Political Geography

World-economy, nation-state
and locality

Fifth edition

Colin Flint and Peter J. Taylor

PEARSON
Prentice
Hall

Harlow, England • London • New York • Boston • San Francisco • Toronto
Sydney • Tokyo • Singapore • Hong Kong • Seoul • Taipei • New Delhi
Cape Town • Madrid • Mexico City • Amsterdam • Munich • Paris • Milan

Pearson Education Limited

Edinburgh Gate
Harlow
Essex CM20 2JE
England

and Associated Companies throughout the world

Visit us on the World Wide Web at:
www.pearsoned.co.uk

———————————————

First published 1985
Second edition published 1989
Third edition published 1993
Fourth edition published 2000
Fifth edition published 2007

© Pearson Education Limited 1985, 2007

ISBN-13: 978-0-13-196012-1
ISBN-10: 0-13-196012-1

British Library Cataloguing-in-Publication Data
A catalogue record for this book is available from the British Library

Library of Congress Cataloging-in-Publication Data
A catalog record for this book is available from the Library of Congress

10 9 8 7 6 5 4 3 2 1
10 09 08 07 06

Typeset in 10/13pt Minion by 35
Printed by Ashford Colour Press Ltd., Gosport

The publisher's policy is to use paper manufactured from sustainable forests.

We dedicate this book to Immanuel Wallerstein for his imagination, inspiration and friendship

Contents

Preface to the fifth edition

A fifth edition – and an interesting and challenging time to be writing a textbook on political geography. If the third edition, written as the established understandings of the Cold War were unraveling, posed difficulties of interpreting a world that seemed to be changing at a rapid rate, then today's background of media reports can be even more bewildering, as well as depressing. In addition, the sub-discipline of political geography is equally fluid, though thankfully lacking the hostility. New theories and approaches are blended into established frameworks as the need to explain a changing world becomes pressing. Pressing, in that the challenges of political violence, ecological disaster, economic inequity, and exclusionary and fundamentalist attitudes to nationalism and religion dominate the news and media commentary. However, it is not *all* bad news. The political geographies of war and difference exist alongside those seeking inter-cultural understanding and reconciliation. In other words, there are political geographies that are attempting to forge a sustainable future.

This edition is the second one jointly authored. Our compatible but different research agendas reflect political geography's consideration of two key processes. On the one hand, Peter Taylor's research studies the integration of the world-economy through the network practices across time and space (currently referred to as globalization). On the other hand, Colin Flint is studying the geographies of war and peace, especially the United States' projection of military power across the globe and into all aspects of society. Both of these topics are to the fore in this edition.

To explain the many political geographies of our world we believe that a historical approach that connects economic and political processes is the most useful. With that in mind, we base the book upon a body of knowledge known as the world-systems approach. This body of knowledge is the product of

the work of many scholars. However, Immanuel Wallerstein has been the driving force behind the world-systems approach, hence our decision to dedicate this edition of the book to him. We explain the world-systems approach in detail, and illustrate its usefulness in explaining and connecting the geography of many different political actions. In addition, we complement the world-systems approach with the perspective of feminist geography. The result is, we hope, an explanation that is able to integrate the complexity of individuals with the complexity of the world-economy.

The five editions of this book may be categorized thus:

1985 *Foundation* text, in which a particular theoretical perspective was brought to bear on the subject matter of political geography.

1989 *Consolidation* text, in which ideas were fleshed out to make for a more comprehensive treatment of political geography (notably in terms of geopolitics and nationalism).

1993 *Post-Cold War* text, in which arguments had to be developed that took account of the traumatic 'geopolitical transition' anticipated by the 1989 (written in 1988) text.

1999 *Globalization* text, in some sense returning to the original theoretical perspective, emphasized the 'global' when it was much less fashionable than it is today.

2007 *Empire and War on Terrorism* text, in which the processes of globalization are discussed in relation to the violent practices of terrorism and counter-terrorism.

Those of you familiar with previous editions will note the new look of this book. We have attempted to retain the rigour of the book with a style that enhances accessibility. The world-systems approach is a historical social science, but one with contemporary relevance. We hope that the integration of text explaining theory and case studies illuminating the theory's relevance enhances the book's usefulness. To keep the book at a reasonable length – each previous edition has been longer than its predecessor – we have cut much of the material describing the theoretical heritages of political geography. With so much to say about the contemporary world and its basis in history we thought that a history of the sub-discipline was best left to specialized books on the topic. But the proof of the pudding is in the eating!

Colin Flint, Champaign, IL, USA
Peter Taylor, Tynemouth, England
22 May 2006

Tips for reading this book

This book contains a number of features designed to help you. The text describes the concepts that we want to introduce to you. These concepts are ideas generated by political geography and world-systems scholars with the intention of explaining events in the world. In addition, we believe that understanding the contemporary world requires consideration of what has happened in the past. Such discussions of the historical foundations of contemporary events are also included in the main text.

Case studies are embedded throughout the book. These are intended to exemplify the concepts we introduce. Mainly, the case studies relate to contem-porary issues. Set off from the text of each chapter in a tinted panel, are short vignettes, gleaned from the media, to show that the news items you come across every day are manifestations of the political geographies we describe in the text.

Finally, each chapter concludes with suggested activities and further reading. As you will see from the text, political geography, as academic subject and real-world practice, is a dynamic affair. Your actions and understandings will maintain existing political geographies and create new ones. The activities and readings are intended to help you plot a pathway.

Acknowledgements

We are grateful to the following for permission to reproduce copyright material:

Figure 1.1 from *The Regional Geography of the World-System: External Arena, Periphery, Semi-Periphery and Core*, Nederlandse Geografische Studies, 144 Utrecht: Faculteit Ruimtelijke Wetenschappen, Rijksuniversiteit Utrecht (Terlouw, Kees 1992); Figure 2.2 from *Britain and the Cold War: 1945 as Geopolitical Transition*, Pinter, London, Guildford Publications, Inc., New York (Taylor, P. J. 1990) By kind permission of Continuum International Publishing Group; Figure 2.3 from *Captain America*: TM © 2006 Marvel Characters, Inc. Used with permission; Figure 2.7 from Linda Panetta at www.opticalrealities.org; Tables 3.2 and 3.3 from 'Industrial convergence, globalization, and the persistence of the North-South divide' in *Studies in Comparative International Development* 38: 3–31, Transaction Publishers (Arrighi, G., Silver, B.J. and Brewer, B.D. 2003) Copyright © 2003 by Transaction Publishers. Reprinted by permission of the publisher; Table 3.4 Copyright (© 2004) from 'Gendered globalization' by S. Roberts in *Mapping Women, Making Politics* (Staeheli, L.A., Kofman, E. and Peake, L.J., eds). Reproduced by permission of Routledge/Taylor & Francis Group, LLC; Table 4.1 from N. Brenner and N. Theodore (2002) 'Cities and the geographies of "actually existing neoliberalism"' in *Antipode* 34: 349–79, Blackwell Publishing; Figures 5.1 and 5.2 reprinted from *Political Geography*, vol. 20, Colin Flint, 'Right-wing resistance to the process of American hegemony', pp. 763–86, copyright 2001, with permission from Elsevier; Table 5.2 from Inter-Parliamentary Union, 'Women in Parliaments', 30 April 2006 http://www.ipu.org/english/home.htm, accessed 3 May 2006; Figure 8.1 reproduced by kind permission of Colin Flint from an unpublished PhD thesis at the University of Colorado.

In some instances we have been unable to trace the owners of copyright material, and we would appreciate any information that would enable us to do so.

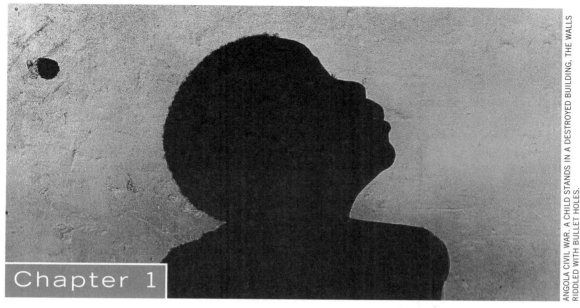

ANGOLA CIVIL WAR. A CHILD STANDS IN A DESTROYED BUILDING, THE WALLS RIDDLED WITH BULLET HOLES.
Source: © Ami Vitale/Panos

Chapter 1

A world-systems approach to political geography

What this chapter covers

Making sense of an age of 'empire'

Political geography
History of political geography

World-systems analysis
Historical systems
  Systems of change
  Types of change
  The error of developmentalism
The basic elements of the world-economy
  A single world market
  A multiple-state system
  A three-tier structure

Dimensions of a historical system
The spatial structure of the world-economy
  The geographical extent of the system
  The concepts of core and periphery
  The semi-periphery category
The dynamics of the world-economy
  Kondratieff cycles
  'Logistic' waves
A space–time matrix for political geography

Power
Types of power
Power geometry
The politics of geographical scale
Scope as geographical scale: where democracy is
  no solution
Ideology separating experience from reality

Power and politics in the world-economy
The nature of power: individuals and institutions
  Power within households
  Power between 'peoples'
  Power and class
  Politics and the state

A political geography perspective on the world-economy
World-economy, nation-state and locality

Making sense of an age of 'empire'

The new and the old are simultaneously vying for our attention. On the one hand, we are told that we live in an age of 'empire', conjuring up images of Victorian era society and elaborate structures and customs deemed necessary to 'civilize' non-Europeans. On the other hand, we are told that we live in a world completely different from the past; a world of terrorism and globalization has, apparently, shattered the established politics of nation-states. The economic and cultural changes of globalization were compounded by the political and military response to 9/11, according to some, to deliver a sense that 'everything has changed'. If everything has changed, it is ironic that some say we have accelerated to an imperial past with all its baggage of racism, colonies, and the like. How can we make sense of this seemingly paradoxical situation?

There are number of ways to negotiate this path. First, it is necessary to discern the relationship between the *material* and the *rhetorical*. Images of the 'real world' are created so that actual political change – the US occupation of Iraq, for example – is seen as 'empire' by some and the growing pains of a 'new world order' by others. Critical commentators and the politicians making the decisions describe the same events in very different ways (see panel below). To understand our world, we must examine the actual causes and nature of current events as well as the way they are portrayed or represented.

Second, to understand the paradox of a new world of 'empire' we must identify the people and institutions (social scientists would say 'actors') that are involved, and then evaluate whether their form and roles have changed dramatically. This second path is one of the *geographical conceptualizing of politics*. To do this we need to reflect on the body of knowledge that political geography has built over the past hundred years or so while also adopting new ideas. How has imperialism, for example, been theorized and described both in the past and by contemporary scholars? What is a state and how has its sovereignty been understood and seen to change over the past decades? Only by being able to conceptualize the actors and identify how they have operated in the past can we evaluate the current situation.

Third, 'making sense' of political changes and the way that they are represented requires *identifying the 'silences' of both analysis and rhetoric*. Gilmartin and Kofman (2004) highlight three such silences, or 'blindspots', in the content of political geography research:

Representing the war in Iraq

On Tuesday, 31 January 2006 the one-hundredth member of the British armed forces died as a result of the US-led invasion of March 2003. The British prime minister, Tony Blair, described himself as 'deeply saddened' and went on to say, 'No life is worth this kind of sacrifice but, in terms of why we are in Iraq, we have now had three democratic elections in a country that was brutalised for decades.' For Prime Minister Blair, the death was a sad but necessary part in the construction of a new democratic Middle East. On the other hand, Reg Keys, an anti-war activist whose son was killed earlier in the conflict, saw the death as an unnecessary loss in an unnecessary conflict. Referring to the milestone of one hundred killed, he was quoted as saying, 'These deaths were 100 per cent preventable. These lads are dying for a falsehood. Their oath of allegiance has been betrayed. This was not what they went to war for. They are not the world's police.'

The reality of the one-hundredth British death was indisputable. What it meant was open to interpretation and became part of the political contest.

Source: All quotes from M. Oliver, '100th British military death in Iraq', *Guardian Unlimited*, 31 January 2006. http://www.guardian.co.uk/print/0,,5388036-103550,00.html. Accessed 31 January 2006.

➤ *Failure to emphasize the persistence of differences in power and wealth.* Despite the persistence of global differences, geopolitics is still focused on state strengths and border issues, for example 'homeland security'.

➤ *The emphasis on elites.* A continued focus upon the state to the detriment of other scales and actors. Hence, there is a need to emphasize the everyday and democratize political geography to include the study of marginalized groups and other non-elites.

➤ *The gendering of geopolitics.* There is a need for a feminist geopolitical approach that focuses upon human security rather than state security.

From a feminist perspective, the challenge is to smash assumptions of which politics and which geographies are the 'most important' and are studied to the detriment of other power relations. In their own words:

> It is important that we acknowledge women's centrality to the day-to-day practice of geopolitics, not just in the documents that tell the stories of geopolitics, but also through their everyday lives that embrace the global.
>
> (Gilmartin and Kofman 2004: 124)

The traditional focus on states has meant an overwhelming concentration upon the elites who control states. If one also acknowledges that it is the 'powerful' states of Europe and North America that have gained the most attention, then focusing upon elites means that it is the geography of the power relations of privileged white men that has constituted the core of political geographical knowledge. Referring back to the panel 'Representing the war in Iraq' on page 2, both quotes are by men, Mr Keys assumes soldiering is a male occupation, and no reference is made to the many more Iraqis who have died in this conflict. Certain power relationships are assumed to be more important; others are marginalized.

Fourth, not only do we need to conceptualize but we must also *contextualize*. Placing current events and changes in historical and spatial context gives them greater meaning and expands our perspective. Gilmartin and Kofman's (2004) call to examine 'differences' is an example of the need for spatial contextualization: wealth, educational opportunities, freedom of expression and movement, for example, vary according to where one lives. Such disparities are local experiences within a global context. Furthermore, such differences are 'persistent'. The broad geography of disparity of wealth, opportunity and security between the global north and south has long been a feature of the world political map.

Fifth, another means to contextualize is to *place events within the process of the rise and fall of great powers.* Competition between states to be the most powerful in the world has been a constant feature of the modern world (Agnew 2003; Wallerstein 2003). Current talk of 'empire' can be understood by looking at the process of the United States' rise to power and the challenges to such power that it is now facing. Comparison with Britain's similar experience in the nineteenth century illuminates commonalities and differences between the two periods. Yet, referring back to the redefinition of 'security' promoted by the feminist approach we must combine a consideration of the power of states and the pursuit of 'national security' with other actors such as multinational businesses, protest groups, families and households.

Sixth, political geography has a tradition of picturing the world as a whole and analysing and *evaluating different localities as components of this larger whole* (Agnew 2003). As we shall see, the global view was an integral part of political geography's role in facilitating imperial conquest. Indeed, seeing the world as a whole is part of the modern *Zeitgeist* and not just an academic exercise. In the panel on page 2, Mr Keys' understanding of a 'proper' use of the British army was in reference to a negative opinion of its current 'global' actions. The predominance of states, national security and global politics are part of the common understanding of the way the modern world works. Academic analysis of political geography has also defined particular global views.

The latter point is important because, for all its 'global heritage', political geography as a sub-discipline has focused its efforts on understanding the modern state and its relations to territory and nation. However, it is important to realize that while contemporary globalization and the idea of 'empire' involve an important 'rescaling' of activities, this is by no means the whole story. Concern for the global should not lead to the neglect of other geographical scales, such as local and national. This is the key point for political geography, and it is relationships *between different geographical scales* that are going to be central to the political geography we develop below. Looking at the scales of the local, the household and the body necessarily requires a study of actors other than states, and of the 'everyday' rather than 'grand events' (Thrift 2000; Hyndman 2004). However, geographical scales and political actors cannot be studied independently of a social theory to inform interpretation and structure the argument. This is where world-systems analysis enters the fray.

Summary

Our political geographic framework to analyse contemporary 'empire' contains seven related elements. They are as follows:

➤ Examine the relationship between the material and the representation of the material.

➤ Geographical conceptualizing of politics.

➤ Challenge long-standing 'silences' to bring in the 'everyday'.

➤ Contextualize politics in spaces and places.

➤ Contextualize politics in the rise and fall of great powers.

➤ Evaluate politics within the larger whole of global politics.

➤ Use geographical scale as a conceptual tool.

Immanuel Wallerstein's (1979, 2004) world-systems approach to social science in general has stimulated a massive literature in recent years. This involves both substantive and theoretical additions to the original ideas and criticisms from a range of alternative perspectives. We do not intend to enter this debate here. In its simplest terms, the choice of Wallerstein's framework is based on the fact that we have found it to be the most useful way to order and understand the subject matter of political geography (Taylor 1982). The proof of the pudding is in the eating, as it were: the remaining chapters of this book are an attempt to illustrate the proficiency of a world-systems political geography. The remainder of this chapter describes the world-systems approach and our particular adaptation of it to political geography.

■ Political geography

Political geographers are interested in how power relations build spaces and places and how, in turn, spaces and places mediate politics and conflict. This simple definition is the essence of all political geography, but the precision of such a definition can be misleading. Contemporary political geography is very diverse, analysing a host of topics through a variety of theories and methods. Power is identified as operating in many different ways, and the relevant spaces range from nation-states, to global geopolitics, the body, networks of migration or capital flows, and the market of cultural products. The oceans and outer space are geographical realms of power struggles and, with the development of cloning, the genome is an embryonic (forgive the pun) arena of politics.

One of the important theoretical foundations for political geography is world-systems analysis: a historical social science approach that is outlined in this chapter. World-systems analysis played a crucial role in the invigoration of contemporary political geography and remains an integral theoretical foundation for the sub-discipline. The vitality of political geography has also meant that other theories have been adopted to explain political geographic phenomena. In this book we cover all the major topics of political geography and, to provide coherence and maintain clarity, emphasize the world-systems approach while illuminating the connections with other theoretical approaches.

A brief history of political geography is necessary to understand the role of world-systems analysis in the sub-discipline.

History of political geography

Political geography was there at the creation of modern geography. Geography was established as a university discipline in Europe at the end of the nineteenth century and on its human side there were initially three sub-disciplines: colonial geography, commercial geography and political geography. The division reflected the discipline's role in the actual politics of imperialism. The two key authors at this time were Friedrich Ratzel in Germany and Sir Halford Mackinder in Britain. Their work defined political geography as geopolitics (see Chapter 2), the key notion being that states operated in competition for territory and resources. Ratzel, in his *Politische Geographie* (1897), claimed that borders were 'naturally' dynamic and expanded with the needs of 'stronger' cultures. His Social Darwinist approach provided a theoretical justification for the expansion of German borders into the Slav regions of eastern Europe. Mackinder produced a grand view of world history, identifying Britain as a commercial maritime state in combat with continental powers based in Eurasia. His pre-scription was the construction of a strong British Empire. We explain such 'classical' geopolitical theories in greater detail in Chapter 2; the point to emphasize here is that political geography emerged as a sub-discipline in order to inform state building and imperialism. In other words we have a skeleton in the cupboard and it is a particularly horrific one.

As part of anti-Nazi propaganda in the Second World War the US media latched on to the term 'geopolitics' and portrayed an army of 'scientists' led by General Karl Haushofer informing Adolf Hitler's plans for world domination (Ó Tuathail 1996). The fictionalization of the facts resulted in a tarnished image for political geography: political geography = geopolitics = Nazism. The sub-discipline suffered. Although it was still taught in many universities, the research agenda became flaccid in the post-war period. Political geography promoted a functional view of the state: an uncritical acceptance of the state that sought geographical ways of maintaining national integration and the accumulation of capital. Colonial or newly independent states were ignored as were the power relationships between rich and poor, or the imperial powers and their dependent states. Political geography became incoherent, atheoretical and uncritical.

The post-war economic boom did not last, and in the late 1960s the Vietnam War and domestic unrest

and economic woes fostered a growing academic engagement with the work of Karl Marx and a radicalization of social science. Following suit, a radical political geography emerged. One strand of political geography was informed by Marx and focused upon dynamics within states (Cox 1973; Harvey 1982). In a parallel development, one of us (Taylor 1982) adopted Immanuel Wallerstein's historical materialist world-systems analysis that culminated in the first edition of this book in 1985. This approach was favoured because it situated political actions at the scale of localities and states within the dynamics of the capitalist world-economy that included the Soviet Union. Thus the demise of the latter in 1991 was accommodated as the collapse of a powerful state and not the defeat of an alternative world-system. Overall, the key point is that politics was contextualized within a hierarchy of geographical scales. Hence the book is subtitled: world-economy, nation-state and locality.

Driven by the Marxist frameworks political geography thrived in the 1980s. In the 1990s human geography became influenced by other social theories and took what has been called the 'cultural turn'. In political geography, there was a decreasing emphasis upon political economy and a new awareness of 'social construction'. For the world-systems stream of political geography this required a re-evaluation of the three scales and also a discussion of new forms of politics. It has been more successful in some of these endeavours than in others, as will be explained in the following chapters. Feminist geographers and queer theorists emphasized the politics of the social construction of scale, or the fact that the actions of individuals and groups 'make' scales. Although these scholars emphasized the scales of the body and the household, the idea that scales are the outcome of political actions rather than pre-existing and theoretical constructs is relevant to all scales. Political geography became increasingly eclectic and the boundaries of the sub-discipline became blurred as cultural geography identified the power relations within cultural constructs and political geography analysed the cultural representation of politics (Painter 1995). A key part of this development was the initiation of critical geopolitics: the examination of the geopolitical practices of states to deconstruct the statements of politicians and theorists in order to show the power relations underlying the 'common-sense' understanding of world politics (Ó Tuathail 1996).

In tandem, the cultural turn and critical geopolitics introduced new forms of politics, or expressions of power, and new political actors, or units of analysis, into the frame of political geography. Focus upon non-state actors, such as social movements, unions, protest groups, indigenous groups and migration diasporas was an outcome of the study of race, gender and sexuality (Kobayashi and Peake 2000; Staeheli et al. 2004). Traditional political geographic analyses of the state and nationalism were enhanced by analysis of the politics of representation and governance (Flint 2003a). The work of philosophers such as Michel Foucault and Pierre Bourdieu challenged the structuralism and economism of Marxist and world-systems analysis, while feminist geographers challenged the binary nature of existing theories and analysis and their exclusion of the analysis of women's activities (Staeheli et al. 2004).

So where are we now? The past hundred years or so have seen a move from a unidimensional political geography in the service of war and imperialism, and subsequent reaction by marginalizing the sub-discipline, through a period of radicalization focusing upon economic explanations to the contemporary eclectic political geography (Mamadouh 2005). Political geography is a rapidly evolving sub-discipline that is the product of a number of creative tensions. Competing theoretical perspectives vie for the spotlight. Moreover, within the social context of the War on Terrorism there is a professional tension between academics wishing to cast a critical eye over foreign policy and university administrators and politicians who see the importance of geographical tools to facilitate counter-terrorism.

A political geography informed by world-systems analysis has a key role to play in contemporary political geography. Its key strength is in being able to place political action in a hierarchy of geographical scales within a historical materialism of economic change in which political competition remains highly relevant. However, it must continually re-evaluate

how its particular framework can include the politics of the everyday. The original focus of world-systems analysis was on the economic structures that created and maintained the persistent differences of wealth and opportunity. Its subsequent concentration on politics discussed the role of states, and then the power of culture. Women, households and the 'everyday' have, according to critics, received inadequate attention.

In this book we use world-systems analysis to provide a critical global and historical view to contextualize contemporary politics. Imperialism, interstate competition, civil war, nationalism, elections and neighbourhood development are all discussed. We also integrate other theories and perspectives when they dovetail with our framework and provide further insight. Especially, the role of discourse or language in framing a dominant understanding of the world is considered, as is the use of feminist theory to explore the variety of non-state politics. In combination we hope to provide global and historical context while also analysing the everyday: the actions of state elites and also the marginalized. In other words, we deal with politics with both a small and a big P (Flint 2003b). The following section begins our journey by outlining the world-systems framework.

■ World-systems analysis

World-systems analysis is about how we conceptualize social change (Wallerstein 2004). Such changes are usually described in terms of societies that are equated with countries. Hence we talk of British society, US society, Brazilian society, Chinese society, and so on. Since there are about two hundred states in the world today, it follows that students of social change will have to deal with approximately two hundred different 'societies'. This position is accepted by orthodox social science and we may term it the 'multiple-society assumption'. World-systems analysis rejects this assumption as a valid starting point for understanding the modern world.

Instead of social change occurring country by country, Wallerstein postulates a 'world-system' that is currently global in scope. This is the modern world-system, also called the capitalist world-economy – Wallerstein refers to them as two sides of the same coin and we use the terms interchangeably in this book. If we accept this 'single-society assumption', it follows that the many 'societies as countries' become merely parts of a larger whole. Hence a particular social change in one of these countries can be fully understood only within the wider context that is the modern world-system. For instance, the decline of Britain since the late nineteenth century is not merely a 'British phenomenon', it is part of a wider world-system process, which we shall term 'hegemonic decline'. The same long-term view can be applied to contemporary debates about the global military presence of the United States, the current hegemonic power. Trying to explain the industrial decline of Britain or the demographics of US Army recruiting policy, for example, by concentrating on Britain or the United States alone will produce only a very partial view of the processes that transcend these particular states.

Of course, the world-systems approach is not the first venture to challenge orthodox thinking in the social sciences. In fact, Wallerstein is consciously attempting to bring together two previous challenges. First, he borrows ideas and concepts from the French Annales school of history. These historians deplored the excessive detail of early twentieth-century history, with its emphasis upon political events and especially diplomatic manoeuvres. They argued for a more holistic approach in which the actions of politicians were just one small part in the unfolding history of ordinary people. Different politicians and their diplomacies would come and go, but the everyday pattern of life with its economic and environmental material basis continued. The emphasis was therefore on the economic and social roots of history rather than the political facade emphasized in orthodox writings. This approach is perhaps best summarized by Fernand Braudel's phrase *longue durée*, which represents the materialist stability underlying political volatility (Wallerstein 1991).

The second challenge that Wallerstein draws upon is the Marxist critique of the development theories in modern social science. The growth of social science after the Second World War coincided with a growth of new states out of the former European colonies. It was the application of modern social science to the problems of these new states that more than anything else exposed its severe limitations. In 1967, Gunder Frank published a cataclysmic critique of social scientific notions of 'modernization' in these new states

The Annales school and the view from below

World systems analysis is based on the principles of the Annales school of thought, a school of French historians linked to the journal *Annales d'histoire économique et sociale* founded by Marc Bloch and Lucien Febvre in 1929. The school advocated 'total history' as a synthesizing discipline to counter the separation of inquiry in disciplines. Most notable was the call for the study of *la longue durée*, or long-term structures and processes, rather than the traditional historical focus on 'big events' and 'great men.'

The ideas of the Annales school are an essential foundation for a political geography approach for the following reasons:

➤ The focus on the big picture offers a global view of structures and processes.

➤ The focus on everyday experience and cultural change means that the structures and processes are identified as the products of social action, and not preordained or immutable.

➤ The focus on everyday experiences means that the view of political geography is 'from below' or democratized. Non-elites are seen as important actors.

➤ In combination, political actions of individuals, groups, and states are placed within the context of large structures and seen to maintain and challenge those structures.

Source: 'Annales school' in *The Dictionary of Human Geography* (2000), ed. R. J. Johnston et al.

which showed that ideas developed in the more prosperous parts of the world could not be transferred to poorer areas without wholly distorting the analysis. Frank's main point was that economic processes operated in different ways in different parts of the world. Whereas western Europe, Japan and the United States may have experienced development, most of the remainder of the world experienced the development of underdevelopment. This latter phrase encapsulates the main point of this school, namely that for the new states it is not a matter of 'catching up' but rather one of changing the whole process of development at the global scale (Wallerstein 1991).

The world-systems approach attempts to combine selectively critical elements of Braudel's materialist history with Frank's development critique, as well as adding several new features, to develop a comprehensive historical social science. As Goldfrank (1979) puts it, Wallerstein is explicitly 'bringing history back in' to social science. And, we might add, with the development of Frank's ideas he is also 'bringing geography back in' to social science: Wallerstein (1991) himself refers to 'TimeSpace realities' as his sphere of interest. Quite simply there is more to understanding the contemporary globalization of the world we live in than can be derived from study of the 'advanced' countries of the world in the late twentieth century, however rigorous or scholarly the conduct of such study is.

Summary

Four key connections between the world-systems approach and our political geography framework:

➤ *La longue durée* facilitates the contextualization of events in long-term historical processes.

➤ The identification of 'one society' adopts a global view.

➤ Critique of development frames the persistence of North–South differences in wealth and opportunity.

➤ The Annales school's call to look at culture and identity promotes the analysis of 'everyday' political geographies.

Historical systems

Modern social science is the culmination of a tradition that attempts to develop general laws for all times and places. A well-known example of this tradition is the attempt to equate the decline of the British Empire with the decline of the Roman Empire nearly two millennia earlier. Similarly, assumptions are often made that 'human nature' is universal, so that motives identified today in 'advanced' countries can be transferred to other periods and cultures. The important point is to specify the scope of generalizations. Wallerstein uses the concept of historical systems to define the limits of his generalizations.

Historical systems are Wallerstein's 'societies'. They are systematic in that they consist of interlocking parts that constitute a single whole, but they are also historical in the sense that they are created, develop over a period of time and then reach their demise. Although Wallerstein recognizes only one such system in existence today, the modern world-system, there have been innumerable historical systems in the past.

Systems of change

Although every historical system is unique, Wallerstein argues that they can be classified into three major types of entity. Such entities are defined by their mode of production, which Wallerstein broadly conceives as the organization of the material basis of a society. This is a much broader concept than the orthodox Marxist definition in that it includes not only the way in which productive tasks are divided up but also decisions concerning the quantities of goods to be produced, their consumption and/or accumulation, and the resulting distribution of goods. Using this broad definition, Wallerstein identifies just three basic ways in which the material base of societies has been organized (for a more complex world-systems interpretation of historical systems, see the work of Chase-Dunn and Hall (1997)). These three modes of production are each associated with a type of entity or system of change.

A *mini-system* is the entity based upon the reciprocal–lineage mode of production. This is the original mode of production based upon very limited specialization of tasks. Production is by hunting, gathering or rudimentary agriculture; exchange is reciprocal between producers and the main organizational principle is age and gender. Mini-systems are small extended families or kin groups that are essentially local in geographical range and exist for just a few generations before destruction or fissure. There have been countless such mini-systems, but none has survived to the present for all have been taken over and incorporated into larger world-systems. By 'world', Wallerstein does not mean 'global' but merely systems larger than the local day-to-day activities of particular members. Two types of world-system are identified by mode of production.

A *world-empire* is the entity based upon the redistributive–tributary mode of production. World-empires have appeared in many political forms, but they all share the same mode of production. This consists of a large group of agricultural producers whose technology is advanced enough to generate a surplus of production beyond their immediate needs. This surplus is sufficient to allow the development of specialized non-agricultural producers such as artisans and administrators. Whereas exchange between agricultural producers and artisans is reciprocal, the distinguishing feature of these systems is the appropriation of part of the surplus to the administrators, who form a military–bureaucratic ruling class. Such tribute is channelled upwards to produce large-scale material inequality not found in mini-systems. This redistribution may be maintained in either a unitary political structure such as the Roman Empire or a fragmented structure such as feudal Europe. Despite such political contrasts, Wallerstein argues that all such 'civilizations', from the Bronze Age to the recent past, have the same material basis to their societies: they are all world-empires. These are less numerous than mini-systems, but nevertheless there have been dozens of such entities since the Neolithic Revolution.

A *world-economy* is the entity based upon the capitalist mode of production. The criterion for production is profitability, and the basic drive of the system is accumulation of the surplus as capital. There is no overarching political structure. Competition between different units of production is ultimately controlled by the cold hand of the market, so the basic rule is accumulate or perish. In this system, the efficient prosper and destroy the less efficient by undercutting their prices in the market. This mode of production defines a world-economy.

Historically, such entities have been extremely fragile and have been incorporated and subjugated to world-empires before they could develop into capital-expanding systems. The great exception is the European world-economy that emerged after 1450 and survived to take over the whole world. A key date in its survival is 1557, when both the Spanish–Austrian Habsburgs and their great rivals the French Valois dynasty went bankrupt in their attempts to dominate the nascent world-economy (Wallerstein 1974a: 124). It is entirely appropriate that the demise of these early modern attempts to produce a unified European world-empire (and therefore stifle the incipient modern world-system at birth) should fail not because of military defeat but at the hands of 'international' bankers. Clearly by 1557 the European world-economy had arrived and was surviving early vulnerability on its way to becoming the only historical example of a fully developed world-economy. As it expanded, it eliminated all remaining mini-systems and world-empires to become truly global by about 1900 (Figure 1.1).

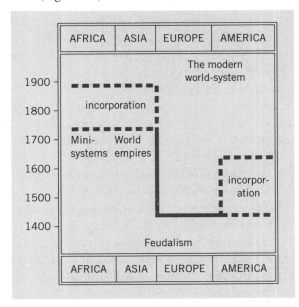

Figure 1.1 Wallerstein's world-system in space and time.

Source: Terlouw (1992) p. 56.

Types of change

Now that we have the full array of types of entity within world-systems analysis we can identify the basic forms that social change can take. It is worth reiterating the key point, that it is these entities that are the objects of change; they are the 'societies' of this historical social perspective. Within this framework there are four fundamental types of change.

The first two types of change are different means of transformation from one mode of production to another. This can occur as an internal process, where one system evolves into another. For instance, mini-systems have begotten world-empires in certain advantageous circumstances in both the Old and New Worlds. Similarly one world-empire, that of feudal Europe, was the predecessor of the capitalist world-economy. We may term this process transition: the most famous example is the transition from feudalism to capitalism in Europe in the period after 1450.

Transformation as an external process occurs as incorporation. As world-empires expanded they conquered and incorporated former mini-systems. These defeated populations were reorganized to become part of a new mode of production providing tribute to their conquerors. Similarly, the expanding world-economy has incorporated mini-systems and world-empires whose populations become part of this new system (Figure 1.1). All peoples of the continents beyond Europe have experienced this transformation over the past five hundred years.

Discontinuities are the third type of change. Discontinuity occurs between different entities at approximately the same location where both entities share the same mode of production. The system breaks down and a new one is constituted in its place. For world-empires, the sequence of Chinese states is the classic example. The periods between these separate world-empires are anarchic, with some reversal to mini-systems, and are commonly referred to as Dark Ages. The most famous is that which occurred in western Europe between the collapse of the Roman Empire and the rise of feudal Europe.

Continuities, the final type of change, occur within systems. Despite the popular image of 'timeless' traditional cultures, all entities are dynamic and continually changing. Such changes are of two basic types – linear and cyclical. All world-empires have displayed a large cyclical pattern of 'rise and fall' as they expanded into adjacent mini-systems until bureaucratic–military costs led to diminishing returns resulting in contraction. In the world-economy, linear trends and cycles of growth and stagnation form an integral part of our analysis. They are described in some detail below.

The error of developmentalism

We have now clarified the way in which world-systems analysis treats social change. In what follows we concentrate on one particular system, the capitalist world-economy, whose expansion has eliminated all other systems – hence our 'one-society assumption' for studying contemporary social change. The importance of this assumption for our analysis cannot be overemphasized. It is best illustrated by the error of developmentalism to which orthodox social science is prone (Taylor 1989, 1992a).

Modern social science has devised many 'stage models' of development, all of which involve a linear sequence of stages through which 'societies' (= countries) are expected to travel. The basic method is to use a historical interpretation of how rich countries became rich as a futuristic speculation of how poor countries can become rich in their turn (Figure 1.2). The most famous example is Rostow's stages of economic growth, which generalize British economic history into a ladder of five stages from 'traditional society' at the bottom to 'the age of high mass consumption' at the top. Rostow uses this model to locate different countries on different rungs of his ladder. 'Advanced' (= rich) countries are at the top, whereas the states of the 'third world' are on the lower rungs. This way of conceptualizing the world has been very popular in geography, where stage models are applied to a wide range of phenomena such as demographic change and transport networks. All assume that poor states can follow a path of development essentially the same as that pursued by the current 'advanced' states. This completely misses out the overall context in which development occurs. When Britain was at the bottom of Rostow's ladder, there was no 'high mass consumption' going on at the top.

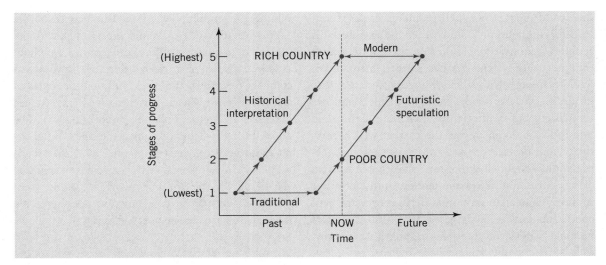

Figure 1.2 Developmentalism.

These developmental models of social change expose the weaknesses of the multiple-society assumption. If social change can be adequately understood on a country-by-country basis then the location of other countries on the ladder does not matter: each society is an autonomous object of change moving along the same trajectory but starting at different dates and moving at different speeds. World-systems analysis totally refutes this model of the contemporary world. The fact that some countries are rich and others are poor is not merely a matter of timing along some universal pathway to affluence. Rather rich and poor are part of one system and they are experiencing different processes within that system: Frank's development and development of underdevelopment. Hence the most important fact concerning those countries at the bottom of Rostow's ladder today is that there are countries enjoying the advantage of being above them at the top of the ladder.

Perhaps more than anything else, world-systems analysis is a challenge to developmentalism: the simplistic world of an international ladder is superseded by the sophisticated concept of the capitalist world-economy.

The basic elements of the world-economy

Now that we have set the study of our world into the overall world-systems framework, we can summarize the basic elements of our historical system, which will underlie all our subsequent analyses. Wallerstein identifies three such basic elements.

A single world market

The world-economy consists of a single world market, which is capitalist. This means that production is for exchange rather than use: producers do not consume what they produce but exchange it on the market for the best price they can get. These products are known as commodities, whose value is determined by the market. Since the price of any commodity is not fixed there is economic competition between producers. In this competition, the more efficient producers can undercut the prices of other producers to increase their share of the market and to eliminate rivals. In this way, the world market determines in the long run the quantity, type and location of production. The concrete result of this process has been uneven economic development across the world. Contemporary globalization is the latest, and in some ways the most developed, expression of the world market.

A multiple-state system

In contrast to one economic market, there have always been a number of political states in the world-economy. This is part of the definition of the system, since if one state came to control the whole system the world market would become politically controlled, competition would be eliminated and the system would transform into a world-empire. Hence the inter-state system is a necessary element of the world-economy. Nevertheless, single states are able to distort the market in the interests of their national capitalist group within their own boundaries, and powerful states can distort the market well beyond their boundaries for a short time. Some interpretations of globalization, for instance, see it as an 'Americanization', a robust expression of US power to stem its relative economic decline over the past couple of decades. This is the very stuff of 'international politics', or 'international political economy' as it is increasingly being called. The concrete result of this process is a competitive state system in which a variety of 'balance of power' situations may prevail. For nearly all of the period since the Second World War, the balance of power was bipolar, organized around the United States and the former Soviet Union. Today, under conditions of globalization, a very different bipolar contest may be emerging between the United States and China.

A three-tier structure

This third essential element is also 'political' in nature but is more subtle than the previous one. Wallerstein argues that the exploitative processes that work through the world-economy always operate in a three-tier format. This is because in any situation of inequality three tiers of interaction are more stable than two tiers of confrontation. Those at the top will always manoeuvre for the 'creation' of a three-tier structure, whereas those at the bottom will emphasize the two tiers of 'them and us'. The continuing existence of the world-economy is therefore due in part to the success of the ruling groups in sustaining three-tier patterns throughout various fields of conflict. An obvious example is the existence of 'centre' parties or factions between right and left positions in many democratic political systems. The most general case is the rise of the 'middle class' between capital and labour since the mid-nineteenth century. Hence, from a world-systems viewpoint, the polarization tendency of contemporary globalization is inherently unstable in the medium term since it is eroding the middle classes. In other contexts, the acceptance of 'middle' ethnic groups helps ruling groups to maintain stability and control in plural societies. The official recognition of Indians and 'coloureds' between the black and white peoples of apartheid South Africa was just such an attempt to protect a dominant class by supporting a middle 'racial buffer'. Geographically, the most interesting example is Wallerstein's concept of the 'semi-periphery', which separates the extremes of material well-being in the modern world-economy that Wallerstein terms the core and the periphery. We define these terms in the next section.

Summary

Political geography and the basic elements of the capitalist world-economy

➤ A single world market is the material basis for our global view.

➤ The multiple-state system contextualizes state politics as one scale of activity in a broader structure.

➤ The three-tier structure is the material basis for analysing the 'persistent differences' in wealth and power.

■ Dimensions of a historical system

If we are bringing history back into political geography, the question obviously arises as to 'what history?' Several recent studies have shared our concern for the neglect of history in geography and have attempted to rectify the situation by presenting brief résumés of world history over the past few hundred years in the opening chapter of their work. The dangers and pitfalls of such writing are obvious: how can such a task be adequately achieved in just a few pages of text? The answer is that we must be highly selective. The selection of episodes to be covered will be directly determined by the purpose of the 'history'. This is nothing new of course; it is true of all history.

It is just that the exigencies are so severe for our purpose here.

We are fortunate that our problem has been made manageable by the publication of *The Times Atlas of World History* (Barraclough 1998), a hugely impressive project already in its fifth edition. Applying Wallerstein's world-systems approach to any subject assumes a level of general historical knowledge that is probably an unreasonable expectation of most students. It is well worth a trip to the library to browse through The Times historical atlas and obtain a sense of the movement of world history. This is recommended to all readers of this book.

Of course, this atlas does not itself employ a world-systems approach. It is divided into seven sections in the following chronological order:

1 The world of early man

2 The first civilizations

3 The classical civilizations of Eurasia

4 The world of divided regions (approximately AD 600–1500)

5 The world of the emerging West (AD 1500–1800)

6 The age of European dominance (nineteenth century AD)

7 The age of global civilization (twentieth century AD).

The product is explicitly global in intent and avoids the Eurocentric basis of many earlier attempts at world history. Nevertheless, it does bear the mark of traditional historiography with its impress of progress from Stone Age to global civilization. Hence Wallerstein's (1980a) conclusion that *The Times Atlas of World History* represents the culmination of a tradition rather than being a pathbreaker. The seven sectors could be termed Stone Age, Bronze Age, classical Iron Age, Dark Ages, age of exploration, nineteenth-century age of trade and imperialism, and twentieth-century age of global society and world wars, with not too much distortion of the flow of ideas. Wallerstein (ibid.) looks forward to a new product in which the waxing and waning of world-empires into and out of zones of mini-systems is gradually replaced after 1450 by the geographical expansion of the capitalist world-economy. That product is not yet

available, but see Chase-Dunn and Anderson (2005) and Chase-Dunn and Hall (1997) for descriptions of such processes. In the meantime, *The Times Atlas of World History* provides an indispensable set of facts catalogued on maps using a traditional model that can be used to bounce world-systems ideas off.

One of the advantages of adopting the world-systems approach is that it enables us to be much more explicit in the theory behind our history. The purpose of this section is to construct just such a historical framework for our political geography that does not simply reflect the weak sense of progress to be found in the other texts referred to at the beginning of this discussion. In place of a linear reconstruction of history, we shall emphasize the ups and downs of the world-economy. Furthermore, the different parts of the world-economy will be affected differentially by these movements. The way in which we present these ideas is as a space–time matrix of the world-economy. This is an extremely poor relation to The Times atlas, but it does provide a succinct description of the major events relevant to our political geography.

The matrix that we generate is not an arbitrary, artificial creation. We are trying to describe the concrete historical entity of the world-economy. Both dimensions of the matrix are calibrated in terms of the system properties of the world-economy. Neither space nor time is treated as in any sense separate from the world-economy. They are not space–time containers through which the world-economy 'travels'. Rather, they are both interpreted as the product of social relations. The time dimension is described as a social product of the dynamics of the world-economy. The space dimension is described as a social product of the structure of the world-economy. Our space–time matrix is a simple model that combines the dynamics and structure to provide a framework for political geography.

The spatial structure of the world-economy

It is unfortunate that the term 'spatial structure' usually conjures up a static picture of an unchanging pattern. Thus we have to emphasize that the spatial structure we are dealing with here is part and parcel of the same processes that generate the dynamics of the system. Spatial structure and temporal cycle are two sides of the same mechanisms that produce a single space–time framework. Space and time are separated here for pedagogical reasons so that in what follows it must always be remembered that the spatial structures we describe are essentially dynamic.

The geographical extent of the system

Our first task is to consider the geographical expansion of the world-economy. We have mentioned that it emerged as a European world-economy after 1450 and covered the globe by about 1900 but have not indicated how this varying size is defined. Basically, all entities are defined in concrete terms by the geographical extent of their division of labour. This is the division of the productive and other tasks that are necessary for the operation of the system. Hence some distribution and trade are a necessary element of the system, whereas other trade is merely ephemeral, and has little relevance beyond those directly participating in it. For instance, luxury trade between the Roman and Chinese Empires was ephemeral, and we would not suggest they be combined to form a single 'Eurasian' system because of this trade. In Wallerstein's terminology, China is part of Rome's external arena and vice versa.

Using these criteria, Wallerstein delimits the initial European world-system as consisting of western Europe, eastern Europe and those parts of South and Central America under Iberian control. The rest of the world was an external arena. This included the ring of Portuguese ports around the Indian and Pacific Oceans, which were concerned with trade in luxury goods. The development of this Portuguese trade had minimal effects in Asia (they merely replaced Arab and other traders) and in Europe. In contrast, Spanish activity in America – especially bullion exports – was fundamentally important in forming the world-economy. For Wallerstein, therefore, Spain was much more important than Portugal in the origins of the world-economy, despite the latter's more global pattern of possessions.

From this period on, the European world-economy expanded by incorporating the remainder of the world roughly in the order Caribbean, North America, India, east Asia, Australia, Africa and finally the Pacific Islands. These incorporations took several forms. The simplest was plunder. This could be only a short-term process, to be supplemented by more productive activities involving new settlement. This sequence occurred in Latin America. Elsewhere, aboriginal systems were also destroyed and completely new economies built, as in North America and Australia. Alternatively, existing societies remained intact but they were peripheralized in the sense that their economies were reoriented to serve wider needs within the world-economy. This could be achieved through political control as in India or indirectly through 'opening up' an area to market forces, as in China. The end result of these various incorporation procedures was the eventual elimination of the external arena (refer back to Figure 1.1).

The concepts of core and periphery

The concept of peripheralization implies that these new areas did not join the world-economy as 'equal partners' with existing members but that they joined on unfavourable terms. They were, in fact, joining a particular part of the world-economy that we term the periphery. It is now commonplace to define the modern world in terms of core (meaning the rich parts of the world in North America, western Europe and Pacific Asia) and periphery (meaning the poor lands of the 'third world'). Although the 'rise' of Japan to core status was quite dramatic in the twentieth century, this core–periphery pattern is often treated as a static, almost natural, phenomenon. The world-economy use of the terms 'core' and 'periphery' is entirely different. Both refer to complex processes and not directly to areas, regions or states. The latter become core-like only because of a predominance of core processes operating in that particular area, region or state. Similarly, peripheral areas, regions or states are defined as those where peripheral processes dominate. This is not a trivial semantic point but relates directly to the way in which the spatial structure is modelled. Space itself can be neither core nor periphery in nature. Rather, there are core and periphery processes that structure space so that at any point in time one or other of the two processes predominates. Since these processes do not act randomly but generate uneven economic development, broad zones of 'core' and 'periphery' are found. Such zones exhibit some stability – parts of Europe have always been in the core – but also show dramatic changes over the lifetime of the world-economy, notably in the rise of extra-European areas, first the United States and then Japan.

How does Wallerstein define these two basic processes? Like all core–periphery models, there is an implication that 'the core exploits and the periphery is exploited'. But this cannot occur as zones exploiting one another; it occurs through the different processes operating in different zones. Core and periphery processes are opposite types of complex production relations. In simple terms, core processes consist of relations that incorporate relatively high wages, advanced technology and a diversified production mix, whereas periphery processes involve low wages, more rudimentary technology and a simple production mix. These are general characteristics, the exact nature of which changes constantly with the evolution of the world-economy. It is important to understand that these processes are not determined by the particular product being produced. Frank (1978) provides two good examples to illustrate this. In the late nineteenth century, India was organized to provide the Lancashire textile industry with cotton and Australia to provide the Yorkshire textile industry with wool. Both were producing raw materials for the textile industry in the core, so their economic function within the world-economy was broadly similar. Nevertheless, the social relations embodied in these two productions were very different, with one being an imposed peripheral process and the other a transplanted core process. The outcomes for these two countries have clearly depended on these social relations and not the particular type of product. Frank's other example of similar products leading to contrasting outcomes due to production relations is the contrast between the tropical hardwood production of central Africa and the softwood production of North America and Scandinavia. The former combines expensive wood and cheap labour, the latter cheap wood and expensive labour.

The semi-periphery category

Core and periphery do not exhaust Wallerstein's concepts for structuring space. Although these processes occur in distinct zones to produce relatively clear-cut contrasts across the world-economy, not all zones are easily designated as primarily core or periphery in nature. One of the most original elements of Wallerstein's approach is his concept of the semi-periphery. This is neither core nor periphery but combines particular mixtures of both processes. Notice that there are no semi-peripheral processes. Rather, the term 'semi-periphery' can be applied directly to areas, regions or states when they do not exhibit a predominance of either core or peripheral processes. This means that the overall social relations operating in such zones involve exploiting peripheral areas, while the semi-periphery itself suffers exploitation by the core.

The semi-periphery is interesting because it is the dynamic category within the world-economy. Much restructuring of space consists of states rising and sinking through the semi-periphery. Opportunities for change occur during recessions, but these are only limited opportunities – not all the semi-periphery can evolve to become core. Political processes are very important here in the selection of success and failure in the world-economy. Wallerstein actually considers the semi-periphery's role to be more political than economic. It is the crucial middle zone in the spatial manifestation of his three-tier characterization of the world-economy. For this reason, it figures prominently in much of our subsequent discussions.

The dynamics of the world-economy

One reason for current interest in the global scale of analysis is the fact that the whole world seems to be struggling to rise out of a period of economic stagnation that has been around for two or three decades – the oil price hikes of the 1970s are often blamed for starting it. What became clear straight away was that the initial slowdown in economic growth was not an American problem or a British problem or the problem of any single state; rather, it is a worldwide problem. More recently interpreted as globalization, despite renewed economic growth, poverty levels are rising in the United States, unemployment is at record levels in Germany, and a banking crisis in Asia threatens the vitality of global trade and finance. Such ambiguity in economic changes makes it impossible to say with any certainty whether or not the world-economy is currently experiencing an upturn. This ambiguity is a manifestation of the polarizing effect of globalization. For example, one of the world's richest people, the international financier George Soros, riding a US$20 billion foreign investment boom in Argentina, now owns the Galería Pacífico, a luxury shopping mall in Buenos Aires housing designer label stores such as Lacoste and Timberland to satisfy the demands of a new class of professionals, while, at the same time, Argentina's unemployment rate is approaching 20 per cent, so that many workers over the age of 40 are being left without either employment prospects or pension. This is a classic 'growth with poverty' example of globalization.

Whether or not contemporary globalization reflects the world-economy emerging from its recent stagnation, it is clear that it would not be the first time the 'world' has experienced such a general stagnation followed by renewed buoyancy. The great post-war boom in the two decades after the Second World War followed the Great Depression of the 1930s. As we go back in time such events are less clear, but economic historians also identify economic depressions in the late Victorian era and before 1850 – the famous 'hungry forties' – each followed by periods of relative growth and prosperity. It is but a short step from these simple observations to the idea that the world-economy has developed in a cyclical manner. The first person to propose such a scheme was a Russian economist, Nikolai Kondratieff, and today such fifty-year cycles are named after him.

Kondratieff cycles

Kondratieff cycles consist of two phases, one of growth (A) and one of stagnation (B). It is generally agreed that the following four cycles have occurred (exact dates vary):

I	1780/90	—A—	1810/17	—B—	1844/51
II	1844/51	—A—	1870/75	—B—	1890/96
III	1890/96	—A—	1914/20	—B—	1940/45
IV	1940/45	—A—	1967/73	—B—	?

These cycles have been identified in time-series data for a wide range of economic phenomena, including industrial and agricultural production and trade statistics for many countries (Goldstein 1988). On this interpretation, we are currently experiencing, perhaps coming to the end of, the B-phase of the fourth Kondratieff cycle.

Whereas identification of these cycles is broadly agreed upon, ideas concerning the causes of their existence are much more debatable. They are certainly associated with technological change, and the A-phases can be easily related to major periods of the adoption of technological innovations. This is illustrated in Figure 1.3, where the growth (A) and stagnation (B) phases are depicted schematically, with selected leading economic sectors shown for each A-phase. For instance, the first A-phase coincides with the original Industrial Revolution, with its steam engines and cotton industry. Subsequent 'new industrial revolutions' also fit the pattern well, consisting of railways and steel (IIA), chemicals (oil) and electricity (IIIA), and aerospace and electronics (IVA). Of course, technology itself cannot explain anything. Why did these technical adoptions occur as 'bundles' of innovations and not on a more regular, linear basis? The world-systems answer is that this cyclical pattern is intrinsic to our historical system as a result of the operation of the capitalist mode of production. Contradictions in the organization of the material base mean that simple linear cumulative growth is impossible, and intermittent phases of stagnation are necessary. Let us briefly consider this argument.

A basic feature of the capitalist mode of production is the lack of any overall central control, political or otherwise. The market relies on competition to order the system, and competition implies multiple decentralized decision making. Such entrepreneurs make decisions for their own short-term advantage. In good times, A-phases, it is in the interest of all entrepreneurs to invest in production (new technology) since prospects for profits are good. With no

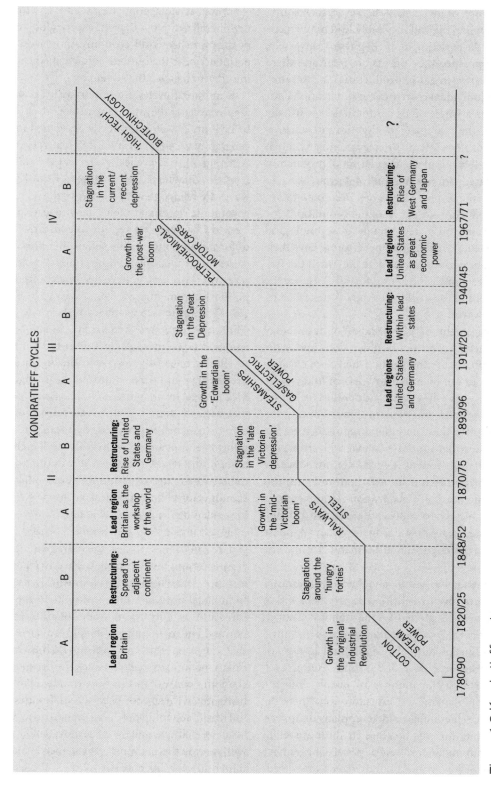

Figure 1.3 Kondratieff cycles.

central planning of investment, however, such short-term decision making will inevitably lead to overproduction and the cessation of the A-phase. Conversely, in B-phases prospects for profits are poor and there will be underinvestment in production. This is rational for each individual entrepreneur but irrational for the system as a whole. This contradiction is usually referred to as the anarchy of production and will produce cycles of investment. After extracting as much profit as possible out of a particular set of production processes based on one bundle of technologies in an A-phase, the B-phase becomes necessary to re-organize production to generate new conditions for expansion based on a new bundle of technological innovations. Phases of stagnation therefore have their positive side as periods of restructuring when the system is prepared for the next 'leap forward'. Hence the ups and downs of the world-economy as described by Kondratieff waves.

The replacement of old bundles of technology with new ones involves political decisions and competition. B-phases are the period when once cutting-edge industries are relocated to areas of lower-wage employment – as witnessed by de-industrialization in the United States and western Europe during the 1980s. To replace these peripheralized industries, the new innovations and industries that will drive production in the subsequent A-phase are introduced – such as today's thriving information and business services at the core of globalization. However, it is not enough merely to reduce the costs of existing industries and create new products. A new A-phase requires increased consumer demand within the world-economy.

Political struggles within and between countries represent a scramble to capture core processes within state borders, as illustrated by the political changes within the former Soviet satellite countries in central and eastern Europe in the late 1980s and 1990s – or the rush to become part of 'Europe'. But if each B-phase increased the number of people enjoying core-like employment and consumption, then the core–periphery hierarchy would eventually disappear. To compensate for this increase in the number of people consuming at core levels, past B-phases have seen an expansion in the boundaries of the world-economy as new populations and territories were peripheralized. Now that the entire globe is covered by the capitalist world-economy, those workers in the periphery bear the burden of intensified exploitation in order to balance the system.

Kondratieff cycles are important to political geography because they help to generate cycles of political behaviour. This link is directly developed in electoral geography (Chapter 6) and local political geographies (Chapter 7), but cyclical patterns pervade our analyses. In Chapter 2, the rhythms of the Kondratieff waves are related to longer cycles of the rise and fall of hegemonic states and their changing economic policies. In Chapter 3, we see how the historical rhythms of formal and informal imperialism follow economic cycles. Such identification of political cycles, regular repetitions of history, have become common among political commentators. For instance, Robert Reich (1998) – former US Secretary of Labor to President Clinton – compares the current climate of political complacency and voter apathy in the United States to similar ones fifty years ago during Eisenhower's presidency and one hundred years ago during McKinley's. In an ominous tone, Reich points out that each of these two previous periods of political calm was abruptly ended by dramatic political reforms and changes, such as the civil rights movement in the 1960s. What we show in this book is that the structure and dynamics of the capitalist world-economy provide a political geography framework for explaining such political actions.

There is a lot more that could be said about the generation of these cycles; the basic geography of the expansion and restructuring is listed in Figure 1.3, for instance. This 'uneven development' is itself related to political processes both as inputs to the mechanisms and as outputs in terms of differential state powers. The main point to make here is to emphasize that the economic mechanisms do not operate in isolation, and we shall consider the broader political economy context in the next chapter. For the time being it is sufficient for us to accept that the nature of the world-economy produces cyclical growth that can be described adequately by Kondratieff waves. This will provide the main part of the metric for the time dimension for our matrix.

'Logistic' waves

What about before 1780? We have indicated that the world-economy emerged after 1450, but we have as yet no metric for this early period. Of course, as we go back in time data sources become less plentiful and less reliable, leading to much less consensus on the dynamics of the early world-economy. Some researchers, including Braudel, claim to have found Kondratieff waves before 1780, but such hypotheses for this earlier period do not command the same general support as the sequence reported above. There is, however, more support for longer waves of up to three hundred years, which have been referred to as 'logistics'. Just like Kondratieff waves, these longer cycles have A- and B-phases. Two logistics of particular interest to world-systems analysis are as follows:

c.1050————A————c.1250————B————c.1450
c.1450————A————c.1600————B————c.1750

The dates are much less certain than for the Kondratieff waves, but there does seem to be enough evidence in terms of land-use and demographic data to support the idea of two very long waves over this general time span.

It will have been noticed that these logistics take us back beyond the beginning of the world-economy. The first logistic is of interest, however, because it encompasses the material rise and decline of feudal Europe, the immediate predecessor of the world-economy. There is a massive literature on the transition from this feudalism to the capitalist mode of production that is beyond the scope of this text. Wallerstein's (1974a) explanation, however, is relevant since it relates to this first logistic wave and the emergence of the world-economy. The B-phase of the first logistic reflects a real decline in production, as indexed by the contraction of agriculture throughout Europe. This is the so-called crisis of feudalism. B-phases terminate when a solution to a crisis is generated. In this case, the solution was nothing less than the development of a new mode of production. This emerged gradually out of European exploration and plunder in the Americas, the development of new trade patterns, particularly the Baltic trade, and technological advances in agricultural production. The result was, according to Wallerstein, a new entity or system: the European world-economy based upon agricultural capitalism. This system itself generates a logistic wave of expansion in its emergence in the 'long sixteenth century' followed by stagnation in the 'crisis' of the seventeenth century. However, Wallerstein emphasizes the fact that this second B-phase in agricultural capitalism is different in kind from the B-phase of late feudalism. Unlike the real decline that occurred in feudal Europe, the world-economy B-phase is more one of stagnation. This involved the reordering of the materialist base so that some groups and areas gained while others lost. There was no general decline like the crisis of feudalism but rather a consolidation of the system into a new pattern. In this sense, the second logistic B-phase is more like the B-phase of the later Kondratieff waves.

Just as there is a dispute over whether Kondratieff waves extend back before 1780 there is a similar disagreement about whether logistic waves can be extended forward to the present. If either set of cycles is extended then we come across the thorny problem of how they relate to one another. For the purposes of our matrix we will avoid this problem by just using the waves generally accepted in the literature and described above. Our time metric for the world-economy therefore consists of ten units, A- and B-phases for the logistic wave after c.1450 and four A- and B-phases in the Kondratieff waves. These two different treatments of time may be thought of as relating to agricultural capitalism and industrial capitalism as consecutive production forms of the world-economy.

A space–time matrix for political geography

The above discussion produces a 10 × 3 matrix involving ten phases of growth and stagnation and three types of spatial zone. In Table 1.1 this framework is used to portray those features of the evolving world-economy that are necessary for an understanding of our political geography. This table should be read and referred to as necessary for subsequent chapters. The historical events in this table are manifestations of the processes that are discussed in depth

Table 1.1 A space–time information matrix

		Core	Semi-periphery	Periphery
Logistic	A	Initial geographical expansion based on Iberia but economic advances based on north-west Europe	Relative decline of cities of central and Mediterranean Europe	Iberian empires in New World; second 'feudalism' in eastern Europe
	B	Consolidation of north-west European dominance, first Dutch and then French–English rivalry	Declining areas now include Iberia and joined by rising groups in Sweden, Prussia and north-east United States	Retrenchment in Latin America and eastern Europe. Rise of Caribbean sugar. French defeat in India and Canada
Kondratieff wave I	A	Industrial Revolution in Britain, 'national' revolution in France. Defeat of France	Relative decline of whole semi-periphery Establishment of United States	Decolonization and expansion – formal control in India but informal controls in Latin America
	B	Consolidation of British economic leadership. Origins of socialism in Britain and France	Beginning of selective rise in North America and central Europe	Expansion of British influence in Latin America. Initial opening up of east Asia
Kondratieff wave II	A	Britain as the 'workshop of the world' in an era of free trade	Reorganization of semi-periphery: civil war in United States, unification of Germany and Italy, entry of Russia	The classical era of 'informal imperialism' with growth of Latin America
	B	Decline of Britain relative to United States and Germany Emergence of the Socialist Second International	Decline of Russia and Mediterranean Europe	Expansion – scramble for Africa. The classical age of imperialism
Kondratieff wave III	A	Consolidation of German and US economic leadership. Arms race	Entry of Japan and Dominion states	Consolidation of new colonies (Africa) plus growth in trade elsewhere (especially China)
	B	Defeat of Germany, British Empire saved. US economic leadership confirmed	Socialist victory in Russia – establishment of Soviet Union. Entry of Argentina	Neglect of periphery. Beginning of peripheral revolts. Import substitution in Latin America
Kondratieff wave IV	A	United States as the greatest power in the world both militarily and economically. New era of free trade	Rise of eastern Europe and Cold War. Entry of OPEC	Socialist victory in China. Decolonization leading this time to neo-colonialism
	B	Decline of United States relative to Europe and Japan End of the Cold War; the War on Terrorism Neo-liberalism	Entry of 'Little Japans' in east Asia. Collapse of communism in eastern Europe, demise of Soviet Union. Rise of debts to core Economic growth of China Pursuit of entry into free-trade regimes	Severe economic crisis and conflict Expansion of poverty Rise of Islamic 'fundamentalism'

in the following chapters. In Chapter 2, the geo-political theories that we discuss reflect the politics of intra-core competition in Kondratieff's waves III and IV. Chapter 3 describes the formation of empires and the maintenance of core–periphery relationships throughout the history of the world-economy. The remaining chapters discuss different expressions of political and economic restructuring within states. For example, Chapter 6 interprets the third world politics of coup and counter-coup as an issue of peripheralization. Table 1.1 is largely self-explanatory, but a brief commentary will illustrate the ways in which this information will be related to our discussion.

The establishment of the world-economy as a system operating from eastern Europe to the New World involved the development of both Atlantic trade and the Baltic trade. The former was initiated from Iberia but gradually came to be controlled financially from the incipient core of north-west Europe on which the Baltic trade was based. Once this core had become established, Iberia was relegated to a 'conveyor belt' for the transfer of surplus from its colonies to the core. It is during the logistic B-phase that the basic elements of the world-economy identified above become consolidated. First, there is a single world-market organized and controlled from north-west Europe. Second, a multiple-state system emerges epitomized by the initiation of 'international law' to regulate relations between states. Third, a three-tier spatial structure clearly emerges and can be identified in the new division of labour for agricultural production: 'free' wage labour developing in the north-west European core; partially free 'share-cropping' arrangements in the Mediterranean semi-periphery; and, in the periphery, two different forms of coerced labour: slavery in the New World and the so-called second feudalism in eastern Europe. Despite massive changes in the world-economy since this time, these three essential features have remained and are just as important today as they were in the seventeenth century.

Following the consolidation of the world-economy, it has grown economically and geographically in fits and starts as described by the four Kondratieff waves. Some degree of symmetry in these changes can be identified from Table 1.1 by designating British and American 'centuries'. These relate to the rise of these two states, the defeat of their rivals (France and Germany, respectively), their dominance of the world-economy (including their promotion of free trade) and finally their decline with new rivalries emerging (including the rise of protectionism and/or imperialism). We consider such patterns in some detail in the next chapter.

In order to illustrate the rest of the matrix, let us highlight the route of today's major states through this matrix. Britain became part of the core by the time of the logistic B-phase, when it restructured its state in its civil war. It has maintained that position despite relative decline since the second Kondratieff B-phase. France's early position is similar to Britain's, but defeat in the periphery and relative decline in the logistic B-phase led to a restructuring of the French state in its revolution. However, subsequent defeat during the first Kondratieff wave led to a second relative decline, but this time within the core. The United States and Germany (Prussia) have had much more volatile histories. Both carved out semi-peripheral positions in the logistic B-phase, but their positions were unstable. In the United States, peripheralization was prevented by the War of Independence. This victory was consolidated by the civil war in the second Kondratieff A-phase, when southern cotton became part of an American periphery (rather than part of a British periphery) in a restructured American state. From this point onwards, the United States prospered to become the major power of the twentieth century. Initially, its major rival was Germany, which also restructured its state in the second Kondratieff A-phase under Prussian leadership. But subsequent economic prowess was set back in the third Kondratieff wave by military defeat. Today, Germany is again a major economic challenger of the United States in the core. The major economic challenger is Japan, which entered the world-economy only in the second Kondratieff wave. It also restructured its state and suffered military setbacks but has now finally come through towards economic leadership. Russia, on the other hand, entered the world-economy earlier but declined in the second Kondratieff wave. This was halted by a reorganization of the Russian state as the Soviet Union, which emerged as a major military power but remained economically semi-peripheral. Finally, China entered the world-economy in the periphery at the end of the first Kondratieff wave and rose to semi-peripheral status with reorganizations of the state in the third and fourth Kondratieff A-phases. Reorganization as the People's Republic of China was successful, and in the current period of restructuring it has manoeuvred to become an economic power based on core processes and activities (such as finance) rather than using its base in agricultural production to ascend through peripheral processes.

This description emphasizes the role of state reorganization in the rise of states to core or semi-peripheral position and their maintenance of that status. Contemporary globalization represents another example of acute state reorganization – the US Republican Party's *Contract for America* in 1994 being a classic example, which we deal with in some detail in Chapter 4. But we should not imply that a state merely needs to reorder its political apparatus to enjoy world-economy success. By only describing the success stories – today's major states – we omit the far greater number of failures: while Germany and the United States were reorganizing in the second Kondratieff A-phase so too was the Ottoman Empire, but to much less effect. In fact, political reorganization has become a way of life in many semi-peripheral states as lack of success in the world-economy brings forth pressure for change again and again. Such a world of winners and losers requires the consideration of power and politics in the modern world-system.

Summary

Political geography and the dimensions of the capitalist world-economy

➤ Identification of Kondratieff cycles places political activity within a context of dynamic economic change.

➤ Recognition of the changing geographical extent of the capitalist world-economy is the foundation for understanding persistent geographical differences in wealth and power.

➤ Core and periphery processes are the theoretical foundation for understanding geographies of inequality.

➤ The semi-periphery highlights the dynamism and social construction of the capitalist world-economy.

➤ The space–time matrix is the foundation for contextualizing political actions in time and space.

■ Power

The value of a political geography approach that adopts world-systems analysis is the insights it offers into the spatiality of power. Politics is an integral component of Wallerstein's approach, despite claims to the contrary (Zolberg 1981). Our framework has a materialist foundation, the capitalist world-economy, but that does not mean that politics is ignored or devalued. In the section on basic elements of the world-economy above, two of the three characteristics described were pre-eminently political in nature: the multi-state system and the three-tier structure. In addition, as we shall see, the construction of geographical scale is a politics of both resistance to and maintenance of the world-economy. It is the purpose of this final section of our introductory chapter to develop the argument concerning politics in the world-economy in order to expand our political geography perspective on the world-economy.

Types of power

Political geography is the study of the spatiality of power, but what is power? Allen (2003) distinguishes three expressions of power that are different but operate together. First is power as an inscribed capacity: power is possessed inherently by individuals, groups and other institutions depending on their relative position to other individuals, groups and institutions. A policeman has an inherent power given his office, the International Atomic Energy Agency has authority to inspect and evaluate nuclear installation in Iran, and refer the country to the United Nations (UN) Security Council which could, in turn, use its inscribed power to make a resolution authorizing military action (see panel at top of next page).

Military action requires consideration of the second conception of power: power is a resource, the capacity to mobilize power to certain ends. The United States had the diplomatic power to force the question of Iran's nuclear capabilities on to the United Nation's agenda. Other members of the UN Security Council (France, China and Russia) have some diplomatic power to block or amend US proposals. Of course, the possession of military technology and

Authority, and the 'right' to enrich uranium

Iran's announcement that it had entered the 'club' of nuclear powers by successfully enriching uranium, an essential step in the ability to make a nuclear weapon, was greeted unfavourably by the United States and the UN. In February 2006, the UN's chief nuclear inspector, Mohamed ElBaradei of the International Atomic Energy Agency (IAEA), called on Iran to freeze its nuclear production for ten years. The IAEA board was urged by the United States to refer the matter to the UN Security Council, and Russia and China (who had previously opposed the referral) consented. The power to inspect and evaluate nuclear operations within a sovereign territory is inscribed to the IAEA. The Security Council has the inscribed capability to make resolutions in response. But what about enforcing any resolutions? That requires power as a resource, and in practice the agreement of the United States to utilize its military capabilities to ensure enforcement.

Identify the different forms of power that this story exemplifies and consider how they relate to each other.

Sources: I. Traynor, 'Tehran urged to agree to 10-year nuclear freeze', *Guardian Unlimited* 3 February 2006 www.guardian.co.uk/print/0,,5930393-111322,00.html. Accessed 3 February 2006. D. Ensor, 'Inspectors to press Iran on claim', CNN.com, 17 April 2006. www.cnn.com/2006/WORLD/meast/04/17/iran.nuclear/index.html. Accessed 20 April 2006.

manpower to attack and occupy another country is an example of power as resource.

The language in the previous paragraph is deliberately gendered: 'policeman' and 'manpower'. The casual use of such terms is an example of the third conception of power: power as strategies, practices and techniques. The French philosopher Michel Foucault stimulated analysis of power in the form of discourse or knowledge that facilitated behaviour that was desired by the powerful, without the powerful seemingly acting. The resource of government was complemented by the techniques of governance. Foucault argues that 'experts' are given authority to create 'knowledge' which then becomes 'common

The relative size of US military spending

In 2005, US military spending amounted to US$420.7 billion. It is hard to visualize that amount of money, but it buys an awful lot of military hardware and maintains hundreds of thousands of military personnel. But it is easier to consider the military might of the United States through a number of comparisons:

➤ The US military spending was almost two-fifths of the global total.

➤ The US military spending was almost seven times larger than the Chinese budget, the second largest spender.

➤ It was more than the combined spending of the next 14 nations.

➤ The US military budget was almost 29 times as large as the combined spending of the six 'rogue' states (Cuba, Iran, Libya, North Korea, Sudan and Syria) who spent US$14.65 billion.

➤ The United States and its close allies accounted for some two-thirds to three-quarters of all military spending, depending on who you count as close allies (typically NATO countries, Australia, Canada, Israel, Japan and South Korea).

➤ The two potential 'enemies', Russia and China, together spent US$139 billion, 30 per cent of the US military budget.

Source: 'Arms trade – a major cause of suffering', *Global Issues that Affect Everyone*, 27 March 2006. www.globalissues.org/Geopolitics/ArmsTrade/Spending.asp?p=1. Accessed 20 April 2006.

Modern monarchy, gender and power

On 3 February 2006 the *Guardian* newspaper carried a story regarding the politics of succession to the throne of the Japanese emperor. Traditionally, the position has been a male preserve, and in 1947 a law was passed legislating that only male descendants of the ruling emperor could ascend to the throne. However, a succession crisis is on the horizon. Crown Prince Naruhito and his wife Princess Masako have only one child, 4-year-old Princess Aiko. The article states: 'No boys have been born into the imperial family since 1965.' Efforts by the Japanese prime minister Junichiro Koizumi to repeal the law have met with opposition, though the majority of the Japanese public favour the change.

The arguments against changing the law to allow for the ascendancy of an empress have taken two angles: patriarchal and racial. Some have argued that the patriarchal line must be maintained, even if it requires reinstating the institution of the concubine. Failure to do so, and the accession of Princess Aiko to the throne, is seen as a risk against Japanese racial purity: Takeo Hiranuma, a former trade minister, is quoted as saying: 'If Aiko becomes the reigning empress and gets involved with a blue-eyed foreigner while studying abroad, and then marries him, their child may be the emperor. . . . We should never let that happen.'

A few days later, another member of Japanese royalty, Princess Kiko, announced she was pregnant, perhaps with a boy who would become the male heir.

Can we identify the three forms of power in this article? Inscribed power is evident in the political powers of the prime minister and his opponents; the ability to mobilize votes or use rhetorical images of Japanese 'tradition' are two examples of power as resource; and the discourses of female promiscuity and the necessity and existence of racial purity are manifestations of power as strategy. Can you find other examples in the article?

Sources: J. McCurry, 'Heirs and races', *Guardian Unlimited*, 3 February 2006. www.guardian.co.uk/print/0,,5390993-105806,00.html. Accessed 3 February 2006. CNN, 'Japan's Kiko is pregnant', CNN.com, 7 February 2006. www.cnn.worldnews.printthis.clickability.com/pt/cpt?action+cpt&title.html. Accessed 7 February 2006.

sense' and normative behaviour. For example, firms and local governments are constantly told they must be 'competitive' in the face of globalization. The discourse of competitiveness becomes taken for granted, or common sense in other words, and catalyses actions such as reduction of labour rights and tax breaks for corporations. In another example, and by referring back to the previous paragraph, patriarchal societies are maintained by 'knowledge' of women's capabilities and roles that promote their marginalization and by reducing access to tasks that are usually deemed masculine. The practice of patriarchy creates gender inequalities in terms of who hold positions of authority, or inscribed power, and hence, who controls powerful resources.

Power geometry

Not only do these different conceptualizations of power complement each other, they also operate in relation to other actors (individuals, groups, and so on). Power is exercised by one entity over another. It is then helpful to think of power geometry, the manner in which individuals are positioned within networks and structures of power, or in other words how they relate to each other (Massey 1993).

There are three ways in which power geometry can be conceptualized geographically. The first is territorially. Inscribed power is often limited by the scope of recognized jurisdiction, and this is especially the case for states. States are sovereign, but this is understood to be over a specific territory. However, state sovereignty was constructed over time, and processes of globalization have altered and eroded the practice of sovereignty. Referring to power as strategy and practice, places are particular pieces of territory within which norms of behaviour are defined. Homosexuals perceive it safe and 'appropriate' to be 'out' in some places but not others, for example (Heibel 2004; Sumartojo 2004). Businesses are

Google and China

On 25 January 2006 Google, the computer search engine company, began operations within China after coming to an agreement with the Chinese government over censorship. Google was eager to enter the Chinese market with its over 100 million internet users and where local competitors and global firms (Yahoo! and Microsoft's MSN) were already established. However, the classic competition for market share within a political territory was complicated for Google by the Chinese government's attitude towards censorship, an expression of its sovereignty over economic activities within its borders. Although Google has agreed that access to websites deemed politically unsuitable by the Chinese government will be blocked, users will be informed whether or not information regarding particular sites has been withheld at the behest of the authorities.

Source: 'Here be dragons', *The Economist*, 28 January 2006, p. 59.

actively seeking market share in particular territories, trying to weaken their competitors by muscling out other products. In the United States, PepsiCo and Coca-Cola often sign contracts with universities so that their products are sold exclusively on campus. Power relations are different in different territories: power is exercised by controlling territory, and different territories interact in power relationships. In this book our discussion of geopolitics, imperialism, states, nations and local politics are all examinations of territory and power.

The second is through networks. Modern political geography operated under the assumption that the compartmentalization of the world into different territorial units – states – was the be all and end all of inquiry. This assumption followed the multi-state assumption of social science. Interaction in the world was seen as bloc against bloc in the case of inter-state war, or as country with country in the case of international trade. Study of the international meant analysing interaction between states. However, this was always an inadequate conceptualization, and within the context of globalization, emphasis has turned instead to transnationalism: a study of networks that transcend state territory.

There are two major ways in which the study of networks has entered human geographical studies: through Castells' (1996) concept of a 'network society', and through actor–network theory. The former aspires to be a new theory of society for an 'informational age' that has now replaced the 'industrial age'.

In other words, information and knowledge are the key commodities of today's world-economy and these are organized through multiple networks. This has led to a change in the nature of social space, the spatial form that we construct in our everyday activities making a living. Historically, this spatial form has been a 'space of places', congeries of different locales. Politically, the classic example is the world political map of countries: nation-states as homelands and territories constitute a space of places. However, the combining of two high-tech industries – computers and communications – in the 1970s has provided opportunities to transcend places and create 'spaces of flows'. Politically, this is the transnationalism that many global thinkers argue is eroding, or even eliminating, the power of states. The prime example is global financial markets that were once controlled by states and that now operate beyond states. At the individual level, the internet best represents the space of flows today. For Castells, our contemporary world is dominated by spaces of flows at the expense of spaces of places. This has implications for the core–periphery concepts we introduced above: we will address this highly contentious position on several occasions as we develop our political geography.

The second theoretical contribution to the study of networks is actor–network theory (Serres and Latour 1995). This rather complex theory is an uneasy collection of a number of contributions. In essence, the world is seen as a combination of diverse and multiple networks of associations. It is the nature of the links and what travels through them that is the focus of study rather than the nodes themselves: what flows rather than the source and terminus of the flow is seen as vital. The actors in the network are constituted by the nature of the flow. This appears somewhat temporary and contingent, but the traffic through the network is quite durable (money, consumer goods, and so forth) and gives a degree of permanency to the network. Power rests in those whose task it is to maintain the network; power is the control of flow or mobility. Some institutions and people have inscribed authority to control flows through control of particular resources (money trading for example), but it is the nature of the flow that holds the disciplining power to form practice; for example immigration authorities control the flow of migrants, but in the act of migrating people will adopt particular behaviours as strategies to be allowed to move and adapt to a new geographical setting.

In a related topical focus, transnationalism has challenged binary understandings of the world. Most important are the binaries of foreign/domestic and developed/undeveloped (or core/periphery in world-systems terminology). A feminist geopolitics (Gilmartin and Kofman 2004; Hyndman 2004) has emphasized mobility (or the lack of it) as a measure of power. In addition, feminists call for the study of people in particular places and within particular power relations to show the way that people challenge and also adapt to structures of national citizenship, patriarchy, racial grouping, and so on. From a world-systems perspective, the study of the flows between world cities has done the most to disrupt the standard view of the territorial geography of nation-states (Taylor 2005). In addition, the world-systems approach to scale may incorporate the feminist approach to view the local scale as the arena of contest within the overarching structure of the world-economy.

The politics of geographical scale

We turn now to the third geographical expression of power geometry: geographical scale. Geographical scale is the main organizing framework for the book and so we will spend more time outlining its usage.

A political geography perspective on the modern world-system is a meaningful project only if it produces something that other perspectives cannot provide. We have hinted above that this is indeed the case; here we attempt an explicit justification. The gist of our argument is that the use of geographical scale as an organizing frame provides a particularly fruitful arrangement of ideas. Specifically, our world-systems political geography framework provides a set of insights into the operation of the world-economy that has not been shown so clearly elsewhere (Flint and Shelley 1996). Such an assertion requires some initial justification. We do this in two ways: first, in terms of a crucial contemporary practical political problem; and second, as a theoretical contribution to our world-systems political geography.

Scope as geographical scale: where democracy is no solution

One way of interpreting geographical scale within politics is to consider the scope of a conflict. By scope we mean who is brought in on each side – we develop this idea further later in this chapter. For the moment, imagine a state that is the weaker party in a political conflict. Given that it will likely lose, it will endeavour to change the scope of the politics. For instance, the Vietnamese in the 1960s attempted to mobilize 'international opinion' on their side and were successful to the extent that there were anti-American demonstrations across the world. Similarly, the anti-apartheid movement was very successful in turning South African domestic politics into an international issue in the 1980s. Kuwait today is not a province of Iraq because it was able to turn a 'local' dispute over its sovereignty into the UN-sponsored Gulf War. Listing such well-known examples should not blind us to the fact that converting the scope of a conflict in this way is actually quite exceptional. We might say that the norm is for the losing side in a dispute to fail to widen the scope. An important reason for this failure can be found in our politically divided world. One key role of state boundaries is to prevent politics overrunning into a global scale of conflict at the whim of every loser. But these state boundaries can themselves be contested politically, and there is

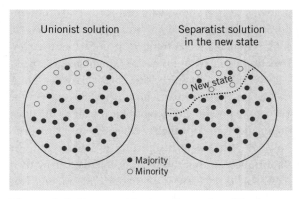

Figure 1.4 Geographical scale and political scope.

no clearer case than contemporary upheavals in eastern Europe following the collapse of communism.

As the former federations of Yugoslavia and the Soviet Union have disintegrated, old federal unit boundaries have been converted into new sovereign state borders. Such a process would be unproblematical if the various national groups formed neat, compact and contiguous spatial units that could be easily enclosed by the boundaries. But cultural and political geography are never that simple; rather, the various ethnic nation groups are typically spatially intermingled. For instance, Russians and Serbians are to be found in sizeable numbers beyond Russia and Serbia, respectively. Figure 1.4 illustrates schematically such a situation, with two groups intermingled in one part of the country. Should this be one country or two? Surely we should let the people say what political structures should prevail. Let us explore such an exercise in democracy.

Assuming that both national groups have been fully mobilized by their respective political elites, we can assume that in a vote at the scale of the original state the unionists in our diagram would win. Most people (that is, the majority national group) do not want their country broken up. 'Foul play!' say the minority, because in the north-west there is a local majority for secession. If the original vote stands, the losers may well take up arms to fight for their 'national independence'. Would such action be anti-democratic? Alternatively, they might be able to marshall international support to get the UN to sponsor

elections held just in the north-west. If this were the case, we can assume that the erstwhile separatists would win and proclaim their independence as a result of a democratic decision of the people. But in doing so they have created a new minority, those who had been part of the majority in the old federation but now find themselves in a state run by their old enemy. If the original unionist solution is abandoned, they in their turn will demand independence for their lands in the new north-west state. Just as surely, the new state will not allow yet another election to divide its newly acquired sovereign territory. So with democracy denied, the only recourse for the newly constructed minority is to resort to arms to win their national independence. Furthermore, these new separatists will be able to draw on the support of their fellow ethnic nationals across the new state boundary.

The above nightmare scenario shows that democracy as a means of solving political disputes is as dependent on the scope of the conflict as any other attempt at resolution. All three solutions to the conflict – unionist, first division, second division – can be legitimated by the democratic will of the people. The question is 'which people?', and in cases of territorial sovereignty this must be a matter of geographical scale. Hence any 'democratic solution' is *not* determined by the voting but by the pre-election geographical decision on the scope of the election: we know the result of the election once we have fixed the boundaries. This is not mere 'fiddling the boundaries' for partisan advantage, for there is no absolutely correct answer to the question of who should vote. In the end, the answer can only be a political decision mediated through the relative power of the participants. But the idea of bringing democracy into the conflict was to prevent the power politics of the elites determining the outcome. We have to conclude sadly that there is no democratic solution in a situation where different geographical scales produce different patterns of national winners and losers.

Although very schematic, most readers will have recognized that the defeat of unionists in Yugoslavia and the Soviet Union has produced situations like the above. Certainly, minority national populations have been produced from what had been the majority; for instance, Serbs, who had been a majority in the original Yugoslavia, are now a minority in Bosnia, and Russians likewise in the Soviet Union and Ukraine. The major losers in this change of scale of politics are these new minorities. It is here that we can expect the most militant nationalism. This has been the case with Serbs in Bosnia and Croatia, and it continues to have the potential to disrupt the former Soviet states. And there is nothing new about this scenario. Who should vote in a referendum to decide whether Northern Ireland joins with the Irish Republic? The scale chosen – all of Ireland, the northern province only – would undoubtedly determine the result. And it is in Northern Ireland that the nationalist cause has most militant support; the nationalist Catholic community is a classic minority that would be a majority. There are no democratic solutions to any of the above because the scope is part of the politics, with choice of geographical scale itself defining victory. As well as illustrating the salience of political geography to understanding the politics of our world today, this example also justifies our choice of geographical scale as the organizing principle for world-systems political geography.

Ideology separating experience from reality

Although globalization is, as Storper (1997: 27) has recently pointed out, a 'fundamentally geographical process labelled with a geographical term', most studies have treated its geography unproblematically. In particular, the basic scale property of globalization has been taken for granted. This is in keeping with the social science tradition, which sees space as merely an inert backdrop to processes of change. Hence the global is treated as a pre-given geographical scale into which modern society and economy have grown. Not surprisingly, such an approach can easily lead to a neglect of other scales of activity, with the global appearing to be almost 'natural'. In contrast, contemporary human geography considers all spaces and places as socially constructed, the results of conflicts and accommodations that produce a geographical landscape. Geographical scale, in particular, is politically constructed (Delaney and Leitner 1997). The argument about democracy and boundaries in the

previous section shows why this is so. Thus contemporary globalization does not represent a scale of activity waiting to be fulfilled; it is part of the construction of a new multi-scalar human geography.

It is perhaps not surprising that it has been political geographers, in particular, who first saw (in the 1970s) the potential of geographical scale as the main organizing frame for their studies. These early works, however, although insightful in identifying the importance of scale, were unproblematical in their treatment of it. Instead of the current singular bias of globalization studies, these political geographies employed a three-scale analysis: international or global, national or state level, and an intra-national, usually urban metropolitan, scale. Although this framework represented a consensus of opinion, it was particularly disappointing that this position was reached with no articulation of theory to justify a trilogy of geographical scales (Taylor 1982). Two questions immediately arise: 'Why just three scales?' and 'Why these particular three scales?' These questions were not answered, because they were not asked. Instead, the three scales are accepted merely as 'given': as one author puts it, these 'three broad areas of interest seem to present themselves' (Short 1982: 1). Well, of course they do no such thing. These three scales do not just happen to appear so that political geographers have some convenient hooks upon which to hang their information. In fact, recognition of the three scales has been implicit in many social science studies beyond political geography (Taylor 1981). It represents a particular way of viewing the world that is subtly state-centric. The scales pivot around the basic unit of the state – hence the international, national and intra-national terminology. Such a position can lead to a separation in the study of geographical scales that destroys the fundamental holism of the modern world-system. For instance, Short (1982: 1) has written of 'distinct spatial scales of analysis', and Johnston (1973: 14) has even referred to 'relatively closed or self-sufficient systems' at these different scales. Obviously, a critical political geography cannot just accept this triple-scale organization as given: the framework must explain why these scales exist and how they relate one to another.

Why three scales? This is not immediately obvious. It is relatively easy to identify many more than three geographical scales in our modern lives. Neil Smith (1993), for instance, argues cogently for a hierarchy of seven basic scales: body, home, community, urban, region, nation and global. Of course, it is easy to add to this; for instance, international relations scholars identify another 'regional' scale between nation-state and the global (western Europe, south-east Asia, and so on). At the other extreme, globalization studies, even when they go beyond their single scale, seem only to see two scales – the local in contrast to the global – for which they have been criticized (Swyngedouw 1997: 159). Environmentalists have been particularly prone to this limited perspective with their famous slogan, 'think global, act local'. Swyngedouw (ibid.) interprets globalization as a 'rescaling' of political economy that moves in two directions away from an institutional concentration of power on the state: upwards to global arenas and downwards to local arenas. With the state remaining in the middle, this represents a construction of a triple-scale organization as pioneered by political geographers but with a theoretical justification. Here we treat the three scales in a more general manner to transcend contemporary globalization by analysing them as integral to the long-term operation of the modern world-system.

From a world-systems perspective, political geographers' triple-scale organization immediately brings to mind Wallerstein's three-tier structure of conflict control (Taylor 1982). We have already come across his geographical example of core/semi-periphery/periphery. We can term this a horizontal three-tier geographical structure. The triple-scale model can then be interpreted as a vertical three-tier geographical structure. The role of all three-tier structures is the promotion of a middle category to separate conflicting interests. In our model, therefore, the nation-state as the pivot becomes the broker between global and local scales. Given that a major political geography aspect of its brokering is to act as a simple buffer, we will treat this arrangement as a classic example of ideology separating experience from reality. The three scales, therefore, can be viewed as representing a national scale of ideology, a local scale

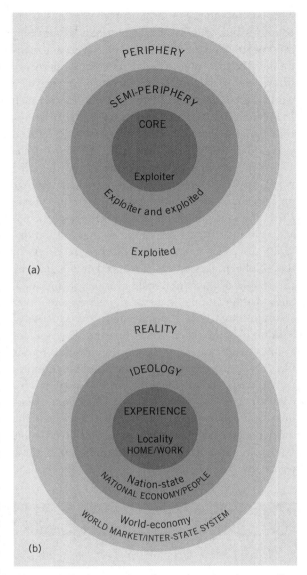

Figure 1.5 Alternative three-tier structures of separation and control: (a) horizontal division by area; (b) vertical division by scale.

of experience and a global scale of reality. This is illustrated schematically in Figure 1.5, where it is compared with Wallerstein's original horizontal geographical structure.

Let us consider this interpretation in more detail. The scale of experience is the scale at which we live our daily lives. It encompasses all our basic needs, including employment, shelter and consumption of basic commodities. For most people living in the core countries this consists of a daily urban 'system'; for most people elsewhere it consists of a local rural community. But the day-to-day activities that we all engage in are not sustained locally. Because we live in a world-system, the arena that affects our lives is much larger than our local community, whether urban or rural. In the current world-economy, the crucial events that structure our lives occur at a global scale. This is the ultimate scale of accumulation, where the world market defines values that ultimately impinge on our local communities. But this is not a direct effect; the world market is filtered through particular aggregations of local communities that are nation-states. For every community, the precise effects of these global processes can be reduced or enhanced by the politics of the nation-state in which it is located. Such manipulation can be at the expense of other communities within the state or at the expense of communities in other states. But the very stuff of Politics (with a capital P) in this framework is in this filter between world-economy and local community.

But why talk of 'ideology' and 'reality' in this context? The notion of scale of experience seems unexceptional enough, but in what sense are the other scales related to ideology and reality? In this model we have very specific meanings for these terms. By 'reality' we are referring to the holistic reality that is the concrete world-economy, which incorporates the other scales. It is, in this sense, the totality of the system. Hence ultimate explanations within the system must be traced back to this 'whole'. It is the scale that 'really matters'. In our materialist argument, the accumulation that is the motor of the whole system operates through the world market at this global scale. In contrast, ideology is a partial view of the system that distorts reality into a false and limited picture. In our model, the reality of the world-system is filtered through nation-centred ideologies to provide a set of contrasting and often conflicting world views. We argue below that such nation-centred thinking has become pervasive in modern politics. This has the effect of diverting political protest away from the key processes at the scale of reality by ensuring that they stop short at the scale of ideology – the nation-state. It is in this sense that we have a geographical model of ideology separating experience from reality. This situation has been summed up well by Nelund (1978: 278):

> [T]he national world picture does not provide us with a language which we can use in our daily life to deal with our concerns. It is a mental burden, and even more it takes us in wrong directions by placing our true concerns beyond our reach, involving us in institutional efforts to reach the issue which we ourselves have displaced.

Shipbuilding: an example of the three-scales approach

Wallsend is a shipbuilding town in north-east England. With the onset of the 1970s' recession there was much local concern for the future of the shipyards, the major employer in the town. Any closure of the industry would have major repercussions throughout the local community. This is the scale of experience. It is at the scale of ideology that policy emerges, however. Following pressure from, among others, the local Labour Party, the British (Labour) government nationalized shipbuilding, including the Wallsend yards. But this was ideological because it reflected only a partial view of the problem. It may have protected jobs in the short term but it did not tackle the problem of Wallsend's shipyards over the long term. These problems derived from the scale of reality – both supply and demand for ships is global in scope. The problem in the industry stemmed from the fall in demand for world shipping in the wake of the 1973–74 oil price rise and the increase in supply with the emergence of new shipyards in other countries such as South Korea. Clearly, a policy of nationalization within Britain is a long way from solving the problems of Wallsend's shipbuilding industry. Rather, it represents a political solution that stops at the state scale, so that there is no challenge to the processes of accumulation at the global scale.

Geographical scale and empire

The construction of imperial control is not simply a matter of one powerful state dominating another. It is about the activities and experiences of individuals operating within groups and institutions. For example, in Chapter 3 we see how British imperialism constructed a particular form of household in colonial Kenya. One key activity of 'empire' is soldiering, and especially the travel into a foreign country to fight for what are represented as 'universal' values of good and morality. For example, an Ohio farmer who served in the Korean War (1950–53) reflects:

> '. . . no war that kills, maims, captures, tortures, and hides so many soldiers and produces so many homeless people, altogether some three million victims, can be justified.
>
> 'Yet, the answer to my question is that I have to justify it. This ex-Ohio farm boy and corpsman must believe the war was worth fighting, for if it was not, then Coy Brewer and all the other casualties of that war died or suffered truly in vain.'
>
> (Quoted in Chappell and Chappell 2000: 142)

The soldier suffers changes in mind and body, serving with a group of comrades, within the hierarchical structure of the armed forces, for country and what was represented as a just cause to protect 'universal' values (see Flint, forthcoming). In turn, his household, family, and community experienced change upon his departure and return.

Of course, there is another side to this story of the destruction of lives, institutions and ways of life of those under the hammer of 'empire' – if not the loss of life, then the loss of property, language, custom, mobility and the like may result. The everyday lives of imperialist and imperialized illustrate that the practice and consequences of 'empire' are enacted at scales from the body to the global.

Consider how your own activity and identity is connected to the local, national and global scales. Are there other scales equally or more important?

This 'national world picture' negates the holism of the modern world-system, keeping most politics away from the scale of the world-economy.

Contemporary globalization represents a classic operation of this scale politics. New state elites are using the global as a threat to redesign national and local politics, and their successes in this new politics show just how limited political resistances to global changes remain. The way politics is legitimated may alter, but the state remains as a buffer between the nationally divided class of direct producers and global capital.

Finally, we must stress that this model does not posit three processes operating at three scales but just one process that is manifest at three scales. In general, this process takes the following form: the needs of accumulation are experienced locally (for example, closure of a hospital) and justified nationally (for example, to promote national efficiency) for ultimate benefits organized globally (for example, by multi-national corporations paying less tax). This is a single process in which ideology separates experience from reality. There is but one system: the capitalist world-economy.

■ Power and politics in the world-economy

The position we take in interpreting political events in the world-economy is based upon the analysis of Chase-Dunn (1981, 1982, 1989). As we have seen, the capitalist mode of production involves the extraction of economic surplus for accumulation within the world-economy. This surplus is expropriated in two related ways. The distinguishing feature of our system is the expropriation via the market, but the traditional expropriation method of world-empires is not entirely eliminated. The use of political and military power to obtain surplus is the second method of expropriation. Of course, this second method cannot dominate the system or else the system would be transformed into a new mode of production. But neither should it be undervalued as a process in the world-economy from the initial Spanish plunder of the New World to today's support of multinational corporate interests by the 'home' country, usually the United States. The important point is that these two methods of expropriation should not be interpreted as two separate processes – or 'two logics' as Chase-Dunn (1981) terms them – one political and the other economic. In our framework, they are but two aspects of the same overall political economy logic. Chase-Dunn (1982: 25) puts the argument as follows:

> The interdependence of political military power and competitive advantage in production in the capitalist world-economy reveals that the logic of the accumulation process includes the logic of state building and geopolitics.

This position has been endorsed and expanded upon by Burch (1994: 52), who sees the 'distinguishing feature of the modern world' as being 'the intimate, inextricable singularity of capitalism and the states system'. In other words, political processes lie at the heart of the capitalist world-economy, and they are not located there separately in isolation.

The argument so far has equated politics with activities surrounding states. While state-centred politics are crucial to an understanding of the world-economy, they do not constitute the sum of political activities. If we equate politics with use of power then we soon appreciate that political processes do not begin and end with states: all social institutions have their politics.

The nature of power: individuals and institutions

We have previously discussed different types of power, here we consider its operation in the modern world-system. For this practice of power, the scope of activities is also brought back into the argument and further elaborated. We can begin by considering power at its most simple level. Consider two individuals, A and B, who are in conflict over the future outcome of an event. Let us suppose that A's interests are served by outcome X and B's by outcome Y. Then simply by observing which outcome occurs we would infer the power ranking of A and B. For instance, if X occurs we can assert that A is more powerful than B. When we inquire why A was able to defeat B, we would expect to find that in some sense A possessed more resources than B. If this were a schoolyard quarrel, we might find that A packed the harder punch.

This model provides us with an elementary feel for the nature of power but only that. The world of politics does not consist of millions and millions of conflicts between unequal pairs of individuals. Potential losers have never been naive enough to let such a world evolve. Let us return to our simple example to show how we can move beyond pairwise conflict. Consider conflict occurring in a schoolyard again. The two fighters will inevitably attract a crowd. B, as we have seen, is losing. What should she or he do? The answer is simple: before defeat occurs she or he must widen the scope of the conflict by inviting the crowd to participate. By changing the scope of the conflict B is changing the balance of power. If B's gang is stronger than A's then it is now in the interests of the latter to widen the scope further – say by bringing in the school authorities to stamp out gang warfare.

This model of power resolution is derived from Schattschneider (1960), who argues generally that the outcome of any conflict will not depend on the relative power of the competing interests but will be determined by the eventual scope of the conflict. Hence it follows that the most important strategy in politics is to define the scope of any conflict. Furthermore, the following basic point now becomes explicit: since every increase in scope changes the balance of power, weaker interests should always be pushing to widen the scope of their conflicts. Historically, this has been illustrated by two opposing political strategies: on the left, collectivist politics are preached; on the right, a much more individualistic approach is favoured. Two important political conflicts of the nineteenth and twentieth centuries show how this has operated in practice (Taylor 1984).

First, the extension of electoral suffrage has gradually increased the scope of national politics, culminating in universal suffrage. This has changed the policies of parties and the practices of governments as politicians have had to respond to the needs of their new clients. Second, the rise of trade unions is an explicit means of widening the scope of industrial conflicts beyond the unequal contest of employer versus individual employee. The basic union strategy of broadening the scope of conflicts has been resisted by employers trying to keep disputes 'local', originally by having trade unions outlawed and subsequently by legal constraints on the scope of their activities. The history of both democratic politics and trade unionism is at heart a matter of changing the scope of conflicts.

The corollary of Schattschneider's ideas is that this 'politics from below' will eventually extend conflicts to a global scale. Such 'internationalism' has a hallowed place in left-wing politics and traces its origins to Marx's 'First International' of 1862. In the late twentieth century, it was noticeable that the poorer countries of the world made most use of the United Nations. But all such internationalism has either come to nothing or has been relatively ineffectual. In fact, globalization can be interpreted as a historical reversal of the politics of scale: today it is capital political interests on a global scale through its neo-liberal rhetoric and practice that is calling the tune (Martin and Schumann 1997: 6–7). However, the scope of most politics is decidedly not global, because a vast array of institutions have been created between the individual and the ultimate scope of politics at the global scale. The main theme of this book is about understanding how the scope of conflicts has been limited. What are the key institutions in this process?

Out of the multitude of institutions that exist, Wallerstein (1984a) identifies four that are crucial to the operation of the world-economy. First, there are the *states*, where formal power in the world-economy lies. States are responsible for the laws that define the practices of all other institutions. We have previously described the importance of this institution, and much of the remainder of this book is devoted to developing themes concerning the power of states.

Second, there are the *peoples*, groupings of individuals who have cultural affinities. There is no single acceptable name for this category of institution. Where a cultural group controls a state they may be defined as a nation. Where they constitute a minority within a state they are sometimes referred to literally as 'minorities' or as ethnic groups. Such minorities may aspire to be a nation with their own state, such as the Tamils in Sri Lanka or the Basques in Spain. To complicate matters, there are some multi-state nations such as the Arab nation. Despite the complexity of this category of institution, the importance of 'peoples' cannot be doubted in modern politics.

The third category of institution is perhaps less complex but no less controversial than 'peoples'. The world's population can be divided into strata based upon economic criteria, which we will term *classes*. Wallerstein follows Marxist practice here and defines classes in terms of location within the system's mode of production. Since the latter is currently global, it follows that in world-systems analysis classes are defined as global strata.

At the other end of the scale are the *households*, Wallerstein's fourth key institution. These are defined not by kin or cohabitation but in terms of the pooling of income. They are, therefore, small groups of individuals who come together to face an often hostile world. The basic behaviour of such a group is the operation of a budget that combines resources and allocates expenditure. Wallerstein considers such households to be the 'atoms' of his system, the basic building block of the other institutions. Hence everybody is first of all a member of a household; that household is subject to the laws of a particular state; it will have cultural affinities with a certain 'people'; and it will be economically located within a specific class.

Wallerstein (1984a) considers these four institutions, as he defines them, to be unique to the capitalist world-economy. They interact one upon another in many ways, forever creating and recreating the temporal and spatial patterns described in the last section. Households are important, for instance, in maintaining the cultural definitions of 'peoples', while 'peoples' fundamentally influence both the boundaries of states and the nature of class conflicts. This is Wallerstein's 'institutional vortex', which underlies the whole operation of our modern world-system.

Institutions both facilitate and constrain the behaviour of individuals by laws, rules, customs and norms. What is and what is not possible will vary with the power of particular institutions. For each category of institution the distribution of power will vary both within and between institutions. For instance, we can ask both who controls a particular state and what the power of that state is within the inter-state system. In this way, we can identify hierarchies of power within and between all four institutions. We illustrate this below by concentrating on particular aspects of informal power distributions.

Power within households

Income pooling may be reduced to daily, weekly or monthly budgets, but it entails a continuity that is generational in nature. Households are frequently changing as some members die and others are born, but they typically show a constancy that allows for the reproduction cycle: it is within households that the next generation is reared. This assumes a pattern of gender relations within households. In the capitalist world-economy, the particular form that gender relations take is known as patriarchy, the domination of women by men.

The notion of income pooling does not, of course, assure equality of access to the resources of a household. The arrangement of work in many different kinds of household across the world provides men with the access to cash and therefore markets, leaving women with 'domestic chores'. In core countries, this has generally led to a devaluing of many women's contributions to the household as 'merely housework'. In peripheral countries, this has often led to the devaluing of food production as 'women's work' relative to male-controlled cash crop production.

This constitutes a very good example of how the scope of a politics has sustained a particular hierarchy of power. In the case of households, we are entering the private world of the family: what goes on between 'man and wife' is not in the public domain. This narrow scope has led to condoning, or at least ignoring, the most naked form of power – physical violence. 'Outsiders', both public officials and neighbours, have been loath to interfere even in the most extreme cases. To the degree that women are confined to the private world of the family they are condemned to political impotence. There are no trade unions for either housewives or food crop producers.

The patriarchy found within households permeates all levels of the world-economy. Where women do enter waged work they typically get paid less and are less well represented as we move to higher levels of any occupational ladder. This endemic sexism is most clearly illustrated in politics, where male politicians dominate the legislatures of all states (refer back to the panel 'Modern monarchy, gender and power' on page 24). Women are under-represented in all legislatures throughout the world. This gender inequality is even more marked in the executive branch of government in all types of regime – liberal democracies, old communist states, military dictatorships, traditional monarchies, and so on.

Power between 'peoples'

'Peoples' reflect the diversity within humanity that has always existed. In the world-economy, this human variety has been used to create specific sets of 'peoples' to justify material and political inequalities. Three types of 'people' have been produced – races, nations and ethnic groups – and each relates to one basic feature of the world-economy.

Race is a product of the expansion of the modern world-system. With the incorporation of non-European zones into the world-economy, the non-European peoples that survived were added to the periphery. In this way, race came to be expressed directly in the division of labour as a white core and a non-white periphery. Until the dismantling of apartheid, the South African government recognized this power hierarchy when it designated visiting Japanese businessmen as 'honorary whites' in their apartheid system. More generally, the ideology of racism has legitimated worldwide inequalities throughout the history of the world-economy.

'Nation' as a concept rose to express competition between states. It legitimates the whole political superstructure of the world-economy that is the inter-state system: every state aspires to be a 'nation-state'. By justifying the political fragmentation of the world, nations play a key role in perpetuating inequalities between countries. The associated ideology, nationalism, was the most powerful political force in the twentieth century with millions of young people willingly sacrificing their lives for their country and its people.

Ethnic groups are always a minority within a country. All multi-ethnic states contain a hierarchy of groups, with different occupations associated with different groups. Where the ethnic groups are immigrants, this 'ethnization' of occupations legitimizes the practical inequalities within the state. In contrast, the inequalities suffered by non-immigrant ethnic groups can produce an alternative minority nationalism to challenge the state.

The concept of 'peoples' covers a difficult and complex mixture of cultural phenomena. We have only scratched at the surface of this complexity here. Nevertheless, it has been shown that peoples are implicated in hierarchies of power from the global scale to the neighbourhood. They remain key institutions both for the legitimization of inequalities and for political resistance. Under contemporary conditions of globalization their salience has increased as groups emphasize their particularities in response to tendencies towards cultural homogenization. We deal with these issues in detail in Chapter 5.

Power and class

All analyses of power and class start with Marx. At the heart of his analysis of capitalism there is a fundamental conflict between capital and labour. In class terms, the bourgeoisie owns the means of production and buys the labour power of the proletariat. In this way, the whole production process is controlled by the former at the expense of the latter. Hence this power hierarchy and the resulting class conflict are central to all Marxist political analyses.

Wallerstein accepts the centrality of class conflict in his capitalist world-economy. However, since his definition of mode of production is broader than Marx's, it follows that Wallerstein's identification of classes diverges from that of orthodox Marxism. For instance, Wallerstein's strata of labour are termed direct producers and include all who are immediate creators of commodities – both wage earners and non-wage producers. Hence the proletariat wage earners are joined by peasant producers, sharecrop-

pers and many other exploited forms of labour, including the female and child labour often hidden within households.

On the other side of the class conflict are the controllers of production, who may or may not be 'capitalists' as owners of capital in the original Marxist sense. For instance, the typical form of capital in the late twentieth century is the multinational corporation. The executive elite who control these corporations need not be major shareholders; certainly their power within the organization does not depend on their shareholding. Although formally employees of the corporations it would be disingenuous not to recognize the very real power that this group of people command. They combine with another group of controllers, senior state officials, who also command large amounts of capital, to produce the 'new bourgeoisie' of the past fifty years or so. First recognized by Galbraith (1958) these have become especially important with economic globalization: Sklair (2002: 98–105) identifies a transnational capitalist class.

Marx recognized the existence of a middle class between proletariat and bourgeoisie but predicted that this intermediate class would decline in size and importance as the fundamental conflict between capital and labour developed. In fact, this has not happened in the core countries of the world-economy. Instead we have had the 'rise of the middle classes' as white-collar occupations have grown and have numerically overtaken blue-collar workers. This large intermediate stratum combines a wide range of occupations with seemingly little connection. Wallerstein interprets persons with these occupations as the cadres of the world-economy. As capitalist production and organization have become more complex there has arisen an increasing need for cadres to run the system and make sure that it operates as smoothly as possible. Originally, such cadres merely supervised the direct producers on behalf of the capitalist controllers. Today, a vast array of occupations are required for the smooth running of the system. These include the older professions, such as lawyers and accountants, and many new positions, such as middle managers within corporations and bureaucrats within state organizations. The end result is a massive middle stratum of cadres between controllers and direct producers. Wright (1997: 15–22) provides a useful classification of class locations by adding 'authority' to 'means of production' to differentiate middle-class 'controllers' in the transnational capitalist class from cadres who have much less authority or expertise. The cadres create a classic example of Wallerstein's three-tier structure facilitating the stability of the world-economy.

As previously noted, since classes are defined in terms of mode of production it follows that in the world-economy today classes are global in scope. We shall term them 'objective' classes since they are derived logically from the analysis. In terms of actual political practice, however, classes have most commonly defined themselves on a state-by-state basis. These subjective 'national classes' are only parts of our larger objective 'global classes'. That is to say, the scope of most class actions has been restricted to less than their complete geographical range. But not all classes have been equally 'national' in scope. While the proletariat have had the internationalist rhetoric, it is the capitalists and the controllers who have been the more effective international actors – in world-systems analysis it is emphasized that capitalist subjective class actions have always been the closest to their objective class interests. At the present time, this is demonstrated by economic globalization, where corporations, guided by the transnational capitalist class, move their production units around to reduce labour costs. The direct producers have no organized strategy to combat the controllers' ability to create new global geographical divisions of labour. The state is clearly implicated in these key constraints on the scope of conflicts in a globalizing world, and this lies at the heart of the political geography that we develop in this book.

Politics and the state

The state is the locus of formal politics. Most people's image of the operation of power and politics comprises activities associated with the state and its government. In this taken-for-granted world the state is *the* arena of politics. Typically, therefore, many political studies have limited their analyses to states and governments. But this is to equate power and politics

in our society with just the formal operation of state politics. Our previous discussion of other institutions has indicated the poverty of such an approach. There is no a priori reason why we should not be equally concerned with questions of power in other institutions, such as households. Marxists, of course, would point to the centrality of classes in any consideration of power.

The way forward from this position is not to debate the relative importance of the different institutions, since it is impossible to deal seriously with any one of them separately. As previously noted, they are interrelated in so many complex ways that Wallerstein (1984a) refers to them as 'the institutional vortex'. Treating them separately as we have done so far can be justified only on pedagogical grounds. In reality, power in the modern world-system operates through numerous combinations of the institutions. From this perspective, one study has enumerated no less than fourteen different types of politics – see Chapter 8. This implies that there are at least fourteen different political geographies we should study. We cannot pretend to do justice to such a range of politics in this book, so there is a need to justify the particular bias in what follows.

Most political geography, like other political study, has been state-centric in orientation. That is to say, it has treated the state as its basic unit for analysis. From a world-systems perspective, the state remains a key institution but is no longer itself the locus of social change. We wish to avoid the state-centric constraints but in no way want to imply that the state is not an important component of our study. In short, the state must be located in a context that maintains its importance but without simultaneously relegating the other institutions. This is what we have attempted to achieve in Figure 1.6, which sets out one of the many relationships that exist between the four key institutions.

Starting with the households: these are the basic social reproducers of the system whereby individuals are socialized into their social positions. In Figure 1.6, we emphasize the transmission of cultural identities that reproduce 'peoples', in particular the nations of the world. These nations then relate to the other two

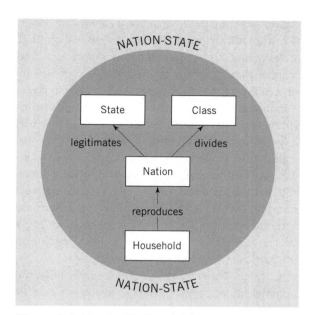

Figure 1.6 Key institutional linkages.

relationship to space. They not only occupy space, as any social institution must, but they also claim particular association with designated places. It makes no sense to have a nation without a 'historical homeland', and states do not exist except through possession of their 'sovereign territory'. In other words, the spatial location of state and nation is intrinsically part of their being. It follows that political geography should focus on the bias highlighted in Figure 1.6 as our particular window on the modern world-system. We develop this argument further in the final section of this chapter; in the meantime, we continue our exploration of the nature of power relations through the familiar activities of states.

institutions in quite contrary ways. For classes, as we have already noted, their global objective status is compromised by subjective organization as national classes. The crucial relation is therefore one of divide. In the case of states, on the other hand, nation and state have been mutually supportive as 'nation-states'. This modern legitimation has become so strong that in our everyday language state and nation are often used interchangeably: by this time, readers should be clear in their own minds that, for instance, it is not nations that compete at the Olympic Games but states. The concept of the nation-state confuses the very important distinction between these two institutions (remember nations are cultural groupings, states are political apparatuses); our purpose in this book is not so much simply to correct this common misconception but to understand how it came about. Hence the particular bias in this political study is towards state and nation but without the neglect of the other institutions that can ensue from a simple state-centric approach.

Our bias can be justified on geographical grounds. As we shall see in the following chapters, as social institutions both state and nation are unique in their

Summary

Political geography and the institutions of the capitalist world-economy

How do the institutions of the capitalist world-economy provide a framework for pursuing political geographic inquiry?

➤ They help us conceptualize the arenas and goals of politics: the struggle to control and manipulate institutions as well as the politics of restricting the scope of particular issues to within institutions.

➤ The institutions both frame and are maintained by everyday activity. For example, the patriarchal household delimits the role of women and is constructed by such constrained roles.

➤ Political activity is contextualized (a) within the scope of institutions and (b) within the way that institutions relate to each other.

➤ Contextualizing everyday activity within institutions illustrates the way that the 'higher' scales of the state and the global are constituted by the activities at the scale of the household and 'peoples'.

➤ 'Difference' may be mapped and understood by analysing how institutions vary in form and function across time and space.

■ A political geography perspective on the world-economy

Power is mediated through institutions in particular places. Institutions and places are expressions of the three types of power we discussed earlier: inscribed, resource and strategy (Allen 2003). In addition, our focus on geographical scale emphasizes that political actions must be contextualized within broader dynamics. In combination we frame political geography as multi-scalar, territorial and networked to complete the power-geometry.

To recap, we have introduced:

➤ one unit of analysis: the capitalist world-economy;

➤ three types of power: inscribed, resource and strategy;

➤ three geographical scales: local, nation-state and world-economy;

➤ four institutions: households, states, 'peoples' and classes;

➤ three silences we hope to assist in breaking: difference, statism and gendered analysis;

➤ a seven-point framework for a political geography adopting the world-systems approach.

In combination we offer a conceptualization that contextualizes politics in a geographical framework of a global structure, and intertwines the material with the discursive. We also wish to provide coherence, a way to negotiate all of these different considerations. Our strategy is to refer predominately to the world-systems framework while emphasizing geographical scale. In doing so, we will continually refer to the different forms of power, the four institutions and the means by which we can provide our particular insights into the silences.

World-economy, nation-state and locality

The model of three geographical scales represents our particular organization of political geography summarized by the subtitle of this book: world-economy,

nation-state and locality. Hence we follow the established political geography pattern of using three scales of analysis but treat them in a more analytical manner than other studies have done. Even though each of the following chapters concentrates largely on activities at one of the three scales, they do not constitute separate studies of each scale. For instance, imperialism is a concept associated with the global scale, but we argue that it cannot be understood without consideration of forces operating within states. Alternatively, political parties operate at the national scale, but we argue that they cannot be understood without consideration of the global scale. In every chapter, discussion ranges across scales depending on the particular requirements for explanation.

Chapters 2 and 3 are devoted to geopolitics and imperialism, respectively. In both chapters we define a framework of political cycles in presenting a dynamic model of politics in the world-economy. After developing a historical analysis we discuss the War on Terrorism and contemporary discussion of 'empire'. The analysis is framed through a world-systems perspective, but we integrate the work of feminist geographers and other scholars to complement and extend our theoretical perspective.

In Chapters 4 and 5 we deal with the classic trilogy of political geography: territory, state and nation. We develop ideas on the state as a mechanism of control and the nation as a vehicle for political consensus. Reinterpretations and new ideas from our world-system logic involve the spatial structure of the state, a theory of states in the world-economy, a materialist theory of nationalism and the representation of gender roles in national identity.

Chapter 6 stays at the same scale in its treatment of electoral geography. We employ world-systems logic to interpret elections and the operation of parties in all parts of the world. We argue that electoral geography must develop its theoretical complexity to match the growth in the output of empirical studies. Especially, questions regarding the global geography of liberal democracy as well as the mismatch between the geography of voting and the behaviour of political and economic elites must be considered.

In the final two chapters we consider the local scale as the localities we experience in our everyday lives.

In Chapter 7 we consider localities as arenas in which politics takes place. We address the geography of capital cities as well as the local politics of citizenship and exclusion. The chapter culminates in a discussion of world cities, thus linking the global and the local, and moving beyond a state-centric view to consider the new networks of global politics. In Chapter 8 we move from space to place considerations, so our localities become more 'lived in'. Here we explore a politics of identities in places and the idea of the emergence of a new politics of identities through the key institutions of the modern world-system. In this concluding chapter we integrate the structures and dynamics of the capitalist world-economy with the concerns of agency, identity and hybridism championed by feminist geographers.

The final product is a political geography that attempts to rethink our studies in world-systems terms. There is some new wine in old bottles but also some old wine in new bottles. Although none of this wine is as yet sufficiently matured, it is hoped that it will not taste too bitter for the discerning reader.

Key glossary terms from Chapter 1

anarchy of production	error of	liberal	protectionism
Annales school of	developmentalism	liberal democracy	racism
history	executive	local government	scope of conflict
apartheid	faction	logistic wave	secession
balance of power	federation	*longue durée*	semi-periphery
boundary	feudalism	Marxism	Social Darwinism
bourgeoisie	formal imperialism	mercantilism	socialism
capitalism	free trade	mini-systems	sovereignty
capitalist world-	fundamentalism	minorities	space
economy	geopolitics	mode of production	spaces of flows
classes	globalization	multinational	spaces of places
Cold War	government	corporations	state
colonialism	home	nation	structural power
communism	homeland	nationalism	suffrage
core	households	nation-state	third world
decolonization	ideology	NATO	unitary state
democracy	imperialism	neo-liberalism	United Nations
development of	informal imperialism	patriarchy	world cities
underdevelopment	inter-state system	peoples	world-economy
economism	Islam	periphery	world-empire
elite	Kondratieff cycles	place	world market
empire	(waves)	political parties	world-system
'empire'	left-wing	power	world-systems analysis
	legislature		

Suggested reading

Agnew, J. (2002) *Making Political Geography.* **London: Arnold.** A concise discussion of the history of modern political geography, key theoretical concepts and future directions.

Hall, T. D. (2000) *A World-Systems Reader: New Perspectives on Gender, Urbanism, Cultures, Indigenous Peoples, and Ecology.* Lanham, MD: **Rowman & Littlefield.** A collection of papers that illustrates the wide range of world-systems studies beyond political economy.

Johnston, R. J. and Sidaway, J. D. (2004) *Geography and Geographers: Anglo-American Human Geography since 1945,* **6th edition. London: Arnold.** A comprehensive analysis of the changing theoretical trends in social science and the manner in which they have been adopted and adapted by human geographers. Political geography is given ample coverage.

Staeheli, L. A., Kofman, E. and Peake, L. (2004) *Mapping Women, Making Politics: Feminist Perspectives on Political Geography.* **New York: Routledge.** A groundbreaking compilation of chapters covering the breadth of political geography subject matter. Shows the costs of ignoring a feminist analysis of political geography, and the insights to be gained by including feminist theory and methods.

Wallerstein, I. (2004) *World-Systems Analysis: An Introduction.* **Durham, NC: Duke University Press.** A concise and accessible discussion of the key concepts and arguments of the world-systems approach.

Activities

1 The documentaries of Michael Moore have been concerned with the experiences of individuals from his home town, the depressed de-industrial town of Flint, Michigan. View his early film *Roger and Me* or the Flint scenes in *Fahrenheit 9/11* and note connections between the plight of the city and its inhabitants on the one hand, and national political decisions, the calculations of firms operating in the world-economy and the global military activity of the United States on the other hand.

2 Consider a news broadcast on the television or the website of the BBC, CNN or a major newspaper. Identify stories that exemplify the three different forms of power discussed in the chapter: capacity, resource and practice. For one of the stories, think about the way one form of power requires the exercise of another form of power. Does considering different types of power lead to consideration of other geographical scales?

3 Select one leading newspaper in your country (non-tabloid) and inspect recent copies (or the newspaper's website). Try to classify news items in terms of our three geographical scales. In fact the stories are likely to be local, national and foreign, with the latter focusing on one country rather than worldwide. This shows the continuing salience of states in the way the world is framed. Nevertheless, search out 'global stories'. Consider ways in which the stories differ across scales. Do the stories exemplify the way we have interpreted the scales – local as concerned with immediate work and life; national and foreign suggesting 'us' together, and 'them' as different or threatening; worldwide on just economic and environmental issues? Are there stories that cut across scales?

IRAN-IRAQ WAR. AN IRANIAN SOLDIER WATCHES AS SMOKE BILLOWS FROM MULTIPLE BURNING OIL REFINERIES IN ABADAN, IRAN.
Source: © Henri Bureau/Sygma/Corbis

Geopolitics revived

What this chapter covers

Geopolitics has played an important role in the history of political geography at two separate times. At the beginning of modern geography, in the 1890s and the following decade, the state-centred strategic calculations of geopoliticians situated political geography as an essential tool of modern state building, imperialism and 'great power' competition. Almost one hundred years later, in the 1980s, critical geopolitics connected political geography to the developments of human geography that, under the broad influence of postmodernism, had taken a 'cultural turn' in which landscapes, media and everyday behaviour were deconstructed and read to uncover power relations. In the first period, geographers aimed at informing the actual practices and behaviour of states. In the second period, geographers were, on the whole, critical of the geopolitical actions of states. Arguably, we are at the beginning of a third period, led by a feminist geopolitics that 'democratizes' the notion of geopolitics by forcing us to consider the 'everyday' and the 'ordinary'. We think the Annales school would approve.

In this chapter we discuss historical and contemporary geopolitics with two aims and beliefs. First, contemporary events – the talk of 'empire' – can only be understood by placing them in historical context. Second, a full understanding of geopolitical actions requires a scalar approach; the 'everyday' can be understood by situating it within structures that require consideration of the nation-state and the capitalist world-economy.

Reading children's stories in camouflage in Afghanistan

In March 2005, the US Department of Defense proudly displayed the details of a programme initiated by a member of the armed forces serving in Afghanistan. The deployment was part of the declared War on Terrorism. The human-interest story was of a programme whereby military personnel could be videotaped, in an army base in Kabul, reading a children's story and have it sent home. The story of the US Army's videotaped reading programme brings together many components of geopolitics that are discussed in this chapter. On the one hand is the 'geostrategic' factor of the United States calculating the level of threat emanating from Afghanistan and prioritizing military control of Afghani sovereign space. This is the stuff of state-centric geopolitical calculation of strategy and territorial control. On the other hand, there is also the story of individuals and the everyday practices that make the grand strategy happen. The soldier featured, Sergeant Tania Steele, is an African-American woman whose daughters are 5 and 7 years old. The reading programme stems from the recognition that the politics of the family is part of the everyday experience of geopolitics. Not only does the programme try to minimize the disruption to the roles of mother and daughter brought on by military service, but it also promotes a normative heterosexual notion of the family and women's roles. Moreover, by tying the defence of the nation to 'family values' the disrupted connection between mother and daughter is used by the Army to illustrate the 'reason' for the military action: to protect 'our way of life' that is somehow captured in the everyday notion of a mother reading to her children at bedtime.

Think back to the nature of power and the institutions of the world-economy discussed in Chapter 1. What forms of power and what institutions are involved in the Army's reading programme? Think of the individual soldier as well as the deployment of the force in your answer.

Source: Lt. Col. Frederick Rice, 'Kabul soldier reaches volunteer program milestone', US Department of Defense. www.defenselink.mil/news/Mar2005/20050222_281.html. 23 March 2005. Accessed 27 March 2005.

Chapter framework

The chapter is framed around four of the seven ideas we introduced at the beginning of Chapter 1:

➤ The practice of geopolitics requires both actual *actions and the representation* of those actions, with critical geopolitics focused upon the latter.

➤ Geopolitics certainly was, and largely still is, seen as the study of elites and the actions of states; new insights stem from *breaking this particular 'silence'.*

➤ World-systems analysis is effective in understanding the *temporal context* (in terms of the rise and fall of hegemonic powers) for the creation of particular geopolitical theories and their adoption by states as geopolitical codes.

➤ A consideration of *geographical scale* situates the everyday within broader structures and shows how everyday actions challenge existing structures.

■ What is geopolitics? Past, present and future

The demise and rise of geopolitics since the Second World War has been quite remarkable. For most of that time, geopolitics was virtually abandoned as an academic discourse. One result of this demise has been the cutting off of political geography from its distinguished heritage of leading geographers such as Friedrich Ratzel in Germany, Sir Halford Mackinder in Britain and Isaiah Bowman in the United States. That political geographers were willing to take such an extreme step is testimony to the profound impact of 1930s German geopolitics on political geography in particular and geography in general. The term 'geopolitics' became an embarrassment to be distinguished from 'respectable' political geography. During geopolitics' 'abandonment' period, Saul Cohen (1973, 2003) was the major exception among political geographers in keeping global thinking alive in political geography. He understood that geopolitical issues were too important a subject for geographers to ignore. And now he has been joined by many

other geographers in an overdue but no less welcome revival of geopolitics.

In outlining a history of geography's engagement with issues of war and peace, Virginie Mamadouh (2005) identifies a broad change from the end of the nineteenth century, when geographers offered practical guidance to states in the pursuit of war and territorial control, to the beginning of the twenty-first century, when geographers were critical of militaristic agendas. Prior to the Second World War, geopolitics could be defined as the representation of territories and the presentation of theories to assist in controlling them. The most notable theories of this time were Sir Halford Mackinder's heartland theory and Friedrich Ratzel's ideas of the organic state and the necessity of border conflicts. A crucial episode in the story of geopolitics is the relationship between the geographer and retired soldier General Karl Haushofer and the leadership of the Nazi Party. During the Second World War the American media portrayed an academy of scientists (the Institute for Geopolitics) and Haushofer as the 'evil genius' behind Hitler's attempt at 'world domination'. The account was highly fictionalized and simply a propaganda tool (Ó Tuathail 1996). However, the consequences for political geography were enormous. Political geography became equated with geopolitics and its 'evil' uses by the Nazis. Geopolitics was a 'hot potato' and most academics dropped it.

The revival of geopolitics has taken three distinct forms. First, and perhaps most intriguingly, 'geopolitics' has become a popular term for describing global rivalries in world politics. Hepple (1986) traces this recent usage of the term to the extensive references to 'geopolitics' in former US Secretary of State Henry Kissinger's widely quoted memoirs. From media discussion of Kissinger's ideas the term re-entered popular language so that today it is not unusual to encounter reference to geopolitics in any serious press article on world political issues. In this usage, geopolitics seems to be a shorthand phrase to indicate a general process of managing global rivalry to produce a balance of power or equilibrium in world affairs. The general implication has been that US politicians need to master this art in the new circumstances of the relative decline of US power. One indication of continued popular acceptance of the term 'geopolitics' it that it has recently spawned a twin: 'geo-economics'. This reflects recognition of changing US foreign priorities after the Cold War, with economic linkages coming to the fore under conditions of globalization.

This popularizing of geopolitics was no doubt important in lessening geographers' inhibitions over their erstwhile wayward child. Hence the second form that the revival has taken is an academic one. Four related research tendencies can be identified. First, there have been revisionist historiographical studies of past geopolitics. Coming to terms with geopolitics' notorious past is an obvious necessity in developing a new 'geographer's geopolitics' (Dodds and Atkinson 2000). Such revisionist histories have included both reassessment of key figures of the past, such as Isaiah Bowman (N. Smith 1984, 2003), and fresh perspectives on German geopolitics (Heske 1986; Bassin 1987; Sandner 1989). Second, political geographers have become particularly involved in researching the geography of traditional topics in international relations. For example, Nijman (1992) illustrated the dynamic geography of the Cold War, while others have studied the geography of trade (Grant and Agnew 1996) and foreign aid (Holdar 1994; Grant and Nijman 1997), conflict (O'Loughlin 2005) and the armed forces (Ek 2000). In addition, Dodds' (1997) study of Antarctica provides fresh insights into a long-standing geopolitical issue. Third, there has been an incorporation of political economy themes into geopolitics, notably hegemony (Agnew and Corbridge 1995; Taylor 1996; Agnew 2005), non-governmental organizations (Mohan 2002) and supra-states (Moisio 2002), as part of a new international political economy. Finally, 'critical geopolitics,' employing post-structural interpretations of geopolitical practices (Ó Tuathail 1996; Ó Tuathail and Dalby 1998), made crucial connections between political geography and developments in human geography and the social sciences. The debates initiated by critical geopolitics have been critiqued and extended by feminist geopoliticians (Hyndman 2001; Dowler and Sharp 2002; Gilmartin and Kofman 2004).

Despite numerous interactions between these recent developments, it is clear that contemporary geopolitical studies lack cohesion. Although our world-systems analysis produces an international political economy bias, in this chapter we attempt to weave together elements of all four current research tendencies in a framework for studying global rivalry in political geography.

Before we embark on this task, however, we need to mention briefly the third form that the revival of geopolitics has taken. This is associated with the neo-conservative, pro-military lobby, which added geopolitical arguments to its 'Cold War rhetoric' in the 1970s and 1980s (Dalby 1990a, b; Sharp 2000). Such studies talk of 'geopolitical imperatives' and treat geography as 'the permanent factor' around which all strategic thinking must revolve. This naive treatment of space as an unchanging stage exposes a basic lack of understanding of the nature of geography. Many years ago, Bowman (1948: 130) provided the answer to such simplistic reasoning:

> It is often said that geography does not change. In truth geography changes as rapidly as ideas and technology change; that is, the meaning of geographical conditions changes.

We shall be concerned with such a 'geographer's geopolitics' that charts the changing meaning of the 'geographical factor'. The simple geopolitics of Cold War rhetoric will be of interest only in so far as it has had an influence on the politics of global rivalry.

Geopolitics is not the only topic in political geography that deals with world politics; there is also the political geography of imperialism. Before we deal with either subject we need to distinguish between geopolitics and imperialism. Although not usually discussed in the same context, both terms relate to political activity at an inter-state scale. Current usage distinguishes geopolitics as concerned with rivalry between major powers (core and rising semi-peripheral states) and imperialism as domination by strong states (in the core) of weak states (in the periphery). Politically, geopolitics describes a rivalry relationship, whereas imperialism describes a dominance relationship. Spatially, they have been described as East–West and North–South spatial patterns, respectively. Although there have been previous definitions, and these are dealt with below, we shall employ these current uses of the terms here. The interesting question for political geography is the relationship between the two concepts in their politics and in their spatial structures. Like so much else in our modern world, the practices that constitute these two concepts are not really separate and autonomous; for instance, the 'founding father' of geopolitics, Halford Mackinder, was simultaneously a great promoter of the British Empire. One advantage of our world-systems approach is that we can treat geopolitics and imperialism together as part of the same story, two faces of one world politics. However, for pedagogical reasons we deal with geopolitics in this chapter and focus on imperialism in the next one.

Summary

This section has shown the important role that geopolitics has played in the history of political geography. Specifically:

➤ geopolitics is both political practice and academic exercise;

➤ geopolitics is a matter of elite foreign-policy makers, academics and also popular culture;

➤ the form and visibility of geopolitics changes over time.

■ Geopolitical codes and world orders

Practical geopolitical reasoning produces what Gaddis (1982) calls geopolitical codes. These are operational codes consisting of a set of political geography assumptions that underlie a country's foreign policy. Such a code will have to incorporate a definition of a state's interests, an identification of external threats to those interests, a planned response to such threats and a justification for that response. There will be as many geopolitical codes as there are states.

Geopolitical codes are closely related to what Henrikson (1980) calls 'image-plans'. Such operational codes involve evaluation of places beyond the state's boundaries in terms of their strategic importance and as potential threats. Geopolitical codes are not just state-centric, they also involve a particular single state's view of the world. They are by definition, therefore, highly biased pictures of the world. Nevertheless, we must come to terms with them and understand them as the basic building blocks of a higher scale: geopolitical world orders.

Geopolitical codes operate at three levels: local, regional and global. The local-level code involves evaluation of neighbouring states. Governments of all countries, however small, must have such a code. Regional-level codes are required for states that aspire to project their power beyond their immediate neighbours. The governments of all regional powers and potential regional powers need to map out such codes. Finally, a few states will have global policies, and their governments will have appropriate world-wide geopolitical codes. Hence, all countries have local codes, many countries have regional codes, and a few countries have global codes.

Some simple examples will help to fix these ideas. Bartlett gives a very clear example of one major power's three levels of concern in the First World War. For Germany, 'the war is one of defence against France, prevention against Russia but a struggle for world supremacy with Britain' (Bartlett 1984: 89). Sometimes a regional code will be in conflict with a local code. The best example of this is the traditional local hostility between Greece and Turkey, which contrasts with their sharing a similar regional code set by membership of NATO. In fact, treaties are a good indicator of codes, especially at the regional level. The change-round of Australia and New Zealand from being part of Britain's global code to having their own regional (Pacific) code is marked by the establishment of the ANZUS pact just after the end of the Second World War. Australian and New Zealand troops fought in Europe in both world wars, but it is unlikely that they will do so again – Europe is now beyond their geopolitical codes.

Although every code will be unique to its particular country, such practical reasoning is not conducted in a vacuum. In creating its geopolitical code a country must take into account the geopolitical codes of other countries. In fact, there has always been a hierarchy of influence within the inter-state system whereby the more powerful impose ideas and assumptions on the less powerful. In particular, the 'great powers' have had an excessive influence on the geopolitical codes of other members of the system, so much so that within any one historical period most geopolitical codes tend to fit together to form a single overall dominant pattern. These are geopolitical world orders.

Ó Tuathail and Agnew (1992) draw on Robert Cox's (1981) concept of world orders and his method of historical structures. The latter is a framework of action combining three interacting forces: material capability, ideas and institutions. The application of this model to world orders can be illustrated by the case of the Cold War, where because the United States had the material capability to dominate the post-Second World War scene, it advanced liberal political and economic ideas, and helped to found institutions such as the United Nations to provide stability to the new world order. For Cox, such world orders combine social, political and economic structures. The Cold War is the political structure of this particular world order. Cox associates his world orders with the hegemony of one state, which imposes and then protects its world order. In short, our specific concern for geopolitical world orders cannot be meaningfully separated from a more general concern for the rise and fall of great powers over the history of the world-economy. We must tackle this topic before we attempt to define particular geopolitical world orders.

Summary

Geopolitical codes

Geopolitical codes:

➤ are the content and assumptions behind the foreign policy decisions of states;

➤ occur at three geographical scales;

➤ combine to form a global pattern of politics we have called a geopolitical world order.

Cycles of international politics

The history of the inter-state system can be characterized by the rise and fall of just a few major powers. Some researchers have identified a sort of 'system within a system' covering just these major powers. Levy's (1983) 'great power system', for instance, includes only thirteen states since 1495 and no more than seven at any one time. The story of these powers has been told by Paul Kennedy (1988) in his best-selling *The Rise and Fall of the Great Powers*. In many ways this is a classic 'book of its times'. Recent decades have witnessed a plethora of studies showing how no one state has been able to maintain its predominance in the world-system. The decline of Britain has long fascinated Americans, and since about 1970 there are signs that even this latest colossus is suffering a relative decline. And therein lies the secret of the remarkable success of Kennedy's book. Understanding 'decline' is both topical and relevant. Can we learn from history? Kennedy is obviously saying yes, but the problem with his answer is that he fails to depart significantly from the 'high politics' of events. There is no sense of Braudel's *longue durée* in Kennedy's analysis and prescriptions. We can rectify this limitation through world-systems analysis (Taylor 1996).

Most studies of the rise and decline of major powers have developed cyclical models of change. Goldstein (1988) described more than a dozen such analyses. We concentrate on two of the most well known here. Modelski provides a model that identified five cycles and four world leaders (Britain 'leads' twice). In contrast, in world-systems analysis just three hegemonic cycles are distinguished with just three hegemonic powers. These differences in numbers of cycles are symptomatic of fundamental differences in the conception of political cycles.

Modelski's long cycles of global politics

Gaddis (1982) has divided writers of history into two groups: 'lumpers' and 'splitters'. The former impose order on the past, the latter revel in differences and discrepancies. All writers on cycles of international politics are lumpers by definition. George Modelski is the arch-lumper of this body of work.

Modelski first presented his long-cycles model in an article in 1978 and has made several modifications since, the most extensive being in his 1987 book *Long Cycles in World Politics*. Basically, Modelski's work is a reaction against two very different traditions in international relations. First, he is at pains to dismiss realist writers, such as Bull (1977), who argue that the international system is essentially anarchic. Modelski discovers a symmetry to international politics that is the antithesis of anarchy. Second, he is a major critic of international political economy, which he believes devalues political processes at the expense of economic factors. He argues for 'two logics' rather than one, with his global politics being autonomous of the world-economy. The original 1978 article was written to provide an alternative political perspective to Wallerstein's (1974b) introduction of his world-systems analysis. This makes Modelski's long-cycle model particularly interesting here, since it enables us to explore the differences between the 'two logics' assumption and our political economy position.

There are many superficial similarities between Modelski's global cycles and the framework we are using. Modelski's global system begins in about 1500 and then proceeds to develop in a cyclical manner. Modelski (1978: 218) creates a new type of cycle of just over one hundred years in length – so we are currently experiencing the fifth such cycle. Each cycle is associated with a world power, which is defined as a state that engages in over half the 'order-keeping' function of the global political system. Four such world powers have existed: Portugal, the Netherlands, Britain and the United States. These have dominated 'their centuries' – Portugal the sixteenth, the Netherlands the seventeenth, Britain the eighteenth and nineteenth, and the United States in the twentieth. Britain is unique in having dominated two global cycles.

The details of these global cycles are shown in Table 2.1. In each case, the rise and fall of the world power is indicated through definite steps. A cycle starts with a weak global organizational structure of severe political competition, which degenerates into global war. Such wars have a wide geographical range and have global pay-offs – the winner is able to order the resulting political system. This phase ends with a legitimizing treaty, which formally sets up the new world order centred on the new world power. As no world power can maintain its control, a decline phase sets in. Initially the world order becomes bipolar and then multi-polar before the system becomes weakly organized again, is ripe for the rise of a new world power, and the cycle starts again. We can be impressed by the symmetry that Modelski finds in international politics.

The geographical scope of Modelski's model is initially larger than the world-economy. His international political system is global from its inception. This is not just a matter of geographical definition: it

Table 2.1 Modelski's long cycles of global politics

Cycles	World powers	World wars	Legitimizing treaties	Key institutions	Landmarks of decline
I	Portugal	Italian wars (1494–1517)	Treaty of Tordesillas (1494)	Global network of bases	Spanish annexation (1580)
II	Netherlands	Spanish wars (1579–1609)	Twelve-year truce with Spain (1609)	*Mare liberum*	The English Revolution
III	Britain	French wars (1688–1713)	Treaty of Utrecht (1713)	Command of the sea	Independence of the United States
IV	Britain	French wars (1792–1815)	Congress of Vienna (1815)	Free trade	Imperialism
V	United States	German wars (1914–45)	Treaties of Versailles and Potsdam (1919, 1945)	United Nations	Vietnam War

relates to how the role of Portugal is interpreted. For Modelski, by taking over the trading network in the Indian Ocean the Portuguese became the centre of a global system and hence a world power. In contrast, Wallerstein places most of the Portuguese activity in the external arena outside the world-economy; Spanish colonization in the Americas is far more important in that it creates a new periphery. For Modelski, Spain operated only on the fringes of his system and never developed a 'world outlook'. Here we come to the crux of the difference between the two approaches, since the notion of a world outlook in the global sense is totally unnecessary for Wallerstein. His world system was originally a European world-economy and only became global in scope in about 1900.

Why is a world outlook an important criterion for Modelski? The original mechanisms for change in his system are twofold. First, there is what he terms 'the urge to make global order' (1978: 224). Once the possibilities of a global order became known, an innate will for power becomes expressed as an urge to shape world order. Only a few people may have such a world outlook, but they respond to the unarticulated needs of the many. Second, the nature of international politics as a system means that structures run down and have to be reconstructed. All systems suffer a loss of order and survive by cyclical development. Table 2.1 describes a particular expression of this general process. It is on the basis of these two mechanisms that change has occurred in Modelski's global system.

Clearly, here we have uncovered the weak point of the model. For all its symmetry, its mechanisms of change are inherently disappointing. We can trace this back to Modelski's starting point: that the global political system must be functionally distinguished from the world-economy. This is not to say that he necessarily underestimates the importance of economic processes – Modelski's politics are about who gets what in the world-system – but he treats the two processes as distinct. In our terminology, this is a classic illustration of the poverty of disciplines. As we have argued in Chapter 1, the state system and the economic system are integral parts of a single process of development incorporated in the concept of the world-economy. There are not two logics but one. The world-economy could not operate except within the political framework provided by a competitive state system. It is a necessary, though not sufficient, definition of a capitalist world-economy (Chase-Dunn 1982). The implication of this is that the mechanisms of change are not economic or political but both. Instead of Modelski's political framework we need a political economy framework for bringing history back into our geopolitics.

Modelski's response to such criticism has been to find an alternative holistic framework in which to situate his model. He has found it in the classic work on social systems by Talcott Parsons. Modelski (1987: 118) finds it 'a matter of some satisfaction' that Parsonian terminology 'corresponds' to his own model. The cycles are broken down into four phases in which each generation of world power elites performs one of the classical Parsonian functions of 'latent pattern maintenance', 'integration', 'goal attainment' and 'adaption' in the changing world-system. The result is that the sequence of global cycles define a 'learning curve', with each cycle building upon those that have gone before. World powers take on the role of teaching agents.

In many ways, this new systems framework is preferable to Modelski's earlier use of general systems concepts in the sense that processes of change are more explicitly specified. But the resulting functionalism would seem to be too much for even the most dedicated 'lumper'. Most important of all, his application of Parsonian analysis does nothing to clarify the interrelationships between his political processes and the massive economic changes occurring at the very same time. Modelski (1981) has flirted with linking his model to Kondratieff waves, but this has proved most unsatisfactory and has been abandoned in favour of Parsons.

In the final instance, Modelski's model must be seen in the context of current US decline. The most unusual feature of the model is the designation of the eighteenth century as Britain's. This sets an important precedent, since at the end of its first cycle Britain lost 'a bungled colonial war' (the American War of Independence), which provided the shock and stimulus to start the processes rolling for a second century of world leadership (Modelski 1987: 87). It is but a short step from one bungled colonial war to another (the Vietnamese War of Independence), and a prediction of a forthcoming second 'American century'. If Britain can have two cycles, why not the United States? In short, Modelski provides a very optimistic model for a progressive world-system and a powerful America. For more recent developments of this approach, see Modelski and Thompson (1995).

Cycles of world hegemony

In world-systems analysis, hegemony in the inter-state system is a rare phenomenon. It has occurred just three times: Dutch hegemony in the mid-seventeenth century, British hegemony in the mid-nineteenth century and US hegemony in the mid-twentieth century. Such hegemonies encompass dominance in economic, political and ideological spheres of activity, but they are firmly based upon the development of an economic supremacy. This has involved three stages. First, the hegemonic state has gained primacy in production efficiency over its rivals. Second, this has enabled its merchants to build a commercial advantage. Third, the bankers of the state have been able to achieve a financial dominance of the world-economy. When production, commercial and financial activities in one state are more efficient than in all rival states, then that state is hegemonic. Such states have been able to dominate the inter-state system, not by threatening some imperium but by balancing forces in such a way as to prevent any rival coalition forming and growing large enough to threaten the hegemonic state's political leadership. Furthermore, the hegemonic states have propagated liberal ideas that have been widely accepted throughout the world-system. Hence hegemonic states are much more than world political leaders.

The rise and establishment of a hegemonic state has been followed by its gradual decline. The very openness of the hegemonic state's liberalism enables rivals to copy the technical advances and emulate the production efficiencies. Soon the hegemonic state's lead over its rivals declines, first in production and subsequently in commerce and finance. In practice,

the two instances of decline have been cushioned to some degree by an alliance between the old declining and the new rising hegemonic state – the Dutch initially became Britain's junior partners, a role that Britain took with respect to the United States after 1945. This may both smooth the transfer of leadership and help to legitimate the new situation.

The rise and decline of hegemonic states defines a hegemonic cycle. Following Gordon (1980), Wallerstein (1984b) has tentatively related such cycles to three logistic waves of the world-economy. Such cycles involve long-term world market control of investments that sustain the existence of hegemonic power. These investments are both political and economic, and produce a system-wide infrastructure. For instance, system-wide transport, communication and financial networks are a necessary requirement for hegemony. There is also a need for diplomatic networks and a pattern of military bases around the world. In this way, each hegemonic state has built up a hegemonic infrastructure through which it has dominated the system. In each case, world wars of approximately thirty years' duration culminated in confirming hegemony and restructuring the inter-state system. Hence the Thirty Years War ending in the Treaty of Westphalia of 1648 marks the coming of the Dutch hegemonic system, the Revolutionary/Napoleonic Wars ending in the Congress of Vienna of 1815 mark the coming of the British hegemonic system, and the twentieth century's two world wars ending in the setting up of the United Nations in 1945 mark the coming of US hegemony.

Cycles of world hegemony are not as neat as Modelski's long cycles. The three cycles do not match the symmetry of Modelski's five 'centuries' of leadership. There are broad parallels of timing with Modelski's model over the Dutch long cycle, the second British long cycle and the current US long cycle. But even here differences arise. Modelski, for instance, does not treat the Thirty Years War in the seventeenth century as one of his world wars (see Table 2.1) but prefers to consider it as a more regional (that is, European) war because of its lack of a major global dimension. The important differences arise, however, in relation to Modelski's Portuguese long cycle and the first British long cycle. We have already discussed the issue of the validity of Portuguese world leadership above. We can merely add here that the additional criteria that world-systems analysis sets for identifying hegemony completely rules out the candidature of Portugal in the sixteenth century.

Differences over Modelski's first British long cycle in the eighteenth century are much more interesting. This cycle is crucial for regularity in Modelski's model. In the hegemony cycles there is no equivalent regularity, since the time span between the first two hegemonies is much longer than that between the last two. This is not only less neat but is also less optimistic from a US viewpoint since it removes the precedent of Britain's double dose of leadership. In Wallerstein's analysis, the eighteenth century represents nothing more than the ending of one logistic, the beginning of another and the intense political rivalry that such circumstances are associated with. The post-Dutch hegemony period of rivalry between Britain and France lasted longer than the post-British hegemony period of rivalry between the United States and Germany, but that does not negate the logic of the model. What is lost in symmetry is more than made up for in the more comprehensive nature of the hegemonic model.

This can be illustrated by showing how the concept of world hegemony denotes much more than a single state dominating world politics. The defeat of the Soviet Union, for example, was due to much more than US military rivalry and threat. Long before its demise, American culture was dominating the black market in eastern Europe in the form of such common consumer items as rock 'n' roll records and denim jeans. The consumer culture that the United States has promoted in the twentieth century is integral to American hegemony and, however powerful the Soviet Union might have been politically, it never found an answer to the American 'good life' exported to western Europeans but denied to eastern Europeans. World hegemons, the Dutch and the British as well as the Americans, are creators of new modern worlds (Taylor 1996, 1998). As well as the 'Americanization' of society in the twentieth century, there was the industrialization of society emanating

from nineteenth-century Britain, and the creation of mercantile society derived from Dutch practices. In each case, the hegemon provides the image of a future world that other countries try to emulate. As the 'most modern of the modern', the world hegemon thus defines the future of other states, and those that resist risk failing to 'catch up', or worse, risk 'falling behind'. That is why by the time it collapsed, the Soviet Union looked so 'old fashioned', almost a nineteenth-century society in the way it emphasized territorial industrialization in an increasingly globalizing world.

World hegemons are thus much more important than a 'world power' in Modelski's terms. They are political leaders to be sure, but they are equally economic, social and cultural leaders, as we have just shown. In terms of political geography, this complex amalgam of power is particularly important in the way that hegemons have used it to define political norms. In the words of the original hegemony theorist, Antonio Gramsci, hegemonic power involves defining the 'ruling ideas' of society. At the scale of the modern world-system, this means inventing and promoting liberalisms. Thus, all three hegemons have been champions of liberalism in their own distinctive ways. We tell that story in the next chapter as we relate world hegemony to cycles of colonization and decolonization.

Summary

In this section we have introduced two theories of the rise and fall of great powers. We have critiqued Modelski's model and used the world-systems approach to introduce the concept of hegemonic power. We shall use cycles of hegemonic power to:

➤ situate geopolitical codes;
➤ situate academic geopolitical theories.

British and American centuries: the paired Kondratieff model

Any model that uses logistic waves as its basis immediately brings forward the question of their relationship with Kondratieff cycles. In Chapter 1, we avoided this question by incorporating both types of cycle into our discussion, but for different time periods. Goldstein (1988) has reviewed this debate in relation to political cycles. Here we consider simply the application of a Kondratieff model to the periods surrounding British and US hegemony. This is important. First, it links the rise and fall of hegemony to the basic material processes of the world-economy as reflected in Kondratieff cycles. Second, it is a necessary addition to the hegemony model, since it brings the other world powers into the model. Finally, it shows that political mechanisms are an integral part of the overall restructuring of the world-economy that occurs within these cycles.

Refer back to Figure 1.3 and note how each Kondratieff A-phase was launched by a new lead sector, a technological innovation that drove a global economic boom and ushered in dramatic societal change. Note too that these innovations were 'captured' within particular territories. For example, in Kondratieff IIIA, at the end of the nineteenth century, Germany and the United States were competing to be the leading economy with the new technologies. At the time, both countries were debating their economic relationships with the rest of the world; simply there was a tension between protectionist policies that would help nurture new industries and secure older ones from competition, and voices for open markets to allow for the new innovations to be sold in the world-economy. War mingled with geoeconomics too. The 'total war' attitude of the two world wars required national economic mobilization. Furthermore, the bombing campaigns of the Second World War flattened much of the industrial base of Britain, Germany and Japan and left the United States as sole economic power. After the Second World War the United States was in a position to call for 'free trade', or the unhindered movement and sale of the products of the lead sector across the world-economy. As we shall see, such an attempt was partially challenged by the Soviet Union.

Figure 1.3 also posits that we are on the cusp of a new A-phase that will be driven by a new lead sector. But this is where things become very interesting and, perhaps, unprecedented. Globalization is inseparable from geopolitics. Many scholars have argued that globalization has undermined the notion of 'national economies'. Instead, the multinational firm is organized transnationally and states are in competition to secure parts of global investment. This 'market access' model is seen as integrating localities located in different states and preventing the identification of anything that could be called a 'French firm' or 'American company'. The 'capturing' of technological innovation, production and profits within a state is, apparently, old hat.

Of course, there is another side to the story, and it is offered by world-systems theory. Giovanni Arrighi's (1994) *longue durée* study of the rise and fall of hegemonic powers identifies a phase he calls 'financialization'. Production is the economic base of the hegemonic cycle, as seen in Figure 1.3. Although the hegemonic state is the initial main producer of the technological innovation it soon faces competition from other countries that emulate, and improve upon, the product and the production process. In Kondratieff IV, Japan soon became the leading producer of cars and lorries, eclipsing what were seen as the older design and poorer quality of American vehicles. In combination the hegemonic state and its emulators flood the world-economy with products and an economic crisis of oversupply ensues. In the past, Arrighi argues, the capitalist world-economy responded by seeking new investment opportunities, which became manifest in increased flows of capital investment. Transnational capital investment (identified as a key element of globalization) is, according to Arrighi, a part of the hegemonic cycle.

The implication of Arrighi's approach is that states may in the future reassert their control over economic flows. Following this scenario it is conceivable that a new round of protection within states could support the 'national' economic development that is the foundation of a hegemonic state. On the other hand, scholars such as Manuel Castells (1996) and John Agnew (2005) lean towards a conclusion that we have crossed a threshold and that transnational economic integration is here to stay and at such levels that hegemonic powers are a thing of the past as national economies have been eroded. The outcome is a matter of ongoing and future political geography.

Global or national steel?

The steel industry is facing a rosier future. Increased demand for steel is, perhaps, a sign of new Kondratieff A-phase period of economic growth. The steel industry was long perceived as weak and in need of state support, but recently it has become the venue for corporate takeovers and agglomeration. The two largest companies are Mittal Steel, run by Indian entrepreneur Lakshmi Mittal, and Arcelor, the new corporate combination of firms located in France, Spain and Luxemburg. In January 2006, Lakshmi Mittal launched an audacious bid to buy Arcelor and create one massive transnational company. Such geo-economics seems the very stuff of the processes that are argued to be undermining national economies.

The steel industry is of particular interest to political geographers because it has been known as a 'strategic industry' since its inception in the nineteenth century. Steel is an essential ingredient for making weapons and so was seen as an industry that states must protect, even at the price of monetary subsidies and nationalization. The contemporary steel industry is, therefore, an interesting window upon the tensions between transnationalism and the practices of the state to 'capture' industry within national borders.

The array of geopolitical actors involved in this struggle over the geography of the steel industry is vast. On the one hand, we have Lakshmi Mittal and other transnational owners of capital seeking to monopolize production within the capitalist world-economy. On the other hand, there are the national politicians such as the French socialists who adopted the language of war in describing Mittal's takeover bid as Europe's 'hour of truth' (*The Economist*, 4 February 2006, p. 12). Another actor is the media. *The Economist*, a consistent advocate of private ownership and free trade stated that, 'If they were wise, European politicians would now be letting Mr Mittal's bid succeed or fail purely on its business merits. An eager buyer while prices are high: surely this is the ideal chance to get European steel out of politics, once and for all' (ibid.).

Not only is Mittal's bid an example of the political tensions between transnational business and states, it also illustrates the important role of representation in geopolitics. *The French socialists and* The Economist *use particular language to promote support for their position. Can you think of other industries in other countries, and even other time periods, that show the interplay between global markets and state protection? Does the behaviour fit into the time–space expectations of Arrighi's argument?*

Sources: 'Age of giants', *The Economist*, 4 February 2006, pp. 55–6. 'Heavy Mittal', ibid., pp. 11–12.

The actions of multinational firms, politicians and people, the latter in their roles as consumers, citizens, migrants, and so forth, will determine the interplay between states and capital.

The geopolitics of the hegemonic cycle

Now that we have conceptualized the dynamics of geopolitical world orders, we can contextualize the construction of geopolitical codes. We begin by discussing political decisions over the form of economic interaction with other states in the world-economy, and then situate the work of classic geopoliticians within our model.

Political activity has always been an integral part of the world-economy. State policies are important processes in the changes observed in the world-economy. Since they are neither independent processes nor mere reflections of economic necessities ('economism') it follows that there is some choice available. If there were no choice, there would be no need for public institutions such as the state. Public agencies have the role of distorting market forces in favour of those private groups controlling the agency. There has never been a 'pure' world-economy, even in periods when free trade has dominated. The power of a public agency and hence its ability to organize the market depends upon the strength of its backers and their material resources. Strong states may promote a 'free' market, while less strong states may favour explicit distortion of the market through protectionism, for instance. In this way, states can act as a medium through which a first set of production processes (upon which the world-economy operates) is translated into a second set of distribution processes and patterns. Since these intermediate processes tend to favour the already strong, it follows that political activity will often increase the economic polarization of the market (that is, helping the core at the expense of the periphery).

The power of a state to organize the market to its own ends is not just a property of that state's resources. The fact that we are dealing with a world-economy and not a world-empire (that is, there is a multiplicity of states) means that relative positions are more important than measures of absolute power. These state positions are relative not only to other states but also to the gross availability of material resources within the world-economy. The cyclical nature of material growth means that opportunities for operating various state policies vary systematically over time. This is not just a matter of different economic environments being suited to alternative state strategies. At any particular conjunction, specifically successful policies can only work for a limited number of agencies. Quite simply, a success for any one state lessens opportunities for other states. There will always be constraints in terms of the total world resources available for redistribution via state activities. Given the 'correct' policies, it is not possible for all semi-peripheral and peripheral states to become core-like. Although this is not a zero-sum game in a static sense, since the available production is always changing in a cyclical fashion, we do have here a sort of 'dynamic zero-sum game'. If state activity is an integral part of the operation of the world-economy, we should be able to model it within our temporal and spatial framework. Just such a model has been proposed by Wallerstein and his associates (Research Working Group 1979). They postulate political activity occurring over a time period covering two Kondratieff waves.

The rise and fall of hegemonic power relates to 'paired Kondratieffs' as follows. If we start with a first growth phase, A_1, we find geopolitical rivalry as core states compete for succession to leadership. With hindsight, however, we can see that new technological advances are concentrated in one country, so increased production efficiency gives this state a long-term advantage. A_1 is associated with the stage of ascending hegemony. In B_1, overall decline of the world-economy leaves fewer opportunities for expansion, but the ascending power now attains commercial supremacy and is able to protect its interests better than its rivals are able to protect theirs. By this stage, it is clear which state is to be the hegemonic power. B_1 is associated with the stage of hegemonic victory. With renewed growth of the world-economy we reach A_2, the stage of hegemonic maturity. By this time, the financial centre of the world-economy has moved to the hegemonic state, which is now supreme in production, commerce and finance (that is, 'true' or 'high' hegemony). Since the hegemonic power can compete successfully with all its rivals, it now favours 'opening' the world-economy. These are periods of free trade. Finally, declining hegemony occurs during B_2, when production efficiency is no longer sufficient to dominate rivals. This results in acute competition as new powers try to obtain a larger share of a declining market. These are periods of protectionism and formal imperialism as each rival attempts to preserve its own portion of the periphery.

According to Wallerstein's research group, the four Kondratieff cycles from the Industrial Revolution can be interpreted as two 'paired Kondratieffs' (Table 2.2). The first pair covering the nineteenth century correspond to the rise and fall of British hegemony, and the second pair describe a similar sequence of events for the United States in the twentieth century. There is no need to consider this table in great detail, except to note how several familiar episodes fit neatly into the model.

In terms of our discussion of state involvement in the operation of the world-economy, phases A_2 and B_2 are particularly important. In A_2, the hegemonic power imposes policies of open trade on the system to reap the rewards of its own efficiency. In the mid-nineteenth century, Britain proclaimed 'free trade' backed up by gunboats, and a century later a new world policeman, this time with aircraft carriers, was

Table 2.2 A dynamic model of hegemony and rivalry

	Britain		USA	
	1790/98		1890/96	
A_1 Ascending hegemony		Rivalry with France (Napoleonic Wars) Productive efficiency: Industrial Revolution		Rivalry with Germany Productive efficiency: mass-production techniques
	1815/25		1913/20	
B_1 Hegemonic victory		Commercial victory in Latin America and control of India: workshop of the world		Commercial victory in the final collapse of British free-trade system and decisive military defeat of Germany
	1844/51		1940/45	
A_2 Hegemonic maturity		Era of free trade: London becomes financial centre of the world-economy		Liberal-economic system of Bretton Woods based upon the US dollar: New York new financial centre of the world
	1870/85		1967/73	
B_2 Declining hegemony		Classical age of imperialism as European powers and United States rival Britain. 'New' industrial revolution emerging outside Britain		Reversal to protectionist practices to counteract Japan and European rivals
	1890/96			

going through the whole process of liberalizing trade once again. These policies certainly contributed to the massive growth of the world-economy in the A_2 phases and were imposed through a mixture of negotiation, bargaining and bullying. Options for non-hegemonic powers were highly constrained, and they largely went along with the hegemonic leadership.

All this changes with the onset of the B_2 phase, however. As production efficiencies spread, economic leadership deserts the hegemonic power. These are key periods because of the opportunities that declining hegemon provides for other core and semi-peripheral states. The imposition of free trade is no longer taken for granted as various states work out new strategies for the new circumstances. In the late nineteenth century, Britain entered the B-phase as hegemonic power and came out behind Germany and the United States in terms of production efficiency. B_2 phases are clearly fundamental periods of restructuring in the world-economy in which geopolitical processes play an important role. We are currently living through just such a phase.

Cycles and geopolitical world orders

We are now in a position to identify geopolitical world orders within our model of world hegemonic cycles. (Hence we continue to concentrate on just the final two hegemonies here, the Dutch cycle being quite different in terms of order.) Following Cox (1981), we have already associated world orders with periods of high hegemony such as the United States and the Cold War. The hegemonic periods of both Britain and the United States are times of relative international stability, and this fact has led to a general hypothesis relating hegemony to world order (Rapkin 1990). However, by geopolitical world order we mean more than these particular periods of stability. Our world orders are a given distribution of power across the world that most political elites in most countries abide by and operate accordingly. This includes hegemonic stable periods to be sure, but there is an order of sorts between the certainties of a hegemonic world. In such times, international

Domestic politics and isolationism?

President George W. Bush, in his annual State of the Union speech in January 2006, identified a new political enemy that he was set to vanquish. The enemy was the idea of 'isolationism', which President Bush blurred with 'protectionism' (see previous panel). President Bush's agenda has been one of global activism, under the title of 'democratization' and most visible in the practice of 'regime change' in Afghanistan and Iraq. In vilifying isolationism, President Bush was trying to neutralize opposition in both the Republican and Democrat parties. Philosophically, some on the right wing of the Republican Party, most notably Pat Buchanan, believe that George Washington's historic warning not to get entangled in messy foreign disputes still applies. Instead of global intervention strengthening the United States, as President Bush argued, Buchanan and his allies believe such engagements drain the United States' power. Some Democrats oppose current intervention in Iraq on an 'anti-war ticket', but there is also a strong economic protectionist tendency led by Congressmen representing industrial constituencies.

President Bush's geopolitical programme and practice of foreign intervention and a belief in the universal application of US-style democracy is what is to be expected of an ascendant or mature hegemonic power. A new hegemonic cycle is built upon material strength as well as the export of an idea to be emulated. On the other hand, the programme has met severe resistance and required a vast amount of military muscle and bloodshed. For the critics, democratization is a codeword for 'empire' and military occupation. The fact that 'regime change' was initiated and maintained by the armed forces suggests that current US geopolitics are the defensive reaction of a hegemonic power in decline. *What then of the 'isolationists'? Could their approach be seen as a recognition that the hegemonic power is weakening and needs to retreat behind national borders for a while and retool?*

Sources: 'President Bush delivers State of the Union address', 31 January 2006. http://www.whitehouse.gov/news/releases/2006/01/20060131-10.html. Accessed 16 February 2006. 'The isolationist temptation', The Economist, 11 February 2006, pp. 27–8.

anarchy has not prevailed; rather, the great powers of the day have accommodated to one another's needs in quite predictable ways. Hence geopolitical world orders transcend the special case of hegemony.

Table 2.3 shows four geopolitical world orders alongside the paired Kondratieff and hegemonic cycles for Britain and the United States. Each world order emerges in a rapid geopolitical transition following a period of disintegration of the previous world order. These transitions are very fluid times, when the old world and its certainties are 'turned upside down' and what was 'impossible' becomes 'normal' in the new order. In other words, they separate distinctly separate political worlds. This will become clear as we describe the geopolitical world orders in Table 2.3.

According to Hinsley (1982), the modern international system only begins with the Congress of Vienna in 1815, which brought to an end the

Revolutionary and Napoleonic Wars. What he means by this is that the political elites of Europe made a conscious effort to define a trans-state political system that would curtail the opportunities for states to disrupt the peace. This was a new departure, and it produced a relatively stable distribution of power, that is to say the first geopolitical world order. In geographical terms, the order consisted of two zones. The Concert of Europe operated as an irregular meeting of the great powers to regulate political disputes across Europe. In the rest of the world there was to be no such regulation. Such a world order was directly complementary to Britain's rising hegemony. It set up the mechanism for Britain to keep a balance of power in Europe so that it could not be challenged by a continental empire such as Napoleon had forged. Equally importantly, it gave Britain a free hand to operate in the rest of the world, where it was now dominant. Hence in Table 2.3 we have termed this

Table 2.3 Long cycles and geopolitical world orders

Kondratieff cycles	Hegemonic cycles	Geopolitical world orders
1790/98 A-phase	BRITISH HEGEMONIC CYCLE Ascending hegemony (grand alliance)	(Napoleonic Wars as French resistance to Britain's ascending hegemony)
1815/25 B-phase	Hegemonic victory (balance of power through Concert of Europe)	Disintegration WORLD ORDER OF HEGEMONY AND CONCERT Transition (1813–15)
1844/51 A-phase	Hegemonic maturity ('high' hegemony: free-trade era)	(Balance of power in Europe leaves Britain with a free hand to dominate rest of the world) Disintegration
1870/75 B-phase	Declining hegemony (age of imperialism, new mercantilism)	WORLD ORDER OF RIVALRY AND CONCERT Transition (1866–71) (Germany dominates Europe, Britain still greatest world power)
1890/96 A-phase	AMERICAN HEGEMONIC CYCLE Ascending hegemony (a world power beyond the Americas)	Disintegration WORLD ORDER OF THE BRITISH SUCCESSION Transition (1904–07)
1913/20 B-phase	Hegemonic victory (not taken up: global power vacuum)	(Germany and United States overtake Britain as world powers, two world wars settle the succession)
1940/45 A-phase	Hegemonic maturity (undisputed leader of the 'free world')	Disintegration COLD WAR WORLD ORDER Transition (1944–46)
1967/73 B-phase	Declining hegemony (Japanese and European rivalry)	(US hegemony challenged by the ideological alternative offered by the Soviet Union)
19??	NEW HEGEMONIC CYCLE?	Disintegration 'NEW WORLD ORDER' Transition (1989–?)

the 'world order of hegemony and concert'. This order lasted until the great transformations of the 1860s, in which political reorganizations across the world – in the United States (civil war), Italy (unification), Russia (modernization), Ottoman Empire (modernization), Japan (modernization) and above all Germany (unification) – showed that the international system was out of hegemonic control. This is the phase of disintegration.

The first geopolitical transition occurred in 1870–71 with the German defeat of France, the subsequent quelling of the Paris Commune and the declaration of the German Empire. The latter was now the dominant continental power, so the British balance-of-power policy in Europe was in shreds. Suddenly a new world order was in place with two major centres of power, London and Berlin. But the 'long peace' instituted at Vienna in 1815 continued in this 'world order of rivalry and concert'. Basically, Germany was

concerned to consolidate its position in Europe, and Britain had the same purpose in the rest of the world, so stability was maintained despite the rivalry. But this world order was relatively short-lived and disintegrated in the 1890s. In Europe, the alliance between France and Russia in 1894 began the threat to Germany on two fronts. In the rest of the world, European dominance was under threat for the first time with the emergence of both the United States and Japan as potential major powers. Rivalry was becoming stronger than concert, and the world order could not survive.

By the end of the century, while the British Empire remained the greatest political power, British hegemony was clearly over. Britain revised its foreign policy and constructed a new world order. The transition took place in the early years of the twentieth century. The first key step was to end British 'splendid isolation' by the naval agreement with Japan in

1901, but the crucial change was the alliances with France and Russia in 1904 and 1907, respectively, which consolidated an anti-German front in Europe. This could hardly be more different from the traditional British policy of non-entanglement in Europe by playing one country off against another. Furthermore, Britain chose its two traditional enemies, France and Russia, as allies – what Langhorne (1981: 85, 93) calls two 'impossible agreements'. But that is the nature of a geopolitical transition: the impossible becomes possible. The pattern of power rivalries set up at the beginning of the century lasted until the defeat of Nazi Germany in 1945. With hindsight, we can interpret this as the 'world order of the British succession'. Although precipitated by British attempts to maintain its political dominance, the two world wars of this era can be seen as the United States preventing Germany taking Britain's place and culminating in the United States' 'succeeding' to Britain's mantle in 1945. For more details on this world order, see Taylor (1993a).

Summary

In the previous sections we have identified Kondratieff waves as the driving force behind the rise and fall of hegemonic powers. In doing so we have:

➤ situated the political decisions made by leaders of hegemonic states and hegemonic challengers within the temporal dynamic of Kondratieff waves;

➤ emphasized the connection between economic and political processes.

■ Situating geopolitical codes and theories

Geopolitical codes are the practices of states. One component of a geopolitical code is the way that such practices are represented to give them meaning and purpose. In other words, they need to be legitimized. In addition, the specific content of a state's geopolitical code is contested. Different groups and interests within a country will have different visions and agendas. For example, different economic sectors will see

different costs and benefits in a protectionist stance. Academics play a crucial role in proposing different geopolitical codes. For example, the US journal *Foreign Affairs* is replete with scholars portraying the world in a particular fashion, and making related policy recommendations. Furthermore, if a state adopts policies that reflect academic theories they are in turn used as justification for state action. The War on Terrorism is often 'explained' by reference to the dubious notion of a Clash of Civilizations disseminated by Samuel Huntington. Huntington's position as Professor at Harvard University gives his treatise authority that, in turn, helps legitimate state actions.

The 'classical' geopolitical theories that were at the core of the founding of political geography are examples of academic input into geopolitical codes. They can be fully understood only by situating them within the dynamics of the capitalist world-economy. The theories of Sir Halford Mackinder and Karl Haushofer were products of, respectively, Britain's hegemonic decline and Germany's attempts to replace it. We situate these foundational texts as partial and subjective knowledge that was constructed within the temporal and spatial contexts of hegemonic decline and challenge.

Mackinder's heartland theory

Mackinder's world model was presented on three occasions covering nearly forty years. The original thesis was presented in 1904 as 'The geographical pivot of history'. The ideas were refined and presented after the First World War (1919) in *Democratic Ideals and Reality*, where 'pivot area' becomes 'heartland'. Then in 1943 Mackinder, at the age of 82, provided a final version of his ideas. Despite this long period covering two world wars, the idea of an Asiatic 'fortress' remains the centrepiece of his model and is largely responsible for its popularity since 1945. Most discussion of Mackinder concentrates on his 1919 work, but here we are particularly concerned with the origins of his ideas just after the turn of the century. For further details, see Parker (1982), Blouet (1987) and Ó Tuathail (1992).

Mackinder's original presentation of his model is a very broad conception of world history. Basically,

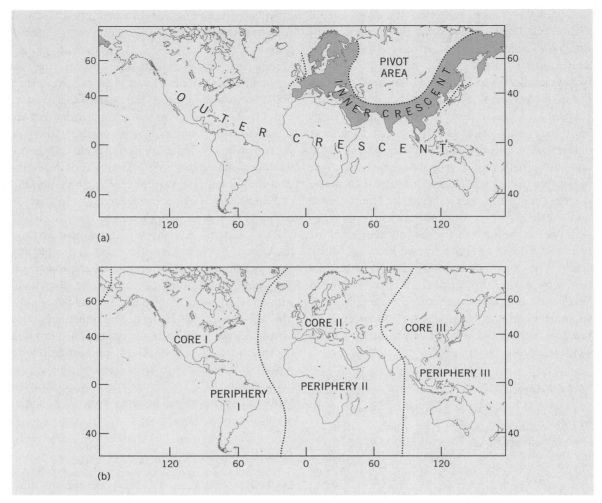

Figure 2.1 Alternative geopolitical models: (a) Mackinder's original model; (b) a model of pan-regions.

he identifies central Asia as the pivot area of history from which horsemen have dominated Asian and European history because of their superior mobility. With the age of maritime exploration from 1492, however, we enter the Columbian era, when the balance of power swung decisively to the coastal powers, notably Britain. Mackinder now considered this era to be coming to an end. In the 'post-Columbian' era, new transport technology, particularly the railways, would redress the balance in favour of land-based power, and the pivot area would reassert itself. The pivot area was defined in terms of a zone not accessible to sea power and surrounded by an inner

crescent in mainland Europe and Asia and an outer crescent in the islands and continents beyond Eurasia (Figure 2.1a).

What had this to do with current (1904) power politics? At its simplest this model can be interpreted as a historical–geographical rationalization for the traditional British policy of maintaining a balance of power in Europe so that no one continental power could threaten Britain. In this case, the policy implications are to prevent Germany allying with Russia to control the pivot area and so command the resources to overthrow the British Empire. Mackinder's message in 1904 was that Britain was more vulnerable than before to the rise of a continental power. British foreign policy needed revision to accord with the new post-Columbian situation to supplement a revised trade policy.

In his 1919 revision of this world model, he redefined central Asia as the 'heartland', which was larger than the original pivot area. This was based on a reassessment of the penetrative capabilities of sea powers. Nevertheless, the same basic structure remains, and the fear of German control of the heartland is still central. In fact, he is much more explicit in his advice as given in his famous dictum:

Who rules East Europe commands the heartland.
Who rules the heartland commands the world-island.
Who rules the world-island commands the World.

(The 'world island' is Eurasia plus Africa, consisting of two-thirds of the world's lands.) This message was specifically composed for world statesmen at Versailles, who were redrawing the map of Europe. The emphasis on eastern Europe as the strategic route to the heartland was interpreted as requiring a strip of buffer states to separate Germany and Russia. These were created by the peace negotiators but proved to be ineffective bulwarks in 1939.

Mackinder's 1943 revision is more comprehensive but far less relevant to our discussion. It reflected the contemporary short-term alliance of Russia, Britain and America and posited them together as heartland and 'midland ocean' (North Atlantic) to control and suppress the German danger between them. This is a long way from the grand history and basic materialist strategic thinking behind his original world model. It is the ability of Mackinder's original broad model to generate specific policy recommendations that has maintained its popularity to the present day.

German geopolitics, 1924–41

As the Second World War approached, German geopolitics was centred on the work of Karl Haushofer. He was professor of geography at the University of Munich from 1921 to 1939, and he edited the flagship journal of geopolitics *Zeitschrift für Geopolitik*. Much of our knowledge of Haushofer has been clouded by reports on his work in the Second World War, when

many myths were created that political geographers have been too slow in correcting. For instance, there never was an Institute of Geopolitics at Munich, and Haushofer never commanded 'a thousand scientists' plotting a German victory. In recent years, there have been several worthy re-evaluations of Haushofer and German geopolitics, which we draw upon here (Heske 1986, 1987; Bassin 1987; Paterson 1987; Sandner 1989; O'Loughlin and Van der Wusten 1990; Ó Tuathail 1996).

Although idealistic ideas dominated most international relations studies in the inter-war years, realist assessments of the world scene prospered in one corner of Europe, defeated Germany. Here idealism was discredited because of its association with what was considered to be the unfair Treaty of Versailles. It is in this context that we must review the rise of German geopolitics. As Paterson (1987) has pointed out, the short-term aim of this geopolitics was the revision of the Treaty of Versailles. The key concept for this challenge was the idea of *Lebensraum* (originally from Ratzel, literally 'living space'), which interpreted Germany's problems as being due to unfair and confining boundaries. The solution was expansion. It is easy to see why such a geopolitics should appeal to Nazi politicians before and after the creation of the Third Reich.

The question of Haushofer's influence on Nazi policy is a controversial one. Certainly, it is now generally concluded that he had much less influence than Second World War reports suggested. Heske (1986) outlines the recent opinion as follows. Haushofer was well known in right-wing political circles for his realist policy prescriptions. He was friendly with Hitler's deputy, Rudolf Hess, from 1919 onwards, meeting him for detailed discussions about once a month throughout this period. He also had contact with other leaders of the Third Reich (Ribbentrop, Goebbels and Himmler), but much less so with Hitler himself. In the 1930s, Haushofer's main link with the political elite seems to have been through his son Albrecht. However, after Hess's abortive peace mission to Britain in 1941 Haushofer lost what influence he had had on the Nazi regime. In 1944, Albrecht was executed for his part in the failed assassination attempt on Hitler.

If we move on from the questions of interpersonal relations to the realm of the ideas themselves, we find further reasons to doubt Haushofer's critical importance at this time. Bassin (1987) has compared geopolitics with National Socialist doctrine and has exposed the fundamental difference in their theories. Whereas geopolitics was derived from Ratzel's scientific materialism, National Socialism promoted ideas of innate human qualities ennobling racial theories of superiority. Despite Haushofer's attempts to avoid this contradiction between the two doctrines (Heske 1987), it is doubtful whether geopolitics could ever have become the leading science of Nazi Germany, as its opponents have claimed. Rather, it represented a set of realist ideas that could be used when convenient. Similarly, Haushofer was a right-wing academic who was used to ease relations between the Third Reich and academia. The result, according to Heske (ibid.), was that geography was implicated more than any other science in the legitimation of the Nazi regime, and Haushofer must bear the main responsibility for this.

With the breakdown of the British-led free-trade system of the mid-nineteenth century, the world gradually moved towards a system of economic blocs behind tariff barriers. In Britain, Mackinder's conversion to tariff reform was to promote the British Empire as an economic bloc – the policy of imperial preference. The ultimate result of such thinking was autarky, or economic self-sufficiency. Since Germany had lost all its colonies after the First World War, autarky for Haushofer and his associates was originally linked to *Lebensraum* and expansion into eastern Europe. But return of the colonies formed an important aspect of the call to revise the Treaty of Versailles, and this led to a more global appreciation of Germany's role in the world. The result was an interpretation of global economic regions as pan-regions.

The idea of economic blocs was not original, of course, but pan-regions were distinctive in their sweeping redefinition of economic patterns. Other proposals of economic blocs were careful to follow the current pattern of colonies and spheres of influence (Horrabin 1942). But pan-regions were more than mere economic blocs. They were based

upon 'pan-ideas', which provided an ideological basis for the region (O'Loughlin and Van der Wusten 1990). The pan-Americanism implicit in the Monroe Doctrine was the classic pan-idea associated with a pan-region.

In German geopolitics, three great pan-regions were finally identified as a new territorial division of the world (see Figure 2.1b) based upon Germany, Japan and the United States. This is an interesting geographical organization in that it involves huge functional regions around each core state cross-cutting environmental-resource regions, which straddle the Earth latitudinally. Hence each pan-region would have a share of the world's arctic, temperate and tropical environments. As political economy units they produce three regions with a high potential for autarky. If it had so evolved, such a world model would have produced three separate world-systems, each with its own core – Europe, Japan and Anglo-America; and periphery – Africa and India, east and south-east Asia, and Latin America, respectively.

The rise to dominance in the world-economy by the United States after the Second World War ended the general drift towards economic blocs and so made the concept of pan-regions temporarily irrelevant. But with the current demise of US dominance of the world-economy, economic blocs and even pan-regions are returning to the world political agenda (O'Sullivan 1986; O'Loughlin and Van der Wusten 1990).

The Cold War as a geopolitical world order

There is no doubt that in 1945 by any reasonable criteria the United States could claim to be hegemonic. Germany, Japan and Italy were defeated, France had been occupied, Russia was devastated and Britain was bankrupt. In contrast, America's economy had expanded during the war. By 1945, the United States was responsible for over 50 per cent of world production. It would seem the US hegemony was even more impressive than the two previous hegemonies. And yet, in geopolitical terms, US hegemony was in no sense as successful as Britain's a century before. US hegemony has been 'spoiled' by the existence of a major ideological and military challenger, the Soviet Union. Whereas Britain's balance-of-power policy involved staying on the outside and diplomatically manipulating the other great powers, the United States was an integral part of the new balance-of-power situation and became continually involved in a massive and dangerous arms race. The Cold War is not what we have come to expect of a hegemonic geopolitical world order. How did this come about?

The short period after the Second World War is a classic example of a geopolitical transition. If we take the world situation on each side of the transition, we soon appreciate the immensity of the change that took place. This is symbolized by two world events only a decade apart and both occurring in German cities. In 1938 in Munich, Britain and Germany negotiated to stave off a world war. In 1948 in Berlin, the United States and Soviet Union confronted each other in what many believed would lead to another world war. In only a decade everything had changed – new leaders, new challengers and a new geopolitical world order.

It surprised many observers at the time how quickly the victorious allies of 1945 split to produce a new confrontation. By 1947, the Cold War was evident for all to see. In fact, we have already observed how previous geopolitical transitions have also involved relatively rapid turnarounds – this seems to be the norm. But for those involved this rapid production of new enemies was a mystery requiring a 'solution'. Hence a huge literature has been produced on the creation of the Cold War. Two main schools of thought have emerged producing contrary conclusions. The so-called orthodox school blames Soviet expansionism, which left the United States with no alternative but containment policies. Responding to the United States' actions in Vietnam, a revisionist school emerged that blamed the United States for the Cold War. In this argument, the Soviet Union was ostracized for not succumbing to the United States' hegemonic world strategy. Today, both of these positions are considered to be too simplistic. There are now numerous post-revisionist positions that analyse the complexity of early post-Second World War international relations. From a geopolitical

perspective, the most interesting are those that emphasize the role of Britain in the making of the Cold War (Taylor 1990).

In 1945 the geopolitical situation was very fluid. The Big Three victors had very different priorities. The United States had clear economic priorities to open up the world for American business. Britain had political priorities to remain a major power. Although it had effectively mortgaged its future in loans to win the war, it remained the largest empire in the history of the world. The Soviet Union's priority was clearly to safeguard its western flank in eastern Europe, through which it had been invaded twice in twenty years. At first, these various interests seemed to be compatible with one another in the goodwill generated by war victory. This was summarized in President Roosevelt's vision of one world and symbolized by the creation of the United Nations. In this idealistic conception of the world, the divisive power politics of the past was banished and replaced by a world of friendship and cooperation.

What went wrong? Figure 2.2 shows the bilateral relations between the victorious allies at the end of the Second World War and the different ways in which the 'Big Three' might be converted into a bipolar world: an anti-hegemonic axis against the United States, an anti-imperial axis against Britain, and an anti-communist axis against the Soviet Union. A

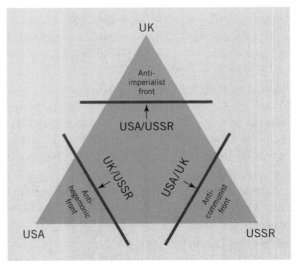

Figure 2.2 Alternative bipolar worlds in 1945.
Source: Taylor (1990).

major reason why an anti-communist front was formed between Britain and the United States can be found in Britian's strategic weakness relative to the other two. Both the United States and Soviet Union were clearly superpowers of continental proportions. Britain, on the other hand, was a small island with a scattered empire whose loyalty could no longer be relied upon. It is not surprising, therefore, that the main priority of British politicians at this time was the devising of policies to maintain Britain in the top rank alongside the United States and Soviet Union. And it was soon appreciated that this would require outside financial aid. The British economy was in such dire straits that bankruptcy could be avoided only by negotiating an emergency loan, for which the United States was the only source. The loan that was successfully negotiated in December 1945 had commercial and financial strings attached to it. In simple terms, the British Empire was being prised open for American business. Despite the support of the latter, the agreement was very difficult to sell to the American public. Why should they foot the bill to prop up the British Empire? In the event, the agreement passed through Congress in the wake of the first post-war bout of anti-communism in the United States. Hence Britain was supported not because it was necessary for US business; rather, it was supported as a bulwark against the threat of expanding communism. According to Kolko and Kolko (1973), US foreign policy moved from negotiation to crusade.

Not unnaturally, Britain encouraged and supported the new crusade. The Soviet Union became isolated and Britain moved into a position of chief ally of the hegemonic power. In the two years after the Second World War Britain undermined the international relations based on the Big Three concept in numerous ways (Taylor 1990). Britain's foreign secretary, Ernest Bevin, worked assiduously against the Soviet Union in the regular meetings of foreign ministers (Deighton 1987). Sir Winston Churchill gave his Fulton speech in 1946, in which he used the famous phrase 'iron curtain' to signify a division of Europe (Harbutt 1986). In 1947, Britain informed the United States that it could no longer maintain its troops in Greece and Turkey. This precipitated President Truman's speech in which the United States promised to support 'free peoples everywhere'. The 'Truman Doctrine' was clearly aimed at the Soviet Union and marks the formal beginning of the Cold War.

From Britain's perspective, this represented a major diplomatic success: the United States had finally come to recognize its hegemonic responsibilities. Britain played a leading role in 1948 in operating the Marshall Plan, giving US aid to western Europe. Then in the next year Britain again played the leading role in the formation of the North Atlantic Treaty Organization (NATO). A new world order was created under US leadership, with Britain having a 'special relationship' as the trusted deputy while leaving the Soviet Union out in the cold. In short, Britain had solved, at least for the time being, its post-war crisis of power. It is not surprising, therefore, that some authorities have concluded that the Cold War was in reality a 'secondary effect' of Britain's prime objective to remain a first-rank power (Ryan 1982).

While covering only part of the massive changes that created the Cold War, this interpretation has the advantage of accounting for the rapidity of change from 'one world' to the bipolar situation of 1947 through highlighting the agency of the parties in a situation of high fluidity. Rapid change provides an opportunity for those in a position to take advantage. For a short time before the Big Three became the Big Two, Britain was in just such a position and took advantage to shore up its faltering world position. The formation of NATO in 1949 confirmed Britain's new position as 'first lieutenant' to the United States in a new world order. This was not such a big reduction in status for a country bankrupt in 1945.

Cold War as superstructure: Great Contest or Great Conspiracy?

How do we interpret this geopolitical world order as a superstructure in world-systems analysis? Basically, the Cold War is a political structure based upon two contrary relations between the superpowers: opposition and dependence (Cox 1986). Most theories of the Cold War emphasize one relation at the expense of the other. We shall try to achieve a better balance between the two.

Theories of opposition are generally concerned to apportion blame, building upon the 'orthodox' and 'revisionist' schools in the making of the Cold War literature. Hence the Cold War is either the result of the Soviet threat or an outcome of US imperialism. In either case, the world is viewed as facing a climactic conflict between two opposing world views: communism and capitalism. These alternative ways of life based upon completely different values are incompatible. In Halliday's (1983) terms this is the 'Great Contest', which may be seen as an ideological battle or, more fundamentally, as global class conflict.

This view of the Cold War interprets the geopolitical world order in the terms set out by the adversaries. It is quite literally the cold warriors' version of the Cold War. The world is divided into just three types of place: 'ours', 'theirs' and various disputed spaces. The conflict is unbridgeable; the end result is either a communist or a capitalist world.

We should, of course, always be wary of those who set the political agenda. We know from Schattschneider that agenda setting can never be a neutral process. There is nothing natural or inevitable about the Cold War. It is a particular world order that favours some issues at the expense of others. We must ask ourselves, therefore, what issues are organized off the political agenda by the Cold War. Presumably these are matters that are not the prime concern of either superpower. In short, they are dependent on each other in maintaining a world order that highlights their superpower politics.

In this sort of argument, the Cold War acts as a diversion from alternative politics. Such diversions have been identified at three geographical scales. At the domestic level for each superpower, the Cold War has been instrumental in mobilizing their populations behind the state in its confrontation with its enemy. It has enabled narrow definitions of loyalty to be used to marginalize alternative politics within each country. In the United States, the main example is the anti-communist hysteria of the early 1950s led by Senator Joseph McCarthy. This produced a consensus on US foreign policy that was maintained for a generation before the Vietnam debacle fractured public opinion. In the Soviet Union the Cold War allowed the persecution and marginalizing of various 'dissident' groups.

Beyond their own countries the Cold War enabled both superpowers to maintain a firm control on their allies. Each superpower led a bloc of countries whose foreign policy options were severely curtailed. The most naked examples of the constraints were to be found in eastern Europe and Latin America. In the former, the Soviet Union intervened militarily to prevent liberal or revisionist regimes appearing, and in the latter the United States intervened to stop radical or socialist regimes appearing. In fact, one of the most remarkable features of the Cold War was the success of both superpowers in maintaining bloc cohesion for so long.

Finally, beyond the blocs the Cold War can be interpreted as deflecting attention from the North–South issue of massive global material inequalities. We have already discussed this briefly in Chapter 1 in terms of the decline of the United Nations as a major actor in the world-economy as its focus has moved away from East–West issues towards North–South issues. Hence the Great Contest is in reality the 'Great Conspiracy'. In this interpretation, US hegemony was not 'spoiled' at all by the Cold War; rather, the latter constituted a great power 'concert' not unlike the world order of British hegemony a century earlier.

There is no doubt that the superpowers used the Cold War to bolster their own positions (Wallerstein 1984a). In concentrating on the military sphere of world politics the United States temporarily side-stepped its economic difficulties and re-established its leadership. There is no doubt that the United States remained number one in the West as a political power in the 1980s. In this sense, the United States used the Soviet Union in its competition with rival friendly states, but, on the other hand, there is no doubt that the Soviet Union represented a genuine challenge to the United States, both militarily and ideologically. It was the Soviet Union that made the Cold War geopolitical world order so unusual. The role of the Soviet Union in the world-economy was a completely novel one – it combined a semi-peripheral economy with a superpower political status. We cannot possibly understand the Cold War without coming to terms with this mammoth mismatch between economics and politics.

Case study

A world-systems interpretation of the Soviet Union

We have seen that the traditional approaches to geopolitics treat the Soviet Union as a land power threat to the traditional dominance of sea powers. To balance the picture, we start our discussion by giving the Soviet view of its world position. In contrast to the expansionary motives assumed in western strategy, the Soviet Union interpreted its position as essentially defensive. This viewpoint is based upon two invasions since its inception in 1917: first, US and western European support for White Russians in the civil war of 1918–21 and second, the German invasion of 1941–44. Hence the Soviet Union saw itself as the protector of the socialist world surrounded by hostile capitalism. One country's containment was indeed another country's expansion.

For our argument here the most relevant part of the Soviet model was the notion of two world-systems operating contemporaneously: a capitalist one and a socialist one. This notion stemmed from Stalin's attempt to build socialism in one country between the world wars. It was expanded with the institution of Comecon after the Second World War as a 'cooperative division of labour' in eastern Europe, in contrast to the capitalist competitive division of labour in the West. An attempt has been made by Szymanski (1982) to integrate this orthodox Marxist position into Wallerstein's framework. He claims that there were indeed two separate world systems, and economic transactions between them constituted luxury trade rather than essential trade. In this sense, the two systems coexisted in the way that other contemporaneous systems (Roman and Chinese world-empires,

for instance) had done before. This conception of two separate economic systems that compete politically was, of course, the mirror image of American geopolitics. We shall show that both positions provide less insight than a world-systems interpretation of a single world-economy.

Charles Levinson (1980) has provided a wealth of evidence to expose what he terms 'the ideological facade' of both US and Soviet geopolitics. The following selection of some of the information he compiled during the Cold War will give an indication of the basis of his argument. He shows that the 40 largest multinational corporations all had cooperative agreements with one or more of the eight eastern European states with communist regimes – 34 of them with the Soviet Union itself. He lists 151 corporations from 15 countries that had offices in Moscow. There were 108 multinational corporations from 13 countries operating in Bucharest alone. In the other direction, Levinson found 170 acknowledged multinational joint ventures by the Soviet Union in nineteen western countries. It is therefore not surprising that by 1977 one-third of Soviet imports and one-quarter of Soviet exports were with western countries. Levinson's conclusion is that although it was international politics that made the news, it was these crucial economic transactions that steered international politics. Hence, détente followed trade, and not vice versa. This is what Levinson calls the 'overworld' of economic dealings, which resulted in the Soviet Union and eastern Europe becoming inexorably integrated into the world-economy. The process is epitomized by the PepsiCo Inc.'s deal to sell its

cola drink in Russia and to market vodka in the West – hence the title of Levinson's book, *Vodka Cola*.

Gunder Frank (1977) has provided further evidence for the same process, which he terms trans-ideological enterprise. He charts the massive rise in East–West trade during the Cold War and discusses the various bartering arrangements and other agreements that made this growth possible. The motives for all this activity were very traditional. For the corporations, there was an extension of the geographical range of their profit making. This was particularly important after 1970 at a time of worldwide recession. The eastern European states provided a source of relatively cheap, yet skilled, disciplined and healthy labour. And there was the vast raw material potential of the Soviet Union. For the latter, the motive was equally straightforward. Cooperation with western corporations was the only solution to a technology lag that the Soviet Union suffered in the wake of the rise of electronic industries in the West. Increasing integration into the world-economy was the price the Soviet Union had to pay for keeping up with its ideological competitors.

Wallerstein interprets all this evidence as placing the Soviet Union and its east European allies in the semi-periphery. Frank (1977) shows that the Soviet Union lay in an intermediate position between 'West' and 'South'. For instance, East–South trade was used to pay for East–West trade in numerous multilateral arrangements. In short, the Soviet Union exploited the South but was itself exploited by the West in terms of trade arrangements. (This process of unequal exchange in trade is explained in Chapter 3 under 'informal imperialism'.) This places the Soviet regime economically on a par with other non-socialist semi-peripheral countries of the period such as Brazil or Iran. Although it can be argued that within Comecon trade was not capitalistic – prices were not set by the world market – trade was still influenced by world market prices. Furthermore, production was for exchange, and production planning was grossly distorted by the world-economy as a whole through its aggressive inter-state system promoting militarization. With our single political economy logic the situation of the Soviet Union and its allies in the Cold War cannot be interpreted as anything other than an integral part of the world-economy.

If the Soviet Union represented an example of an aggressive upward-moving semi-peripheral state, where does this leave its socialist rhetoric? According to Brucan (1981), communist eastern Europe provided a model for development rather than for socialism. From the very beginning, with Lenin's New Economic Policy of 1921, the essence of Soviet policy has been to 'catch up', and this involved all-out mobilization of national potential. The original heavy industry 'import substitution' phase of protectionism or autarky gave way to the export orientation in the 1970s, which Frank and Levinson

have charted (Koves 1981). In fact, the original ideology of the revolution had to be developed to invent the intermediate state of 'socialism' between capitalism and communism to cover the period when the Soviet Union was 'developing the forces of production'. All this seems very much like mercantilism in new clothing (Frank 1977; Chase-Dunn 1982; Wallerstein 1982). The United States, Germany and Japan have in their turn been aggressive, upwardly mobile, semi-peripheral states which have used political means (trade protectionism, state investment in infrastructure and other support) to improve their competitive position in the world-economy. Soviet 'socialism' seems to have been a classic case of a modern semi-periphery strategy.

But the Soviet Union was always more than just another rising semi-peripheral state. The establishment of the Soviet state in 1917 was the culmination of a revolutionary movement whose internationalism was stemmed but which nevertheless represented an ideological challenge to the capitalism of the world-economy. With the revolution initially limited to Russia, Stalin had no option but to build 'socialism in one country', in other words to catch up before the new state was destroyed by its enemies. From this point onwards, the logic of the world-system placed the Soviet Union in a 'Catch 22' situ-

ation. In order to survive, the state needed to compete with other states, but this competition involved playing the world-economy game by capitalist rules. This came to a head in the 1980s. There have always been policy conflicts within the Soviet bloc between fundamentalists, who emphasize their socialist credentials, and the technocrats, who emphasize efficiency. In times of recession we can expect this 'red versus expert' conflict to be resolved in the latter's favour, and this happened throughout the communist world in the 1980s. In China, it led to economic liberalization but political repression after 1989. In the Soviet Union, the resolution of this conflict under President Gorbachev produced attempts at both economic reforms (*perestroika*, meaning 'restructuring') and political reforms (*glasnost*, meaning 'openness'), which had effects far beyond the Soviet Union. By signalling his intention not to support unpopular communist governments in eastern Europe, Gorbachev precipitated the revolutions of 1989 that ended the Cold War. This shows the particular nature of the Soviet Union as semi-peripheral state and superpower. Other semi-peripheral states suffered severe economic difficulties in the 1980s (symbolized by debt crises), just as the Soviet Union suffered, but only the latter, through tackling its problems, brought a world order to an end.

Summary

In this section we have:

➤ situated academic geopolitical theories within the dynamics of the rise and fall of hegemonic powers;

➤ shown the dynamic nature of academic theories as their utility waxes and wanes within the changing context.

■ The War on Terrorism: geopolitical transition and world order

The end of the Cold War has been interpreted in many ways. But although the interpretations are diverse they share a common belief that we live in a period of dramatic geopolitical change. Using our paired Kondratieff model we would expect current challenges to the United States as hegemonic power, and so there is a temptation to look for potential challengers. Yet there is more to the dynamics of the capitalist world-economy than cyclical change. In addition, there are linear changes over time, some of which are catalysed by hegemonic powers. One such linear change is an increased integration of the world-economy. The degree of integration has provoked the identification of a qualitatively new form of social organization in the capitalist world-economy that falls under the broad title of 'globalization'. Some have claimed that as a result of globalization it is more accurate to say we live in a 'network society' instead of a world of states (Castells 1996).

It is clear that states are still an integral part of the capitalist world-economy, and key geopolitical actors. However, networks and states interact so that the form of the geopolitical transition is likely to be a function of three different processes and actors. In no order of preference or importance they are:

➤ the continuation of states as territorially based actors in competition with one another;

➤ transnational actors, both economic and civil society, who propel a greater integration between states and undermine state sovereignty;

➤ movements resisting both the processes of 'globalization' (or integration) and the power of the state.

Although the ending of the Cold War was precipitated by the economic and then the political collapse of the Soviet Union, we should not forget that it also occurred in the period of relative US economic decline. Hence most of the literature of the 1980s focused on the latter, especially in relation to the rise of Japan. Certainly the most discussed scenario was derivative of our paired Kondratieff model predicting a 'Japanese century'. As US hegemony winds down, the question of a successor naturally arises, and on almost all economic criteria Japan seemed to be the most likely hegemonic candidate. This is certainly the reason for the phenomenal success of Paul Kennedy's (1988) book just before the end of the Cold War. His model postulates that all great powers overstretch themselves militarily, and this becomes a particularly

acute problem when economic decline sets in: hence the demise of great powers from the Habsburgs to the British Empire. Geo-economics was to the fore as the 'threat' to US well-being (Ó Tuathail 1992).

However, the nature of 'threat' changed dramatically on 11 September 2001. Despite the fact that al-Qaeda had been accused of targeting the United States for a number of years and conducted terrorist attacks on the World Trade Center (1993), US embassies in Tanzania and Kenya (1998), US military personnel in Saudi Arabia (1995 and 1996), and the battleship *USS Cole* in Yemen (2000) – the magnitude of the September 11 attacks and their targeting of the US 'homeland' marked a change in the global geopolitical landscape. Since then the world has been at war. The United States declared War on Terrorism. It has resulted in the invasions of Afghanistan and Iraq, domestic political debate in the United States and its allies over the form of the military reaction, terrorist attacks in Spain and Britain, and the establishment of new US bases in eastern Europe and central Asia – just recently the geopolitical domain of the Soviet Union.

The world-systems approach allows us to conceptualize the timing and meaning of al-Qaeda's terrorist campaign against the United States within the temporal dynamics of hegemonic cycles and the core–periphery structure of the capitalist world-economy. In addition, we must consider new geographies of power – networks – to conceptualize the War on Terrorism. Dispute over the necessity, legality and morality of the War on Terrorism has highlighted the role of representation, and for this we turn to critical geopolitics. Finally, we move into the next chapter through an examination of resistance to the geopolitics of hegemony and the call for a politics of solidarity across the persistent differences of the capitalist world-economy; for that task we utilize feminist geopolitics.

The structural geography of the War on Terrorism

Extrapolating the paired Kondratieff model was always too simple, and one of the features of the post-Cold War period is the recognition of the complexity of contemporary geopolitics. It is essential to move beyond the 'classical' geopolitical trait of identifying 'great powers' as the only subjects worthy of attention: the demonization of the United States, and the West in general, by al-Qaeda highlights the role of the core–periphery structure in contemporary conflict. Johan Galtung (1979) posited the core–periphery divide as the line of future conflict in his world classes model. Disparities in wealth and opportunity were identified as the basis for a conflict between the rich and poor at a global scale. Taylor (1992b) resurrected Galtung's world classes model and gave it a new basis that foresaw the rise of al-Qaeda. Taylor's third-worldist position emphasizes the rise of Islam as a world political force. The mobilizing potential of Islam was briefly revealed in the Gulf War, where support for Iraq was widespread among the Muslim population of the world although most Muslim states supported the UN position. Since then, the intensity and breadth of Muslim political mobilization has risen, with al-Qaeda the high-profile and violent manifestation. Of course, Taylor's argument and the existence of al-Qaeda do not mean that there is a homogeneous Islamic geopolitical bloc. The majority of Muslims renounce terrorist violence. However, it is clear that Islam has the potential to replace the now defunct Marxism–Leninism as the spearhead of the political mobilization of the periphery.

The world-systems approach emphasizes the role of hegemonic decline and the core–periphery structure in the current geopolitical transition. The United States, in its role as hegemonic power, has been challenged by the rhetoric and actions of al-Qaeda, and it is attempting to redefine and reinvigorate its global role through the self-constructed War on Terrorism. The difference between the current situation and previous analogous phases defined by the paired Kondratieff model is that the challenge to the hegemonic power is not another strong state – strong through the dominance of core processes within its borders – but a non-state geopolitical actor, al-Qaeda, that is based in the peripheral regions of the capitalist world-economy. There is one more complicating factor; the mode of conflict has changed too. To understand the War on Terrorism it is necessary to consider the geography of networks.

Netwar: a new political geography?

The War on Terrorism has played its part in the invigoration of the discussion of networks in geography and other social sciences. Researchers at the RAND organization in the United States, long related to the military establishment, disseminated the term 'netwar' to describe the new challenges facing the country. The term was originally coined to highlight the potential for states as well as terrorist or criminal groups to exploit the interconnectivity of computer networks: computer-hacking and the destruction of information technology infrastructure to cripple the military capability and economic activity of a state. The concept of netwar was quickly translated into the War on Terrorism. Al-Qaeda was portrayed as a global terrorist network, despite the fact that scholars found it hard to identify an organization with any such coherence. Netwar is defined as the use of network forms of organization, doctrine, strategy, and technology to engage in conflict (Arquilla et al. 1999; Arquilla and Ronfeldt 2001; Hoffman 2001; Ronfeldt and Arquilla 2001). The identification of netwar parallels calls by political geographers to study the flows of information, people and goods across the globe (Ó Tuathail 2000; Hyndman 2004).

However, in the study of terrorism as netwar attention must also be given to the nodes that constitute the network. Nodes are categorized by their function: organizational, narrative, doctrinal, technological and social (Arquilla and Ronfeldt 2001). Also of importance is the likely degree of permanence of the node and its degree of connectivity. Of interest to counter-terrorism is the hypothesis that nodes with more connections would require greater permanence (Flint 2003c). For example, the node that is a suicide bomber, or a cell of bombers, is connected to only a few other cells in order to maintain the security of whole network if they were to be captured and is, by definition, temporary. By way of contrast, key decision makers in terrorist networks would require connections to subordinates that have a degree of permanence.

But the network of the terrorist organization is not the only political geography that must be considered.

Case study

Hegemonic decline, semi-peripheral reaction and international terrorism

Hegemonic decline is characterized as a phase of political disorder; it is the obverse of the relative peace that comes when hegemony is attained. The new politics of upheaval may take several forms, many of them violent, most notably the world wars. The latter inter-state conflicts pit the strong against the strong, but what of the weak? Terrorism is a weapon that is used in certain circumstances by non-state actors disputing the claim of states to sole legitimate use of violence. Such 'private' political coercion has been undertaken at all geographical scales from the local through the national to the international. Here the focus is on international terrorism.

Is international terrorism cyclical? Several world-systems analysts have tried to show this to be the case. The basic model is that international terrorism originates in failing countries of the semi-peripheral zone. During hegemonic decline, the semi-periphery is a zone of new political possibilities but not all semi-peripheral states can take advantage of the opportunity. Thus during Britain's hegemonic decline, the opportunity to rise in the inter-state system was grasped by the United States and Germany. In the semi-periphery of eastern and southern Europe, however, this was not the case. Despite political moves towards 'modernization' in tsarist Russia, the Austro-Hungarian Empire, Italy, Spain and the Ottoman Empire, these parts of Europe did not break into the core of the world-economy before 1914 (the

three empires disintegrated as a result of the 1914–18 war). It is through political disillusionment towards these failing states that anarchism developed as a threat to these regimes. The tactic of propaganda by deed involved assassination of political leaders in order to 'awaken the people'. The most famous victim was Tsar Alexander II in 1881; in 1878 there was an attempt to assassinate the Spanish king; in 1898, the Austrian empress was killed; and in 1914 the Austrian archduke was assassinated, precipitating the First World War. The movement spread beyond the semi-periphery: in 1878 there was an attempt to kill the German emperor; the French president was assassinated in 1894, and US president McKinley was assassinated in 1901. The anarchist with a bomb under his cloak became a figure of fear in a reign of international terror from 1879 to 1914. This political phase ended only when anarchism triggered world war.

The parallels with contemporary Arab–Islamist terrorism are striking. A century on, it is Pacific Asian countries that have taken the opportunities available to semi-peripheral states during US hegemonic decline. Thus the 'greater Middle East' (from north Africa to central Asia) is largely composed of states that are failing their populations despite some political 'modernizations', or oil wealth that has allowed eschewing such 'modernization'. It is through political disillusionment towards these failing states that Islamism has developed

as a threat to these regimes. Assassinations (for example of the president of Egypt, Anwar Sadat) and bombings (for example in Saudi Arabia) in this region have spread beyond the semi-periphery to western Europe (France, Spain, Britain) and the United States. The Islamist with a bomb in a bag has become the figure of fear in a reign of international terror since 1979. If there is to be another world war, most commentators agree that the trigger will come from the Middle East.

However, as with all historical parallels we have to be careful not to over-interpret. Attempts to apply this model to the first hegemonic cycle are less successful – the 'Terror' of the failing French state in 1792 preceding the Napoleonic world war is not a good fit, and in earlier periods the distinction between 'private' and state coercion was less clear-cut. But as with applying all structural models to political process, there is the vital question of contingency. In this case, the decision of the United States after the Second World War to support the creation of the state of Israel in the semi-peripheral Middle East, the ensuing Israeli–Arab wars, and the Israeli–Palestinian conflict are all obviously critical to Islamist popular disillusionment. Here there is no parallel whatsoever with the rise of anarchist politics in outer Europe during British hegemonic decline.

The interested reader is referred to Bergesen and Lizardo (2005) for further discussion of the issue.

The nodes must be located somewhere, and the contextual setting and connectivity of the various nodes in a terrorist network is crucial. Netwar is not separate from the political geography of territorial nation-states, but takes place over the existing terrain of territorial nation-states. Terrorist nodes are located with regard to their function. For example, Osama bin Laden, as the central node of al-Qaeda, is located in a territory (probably northern Pakistan) where it is very hard for the US armed forces to find him. On the other hand, bombers must be near their targets and therefore, most likely in situations where counter-terrorist organizations have greater access and power (Flint 2003c). The layering of networks and states has an implication for the legitimacy of the War on Terrorism. Security forces are still state organizations restricted by the territorial jurisdictions of nation-states. However, to counter terrorist networks, nation-states must enter the sovereign territory of other states. Such actions risk being cavalier with other people's sovereignty, and require a large military presence to cover all the territory where nodes of the terrorist network may be located. In sum, the 'global reach' of the United States in its War on Terrorism is easily perceived as invasion and occupation. The implication is that countering terrorist networks through territorial occupation is the very expression of hegemonic force and power to which terrorists and their supporters are reacting in the first place. It is, arguably, counter-productive counter-terrorism.

Geopoliticians such as Sir Halford Mackinder and Friedrich Ratzel theorized the political construction of spaces: imperial blocs and *Lebensraum*. The War on Terrorism constructs its own spaces too: terrorist networks, networks of US military bases, and the post-Saddam Hussein Iraq. But geopolitics, old and contemporary, is also a matter of representing political space, or imagining certain parts of the world in a manner that justifies foreign policy. Scholars who investigate such representations are identified as practitioners of critical geopolitics.

Summary

In this section we have emphasized two aspects of the current geopolitical transition:

➤ the actions of non-state actors, such as al-Qaeda, can also be understood within the concept of geopolitical codes;

➤ the metageographical context of geopolitics may be changing, with a growing emphasis upon networks rather than states.

Critical geopolitics: representations of the War on Terrorism

Critical geopolitics is part of the post-structural turn in human geography. As such, these political geographers are suspicious of any general framework for ordering knowledge, including the world-systems analysis we use here. They do not see their research as creating a new school of thought but rather it is represented as a loose 'constellation' of ideas (Dalby and Ó Tuathail 1996: 451–2), 'parasitical' on other knowledge creation by making tactical interventions in other work rather than indulging in any broad strategic thinking of their own (Ó Tuathail 1996: 59). Of course, such perennial critics are indispensable for any discipline, and there is no reason why we should not turn the tables and use their fresh insights to inform our world-systems political geography.

Part of the contested construction of the next world order will be a battle over how geographical space is represented. The question of representation is at the heart of critical geopolitics and relates directly to what we have previously referred to as the construction of geopolitical codes. Critical geopolitics aims to interrogate the implicit and explicit meanings that are given to places in order to justify geopolitical actions. For example, the debate in 1990 and 1991 within the United States over whether or not troops should be sent to Bosnia entailed the manipulation of two competing images (Ó Tuathail 1996: 196–213). First, the Bush administration, which was opposed to sending US troops, conjured up the image of Bosnia as a 'quagmire' or 'swamp'. This representation was intended to awaken images of Vietnam to generate support for a policy that would not endanger US troops. On the other hand, proponents of military intervention referred to the genocide in Bosnia as a 'holocaust' to stimulate images of the Nazis' atrocities against the Jews. Both sides were drawing competing images of a little known part of the world to influence and determine international political and military policy. This is a classic example of the fluidity of a geopolitical transition in which a new geopolitical code was being constructed. (During the Cold War, few people had even heard of Bosnia; it was an uncontested part of a communist country; no more needed to be said.) The importance of the critical geopolitical research is to show explicitly that the very construction of the images used in foreign policy making is itself a key geopolitical act.

Critical geopolitics entwines a number of intellectual strands to illustrate the importance of space in geopolitics. As well as the importance of representations of space, spatial practices and the importance of a spatial Other are also seen as components of the way the geography of the world is constructed (Ó Tuathail 1996). Spatial practices refer to the often unquestioned ways in which particular institutions and scales constrain political activity. For example, the dominance of the state in both political practice and intellectual inquiry has inhibited the exploration of alternative politics at both the global and local scales (Walker 1993). The logic of the critical geopolitics position is not to privilege states but to examine the geopolitical actions of social movements, environmental politics and gender as well. The spatial Other refers to Edward Said's (1979) classic *Orientalism*, in which dark images of foreign cultures are painted to make one's own appear in a better light. One of the early studies of critical geopolitics, for example, employed this way of thinking to categorize the Cold War with the Soviet Union as the United States' Other (Dalby 1990b). The approach informs the politics of 'us' by showing how it defines its nemesis. The case of the criticism of Saddam Hussein does more than demonize the man and legitimate military

action against Iraq, it also reflects a discourse that promotes an uncritical view of western democracy.

Captain America: Defender of Hegemony

Al-Qaeda was a challenge to the universal pretensions of the United States as hegemonic power. Following the world classes idea of Johan Galtung, al-Qaeda used the rhetoric of material deprivation and cultural oppression to justify its actions. To counter this rhetoric and to represent its military actions as just, the United States represented its own actions as being in the name of the good of all humanity – the ultimate expression of the liberal universal geopolitical code of the hegemonic power (Flint and Falah 2004). It attempted to portray its actions as something other than national self-interest, or actions that critiques could paint as the construction of 'empire'. It did so by changing the geographical scale of the rhetoric. It eschewed the language of states, and instead concentrated upon human rights – a scale that was simultaneously individual and universal (Flint and Falah 2004).

It is not just the language of press statements and political speeches that is a crucial component of the representation of geopolitical actions. As hegemonic power, the United States must address global public opinion when justifying its actions, including military invasion. It must also gain the support of the American public. Although public support for President George W. Bush and the War on Terrorism waned through 2005 and 2006, what is of more importance is the general common-sense understanding held by the American public that its country has particular rights and responsibilities in the inter-state system. In other words, although political debate occurs over whether the United States should have invaded Iraq (there was practically no dissent in Congress) and what it should do in the face of insurgency, the general belief that the United States should undertake political and military interventions is not even questioned. It has become an assumed part of the US national identity, what E. P. Thompson (1985) called 'hegemonic nationalism'.

Such ideas are constructed and maintained through popular culture: films, books, songs, TV series and so on. One avenue is cartoon books, and one obvious character is Captain America (Dittmer 2005). Captain America first appeared in 1940, prior to the United States' entry into the Second World War. Interestingly, the character is a citizen who is too physically weak to become a regular soldier but volunteers for a dangerous scientific experiment. He is injected with a serum that transforms his physique into the epitome of athleticism. In other words, Captain America, a representation of the nation itself, is transformed through the application of the science and industry of the military–industrial complex. The nature of his mission also transmits a national myth: he only acts violently to defend the nation. The actions of Captain America are not an aggressive usage of his extraordinary strength (Dittmer 2005).

Captain America comic books went through a number of manifestations reflecting different periods of, and challenges to, US hegemony: Second World War hero as the United States assumes its global role; 'Captain America . . . Commie Smasher!' at the beginning of the Cold War; and his reinvention in 1964 since when he has fought issues such as poverty, racism and pollution (Dittmer 2005). Significantly, in the wake of public discomfort with the Vietnam War, Captain America battled for 'American values' mainly within the borders of the United States. But then came the terrorist attacks of 11 September 2001 and Captain America had a new mission.

Dittmer's (2005) critical geopolitical analysis of Captain America reveals a continued focus upon battles within the United States. However, the geopolitical context has changed and the values that

Captain America now defends are shown to be Christian-American in the face of the threat of Islamic fundamentalist violence. The scene is Centerville, a geographical landscape that represents the socio-political ideal of 'Middle America'. It is Easter Sunday morning, and most of the town's inhabitants are at the church, portrayed in such a way as to show how Christian values are a part of the social landscape (Figure 2.3). The church congregation is attacked by Islamic terrorists, and the story unfolds through a dialogue that pits American values (as the epitome of Christian values) against Muslim ones. Captain America is fighting the Clash of Civilizations, Samuel Huntington's (1993) representation of the new form of global conflict.

The imagery used reveals the different representations of the two religions. Outside, the church sign says 'Easter Service: All Are Welcome' (Dittmer 2005: 639, quoting Rieber and Cassaday 2002a: 3), expressing the 'openness and tolerance of Christianity' (Dittmer 2005: 639). On the other hand, Islam is portrayed as irrational, intolerant and violent. One of the terrorists in the story refers to a woman as a 'whore with a painted mouth' (Dittmer 2005: 639, quoting Rieber and Cassaday 2002a: 8), and another declares, 'Death is peace for me' (ibid.). The nation that Captain America, the ordinary guy empowered by the United States military and scientific might, must defend is represented as a tolerant Christian nation, and the new enemy is Islam.

However, the story is not as clear as that. In a controversial stance, the storyline took a nuanced, even overtly critical, attitude towards the overseas

Figure 2.3 Centerville, USA. Venue for Captain America's battle with Islamic terrorists.
Source: © 2006 Marvel Characters, Inc. Used with permission.

military interventions of the United States. Fictional Centerville is the location of a bomb manufacturing plant, and one of the terrorists, al-Tariq, links that to the fate of the captive congregation: 'Some of you are asking your God why you will die today. Some of you know – those of you who work at the bomb manufacturing facility at the edge of this peaceful town. Today you learn what it means to sow the wind and reap the whirlwind' (Dittmer 2005: 640, quoting Rieber and Cassaday 2002b: 1). In a conversation between al-Tariq and Captain America, the hero learns that the children fighting with the terrorist have prosthetic limbs; they are the casualties of American bombs. It forces Captain America to reflect: 'Are we hated because we're free – free and prosperous and good? Or does the light we see cast shadows that we don't – where monsters like this al-Tariq can plant the seeds of hate?' (Dittmer 2005: 639, quoting Rieber and Cassaday 2002b: 15).

Although it is al-Tariq who is labelled the monster, and not the US military machine, the violence of US military actions is questioned. Although American values are to be defended, it is the 'homeland' and its values that must be saved. In this story Captain America raises the morality and efficacy of military actions in other sovereign spaces. It was this self-same theme that we can see was adopted by political cartoonists in Arab newspapers: a very different audience but demonstration of the same political geographic imagination.

Arab political cartoons and the critique of 'empire'

Falah et al. (2006) analysed the political geographic themes in cartoons published in Arab newspapers in the months leading up to the US invasion of Iraq in March 2003. Political cartoons are an important component of popular culture in Arab countries as they are able to disseminate views that are likely to be censored if they appeared in the form of text. Cartoons from all Arab countries were analysed and five common themes were identified (Falah et al. 2006): imperialist intent; the arrogance of power; realist power politics; double standards towards the United Nations; and US support for Israel as an immoral act.

Figure 2.4 'Uncle Sam' and the control of Middle East oil.

Source: Ad-dustour Newspaper (15 November 2002).

The cartoons portrayed President George W. Bush as a 'muscle-man', a powerful geopolitical actor who could throw his weight around for his own purposes and not the good of humankind. The cartoons identified US actions as the classically 'realist' politics of self-interest and not the dissemination of 'good'. Moreover, these actions were seen to violate the political geography of territorial and sovereign nation-states. Instead, the United States was portrayed as 'imperial', able to invade countries at will in order to control natural resources (Figure 2.4). Figure 2.5 shows an Uncle Sam figure with a 'to-do' list of Arab countries: Palestine and Iraq have already been accounted for, and Syria, Saudi Arabia, and Egypt remain on the agenda. Other cartoons portrayed the

Figure 2.5 'Uncle Sam' and military intervention in the Middle East.

Source: El-Osboa Newspaper (6 November 2002).

Figure 2.6 US alliance with Israel portrayed as a barrier to peace.

Source: Alwatan Newspaper (November 2002).

President George W. Bush has justified its activity in a way common to all hegemonic powers: the duty to disseminate the benefits of liberalism across the globe (Taylor 1998).

Summary

In this section we have shown:

➤ the important role of representation in geopolitics;

➤ thereby, illustrating that geopolitics is as much a matter of everyday popular actions as it is the behaviour of state elites or academics.

United States as arrogant and uninterested in Arab viewpoints: a hegemonic power acting for its own self-interest and not a 'world leader'. Furthermore, the United States was portrayed as contemptuous of United Nations actions, with the UN seen as the guardian and adjudicator of territorial disputes. As with the themes of realist power and imperial intent, the United States was critiqued for violating the norms of territorial sovereignty. In addition, such a US stance was portrayed as hypocritical because it supported Israel, whose actions are vilified across the Arab world for its occupation of Palestinian territory and its violation of international law. Figure 2.6 provokes dissent towards the United States by integrating two of these themes: support for Israel prevents the establishment of peace in the Middle East built upon UN resolutions.

Captain America and Arab political cartoons are both illustrations of the role of representations in popular culture in justifying and challenging the geopolitical practices of states. At the centre of the representations is a geopolitical imagination of the propriety of certain geopolitical actions, namely the dubious morality of intervening in the sovereign spaces of other countries (Flint and Falah 2004). The United States is facing a geopolitical challenge in the form of terrorism, portrayed as netwar. Its response, the War on Terrorism, is a practice of global military intervention. Popular culture in the United States and in Arab countries has questioned the moral basis of such activity. In turn, the administration of

■ Feminist geopolitics and the interrogation of security

Critical geopolitics plays an essential role in contemporary political geography. It facilitates the ability to 'unpack' policy statements and academic theories to see their role in maintaining power relations. The state-centric, white, and male voice of geopolitics is made transparent. However, critical geopolitics has recently come to be critiqued itself. Feminist scholars have advanced a feminist geopolitics that challenges critical geopolitics, and other frameworks including the world-systems approach, in two ways. First, the content of critical geopolitics is critiqued for remaining state-centric: the focus is still upon 'old white guys' speaking for the government. Second, the practice of critical geopolitics is seen as stopping at academic critique, rather than doing something. In other words, no normative geopolitical agenda is promoted.

Feminist geopoliticians do not present a single well-defined agenda (see Koopman 2006 for a clear discussion of the difficulties of academics speaking of the political goals of others). However, the feminist approach involves an understanding that academics are themselves activists and that research and teaching is, in part, a political act. Much feminist scholarship is participatory research that attempts to publicize and understand 'the voices of others and the political

'Empire' and participatory research

November 21, 2004: Columbus, Georgia, USA. Neris González, a Salvadoran woman, stands on stage. Before her: 16,000 mostly white middle-class U.S. protestors, over half of them women. Behind her: two long, tall fences topped with huge rolls of barbed wire. Behind that: Fort Benning, home of the School of the Americas, a U.S. Army training academy for Latin American military officers. Neris tells the crowd how a Salvadoran army official trained at this U.S. Army school tortured her for weeks, raped her, and beat her seven-month unborn child to death inside her. Above her a U.S. military helicopter hovers low in the sky, the sound of its wings thumping hard. Her words are followed by a liturgical reading of the names of the dead at the hands of graduates of the School. 'Ignacio Ellacuría, 59 years old'. We all cry, 'Presente!' and lift our crosses, each with a name, a body. With each name we take a slow step forward towards the base in a slow funeral procession. We weave the crosses into the Army's chain-link fence. We bring the dead from other worlds, other times, and lay these murders at the feet of the U.S. Empire's trainers of torture and death. We fill the fence and turn it into an altar. Liz Deligio is one of fifteen who cross over the fence and on to the base, in an act of civil disobedience for which she serves ninety days in a federal prison.

Source: Koopman, S. (2006).

The preceding quote illustrates Sara Koopman's position as both academic and activist. The scene described is of a protest at Fort Benning, Georgia (Figure 2.7), a military installation that has trained many people from foreign armies, Colombia being a particularly frequent customer, in the skills of surveillance and torture. The US government claims that torture is no longer on the curriculum, but the graduates of the school work in forces with appalling human rights records. Fort Benning is part of the infrastructure of 'empire' – the bases and facilities that allow the United States to project its military might across the globe, either directly or in conjunction with allied armies.

The other insight from the quote is the tension Koopman explores between being a white middle-class academic from North America (a relatively privileged position) and writing about marginalized groups. How does an activist scholar avoid speaking for people in less privileged situations while still writing and teaching of their plight? This is a difficult balancing act – and it is feminist scholars who have led the way in trying to negotiate the awkward position of the researcher. Koopman's solution is to speak with rather than for the political movement with which one is engaged, or a collaborative rather than dominant and exploitive relationship.

project of facilitating the empowerment of others through the research process' (Sharp 2004: 97).

Concentration on 'others' incorporates many other actors into the analysis of geopolitics. It extends the topic beyond the state. For example, Jennifer Hyndman's (2000, 2004) research on refugees explores the geopolitics of mobility. In the process the intersection of racist and sexist practices with state institutions and borders are highlighted. The family, the cultural norms of child care and other manifestations of gendered divisions of labour, the fear of travelling into particular regions and places because of sexist and racist violence, and the way movement is inhibited or facilitated by wealth, are all seen as components of geopolitics. Feminist geopolitics requires a change in the conceptualization of security by switching the scale of analysis. The proposed move is from national security to human security, in a manner that gives voice to people (especially the marginalized) and broadens the processes and structures of geopolitics beyond states to include access to food, water, land, education and health care, for example, and protection from violence motivated by sexism and racism.

The world-systems approach to geopolitics contextualizes political activity within the structure and dynamics of the capitalist world-economy. Research topics have, to date, been focused upon the

Figure 2.7 Demonstration at Fort Benning, Georgia.
Source: Linda Panetta (www.soawne.org).

geopolitical codes of states within the broader system. Feminist geopolitics is a catalyst for political geographers to consider how the structure and institutions (introduced in the previous chapter) of the capitalist world-economy interact to create varying contexts of human security for different groups. Both the world-systems approach and feminist geopolitics advocate a political agenda of emancipation. It will be interesting to see whether the methodology and subject matter of feminist geopolitics can and will be integrated with the historical and structural account provided by the world-systems approach.

It is unlikely that the serious study of geopolitics can ever return to the taken-for-granted world of power politics from which it emerged. After the Cold War, the general state custom of separating political relations from economic relations, always adhered to more in the theory than the practice, has finally disintegrated. State security services are now expected to pay at least as much attention to 'geo-economics' as to geopolitics. This was presaged in political

geography by the introduction of political economy perspectives as, for instance, in the first edition of this book. But currently it is the adherents to critical and feminist geopolitics who, more than any other research group, are interrogating the assumed worlds of formal and practical geopolitical reasoning.

Possibly the most valuable contribution of critical and feminist geopolitics is to point the way forward for how political geography copes with globalization and 'empire'. The various trans-state positions of globalization undermine the state-centric assumptions of a century of geopolitics, whether British, German, American or from any other political base. In terms of geographical globalization, this can be represented as a 'de-territorialization' of world politics (Ó Tuathail 1996). Critical geopolitics puts us on the alert for crude 're-territorializations' that try to reconstruct simple stable representations in a fluid world of massive social change. But new simple spatial imagery (such as the 'Axis of Evil') is at odds with the material changes that are globalization. Feminist geopolitics catalyses a questioning of accepted notions of 'security' during a time of war. In the face of military invasions and occupations in the name of counter-terrorism and national security, feminist geopolitics asks whether that is the sole or most appropriate form of security: security from what and for whom? It is our contention that the valuable contributions of critical and feminist geopolitics complement the world-systems approach. The latter offers a temporal and structural contextualization of political actions, while the former point to the agency and struggles of geopolitical actors within those contexts.

Summary

In this section we have introduced feminist geopolitics and:

➤ illustrated its role in challenging established lines of inquiry;

➤ suggested ways in which feminist geopolitics and the world-systems approach may complement each other.

Chapter summary

In this chapter we have:

➤ investigated the changing meaning of the term 'geopolitics';

➤ introduced the concepts geopolitical codes and geopolitical world orders;

➤ used the paired Kondratieff model of hegemony to show the interconnections in our political economy approach;

➤ used the paired Kondratieff model of hegemony to situate geopolitical codes and explain their dynamism;

➤ situated the contemporary War on Terrorism within our political economy approach of hegemonic cycles;

➤ illustrated the importance of geopolitical representations through the perspective of critical geopolitics;

➤ interrogated the mainstream ideas of what constitutes 'security' through the lens of feminist geopolitics.

Together, the concepts and approaches introduced in this chapter provide a means for understanding the dynamism of geopolitics, and the relationship between economic and political processes. In addition, the contested nature of geopolitics is revealed as different 'common-sense' views of 'threats', and appropriate responses are advocated by different states or political interests within them. The most radical stance is provided by feminist geopolitics and its aim to overhaul our taken-for-granted notions of foreign policy.

Key glossary terms from Chapter 2

administration	elite	inter-state system	place
Annales school of	empire	Islam	power
history	'empire'	isolation	protectionism
ANZUS	First World War	Kondratieff waves	racism
autarky	free trade	(cycles)	right-wing
balance of power	functionalism	*Lebensraum*	Second World War
bloc	geopolitical code	liberal	semi-periphery
capitalism	geopolitical transition	logistic wave	socialism
capitalist world-	geopolitical world order	*longue durée*	sovereignty
economy	geopolitics	mercantilism	space
civil society	globalization	Monroe Doctrine	state
classes	government	multinational	superpower
Cold War	heartland–rimland	corporations	Thirty Years War
Comecon	thesis	Napoleonic Wars	Treaty of Tordesillas
communism	heartland theory	nation	Treaty of Versailles
conservative	hegemony	nationalism	Truman Doctrine
containment	homeland	nation-state	United Nations
core	human rights	NATO	world-economy
decolonization	idealism	neo-conservative	world-empire
democracy	ideology	opposition	world market
détente	imperialism	pan-region	world-system
economism	informal imperialism	periphery	world-systems analysis

Suggested reading

Agnew, J. (2003) *Geopolitics: Re-visioning World Politics*, 2nd edition. New York: Routledge. Defines and discusses the features of the modern geopolitical imagination and the way that it has dominated foreign policy, before positing that a new form of geopolitics is emerging.

Enloe, C. (2001) *Bananas, Beaches and Bases: Making Feminist Sense of International Politics*. Berkeley: University of California Press. Seminal feminist text on international politics that was the foundation for feminist geopolitics.

Ó Tuathail, G. (1996) *Critical Geopolitics*. Minneapolis: University of Minnesota Press. Seminal text introducing the field of critical geopolitics with arresting examples.

Ó Tuathail, G., Dalby, S. and Routledge, P. (2006) *The Geopolitics Reader*, 2nd edition. New York: Routledge. A collection of key geopolitical statements from the late nineteenth century to the present, featuring statesmen, academics, journalists, and dissidents.

Activities

1 The United States government is well established on the internet. Some relevant sites are listed below. Use the sites to find recent policy documents or press releases and then interpret them through the lens of the paired Kondratieff model. In what way do the policies speak of active engagement with the rest of the world, in terms of free trade, investment, immigration or military action? On the other hand, in what way do the policies emphasize territorial security and sovereignty?

http://www.whitehouse.gov/   The White House
http://www.commerce.gov/   Department of Commerce
http://www.commerce.gov/   Department of Defense
http://www.state.gov   State Department
http://www.dhs.gov   Department of Homeland Security

2 Consider the representation of geopolitical events in cinema films currently on release. Who or what is the threat? What was the nationality and gender of the 'hero'? Reading between the lines, did the storyline imply particular foreign policy actions or inactions by the United States or other countries?

3 Refer to the webpage of the US Department of Defence, the British Ministry of Defence, or the defence department of another country and read its discussion of an ongoing mission. Then consider the same situation from the perspective of feminist geopolitics. What 'silences' in the official commentary does the feminist perspective help to illuminate?

Chapter 3

Geography of imperialisms

What this chapter covers

The rise of geography as a university discipline in the late nineteenth century was closely related to imperialism (Hudson 1977). The systematic parts of the discipline that were developed at this time – political geography, commercial geography and colonial geography – all covered fields of study that were useful to those engaged in imperial activities, be they politicians, soldiers, traders or settlers. Outside the universities, geographical societies flourished as vehicles for providing advice to would-be imperial actors (McKay 1943). Contemporaries who wanted to find out about the imperialism of their age turned to geography for the facts. Lenin, for instance, used a work of the German geographer A. Supan, *The Territorial Development of the European Colonies*, for information on the extent of late nineteenth-century European expansion. Geography and imperialism were intimately entwined (Godlewska and Smith 1994).

Just as geopolitics was found to be an embarrassment to geography after the Second World War, so a generation earlier links to imperialism were found to be an embarrassment to geography in the aftermath of the First World War. In the inter-war period (1919–39), according to Bowle (1974: 524), 'the main intellectual fashion [was] set strongly against Empire'. And this intellectual trend was confirmed by the disintegration of the European empires after 1945. As a result, imperialism was largely removed from both the political and academic agendas as a matter of little or no contemporary relevance. Beginning in the 1970s, world-systems analysis was part of a new trend that reversed this agenda setting. Imperialism is a central concept in our political geography. As with geopolitics, we are not in the business of endorsing this form of global politics; rather, we treat it as an object of study necessary for understanding our modern world.

Imperialism is an obvious combination of political and geographical characteristics. Yet the topic of imperialism was a neglected theme in political geography. This was not just a problem of political geography, however, but more to do with the nature of modern social science as a whole. It relates directly to what is referred to as the poverty of disciplines. The term 'imperialism' is a classic political economy concept that cannot be properly defined in either

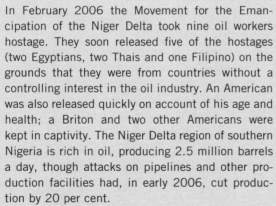

Western oil workers taken hostage in Nigeria

In February 2006 the Movement for the Emancipation of the Niger Delta took nine oil workers hostage. They soon released five of the hostages (two Egyptians, two Thais and one Filipino) on the grounds that they were from countries without a controlling interest in the oil industry. An American was also released quickly on account of his age and health; a Briton and two other Americans were kept in captivity. The Niger Delta region of southern Nigeria is rich in oil, producing 2.5 million barrels a day, though attacks on pipelines and other production facilities had, in early 2006, cut production by 20 per cent.

The claims of the militant groups were simple. They wanted a greater share of the wealth generated by the oil industry. The local population was impoverished. In the words of Macon Hawkins, the released American, 'they're dirt poor – poor as field mice'. The wealth generated by oil extraction in the region has bypassed the local population. It benefits workers hired from foreign countries, and the profits flow through the national government and into the coffers of international oil companies.

The Movement for the Emancipation of the Niger Delta believes it is fighting against an imperial relationship that extracts resources from 'their' land but allows the profits to flow into bank accounts outside of Nigeria and even Africa. There is clearly a differential in the ability to profit from Nigerian oil and it is one means by which differences in wealth and life opportunities are manifest in today's world. *But should such a situation be identified as an expression of imperialism? Consider how and why your response to this question changed, or not, after reading this chapter.*

Source: 'Six oil hostages released in Nigeria', *Guardian Unlimited*, 2 March 2006. www.guardian.co.uk/print/ 0,,329425308-1130373,00.html. Accessed 2 March 2006.

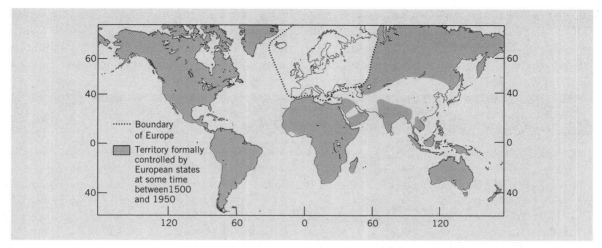

Figure 3.1 The geographical extent of European political control of the periphery.

political or economic categories alone (Barratt Brown 1974: 19). Hence the neglect of imperialism spread far beyond political geography. One of the most cogent criticisms of the whole 'modernization' and 'development' schools of modern social science, for instance, is that they seem to conveniently 'forget' or at least 'ignore' the contribution of imperialism to the modern world situation.

The geographical extent of European political control in the periphery is shown in Figure 3.1. All areas that were at some time under core control are shown and can be seen to include almost all the periphery. The major exception is China, but even here the leading core states delimited their 'spheres of influence'. In geographical terms, the result of this political control was a world organized as one huge functional region for the core states. This has been, and, we shall argue, remains, the dominant spatial organization into the twenty-first century. And yet until recently this subject was considered so unimportant that it was largely left to historians to debate it outside the social sciences.

In our world-systems approach, imperialism is much more than a problem of history. The constant and necessary core–periphery hierarchy of the capitalist world-economy requires imperialism to be a continual political project. However, the dynamism of the capitalist world-economy means that the form of imperialism will change in relation to cyclical and linear trends: 'bringing history back in' means

opening up the issues of imperialism once again. In the wake of the terrorist attacks of 11 September 2001 the word 'empire' has reappeared with a vengeance. Now it is almost impossible not to see the word when perusing the current affairs section of the bookshop. But we must remain critical of its current usage. In this chapter we place imperialism in the core–periphery relationship and relate it to the current actions of the United States as hegemonic power: the military expeditions that have generated the renewed focus on 'empire'.

Our discussion of imperialisms will emphasize the material construction of functionalist spaces of formal empire, and conceptualize imperialism within the structure and dynamics of the capitalist world-economy. From this discussion we can consider the construction of localities in the periphery with regard to the role they played in a global imperial system. Clearly, this requires seeing the connection of global and local processes as the dynamics of the capitalist world-economy are manifested in local spaces and social relations. In the final section of the chapter, contemporary definitions of 'empire' are placed within the temporal context of hegemonic cycles and their interaction with the changing nature of capitalism, or 'globalization'.

One of the achievements of neo-Marxist perspectives on social science has been the rediscovery of the revolutionary heritage of imperialism. Of course, the concept has not been neglected just because it is difficult to fit into the modern disciplines of political science and economics; imperialism has been neglected because it is part of a classical revolutionary theory that modern social science was designed to circumvent. We cannot understand imperialism, in a world-systems or any other framework, without first understanding its revolutionary heritage. The first

Formal and informal imperialisms in contemporary global sports

Curiously, one of the easiest ways to discern past formal and informal imperialisms in today's world is to look at which sports are popular in which countries.

The contemporary world geography of sport has a direct link to past patterns of imperialisms as cultural dominations. Although sports and pastimes were a mainstay of traditional rural life, the urbanization consequent upon industrialization in the nineteenth century changed the nature and character of sports. In the second half of the nineteenth century sports became major events for urban spectators adding excitement to hard lives in difficult times. Initially focusing on betting to fund sport (horse racing, road running, boat racing, boxing), towards the end of the century there was the widespread development of team games paid for by attracting spectators to dedicated club stadia. National organizations were formed with their leagues and cup competitions leading to popular rivalries between clubs from different cities.

But which sports would catch on in which countries? By the turn of the nineteenth and twentieth centuries there was a real smorgasbord of different games available with, for instance, English cricket tourists stopping off in New York for a game on their way to Canada while professional baseball was beginning in England. But we know these early initiatives were not to bear fruit as major sports in the United States, Canada and England. Why? Basically the world geography of sport evolved through four main zones:

➤ In old British colonies – the formal British Empire – English political elites diffused the game of cricket. Today cricket remains a major sport in south Asia (India, Pakistan, Sri Lanka and Bangladesh), Australasia (Australia and New Zealand), southern Africa (South Africa and Zimbabwe), and English-speaking West Indies as well as in the sport's country of origin, England.

➤ In the 'informal empire' – countries with strong trading links to Britain – British economic elites introduced football (soccer), sometimes founding football clubs. Today, this is the most global of sports but all World Cup winners have been countries that come from three of Britain's nineteenth century prime trading zones, and who still provide the leading contenders: South America (Argentina, Brazil, Uruguay), southern Europe (Italy, Spain, Portugal), western Europe (France, Germany, the Netherlands) as well as England, the sport's country of origin.

➤ In the late nineteenth century direct British influence, both political and economic, was least in the United States, resulting in neither cricket nor soccer gaining a foothold against home-grown sports, notably baseball and grid iron football. US professional sports spread to Canada in the twentieth century so that it became the only part of the old British Empire/Commonwealth not to have either cricket or soccer as a major sport. In the early twentieth century there were debates about whether Canada was more influenced by Britain or America: sport has provided the answer.

➤ In other parts of the world that Britain was unable to dominate in the late nineteenth century, the sports menu remained open to US influence in the twentieth century. This resulted in baseball diffusing to more than just Canada: it became a major sport in Spanish-speaking West Indies, Central America and parts of Pacific Asia, notably Japan. In some of these countries baseball vies with soccer as the national sport.

Like so many other popular cultural practices, contemporary world geographies of sport have been shaped through activities of Britain and the United States during their hegemonic cycles.

section of this chapter describes this heritage and the subsequent interpretations of historians who could safely label these issues as concerns of the past.

We then order, or conceptualize, an understanding of imperialism by mapping periods of growth in relation to the dynamic model of hegemony and rivalry presented in the previous chapter. These descriptions are divided into two sections, the first dealing with formal imperialism and the second with informal imperialism. Both imperialisms are characterized by the dominance relation between core and periphery, the distinction between them being that the former involves political control of peripheral territory in addition to economic exploitation. In the concluding section we discuss the interaction of globalization and war to explain the current identification of 'empire'.

Chapter framework

This chapter addresses the following components of the political geography framework that we introduced in Chapter 1:

➤ The chapter *conceptualizes* the *persistent differences* of wealth and opportunity that categorize the capitalist world-economy.

➤ The *concepts* of formal and informal imperialism are situated within the *temporal context of the rise and fall of great powers*.

➤ The *spatial interconnections* that facilitate imperialism as political practice are elucidated.

➤ The experiences of people within *localities are understood within the wider whole* that is the capitalist world-economy and its core–periphery structure.

➤ In our discussion of contemporary 'empire' we discuss the interaction between the *material world and the way it is represented*.

■ The revolutionary heritage

One of the problems in dealing with a concept like imperialism is that its meaning has changed over time. We are using it in this study in its current meaning as indicating a dominance relation, but in considering the heritage of our study we must understand past meanings. In fact imperialism, unlike the terms 'imperial' and 'empire' from which it derives, is of fairly recent origin. Its initial use was as a term of abuse to describe French Emperor Louis Napoleon's foreign policy in the mid-nineteenth century, the implication being that the policy was aggressive and reckless. By the end of the century, imperialism was associated with aggressive expansion by core countries and was a major concern of domestic politics – both the British general election and the American presidential election of 1900 had imperialism as a major campaign issue, for instance. From this contemporary debate one contribution stands out as being of continuing significance: J. A. Hobson's (1902) *Imperialism: A Study*, written in the aftermath of the British imperial war in South Africa. As a polemic against imperialism this was to become a source for Lenin's theory, and we consider these surprising links between English liberalism and Russian Marxism in our outline of the revolutionary theory below. This is followed by a description of how the subject matter became despatched into history before we consider the development of new theory that has become part of the world-systems approach.

The rise and fall of the classical theory

Imperialism had a central role to play in the theories of the generation of Marxists after Marx and Engels. Whereas Marx did not use the term 'imperialism' and made no general study of the effects of capitalism in the periphery, for Lenin and his associates imperialism had become the essence of the capitalism of their time. Hence when we refer to the classical theory of imperialism it is to Lenin not Marx that we refer. However, we must be careful not to equate the classical Marxist concept of imperialism with the modern concept. The classical concept was both broader and more akin to our geopolitics than is often realized. The various meanings of the concept will become clear as we look at the different theories.

Imperialism as geopolitics

Lenin's famous pamphlet on imperialism was written during the First World War as part of an ongoing debate in Marxist circles on the meaning of the war. It was avowedly polemic and popular and represents a position that subsequently became orthodox Marxism rather than an original contribution (Brewer 1980: ch. 5). That is to say, Lenin draws upon the work of other authors, reorders some ideas, changes the emphases to suit his debating position but adds little to the argument. He particularly draws on the work of Hilferding, an Austrian Marxist, and the British liberal writer Hobson.

According to Brewer (1980), Hilferding is the real founder of the Marxist theory of imperialism. He used the term 'imperialism' to mean inter-core rivalry that incorporated dominance over the periphery but did not make this its primary characteristic. Rather, he emphasized the rise of finance capital in a new monopoly era when industrial and financial capital are fused into one system. He was particularly impressed by the power of the banks and their relations to industry and the state in pre-1914 Germany. He concluded that finance capital needed strong state support in terms of economic protection and obtaining territories for investment, markets and raw materials, but he remained concerned largely with the internal developments of core countries rather than the 'spillover' effects into the periphery. These ideas are developed in Bukharin's (1972) *Imperialism and the World Economy* of 1917, in which the idea of imperialism as the geopolitics of a particular stage of capitalism – Bukharin's finance monopoly capitalism – is proposed. Lenin takes this argument one step further and defines imperialism as a stage of capitalism, the 'highest', in fact. By this he meant that the competitive capitalism that Marx had described in the mid-nineteenth century had been replaced by a monopoly capitalism. This was the final stage of capitalism, because the contradictions of the system were reflected in inter-state rivalries that had produced a world war as the beginning of the revolution. For Lenin, the First World War represented the death throes of capitalism.

The Hobson–Lenin paradigm

The other major source for Lenin's ideas was, as we have indicated, Hobson's non-Marxist anti-imperialism. For Hobson, the 'taproot of imperialism' was surplus capital generated in the core looking for investment outlets in the periphery. Hobson collected data on overseas investment, areas and populations of colonies and dates of acquisitions and concluded that 1870 was a vital turning point in European history. Lenin reproduced this information in his own pamphlet, and Wallerstein (1980a) feels that it is reasonable to talk of the Hobson–Lenin paradigm. This has three basic positions: (1) within states there are different interests between different sectors of capital; (2) a monopoly finance sector was emerging as the dominant interest that could steer the state into imperial ventures for its own interests but against the interests of the other sectors; and (3) despite much popular support, these ventures were against the real interests of the working classes in these countries. Wallerstein terms this a paradigm because of its influence in setting the agenda for all subsequent studies of imperialism. In this sense, whether we agree with Hobson's and Lenin's theses or not, we remain a 'prisoner' of their ideas as long as they continue to be the starting point of discussions on imperialism.

Wallerstein (ibid.) emphasizes the similarities between Hobson and Lenin on imperialism. This is important, because their association has given Lenin's revolutionary theory a respectability in western thought that it might not otherwise have received. But there are fundamental differences between the two men. As an English liberal, Hobson's anti-imperialism pointed towards reformist solutions of free trade and raising domestic consumption. Lenin's theory was intended as a contribution to overthrowing the capitalist system. By understanding the stage that capitalism had reached, revolutionary forces could seize the opportunity that the world war offered and create a new system. This was not a prediction but a statement of the theoretical rationale for revolutionary strategy and practice. In the end, the practice was only partially successful, producing a successful revolution in Russia alone. But the theory remained to be refined and updated by sympathizers and criticized and dismissed by opponents. The best way of achieving the latter is to take the subject off the political agenda. This was done by despatching the topic to history, where the final stage of capitalism becomes converted into the harmless 'age of imperialism'.

Despatching imperialism to history

Since the Second World War, there has been a reaction against narrow historical interpretations. In a series of studies, Robinson and Gallagher have attempted to rethink the whole debate. Their critique is of particular relevance to our study on two counts. First, they challenge the whole idea of an 'age of imperialism' in the late nineteenth century (Gallagher and Robinson 1953). They emphasize the continuity of policy throughout the nineteenth century and do not accept that 1870 represents a fundamental watershed in modern history. After all, India, 'the jewel in Britain's imperial crown', was obtained before the age of imperialism and continued to be by far the most important imperial possession throughout the age of imperialism. Second, they queried whether the causes of European expansion could be found solely in processes – economic or political, or both – operating in the core (Robinson et al. 1961). As studies of imperialism began appearing from the periphery, it became clear that the timing, nature and form of dominance relations were often conditioned by local circumstances in the periphery. Robinson (1973) has elaborated this into a theory of collaboration whereby certain peripheral elites interact with core states to help to produce imperialism. This explains why European powers could control so much of the periphery with relatively little military involvement. Clearly, British control of India would have been impossible without collaboration. In short, the classical debate was hopelessly Eurocentric. Hence we can say that Robinson and Gallagher were steering the temporal and spatial coordinates of the debate in the direction of the framework adopted for this study.

A world-systems interpretation of imperialism

Robinson and Gallagher help us to break out of the Hobson–Lenin paradigm but they do not chart a new theory. By extending the debate beyond the 'age of imperialism' they leave the way open for other historians to commit what we have termed the myth of universal law whereby imperialism is a general political process based on motives of expansion and conquest. Lichtheim (1971), for instance, considers the 'imperialism' of the Roman Empire and other world-empires alongside modern imperialism. Within our world-systems framework – the time limit is quite explicit – imperialism is a dominance relation in the world-economy and therefore is not found prior to the sixteenth century. Previous political expansions are based on fundamentally different political economy processes and require a separate term to describe them.

Of course, if we can extend Robinson and Gallagher's concern for the nineteenth century backwards to the origins of our system we can also bring the concept forward to the present. It was not just historians who were taking note of new studies of the periphery in the period since the Second World War. As pointed out in Chapter 1, dependency theory was developed and extended to become the world-systems approach. In political terms, decolonization put imperialism back on the agenda with theories such as Ghanaian leader Kwame Nkrumah's concept of 'neo-colonialism' as another stage of capitalism. The time was ripe for a second set of revolutionary theories to emerge to inform and direct the periphery in their relations with the core (Blaut 1975). Here we briefly describe the world-systems theory of imperialism.

Transcending Robinson and Gallagher

Wallerstein (1980a) follows Robinson and Gallagher in breaking out of the Hobson–Lenin paradigm. In world-systems terms, this paradigm is a classic example of the error of developmentalism, taking as it does countries as units of change and identifying stages through which they pass. But we go beyond Robinson and Gallagher's theses by transcending their dichotomies of continuity versus discontinuity and core versus periphery.

The continuity–discontinuity debate is subsumed under our cyclical property of the world-economy. Imperialist activity will vary with the political opportunities afforded to states during the uneven growth of the world-economy. Our paired Kondratieff model in the previous section illustrates how formal imperialism is part of an unfolding logic interacting with periods of hegemony when informal imperialism is prominent. Hence imperialism is a relation that has occurred throughout the history of the world-economy. But this continuity does not preclude identification of particular phases when different strategies prevailed. Hence the very real differences between mid-nineteenth-century British hegemony and the late-nineteenth-century 'age of imperialism' are incorporated but without any suggestion of a particular 'stage' in a linear sequence.

In a similar manner, the world-systems approach can incorporate both sets of arguments in the 'core versus periphery' debate on causes of imperialism. We shall use part of Johan Galtung's (1971) 'structural theory of imperialism' to model a range of sub-relations through which the overarching dominance relation of imperialism operates. In Figure 3.2, we have simplified the argument to basics with just two types of state, core (C) and periphery (P), and two classes in each state, dominant (A) and dominated

(B). This produces four groups in the world economy: core/dominant class (CA), periphery/dominant class (PA), core/dominated class (CB) and periphery/dominated class (PB). From this, we can derive four important relations: collaboration, CA–PA, whereby the dominant classes of both zones combine to organize their joint domination of the periphery; social imperialism, CA–CB, in which the dominated class in the core is 'bought off' by welfare policy as the price for social peace 'at home'; repression, PA–PB, to maintain exploitation of the periphery by coercion as necessary; and division, CB–PB, so that there is a separation of interests between dominated classes, that is, the classic strategy of divide and rule. In our approach therefore, Robinson's (1973) collaboration is just one sub-relation of a broader set of relations of imperialism.

Transcending the classical Marxist theory

In transcending Robinson and Gallagher, we have simultaneously highlighted a key difference between world-systems analysis and classical Marxist theories of imperialism. Whereas the latter consists of various theories of linear change (stages), Wallerstein postulates long-term cyclical change.

There are two further fundamental differences that should be understood. First, world-systems analysis implies a new geography of revolution. This is partly

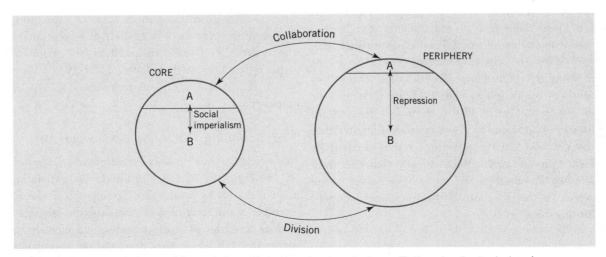

Figure 3.2 Four relations of imperialism (A is the dominant class, B the dominated class).

5

expressed in the different definitions of imperialism, with the traditional theory incorporating much more than core–periphery relations. But it is much more than a matter of definitions. Brewer (1980) shows that the classical theory's relative neglect of the periphery was no accident but a reflection of a theory of revolution that expected transformation to occur in the core, where the forces of production, and hence the contradictions of capitalism, were most developed. Indeed, in the classical view penetration of the periphery by core countries could be seen as beneficial, since 'progressive' capitalism would free the area from the shackles of feudalism, just as it had done earlier in Europe. As we have seen, the neo-Marxist view interprets such penetration in a completely different manner, with capitalism in the periphery never having a progressive liberating role but rather being regressive from the beginning – what Frank terms the development of underdevelopment. The so-called progressive elements of capitalism, therefore, are geographically restricted to the transformation from feudalism to capitalism in the core. From the world-systems perspective, this results in a geographical realignment of revolutionary forces in terms of core versus periphery, with the latter becoming a major focus of future revolt and change. This is the political argument of the radical dependency school and is usually termed 'third-worldist' because of its geographical emphasis. It is most closely associated with Mao Zedong and his theory of global class struggle.

The second difference between classical Marxist theories and world-systems analysis is related to arguments about the reasons for imperial expansion. Viewing this activity over the long term, Wallerstein comes to a different conclusion concerning its purpose. Many non-Marxist critics of imperialism, Hobson for instance, have argued that formal imperialism is uneconomic for the core states concerned, benefiting only that small group directly involved in the imperial ventures. As we have seen, this line of argument leads on to the notion of economic dominance by large monopoly or finance capital and the identification of imperialism as a 'stage'. But formal imperialism existed as a major aspect of the world-economy for over four hundred years, starting with

the activities of state-licensed charter companies. The problem with such strict economic evaluations of formal imperialism is that they have emphasized trade and the search for new markets for core production. But, as Wallerstein (1983: 38–9) points out, this explanation does not accord with the historical facts: 'by and large it was the capitalist world that sought out the products of the external arena and not the other way round'. In fact, non-capitalist societies did not need products from the core states, and such 'needs' had to be 'created' after political takeovers. Certainly, the search for markets cannot explain the massive imperialist effort on the part of core states over several centuries. Instead, Wallerstein suggests that the search for low-cost workforces is a much better explanation. This switches emphasis from exchange to production. Incorporation of new zones into the periphery invariably led to new production processes based upon the lowest labour costs. In the first instance, therefore, imperial expansion is about extending the division of labour that defines the boundaries of the world-economy. Imperialism, both formal and informal, is the process that created and continues to recreate the periphery.

Summary

In this section we have discussed the theoretical foundations of contemporary understandings of imperialism. In so doing we have:

➤ shown the interrelationship between classes;

➤ shown the interrelationship between the core and periphery of the world-economy;

➤ built upon existing theories to demonstrate a world-systems approach to imperialism.

■ Formal imperialism: the creation of empires

The formal political control of parts of the periphery has been a feature of the world-economy since its inception. From the early Spanish and Portuguese Empires through to the attempt by Italy in the 1930s to forge an African empire, formal imperialism has been a common strategy of core domination of the periphery. This process must not be confused with the concept of the world-empire as an entity with its own division of labour. Even the British Empire, on which the sun did not set for over a century, was not a world-empire in our terms but rather a successful core state with a large colonial appendage. In this section, we describe the rise and fall of such 'appendages' within the framework of the dynamic model of hegemony and rivalry described in Chapter 2. This description is organized into four parts. First, we look at imperialism at the system scale to delineate the overall pattern of the process. Second, we consider the imperialist activities of the core states that created the overall pattern. Third, we turn to the periphery and briefly consider the political arenas where this dominance relation was imposed. Finally, we look at how the two sides of the imperial relation fitted together in a case study of the British Empire.

The two cycles of formal imperialism

If we wish to describe formal imperialism, the first question that arises is how to measure it. Obviously, figures for population, land area or 'wealth' under core political control would make ideal indices for monitoring imperialism, but such data are simply not available over the long time period we employ here. Instead, we follow Bergesen and Schoenberg (1980) and use the presence of a colonial governor to indicate the imposition of sovereignty of a core state over territory in the periphery. These personages may have many different titles (for example, high commissioner, commandant, chief political resident), but all have jurisdictions signalling core control of particular parts of the periphery. Bergesen and Schoenberg obtain their data from a comprehensive catalogue of colonial governors in 412 colonial jurisdictions compiled by Henige (1970).

We rework the data to portray them in a way that is relevant to our space–time matrix. The metric we have used is a simple 50-year sequence from 1500 to 1800, and a 25-year sequence from 1800 to 1975. This provides for some detail in addition to the long A-

and B-phases in the original logistic wave plus an approximation to the A- and B-phases of the subsequent Kondratieff waves. Second, we record more than merely establishment and disestablishment of colonies. From the record of governorships we can also trace reorganizations of existing colonized territory and transfer of sovereignty of territory between core states. Both of these are useful indices in that they are related to phases of stagnation (and hence the need to reorganize) and core rivalry (expressed as capturing rival colonies). In the analyses that follow, the establishment of colonies is divided into three categories: creation of colony, reorganization of territory and transfer of sovereignty.

The cumulative number of colonies

By cumulating the number of colonies created and subtracting the number of decolonizations, the total amount of colonial activity can be found for each time period. The results of this exercise are shown in Figure 3.3, which reproduces Bergesen and Schoenberg's two long waves of colonial expansion and contraction. There is a long first wave peaking at the conclusion of the logistic B-phase and contracting in the A-phase of the first Kondratieff cycle. This largely defines the rise and fall of European empires in the New World of America. The second wave rises through the nineteenth century to peak at the end of the 'age of imperialism' and then declines rapidly into the mid-twentieth century. This largely defines the rise and fall of European empires in the Old World of Asia and Africa. Hence the two waves incorporate two geographically distinct phases of imperialism. This simple space–time pattern provides the framework within which we investigate formal imperialism more fully.

Establishment: creation, reorganization and transfer

When governors are imposed on a territory for the first time we refer to the creation of a colony. Since this is one political strategy of restructuring during economic stagnation, we expect colony creation to be associated with B-phases of our waves. This is generally borne out by Figure 3.4. The major exception is the imperialist activities of Spain and Portugal in the

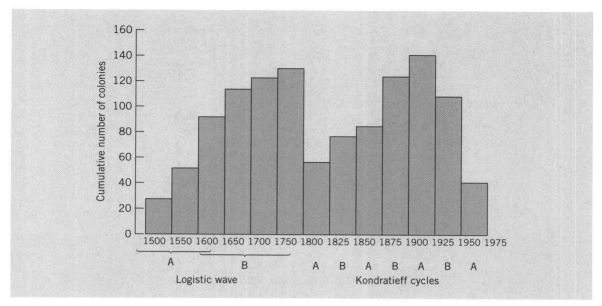

Figure 3.3 The two long waves of colonial expansion and contraction.

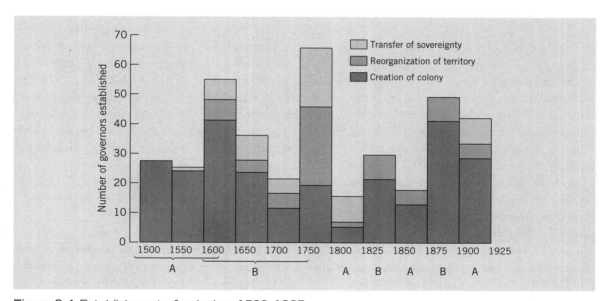

Figure 3.4 Establishment of colonies, 1500–1925.

original A-phase in the emerging world-economy. This followed the Treaty of Tordesillas, when the Pope divided the non-European world between these two leading states. This curtailed the usual rivalry associated with colony creation, and it seems that the world-economy was not yet developed sufficiently to enable informal imperialism to operate outside Europe. With the onset of the seventeenth-century stagnation phase, colonial creation expanded with the entry into the non-European arena of north-west European states. From this first peak of colony creation, the process slows down until a minor increase

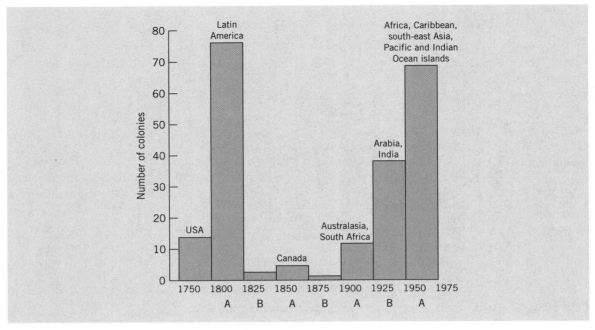

Figure 3.5 Decolonization, 1750–1975.

during the period of British–French rivalry at the end of the logistic B-phase. During the Kondratieff cycles, colony creation goes up and down with the A- and B-phases, but the most notable feature is the 'age of imperialism', clearly marked by the late-nineteenth-century peak. Hence we incorporate both the continuity and the discontinuity arguments by viewing colonial activity as a cyclical process.

Reorganization of territory should be particularly sensitive to periods of stagnation. Such periods involve pressures on states to cut back public expenditure. In the time-scale we are dealing with here this is reflected in attempts to make colonies more 'efficient'. Hence the reorganizations shown in Figure 3.4 are generally associated with B-phases. The major peak here is at the end of the logistic B-phase, when the relative decline of Spain and Portugal meant that their colonies were becoming acute burdens on the state exchequers.

Transfer of sovereignty is a direct measure of inter-state rivalry in the periphery. This is mainly a feature of the logistic B-phase, when this type of activity was relatively common. With the onset of the Kontratieff cycles, such 'capture' is quite rare and is concentrated into just two periods, both A-phases. What these actually represent is the sharing out of the colonial spoils after two global wars. The first relates to the defeat of France and the confirmation of British hegemony. The second relates to the defeat of Germany and the confirmation of US hegemony. In both cases, the losing powers were deprived of colonies.

Decolonization: geographical contagion and contrasting ideologies

The pattern of decolonization is a much simpler one (Figure 3.5). There have been two major periods of decolonization, and these are directly responsible for the troughs in Figure 3.3 and hence generate the two-wave pattern.

Although both peaks occur in A-phases, they do not correspond to equivalent phases in our paired Kondratieff model: the second A_2 phase of American hegemony is a period of major decolonization, but the first A_2 phase of British hegemony is clearly not such a period. The first decolonization period occurred earlier during an A_1 phase of emerging hegemony. This is best interpreted as the conclusion of the agricultural capitalism of the logistic wave. The decolonization involved the termination of Spanish and Portuguese colonies in Latin America; by this time, the colonizing powers had long since declined to semi-peripheral status. For the competing core powers in this period of emerging British hegemony there were obvious advantages to freeing these anachronistic colonies. This was particularly recognized by George Canning, the British foreign secretary, in his oft-quoted statement of 1824 that a free South America would be 'ours' (that is, British). This decolonization set the conditions for British 'informal imperialism' in Latin America in the mid-nineteenth century, which is discussed in the next section.

One feature of the decolonization process is its geographical contagion. Decolonization is not a random process but is spatially clustered at different periods. The decolonization of the Americas, for instance, did not affect existing colonies in the Old World. This spatial contagion shows the importance for decolonization of processes operating in the periphery. The American War of Independence served as an example for Latin America in the first decolonization phase; Indian independence led the way for the rest of the periphery in the second decolonization period. At a regional level, political concessions in one colony led to cumulative pressure throughout the region. In this sense, the independence awarded to Nkrumah's Ghana in 1957 triggered the decolonization of the whole continent in the same way that Simon Bolivar's Venezuelan revolt of 1820 led the way in Latin America.

One final point may be added to this interpretation. The rhetoric of the colonial revolutionary leaders in the two periods was very different. The 'freedom' sought in Latin America was proclaimed in a liberal ideology imitating the American Revolution to the north, whereas a century and a half later the ideology was socialist, mildly at first in India, but much more vociferous in Ghana and culminating in many wars of national liberation. It is for this reason that the first decolonization (Latin America) was more easily realigned in the world-economy under British liberal leadership than the second decolonization, with its numerous challenges to American liberal leadership. In broad terms, we can note that the

first decolonization period is a liberal revolution at the end of agricultural capitalism, whereas the second period encompasses socialist-inspired revolutions at the end of industrial capitalism, which have been much more anti-systemic in nature.

The geography of formal imperialism

Imperialism is a dominance relation between core and periphery. Our discussion so far has stayed at the system level, and we have not investigated the geography of this relation, that is, we have not asked who was 'dominating' whom where. We answer this question below by dealing first with the core and then with the periphery.

Core: the imperial states

Who were these colonizing states? In fact they have been surprisingly few in number. In the whole history of the world-economy there have been just thirteen formal imperialist states, and only five of these can be said to be major colonizers. Figure 3.6 shows the colonial activity of these states in graphs that use the same format as Figure 3.4. Seven graphs are shown: separate ones for Spain, Portugal, the Netherlands, France and Britain/England; and combined ones for the early and late 'minor' colonizing states. The early states consist of the Baltic states of Denmark, Sweden and Brandenburg/Prussia; the 'latecomers' are Belgium, Germany, Italy, Japan and the United States. The graphs show the individual patterns of colonial activity of the states that created the total picture we looked at above.

In the logistic A-phase before 1600, all colonial establishment was by Spain and Portugal. The onset of the B-phase brings the Netherlands, France, England and the Baltic states into the fray. This is complemented by a sharp reduction in colony creation by both Spain and Portugal – the location of the core of the world-economy had moved northwards, and this is directly reflected in the new colonial activity. As the B-phase progresses, all of these new states continue their colonial activities but at a reduced scale except for England/Britain, for which colony capture is as important as, or more important than,

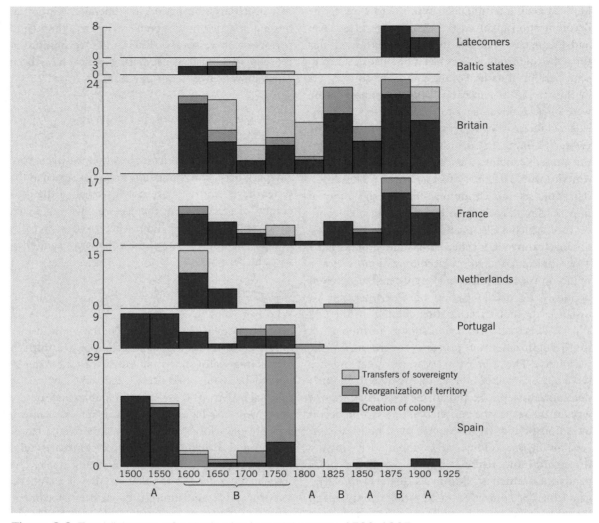

Figure 3.6 Establishment of colonies by imperial states, 1500–1925.

colony creation. As we move into the Kondratieff cycles, the Netherlands almost totally ceases establishment of colonies, so that in the mid-nineteenth century all colonial creation is either British or French. In the classical 'age of imperialism', these two old-stagers are joined by the five latecomers.

Figure 3.6 enables us to define four periods of colonial activity by imperial states:

1 The first non-competitive era occurs in the logistic A-phase, when only Spain and Portugal are imperial states.

2 The first competitive era occurs in the logistic B-phase, when eight states are involved in imperialist expansion.

3 The second non-competitive era of the mid-nineteenth century coincides with the rise and consolidation of British hegemony. In this period, only two states are involved in imperial expansion: Britain and France.

4 The second competitive era is the 'age of imperialism' and coincides with the decline of British hegemony. In this period, eight states are involved

in imperial expansion. We shall use this division of core-state activity in our discussion of peripheral arenas.

Periphery: the political arenas

Fifteen separate arenas can be identified in which colonial activity occurred in the periphery. The first arena we can term Iberian America; it includes Spanish and Portuguese possessions in America obtained in the first non-competitive era. The other fourteen arenas are shown in Figures 3.7, 3.8 and 3.9, which cover the other three periods of colonial activity. The arenas have been allocated to these periods on the basis of when they attracted most attention from imperial states. We shall describe each period and its arenas in turn.

The dominant arena of the first competitive era was the Caribbean (Figure 3.7). This was initially for locational reasons in plundering the Spanish Empire, but subsequently the major role of the greater Caribbean (Maryland to north-east Brazil) was plantation agriculture supplying sugar and tobacco to the core. Of secondary importance were the North American colonies that did not develop a staple crop and effectively prevented themselves becoming peripheralized. This was to be the location of the first major peripheral revolt. The other important arena for this period was the African ports that formed the final apex of the infamous Atlantic triangular trade. It is this trade and the surplus value to be derived from it that underlay the colonial competition of this era. The final two arenas were much less important and related to the Indies trade, which Wallerstein doubts was integral to the world-economy until after 1750.

In the second non-competitive era colonial activity was much reduced, but four arenas did emerge as

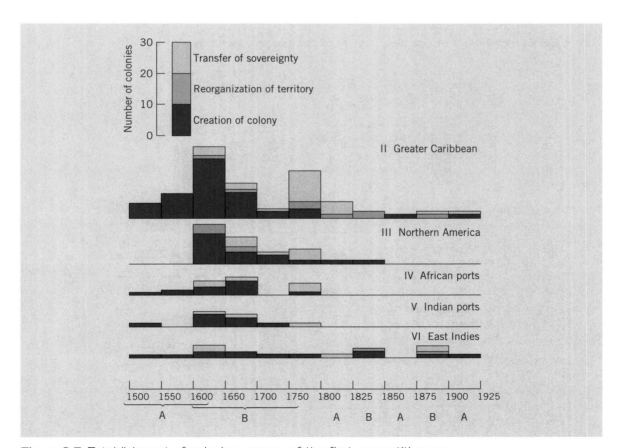

Figure 3.7 Establishment of colonies: arenas of the first competitive era.

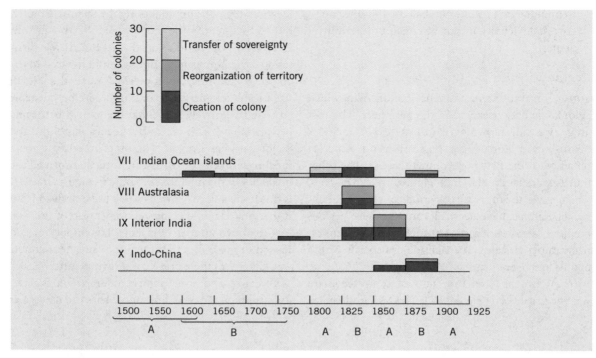

Figure 3.8 Establishment of colonies: arenas of the second non-competitive era.

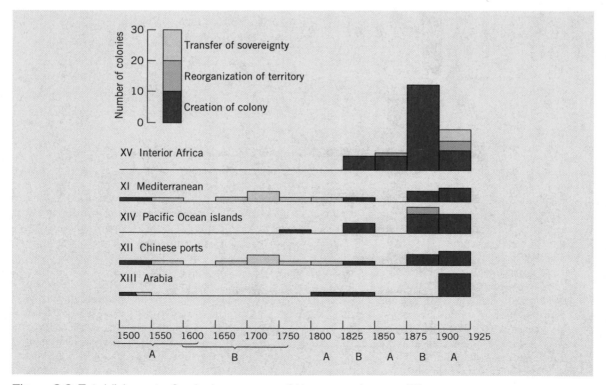

Figure 3.9 Establishment of colonies: arenas of the second competitive era.

active in the mid-nineteenth century (Figure 3.8). There was no competition between core states within these arenas, which were consequently divided between France and Britain. Although without the authority of the papal bull legitimizing the earlier Spain–Portugal share-out, Britain and France managed to continue some colonial activity while avoiding each other's ambitions. The Indian Ocean islands (including Madagascar) and Indo-China were 'conceded' by Britain as French arenas, and the latter left India and Australasia to the British.

This peaceful arrangement was shattered in the next competitive period during a series of 'scrambles', the most famous being that for Africa, although similar pre-emptive staking out of claims occurred in the Mediterranean arena, Pacific islands and for Chinese ports (Figure 3.9). With the collapse of the Ottoman Empire, there was a final share-out of Arabia after the First World War. The brief takeover of this last arena completes the pattern of formal core control of periphery as depicted by Figure 3.1.

The economics of formal imperialism

The pattern of formal imperialism is now clear. In two major cycles over four hundred years a small group of core states took political control in fifteen separate arenas covering nearly all the periphery. Why? Because of the dominance of the Hobson–Lenin paradigm in this debate, most discussion has focused on the second cycle's 'age of imperialism'. This has produced a rather unnecessary debate pitting economic causes against political ones. Foeken (1982) has reviewed these arguments in relation to the partition of Africa and shows that neither set of causes can be sensibly dismissed. What is required is a political economy view of the situation, where complementary aspects of the various theories are brought together to produce a more comprehensive account (ibid.: 140). This would bring analysis of the second cycle into line with that of the first cycle, where it has never been doubted that economics and politics are intertwined in the era of mercantilism.

Case study

'Islands of development' in Africa

At the African end of the slave trade the European states secured 'stations' on the west African coast (see Figure 3.7), where they exported slaves received from local slave-trading states. Originally, this trade was only a luxury exchange with an external arena. By 1700, however, Wallerstein (1980b) considers it to have become integral to the restructured division of labour occurring during the logistic B-phase. However, as west Africa became integrated into the world-economy, the production of slaves became a relatively inefficient use of this particular sector. Hence British abolition of the slave trade in 1807 reflects long-term economic self-interest underlying the moral issue. Gradual creeping peripheral-ization of west Africa throughout the nineteenth century is accelerated by the famous 'scramble for Africa' in the age of imperialism (Figure 3.9). In the final quarter of the nineteenth century, colonies were being created in Africa at the rate of one a year. This enabled the continent to be fully integrated into the world-economy as a new periphery.

The spatial structure of this process was very simple and consisted of just three major zones (Wallerstein 1976). First there were the zones producing for the world market. Every colony had one or more of these, and the colonial administrators ensured that a new infrastructure, including ports and railways, was developed to facilitate the flow of commodities to the world market. This produced the pattern that economic geographers have termed 'islands of development'. These 'islands' were of three types. In west Africa, peasant agricultural production was common – the Asante cocoa-growing region is a good example. In central Africa, company concessions for forest or mineral production were more typical, such as the concessionary companies in the Congo, which devastated the area in what was little more than plunder. In east Africa and southern Africa, production based on white settler populations was also found. In all three cases, production was geared towards a small number of products for consumption in the core.

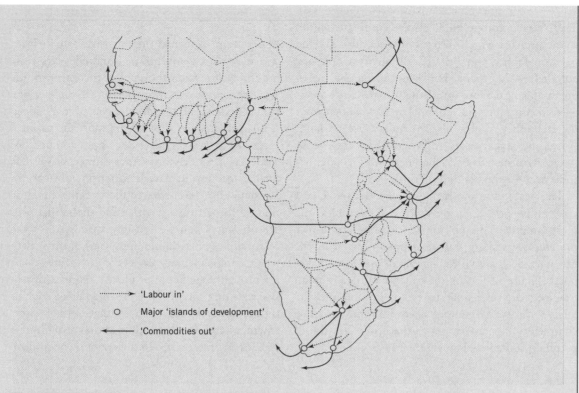

Figure 3.10 Africa south of the Sahara: the imperial economic structure.

Surrounding each 'development island' was a zone of production for the local market. These were largely peasant farming areas producing food for the labour attracted to the first zone. The remainder of Africa became and still is a large zone of subsistence agriculture that is integrated into the world-economy through its export of labour to the first zone (Figure 3.10). One of the key processes set in motion by the colonial administrators was a taxation policy that often forced peasants outside the two market zones into becoming wage labourers to provide for their new need for money. Labour migration continues today with the massive flows of labour from the Sahel to the coast in west Africa and from central Africa to southern Africa. Every island of development has its own particular pattern of source areas for labour (see Figure 3.10). This international migration incorporates all the advantages to capital of alien labour. It is cheap; it has few rights; the cost of reproduction is elsewhere; and it can be disposed of easily as necessary in times of recession. The third zone, the labour zone, is at the very edge of the world-economy – the periphery of the periphery. This is the most vulnerable zone of the world-economy and has been marked during the current Kondratieff B-phase by widespread destitution and famine.

It is ironic but not surprising that huge tracts of Africa have been merely producers of cheap labour for the world-economy. This is where Africa entered the story, and although the legal status of this labour has changed, it is the same basic process that relegates most of Africa to the very bottom of the world order. This has continued despite the granting of independence to nearly all of Africa in the post-Second World War era. The migration goes on unabated, although now it is between independent states instead of between colonies. Formal imperialism may have been an effective strategy for setting up this situation, but it is clearly not a necessary criterion for its continuation: imperialism is dead; long live imperialism! The alternative imperialism, the informal variety, is the subject matter of the final section of this chapter, but before we unravel the intricacies of this 'hidden imperialism', we need to consider how our previous discussions of core states and peripheral arenas, and the geography and the economics of imperialism, fit together in a single empire.

In world-systems analysis, as we have seen, formal imperialism is interpreted as the political method of creating new economic production zones in the world-economy. From the original production of bullion in the Spanish colonies in the sixteenth century to the production of uranium from Namibia, which became independent only in 1990, formal imperialism has been the prime means of ensuring the transformation of external arenas into the world-economy's division of labour.

Hegemony and the British Empire

Hegemony implies an open world-economy; empire involves the closing off of part of the world-economy from rivals. Hence, as we have seen, imperial expansion is not associated with hegemonic strategies of dominance. But Britain did not give up its empire on achieving hegemony and in fact continued to extend it, albeit at a slower rate (see Figure 3.6). India became the great exception to British foreign policies of the mid-nineteenth century. The crude mercantilist restrictions of the eighteenth century continued in India unabated. As Hobsbawm (1987: 148) has noted, India was 'the only part of the British Empire to which laissez faire never applied. . . . And the "formal" British Empire expanded in India even when no other part of it did. The economic reasons for this were compelling.' India imported 50 per cent of Lancashire cotton textiles and accounted for 50 per cent of Chinese imports in the form of opium; with government charges and debt interest, 40 per cent of Britain's deficit with the rest of the world was covered by payments from India. No wonder it was called the jewel in the crown!

Despite hegemony, therefore, Britain was the major imperial power in the world before the general expansion of imperialism in the late nineteenth century that marked the decline of that hegemony. Since Britain remained politically the world's most powerful state, it gained most from the 'new imperialism', thus creating the mammoth empire so lavishly celebrated in 1897 (see Case study overleaf).

But what was this imperial creation? Great empires of the past had usually been compact land empires, not disconnected pieces of territory spread all over the globe. There was certainly never any overall imperial strategy to produce the particular pattern of lands that ended up coloured pink on the world map. That is why people joked that the empire had been acquired 'in a fit of absence of mind' (Morris 1968: 37). But it was much more than that. Although not centrally directed, the empire was created in a series of minor and major conflicts with other European powers and local populations over a period of four hundred years. Hence it was fragmented because it reflected events in the periphery as much as those of the core.

The problems of geographical fragmentation were exacerbated by the ad hoc nature of the administration of the empire. The Colonial Office directly administered the crown colonies. Other colonies were self-governing (the white-settled territories). Much of the empire consisted of protectorates, which were technically still foreign countries and were controlled from the Foreign Office, which also ran the Egyptian government although the country was never formally British. India was again the exception, having its own department of state, the India Office. No wonder Morris (1968: 212) has concluded: 'It was all bits and pieces. There was no system.'

For the imperialist politicians in London, geographical fragmentation could be overcome by the new technologies of travel and communications. Contemporary geographers emphasized the decline of distance as an obstacle of separation and discovered the concept of 'time–distance' to replace simple physical distance (Robertson 1900). In the new analysis, the world was shrinking and the far-flung empire was converted into a viable political unit. In particular, the steamship and the imperial postal service were seen as bringing the empire together. Even more dramatic, by the 1890s the electric telegraph was producing almost instantaneous links between all components of the empire. Britain had laid thousands of miles of submarine cables in the late nineteenth century, and in 1897 Queen Victoria's jubilee message was sent in a mere few seconds to all corners of the empire by telegraph. In this new world, it seemed that Britain was in the process of producing a new sort of world state. Instead of 'old-fashioned' laissez-faire, new institutions like the Imperial Federation League and the United Empire Trade

Case study

Where the sun never set

Although there have been five major imperialist states in the modern world-system, one stands out as pre-eminent (see Figure 3.6). By the late nineteenth century, the British Empire was not only viewed as the most powerful state in the world, it was widely regarded as the greatest empire the world had ever known. One-quarter of the world's lands and population were formally controlled from London (Figure 3.11). The empire reached its widest extent after the First World War, when it took over former German colonies and Ottoman territories as mandates from the League of Nations. But they merely represented some late spoils of military victory. The empire reached its true zenith at the end of Queen Victoria's reign. As 'queen-empress', she was a symbol of Britain 'winning the nineteenth century'. The celebration of the Queen's diamond jubilee in 1897 centred on a great imperial pageant through the streets of London, and this is usually taken as the occasion that marks the high point of British imperialism (Morris 1968). Within two years, Britain was embroiled in the Boer War, and British self-confidence in its right to rule over so much of the world began its long decline.

Although there is nothing typical about British imperialism in relation to other imperialisms, we describe it in this section as our case study of formal imperialism because it contributed so much to the making of our modern world. We deal with three themes: the relation of the empire to British hegemony, the geopolitical codes that were created for empire and the ideology that attempted to hold the whole structure together.

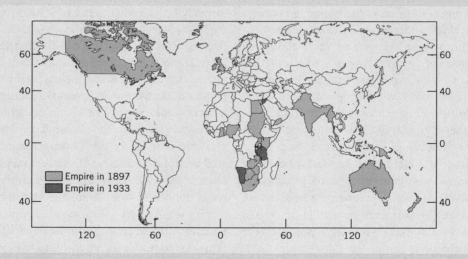

Figure 3.11 The British Empire in 1897 and its extension in 1933.

League, supported by geographers such as Mackinder, were formulating and promoting plans for a British imperial federated state. Since it would cover all parts of the world, the 'imperial harvest' would be a continuous one. Hence self-sufficiency was a practical goal that was both feasible and that might become necessary as inter-state rivalries heightened. Britain may have lost its hegemonic superiority, but here was a chance to create an alternative way of maintaining world dominance. For Britain's imperial politicians, there was one important black cloud on the horizon of their new world. The strategic defensive problems of such a fragmented world state were immense. This was

Mackinder's major concern, as we have seen. Imperial geopolitical codes were required to make sense of this problem.

Imperial geopolitical codes

Britain's haphazard creation of an empire had required a whole series of local geopolitical codes. In every arena, local British military and civilian officials had to compete with other European states and come to terms with local populations. After the defeat of, or accommodation with, the former, strategies of control had to be devised for the latter. This meant the identification of collaborators. The basic British strategy was divide and rule. It was at this time that British governors throughout the world 'officially' recognized various cultural groups in order to play one off against another. Official designation in administrative documents such as the census turned these groups into political strata competing for the favours of the empire. Quite literally, the British Empire was the great creator of 'peoples' throughout the world. The legacy of this policy remains with us today in such political rivalries as Hindus, Muslims and Sikhs in India, Tamils and Sinhalese in Sri Lanka, Greeks and Turks in Cyprus, Indians and indigenous Fijians in Fiji, Jews and Palestinians in Israel/Palestine, Chinese and Malays in Malaya, plus numerous ethnic rivalries in the former British colonies of Africa.

Divide and rule could help to keep control of particular colonies or protectorates, but it did not help in global strategy required to defend the empire as a whole. As a sea empire, Britain's strategy was based on its navy. In the nineteenth century, government policy operated a two-to-one rule whereby the Royal Navy was kept twice as large as the combined fleet of its two main rivals. Furthermore, the Royal Navy was the only navy with truly global range based upon coaling stations on islands and key ports at intervals of 3,000 miles or less on all the major shipping routes.

The route to India was the 'main street' of the empire. Originally around Africa via the Cape of Good Hope, after the opening of the Suez Canal in 1871 the new route to India became the focus of Britain's imperial strategy. The Mediterranean and the Red Sea became vital to British interests, with the canal viewed as the 'jugular' of the empire – hence the need to oust France from Egypt and take over the government of Egypt. But even more important than rivalry with France in the Mediterranean was the threat from Russia to the north-west of India.

At the same time as Britain had been extending its position in India, Russia had been expanding its land empire into central Asia, inevitably provoking a clash of interests between the two powers. The result was the 'Great Game' of the nineteenth century, the intrigue and threats between Russia and Britain in a zone extending from Turkey through Iran to Afghanistan and the north-west frontier of India. This Victorian cold war, as Edwardes (1975) has called it, rarely erupted into 'hot war' but was an underlying concern of Britain's global imperial geopolitical code. It was this Great Game that Mackinder extended to become the heartland theory and that the United States finally inherited as containment in the 'real' Cold War.

The imperial ideology

Finally, it must be emphasized that the supposed unity of the British Empire was not based solely upon physical facilities such as coaling stations and telegraph offices. Imperialism was not a derogatory term in the late nineteenth century; on the contrary, it was an official and popular ideology. Although we can agree with Morris (1968: 99) that 'one imperial end was basic to another – profit', and that this was always 'profit for the few', the glory that was empire could be shared by the many. All Britons, both high and low, could be proud to be members of the country with the greatest ever empire. They were an 'imperial people' with a 'civilizing mission' for the world. It is here that we can find the ultimate contradiction of formal empire, which was to signal its end within a couple of generations of its zenith.

Basically, Britain's imperial ideology incorporated two incompatible principles. First, there was the 'imperial philosophy of equality' (Huttenback 1976: 21), sometimes known as 'the High Victorian concept of fair play' (Morris 1968: 516). In theory, all the peoples of the colonies were subjects of the Queen and therefore enjoyed the Queen's justice irrespective of colour or creed. Hence 'the rule of law was the one convincingly unifying factor in imperial affairs' (ibid.: 195). But alongside this principle was a second principle of racial superiority – Cecil Rhodes' 'visionary project' of 'the world supremacy of the Anglo-Saxon peoples' (Bowle 1974: 359). These two principles clashed most acutely where coloured immigrant labour came into contact with white labour in the settler colonies.

In the creation of new production zones for the world market, large numbers of people were moved to form new workforces. For tropical crop production, Indians and Chinese were transported all around the world from Fiji to Trinidad. In the more temperate parts of the empire, European immigration had produced another workforce that was organized into unions and was relatively highly waged. Such 'labour problems' could be 'solved' by using coloured contract labour to replace expensive white labour. In the self-governing colonies, the latter were powerful enough to prevent this happening. But if all the Queen's subjects were equal under the law, how could this cheap labour be kept out of Australia, Canada, New Zealand and South Africa?

This political problem confronted the chief ministers of the self-governing colonies, who met in London at the jubilee celebrations under the chairmanship of Joseph Chamberlain, the colonial secretary. He advised adoption of what had come to be known as the 'Natal formula' (Huttenback 1976: 141). This involved the use of a European language test to prevent entry to a colony. Since all immigrants were liable to take the same test, it preserved the first principle of equality while enacting the second principle of racial discrimination. Huttenback (ibid.: 194) sums it up thus: 'The "Natal formula" was simply a means for keeping unwanted immigrants out of a colony through the use of a mechanism which seemed innocuous in legislation.' In short, it provided a veil of decency to cover the racialism. Within a year, the Australian colonies began enacting the required legislation to produce the 'White Australia' policy while retaining the 'principles of empire'.

Imperialism's claim to the high moral ground – the European 'civilizing mission' – and Britain's particular reputation for fair play were disputed and soundly defeated in the twentieth century. Gandhi in India more than any other individual exposed the contradictions of empire and won the high moral ground for the anti-imperialists. Imperialism became unfashionable and doomed. Political freedom still had to be won, by physical resistance in many colonies, but after 1945 no European power, not even Britain, could stem the flood tide of independence. In the new world of US hegemony and the Cold War, there was no room for an anachronism such as the British Empire.

Summary

In this section we have:

➤ introduced the concept formal imperialism;

➤ related the rise and fall of formal imperialism to economic cycles;

➤ related the practice of formal imperialism to hegemonic cycles;

➤ shown that the practice of formal imperialism demands a related representation of state behaviour to justify such actions.

■ Informal imperialism: dominance without empire

In *The Geography of Empire*, Keith Buchanan (1972) does not discuss the formal imperialisms of the past but concentrates solely on the contemporary dominance of the United States in the world-economy. In what he terms 'the new shape of empire', he points out that whereas the decolonization process has provided formal independence for colonies from a single imperial state, it has not provided independence from the imperial system as a whole (ibid.: 57). In world-systems terms, what we have is a change of strategy by core states from formal to informal imperialism. This is not a new phenomenon. In our model of hegemonies and rivalries the former were associated with informal imperialism. Hence we expect the rise of each hegemonic power to lead to a period of informal imperialism similar to that described by Buchanan for American hegemony. In fact, this is exactly what we find. There have been only three hegemonic powers in the history of the world-economy, and each is associated with one of the three classic examples of informal imperialism. First, in the mid-seventeenth century Dutch hegemony was based, in large part, on the Baltic trade, whereby eastern Europe remained politically independent while becoming peripheralized. Dutch merchants dominated the trade, but there was no Dutch political control. Second, in the mid-nineteenth century Britain employed the 'imperialism of free trade', when Latin America became known as Britain's 'informal empire'. Finally, in the mid-twentieth century American hegemony has been associated with decolonization, to be replaced by neo-colonialism – political independence of the periphery tempered by economic dependence.

Informal imperialism is a much more subtle political strategy than formal imperialism. For this reason, it is much less amenable to the descriptive, cataloguing approach employed in the previous section. Buchanan (1972) produces numerous interesting maps on topics such as US support for indigenous armies and police to control their own populations, which he terms 'Vietnamization', but he is unable to capture the basic mechanism of informal imperialism through this empirical approach. Here we develop our argument in two stages. In the first place, we show that informal imperialism is no less 'political' despite its emphasis on 'economic' processes. This involves a discussion of trade policy not as a part of economic theory but as alternative state policies within different sectors of the world-economy. But political intervention in the world market cannot change the structural constraints of the world-economy. In the final part of the chapter, we describe the basic mechanism of unequal exchange that generates and maintains uneven development throughout the world.

The international relations of informal imperialism

The mainstream of economic thought traces its origins back to Adam Smith's *Wealth of Nations*, written in 1776. This book criticized the policy of mercantilism as generally practised at that time and instead advocated a policy of laissez-faire. Ever since Smith, free trade has been a basic principle of orthodox economics. In the early nineteenth century, David Ricardo added the idea of comparative advantage to the theory. This claimed that with each state specializing in what it could best produce, free trade would generate an international trade equilibrium to everyone's mutual advantage. Free trade was therefore the best policy for all states, and any political interference in the flow of commodities into or out of a country was not in the interests either of that particular country or of the system as a whole.

There are two related paradoxes in this orthodox economics. The first is that it simply does not work in practice. In the three cases that we have identified as informal imperialism, the peripheral states did not gain from the openness of their national economies – eastern Europe still lags behind western Europe; Latin America is still a collection of peripheral or semi-peripheral states; and Africa and most of Asia are part of a 'South' periphery in which mass poverty has shown little sign of abating in recent decades. As we shall see, states that have 'caught up' have employed very different policies. This leads on to the second paradox concerning free trade, which is that most politicians in most countries at most times have realized that it does not work. Although they have not always had theoretical arguments to back up their less than orthodox economics, most politicians have found that the interests of the groups they represent are best served by some political influence on trade rather than simply leaving it all to the 'hidden hand' of the market. We could ask: who is right – 'economic theorists' or 'practical politicians'? The answer is that they both are – sometimes. It all depends on the world-economy location of the state in question. In Table 3.1, different trade policies are related to different zones of the world-economy through the three hegemonic cycles that we described in Chapter 2. Below we outline the politics of these policies for each zone.

Free trade and the hegemonic state

We can interpret the orthodox economic advocacy of free trade as a reflection of the structural advantage of core powers, in particular hegemonic core powers, in

Table 3.1 Trade policies through three hegemonic cycles

Cycle	Core: 'universal' theory	Semi-periphery: political strategy	Periphery: dilemma and conflict
Dutch	Grotius's *mare liberum*	England: Mun's mercantilism France: Colbertism	Eastern Europe: landowners vs burghers
British	Smith's laissez-faire Ricardo's comparative advantage	Germany: List's protectionism United States: Republican tariff policy	Latin America: 'European party' vs 'American party'
American	Modern economics orthodoxy's free enterprise	Soviet Union: Stalin's 'socialism in one country' Japan: 'hidden protectionism'	Africa and Asia: 'capitalism' vs 'socialism'

the world-economy. Hence we would expect to trace such ideas back beyond Adam Smith to the original Dutch hegemony. Not surprisingly, the first great trading state of the modern world-system was concerned for freedom of the seas, and this was expressed in the work of the Dutch political writer Hugo Grotius. He wrote his *Mare Liberum*, which became the classic statement in international law justifying Dutch claims to sail wherever there was trade to be had, in 1609. As the most efficient producers of commodities, hegemonic core states promote 'economic freedom' in the knowledge that their producers can beat other producers in any open competition: the market favours efficient producers, and the efficient producers are concentrated in the hegemonic state by definition. In such a situation, it is in the interests of the rising hegemonic power to present free trade as 'natural' and political control as 'interference'. Hence from Grotius through Adam Smith to modern economics, economic freedoms are presented as universally valid theory masking the self-interest of the economically strong (see Table 3.1). But there is nothing natural about free trade, the world market or any other socially constructed institution. 'All organisation is bias' is Schattschneider's (1960) point, as we have discussed in Chapter 1, and orthodox economics represents a classic case of attempting to organize non-hegemonic interests off the political agenda. The question we ask of any institution, however, is: what is the organization of bias in this institution? (Bachrach and Baratz 1962: 952). In the case of the world market, it is clear that the bias is in

favour of core states, and hegemonic core states in particular. The whole history of the world-economy is testimony to this fact. The purpose of inter-state politics is either to maintain this bias or to attempt to change it. The former political strategy is the free-trade one, which is associated with informal imperialism. This is neither more nor less 'political' than protectionism, mercantilism or formal imperialism, which attempt to change the status quo. The former is political non-decision making, the latter political decision making in Bachrach and Baratz's terms. Of course, politics is so much easier when the system is on your side.

Protectionism and the semi-periphery

The practical politicians, who have generally failed to adhere to the orthodox prescriptions for trade, have not been without their economic champions. The most famous is the mid-nineteenth-century German economist Friedrich List. The world-systems approach is much closer to his analysis than Adam Smith's. For List, there was no 'naturally' best trade policy; rather tariffs were a matter of 'time, place and degree of development' (Isaacs 1948: 307). List even admitted that had he been an Englishman he would have probably not doubted the principles of Adam Smith (Frank 1978: 98). But List realized that free trade was not a good policy for the infant industries of his own country, Germany. Hence he advocated a customs union – the famous Zollverein – with a tariff around the German states under Prussian leadership. List rationalized his unorthodox position by arguing that there are three stages of development, each of which requires different policies. For the least advanced countries, free trade was sensible to promote agriculture. At a certain stage, however, such policy must give way to protectionism to promote industry. Finally, when the latter policy has succeeded in advancing the country to 'wealth and power', then free trade is necessary to maintain supremacy (Isaacs 1948). In world-systems terms, List's theory can be translated into policies for periphery, semi-periphery and core countries, respectively. Since the Germany of his time was semi-peripheral, he advocated protectionism. In fact, we can identify protectionism or, more generally, mercantilism as the strategy of the semi-periphery. Both modern champions of free trade – Britain and the United States – were major advocates of mercantilist policies before their hegemonic period: England against the Dutch, the United States against Britain (see Table 3.1). In fact, the early classic mercantilist tract is by an Englishman, Thomas Mun in 1623, who advocated mercantilist measures to protect England from the superior Dutch economy (Wilson 1958). Similarly, US Secretary of State Alexander Hamilton's famous Report on Manufactures in 1791 remains a classic statement on the need to develop a semi-peripheral strategy as defined here (Frank 1978: 98–9), although the policy was not consistently pursued until after the pro-tariff Republican Party won the presidency with Abraham Lincoln in 1861. More recently, Soviet autarky – 'socialism in one country' – and subsequent controlled trade can best be interpreted as an anti-core development strategy, as we argued in the previous chapter. And the 'hidden protectionism' of post-Second World War Japan is still an issue of contention with the United States today.

The dilemma for the periphery

Friedrich List advocated free trade as the tariff policy of the periphery. In fact, there have been and continue to be disputes within peripheral countries on the best policy. Gunder Frank has described this for mid-nineteenth-century Latin America as a contest between the 'American' party and the 'European' party. The former wanted protection of local production and represented the local industrialists. The latter were liberals who favoured free trade and were supported by landed interests, who wished to export their products to the core and receive back better and cheaper industrial goods than could be produced locally. Generally speaking, the 'European party' won the political contest and free trade triumphed. It is in this sense that Frank talks of local capital in allegiance with metropolitan capital underdeveloping their own country. This is the collaboration relation in informal imperialism epitomized by nineteenth-century Latin American liberals. In contrast, in the United States the 'American party' (notably pro-tariff Republicans) was triumphant, and the country did not become underdeveloped.

Frank's political choices for Latin America in the second half of the nineteenth century can be identified in the two other classic cases of informal imperialism (see Table 3.1). His terminology is no longer appropriate – we shall rename his positions peripheral strategy (the European party) and semi-peripheral strategy (the American party). In eastern Europe, the Counter-Reformation represents the triumph of Catholic landed interests over local burgher interests. In our new terms, the landed interests of eastern Europe adopted a peripheral strategy and opened up their economy to the Dutch.

The current pattern of informal imperialism provides modern political leaders in the periphery with the same basic choice. In any particular state, which strategy is adopted will vary with the internal balance of political forces and their relation to core interests. This has been somewhat obscured, however, by the same ideological facade that confuses the geopolitics previously described. In Africa, for instance, Young (1982) distinguishes between states in terms of the self-ascribed ideology of their governments. The two most common categories are 'populist socialism' and 'African capitalism'. In our framework, these represent semi-peripheral and peripheral strategies, respectively. Ghana provides a good example of a country where both options have been used. Kwame Nkrumah's development policy of using cocoa export revenues to build up the urban industrial sector is a typical semi-peripheral approach that only officially became 'socialism' towards the end of his regime. Nkrumah's great rival, Busia, on the other hand, led a

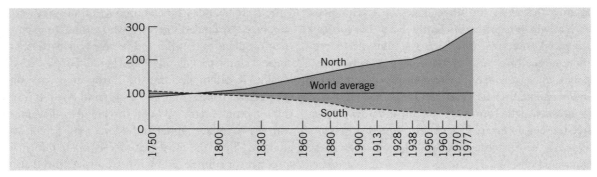

Figure 3.12 The growing 'North–South' gap: relative wage levels, 1750–1977.

government that adopted a liberal trade policy, to the advantage of the landed interests behind the cocoa exports – what we would term a peripheral policy. Hence the overthrow of Nkrumah was not a defeat for socialism, and the overthrow of Busia was not a defeat for capitalism. In world-systems terms, in the case of Ghana both semi-peripheral and peripheral strategies failed politically and economically.

Informal imperialism as a structural relation

The general argument relating to international relations can be summarized as saying that core states, especially hegemonic core states, have a structural advantage in the world-economy. By 'structural', we mean that the advantage is built into the whole operation of the world-economy. This is more than a mere cumulative advantage – the system relies on this inequality as part of its functioning. Hence there are no solutions to overcoming world inequalities within the world-economy, but there are state strategies that can aid one state at the expense of the others. Wallerstein (1979) uses Tawney's tadpole philosophy to illustrate this. Although a few tadpoles will survive to become adult frogs, most will perish not because of their individual failings but because they are part of an ecology that limits the total number of frogs. Similarly, if all countries adopt 'perfect' policies for their own economic advance, this does not mean that all will rise to membership of the core. To have a core you need a periphery, and without both there would

be no world-economy. In this situation, it is easier to maintain a core position than to rise upwards.

But what is the mechanism that maintains the core–periphery structure? The exact process has changed over the history of the world-economy, and here we will concentrate upon the period of industrial capitalism. Our discussion is based loosely on Emmanuel's (1972) concept of unequal exchange, in which we shall emphasize the political process. Emmanuel's work is an attempt to explain the massive modern inequalities of the world-economy. Whereas before the mid-nineteenth century wages for direct producers were not very different across the various sectors of the world-economy, wage differences became very large (Figure 3.12). Why this change in the intensity of the core–periphery structure? The answer to this question provides us with the basic mechanism of informal imperialism.

The rise of social imperialism

Emmanuel's starting point is the concept of a labour market. The rise of the world-economy initially produced 'free' labour in the core countries, where men and women were able to work for whom they pleased. But this freedom was a very hollow one when there were insufficient jobs or when wages were set by employers. In fact, 'free' labourers were no better off than their predecessors in feudal Europe, and their lack of security might mean that they were worse off. The labour market operated initially on an individual basis, with the result that the more powerful party to the agreements – the employer – could force the lowest wages on the worker. In this situation, subsistence wages were the norm, with wage levels reflecting the price of bread, which would constitute up to half of a household's expenditure. The purpose of wages was to sustain and reproduce the worker and no more. In classical economics, subsistence wages were just as 'natural' as free trade but, unlike the latter, this part of our economics heritage has not remained orthodox, at least in the core countries. Quite simply, economists were not able to keep wages off the political agenda.

Wages could rise above subsistence levels under certain circumstances. For instance, relative scarcity of labour would tip the balance of negotiations in favour of labour. Hence in the mid-nineteenth century the highest wages were not in the European core but in the new settler colonies – notably Australia – with their labour shortages. Marx also mentioned a 'historical and moral element' beyond the market, and it is this idea that Emmanuel develops. Marx had used this concept to cover such things as differences in climate and in consumption habits that lead to different levels of subsistence wage. Emmanuel adds a political dimension: where workers combine, they can negotiate from a position of strength in the labour market and obtain more than subsistence wages. This was recognized by politicians, who legislated against unions – in Britain through the Combination Acts of the early nineteenth century. Thompson (1968) argues that after 1832 in England a working-class politics emerged to challenge the state. Although initially unsuccessful, in the mid-nineteenth-century period of economic growth, unions consolidated their position and made economic gains for their members. Subsistence wages were no longer 'natural'; the issue of wage level was negotiable. Although originally restricted to skilled workmen – Lenin's labour aristocracy – unionism gradually spread to other workers. With the extension of the voting franchise, governments began to make further concessions to workers, culminating in the establishment of the welfare state in the mid-twentieth century. This process was also going along in different ways but essentially the same direction in other countries, but only in core countries. Political pressure to increase the well-being of the dominated class has been successful only in core and a few semi-peripheral countries. The end result has been a high-wage core and a low-wage periphery, reflecting both the 'social imperialism' and 'division' relations described previously (see Figure 3.2).

The key mechanism: unequal exchange

Modern massive material inequalities at the world scale reflect relatively successful political pressure from the dominated class in the core and the lack of any such success in the periphery. But how does this contrast help to maintain the current structure of core and periphery? This is where unequal exchange comes in. Every transaction between core and periphery is priced in a world market that incorporates these inequalities into its operation. Hence peripheral goods are 'cheap' and core goods are 'expensive'. When a German consumer buys Ghanaian cocoa, low Ghanaian wages are incorporated into the price. When a Ghanaian consumer buys a German car, high German wages are incorporated into the price. This is not just a matter of different technology levels, although these are interwoven with unequal exchange; the essential difference is in social relations at each location – the relative strength of the German worker compared with his or her Ghanaian equivalent. For 1966, for instance, it has been estimated that the peripheral countries' trade of US$35 billion would have been 'worth' US$57 billion if produced under high-wage conditions (Frank 1978: 107). The shortfall of US$22 billion is the result of unequal exchange. Needless to say, this was very much greater than all aid programmes put together. It is the difference between social imperialism and subsistence wages.

We have now reached the crux of our argument. The interweaving of class conflict at a state scale and the core–periphery conflict at the global scale through the process of unequal exchange produces the uneven development so characteristic of our world. And the beauty of this process is that it goes on day after day unrevealed. Unlike free trade and subsistence wages, which have been victims of political action, the world market remains off the political agenda. It cannot be otherwise in a world divided into many states where each has its own separate politics. Unequal exchange is an integrated mixture of inter-state and intra-state issues that conventional international politics cannot deal with. The world market appears to be based upon the impersonal forces of supply and demand, which determine prices. The only issues that arise are the terms of trade or the balance of prices between core and periphery goods. The fact that these terms do not reflect the hidden hand of the market but are defined by centuries of imperialism producing global differentials in labour costs is conveniently forgotten. This non-decision making represents a major political achievement of the dominant interests of the modern world-economy.

Technological shift and the persistence of global inequality

The core–periphery hierarchy has been a constant feature of the capitalist world-economy. However, that does not mean that it lacks dynamism. Indeed, it is expected that as new industries are created in the world-economy they will replace the previous innovations as the most profitable. To maintain the profitability of the former lead-sector industries, production costs must be reduced. One method of doing this is to reduce labour costs. In world-systems language, these older industries will be in the economic sphere of peripheral processes and their low wages.

Often, such a transition will entail a geographical move too. The consequent industrialization of countries is then argued, by scholars in the developmentalist school, to be evidence of development or progress.

However, world-systems analysis focuses more upon the maintenance of relative economic position within the core–periphery structure. Industrialization by a process of gaining industries that are less profitable than newly emerging industries is unlikely to close the gap between rich and poor zones of the world-economy. What is necessary is a focus upon wealth rather than industry.

We utilize recent work by Arrighi et al. (2003) to show how this process works. Table 3.2 shows the geographical shifts in the manufacturing industry from 1960. The numbers represent a region's share of manufacturing in relation to the 'first world', or those countries of the world where core processes predominate. All the regions of the 'third world' (or countries where peripheral processes predominate) have increased their relative share of manufacturing. Notably, in 1998 Latin America, east Asia, and China had surpassed the first world as sites for manufacturing industry. In the first world all regions except Japan had decreased the proportion of the manufacturing industry in their economy.

Table 3.2 Region's manufacturing GDP as a percentage of first world GDP

Region	1960	1970	1980	1990	1998
Sub-Saharan Africa	53	63	71	88	78
Latin America	98	95	115	113	105
West Asia and north Africa	38	43	41	70	71
South Asia	48	51	71	82	79
East Asia (excl. China and Japan)	49	68	95	115	130
China	82	107	166	150	190
Third world	**75**	**78**	**99**	**108**	**118**
North America	96	88	88	84	93
Western Europe	102	101	97	97	97
Southern Europe	91	92	111	100	96
Australia and New Zealand	87	86	80	68	67
Japan	120	127	120	128	120
First world	**100**	**100**	**100**	**100**	**100**

Source: Arrighi et al. (2003).

Table 3.3 Region's per capita GNP as a percentage of first world's per capita GNP

Region	1960	1970	1980	1990	1999
Sub-Saharan Africa	5	4	4	3	2
Latin America	20	16	18	12	12
West Asia and north Africa	9	8	9	7	7
South Asia	2	1	1	1	2
East Asia (excl. China and Japan)	6	6	8	10	13
China	1	1	1	1	3
Third world	**5**	**4**	**4**	**4**	**5**
North America	124	105	100	98	101
Western Europe	111	105	105	100	98
Southern Europe	52	59	60	59	60
Australia and New Zealand	95	83	75	66	74
Japan	79	126	134	150	145
First world	**100**	**100**	**100**	**100**	**100**

Source: Arrighi et al. (2003).

Table 3.3 tells a very different story. The wealth of the same regions is represented as a share of the first world's wealth. Instead of wealth becoming more equitable as manufacturing industry moved from first to third world, disparities actually increased. The disparity with the first world grew in all of the third world regions except east Asia and China. Notably, in the first world, the relative wealth of North America and western Europe declined as Japan's rose.

The overall core–periphery structure of the world-economy remains constant. The temporal dynamism of the system produces what on the surface are dramatic changes – the industrialization of Asia for example. The structural approach does not deny the widespread impact that such change will have on the politics and sociology of particular countries, the environmental impacts and the improvements in material well-being for many people. However, it is also important to note that the world-system does not become more equitable as a result of such industrial shifts.

Informal imperialism under American hegemony

The main economic agents of today's informal imperialism are the 'multinational corporations' that produce and trade across several states. These have been the major economic feature of US hegemony, and the decline of that hegemony has been marked by a consequent rise of European and Japanese corporations. The relationship between these economic enterprises and the states they operate in is a very important issue, which we deal with in some detail in Chapter 4. Here we concentrate on the way in which informal imperialism operates below the scale of the state and corporation. Individuals organize their lives through households, and the question arises as to how this institution relates to informal imperialism. In fact, households are integral to the operation of unequal exchange.

For unequal exchange to occur, we require two zones where direct producers obtain very different levels of remuneration for their labours. We have already considered how high-wage and low-wage zones come about, but why do they still exist? What are the mechanisms that enable core and periphery to be reproduced in the day-to-day activities of individuals? The answer is that in each zone different types of household have been created to accommodate different levels of resources. In this way, the households become part of the structure through which imperialism continues.

Wallerstein (1983) has introduced the concepts of proletarian and semi-proletarian households to describe the different institutions in core and periphery. Proletarian households derive most of their income from waged work. They evolved out of the social imperialism and welfare state developments in the core states in the first half of the twentieth century. As the direct producers obtained higher wages, a new form of household, centred upon the nuclear family, was created. Older forms of household covering a larger extended family could be displaced when a single wage became large enough to maintain the immediate family of the wage earner. In the ideal form of this new household, the husband becomes the sole 'breadwinner', the wife becomes a 'housewife' and the children are full-time at school. This produces a patriarchy where women are relegated to the private sphere of the home, so their work is unwaged and largely unrecognized. This leaves men with their wages as the providers of all necessary items from outside the household. Their role as head of household is therefore predictable and almost seems 'natural'.

This household form expanded with US hegemony. The 1947 General Motors–Auto Workers Union deal is usually seen as symbolic because the corporation conceded the high wages to sustain the new proletarian household. The suburban way of life for direct producers was born, and J. K. Galbraith (1958) announced that we had now entered a new sort of society, the 'affluent society' no less. As the Kondratieff boom spread to Europe after 1950, so too did the affluent society and its associated consumerism. But this is only the first part of only half the story.

Meanwhile, in the periphery economic changes were reinforcing a very different form of household. These are termed semi-proletarian because wages constitute only a minority of household income. In this low-wage zone, it is not possible for one person to be the sole breadwinner. Hence other members of the household have to contribute various forms of income for their survival. An example will help to clarify the situation. In the spatial division of labour we have described in Africa (see Figure 3.10), for instance, households straddle the different geographical zones, producing a distinct sexual division of labour. The different zones provide contrasting opportunities for men and women. In the 'islands' producing commodities for the world-economy, for instance, much of the labour is carried out by male migrants who originate from the subsistence zone. These direct producers provide the main wage component of a household income. The other members of the household remain in the subsistence zone, where most of the labour is female and unwaged. This form of patriarchy is superficially similar to that described for the proletarian household, with the man controlling the cash, but in this case the cash is much less important to the household. Money from the migrants is necessary for payment of taxes and buying a few items in the market, but the bulk of a household's day-to-day needs are produced within the household. It is this subsistence activity that enables low wages to be paid to the migrant males. In effect, the women of the subsistence zone are subsidizing the male labour of the world-economy production zones.

Such migrant-labour households are common throughout the periphery, but they constitute just one of several semi-proletarian household types. Their common feature is that they transfer reproduction costs away from the production costs for the world market. Hence necessary activities, such as rearing children for the next generation of labour and the looking after of the previous generation of labour after their working lives, are not costed in the pricing of commodities originating from the periphery to the same degree as for commodities from the core. Hence buyers of core goods in the periphery pay a price that contributes to the welfare of direct producers in the core, while buyers of periphery goods in the core do not make the same contribution to the welfare of direct producers in the periphery. Thus patriarchy has been moulded in different contexts to generate unequal exchange.

There have been important changes in recent years that have modified the simple pattern of household structures we have just described. In the core, the 'ideal' proletarian household based upon one wage has been severely eroded by the massive increase of women in the labour force. The patriarchy represented by the notion of a male breadwinner has been undermined by the ideas generated by the women's movement from the late 1950s onwards. Simultaneously, the spread of mass-production techniques created a need for additional labour that women could supply. Hence proletarian households in the core have become even more 'proletarian' as they typically rely on more than one wage to maintain their standard of life. This has meant even higher levels of commodity consumption by core households, which has contributed to the maintenance of the vast material differences between core and periphery. Never doubt that the ubiquitous suburban shopping malls of the rich countries are political symbols of the continuing world victory of the core.

Meanwhile, in parts of the periphery new developments have also been occurring, involving bringing more women into the waged workforce. Since the 1960s, there has been appreciable growth of industrial production outside the core (Table 3.2). This is sometimes referred to as the new international division of labour. In south-east Asia, for instance, a massive electronics industry has grown up employing large numbers of women. Interpretation of this industrial growth has been confused by a popular assumption that industry is a property of core countries, leaving the periphery to produce agricultural goods and raw materials, as we dismissed earlier (Arrighi et al. 2003). If this assumption is accepted, then the new international division of labour represents a genuine de-peripheralization process. But we have disposed of the 'industry equals core' argument. In world-systems analysis, core production processes involve relatively high-wage, high-technology activities irrespective of the commodity produced. Throughout its hegemony and beyond, the United States has been the major agricultural exporter on the world market, for example. The important point is how the production is organized – the social relations of production – not what happens to be produced. Hence peripheral production processes are compatible with industrial activity where the latter is relatively low wage and low technology. Electronics components production can be either core-like or periphery-like, depending upon the social relations of production. In south-east Asia, the components are produced in a periphery production process that uses the patriarchy of the region in a new way. A workforce of young

women has been created whose gender subordination is translated into 'docile, subservient and cheap' labour (Momsen and Townsend 1987: 79). Hence the increased proletarianization of households is not resulting in appreciable increases in the affluence of the households. The low wages ensure that, despite industrialization of the periphery, unequal exchange continues unabated.

Gender, globalization and scale

Globalization is a term that seems to imply a blanket coverage of the world, processes leading to homogenization. This has been expressed, in terms of finance, as 'the end of geography' (O'Brien 1992). Michael Storper (1997: 27), however, sees it as 'quite curious that a fundamentally geographical process labeled with a geographical term – "globalization" – is analysed as a set of resource flows largely without considerating their interactions with the territoriality of economic development.' We are at one with this position. Quite simply: 'Higher systemic integration has not replaced core/periphery structures or core rivalry' (Marshall 1996: 886). Hence the question becomes not whether globalization has replaced imperialism but how does imperialism operate under conditions of globalization? It might be much more complex than earlier spatial structures, but even researchers who doubt the continuing salience of core–periphery concepts nevertheless fall back on just such analysis in their comparative studies (Castells 1996: 108).

The identification of households as an arena of politics is a key conceptual tool that achieves two related tasks. Households are a particular construction of geographical scale that is simultaneously local and global. The gender relations that are evident in semi-proletarian households are everyday manifestations of gender roles that are proscribed and resisted. At the same time, they play a functional role in maintaining inequalities in wage levels that is an essential component of the core–periphery hierarchy. A feminist approach to globalization and North–South differences is effective in showing how women have usually been excluded from analyses of globalization through a variety of binary frameworks (Table 3.4). In sum, these frameworks separate the global from

Table 3.4 Binary understanding and the
marginalization of women

Global	Local
Economy/market	Culture/nonmarket
Theory	Ethnography
Production	Consumption
Formal sector	Informal sector
Public	Private
Macro	Micro
Modernity	Tradition
Reason/logic	Affect/emotion/belief
Cause	Effect
Agent/action	Victim/passive
History	Everyday life
Space	Place
Abstract	Grounded
Universal	Particular

Source: Roberts (2004).

the local and relegate women to the latter. The twin
consequences of these dichotomies are that women
are conceptualized as passive victims on the margins
of theoretical understandings of the capitalist world-
economy and also 'bit players' in the grand and
abstract processes of global finance and economic
restructuring (Roberts 2004).

However, the gendered relations within the semi-
proletarian household clearly show the importance
of women in the sphere of production and how
women's labour is an integral component of the
structure of the world-economy. 'Local' and 'global'
are not oppositions, but rather a nexus of activity.
In combination, such activity creates and maintains
the capitalist world-economy requiring a hybrid
combination of production and consumption, and
formal and informal labour that shows the essential
nature of gendered roles in all activity, rather than
activities that can be marginalized through binary
distinctions (Roberts 2004).

Moving from a critique of binary frameworks
that have conspired to exclude women from studies
of global economic relations, Roberts (2004) explores
how the contemporary neo-liberal rhetoric and practices of globalization aim to construct gender roles
in a particular way. In Chapters 1 and 2 we identified

the current period of globalization as a moment
on the cyclical dynamics of the capitalist world-
economy. Roberts takes a different approach and sees
globalization as a label for a new form of economics
known as neo-liberalism: 'the bundle of discourses
and social practices that in large part animate the
dynamics of the contemporary global economy'
(Roberts 2004: 135). Neo-liberalism *opens* markets
by reducing or eradicating barriers to trade and finance erected by states, the World Trade Organization
being the key site of agency. Neo-liberalism also
extends the economic scope of markets by privatiz-
ing what were once communal assets such as water
and land.

To undertake neo-liberal projects dominant
representations of the world are offered, that fall
under the broad rubric of 'development' and
'progress' that Wallerstein has identified as a constant ideology of the capitalist world-economy.
Following the feminists' identification of the local
and the global as part of a continuum, neo-liberalism
identifies women as objects in four ways that simultaneously promote certain gender roles within the
much larger neo-liberal project. Roberts (2004: 135)
identifies four ways in which neo-liberal rhetoric
creates women as objects in contemporary economic
restructuring:

➤ Women are seen as individual market actors
rather than as a group.

➤ Women are seen as 'human capital' to be 'developed' through education and training.

➤ Women are seen as political subjects with human
rights.

➤ Women are seen as 'social capital' active in civil
society, especially non-governmental organizations.

In general, these four ways of trying to define women
relate to the drive to maximize profit in a moment of
global economic restructuring. From our discussion
of 'islands of development' in colonial Africa to
present-day neo-liberalism we can see that the household is a key site of the politics of imperialism.
Women are essential actors in the processes of imperialism, but, as we shall see in the next section, have
been active in a politics of resisting imperialism.

The post-colonial voice

Post-colonial studies are concerned with both colonial discourses, or the way colonized countries and people have been represented, and neo-colonial processes that maintain inequalities of wealth and power. Post-colonial studies are a challenge to the way we 'see' the world, or construct academic frameworks, as well as the identification and support of political projects to address structural inequalities (Laurie and Calla 2004). The world-systems approach is, in some senses, a kindred spirit. On the other hand, its structural approach can, falsely we would argue, be critiqued by post-colonial scholars as maintaining a privileged male gaze. Post-colonialism has identified the Eurocentric nature of social science, perhaps best reflected in the 'error of developmentalism' discussed in Chapter 1. The North–South/modern–undeveloped dichotomies of mainstream social science have created a cultural and scientific hierarchy by which scholars from European and American backgrounds are privileged to speak about and on behalf of the rest of the world.

The struggle of marginalized groups is a central component of world-systems analysis as the nature of political change is seen to emanate from the periphery of the capitalist world-economy rather than from the core (Amin et al. 1990). This view is in contrast to orthodox Marxism that identifies the most industrialized parts of the world as the venues for socialist revolution. Instead, the world-systems approach views the material exploitation of the periphery (as profits flow to the core) and its marginalization through racist structures and representations as generating an opposition to the ideology of capitalist 'development' that may provoke new post-capitalist politics and society. Although the efficacy and likelihood of such actions may be debated, the world-systems view of the world echoes the post-structural belief that change emerges from the margins of society (Cope 2004).

The world-systems approach identifies the power relationships within the capitalist world-economy and so helps situate or conceptualize the 'Anglo gaze' of social science. However, its structural approach can also be identified as a hegemonic white-male Anglocentric perspective. The balance is for the world-systems approach to give temporal and spatial context the voices of those marginalized within the capitalist world-economy without forcing their words into a pre-established theoretical framework. We acknowledge the theoretical tensions between post-colonial studies and the world-systems approach, but think they are 'creative tensions'. A broad geographical and temporal framework is, we believe, a productive complement to attempts to give voice to political alternatives and insights voiced by those marginalized within the capitalist world-economy.

Summary

In this section we have:

➤ introduced the concept informal imperialism;

➤ related the practice of informal imperialism to the actions of the hegemonic power;

➤ related the practice of informal imperialism to the maintenance of the core–periphery structure of the capitalist world-economy;

➤ situated political decisions made in the core and periphery to the dynamics of informal imperialism and the persistent core–periphery structure;

➤ illustrated how informal imperialism is constructed by, and partially determines the actions of, households;

➤ shown the role of the feminist perspective in understanding the practice of informal imperialism.

■ 'Empire' in the twenty-first century

'Empire' is the word of the moment. After being despatched to history as a positive politics in the wake of the First World War, the War on Terrorism has brought it back as a central theme. However, a few strong words of caution are necessary. The usage of 'empire' is as much polemic as it is academic. Some neo-conservative commentators in the United States have embraced the term 'empire', seeing it as the current form of an unquestioned right and duty of the United States to do good in the world (Boot 2002; Ignatieff 2003). At the same time the negative connotation is also being pushed: critics of the War on Terrorism have noted the disapproving baggage that is attached to the word 'empire' and so are eager to paint the foreign policy of President George W. Bush in colours that evoke images of racist colonial officers, military expeditions to pacify swathes of territory and their indigenous populations, all in the name of economic exploitation.

Such literature is careless in its conceptualization of what imperialism actually has been and is. However, in a parallel development, social scientists have begun to think about 'empire' in a more systematic manner. Political scientist Joseph Nye has focused on the role of 'soft power' in the projection of US influence across the globe. Soft power, 'the ability to get what you want through attraction rather than coercion or payments' (Nye 2004: 256), is the geopolitical role played by a country's culture and political ideals. It is not just a matter of rhetoric. The power of the United States is judged by its policies and actions, and Nye (2004) laments that coercion, or 'hard power' is the dominant form of US presence in the world, resulting in increasing anti-Americanism and declining appreciation of American cultural products.

The term 'empire' has increased in popularity because of the military dominance of the United States, its declaration of the War on Terrorism, and the invasions of Afghanistan and Iraq. The manifestation of coercive power is clear and present. Nye (2004) argues that the form of intervention is overly militaristic and is counter-productive. He claims that

we are witnessing 'imperial understretch' (Nye 2004: 264): 'The United States has designed a military that is better suited to kicking down the door, beating up a dictator, and then going home rather than staying for the harder imperial work of building a democratic polity' (Nye 2004: 264). The United States' military expenditures are nineteen times the size of its budget for international affairs (including aid donations) (Nye 2004). In other words, 'empire' is primarily a matter of military conquest rather than creating a set of countries with the political and cultural values of the United States and who show allegiance to the dominant power.

Related work emphasizes the militaristic stance of the United States and its reliance upon coercive power (see, for example, Bacevich 2002; Johnson 2004; Klare 2004; Boggs 2005). In this body of work the United States is identified as resorting to a militaristic foreign policy in the face of increasing resistance to its authority abroad. However, and agreeing with Nye (2004), we argue that military power is not a sign of strength, but a sign that its hegemonic ability to set the global agenda is waning (Boulding 1990). Such a drift into military expeditions makes sense from our cyclical world-systems approach. We saw that Britain's hegemonic decline was related to increasing military expeditions abroad and the need for a greater colonial presence.

There are a number of ways of interpreting the current focus on 'empire'; one is to place it within the dynamics of hegemonic cycles discussed in the previous chapter. It is a policy and representational response to hegemonic decline. The other approach is to see contemporary changes as ushering in a new form of political geography. Hardt and Negri's (2000) book *Empire* provoked a bout of serious social scientific consideration on the topic. We begin by outlining Hardt and Negri's contribution, but then move on to critical considerations that force us to consider this round of empire within the dynamics of hegemonic cycles, just as we did for earlier periods of core–periphery relations.

For Hardt and Negri a new form of sovereign power is now dominant. This is not one based on territorial sovereign states and their relative ability to colonize other territorial entities; instead, 'empire' is a decentered and deterritorializing apparatus of rule that progressively incorporates the entire global realm within its open, expanding frontiers. Empire manages hybrid identities, flexible hierarchies, and plural exchanges through modulating networks of command. The distinct national colors of the imperialist map of the world have merged and blended in the imperial global rainbow.

(Hardt and Negri 2000: xiii)

They go on to say that there are no territorial limits to 'empire' – it encompasses the whole '"civilized" world' (ibid.: xiv) in their terms. It is also portrayed as timeless, meaning that no ideological space is given for alternatives. Empire pervades all social relations, regulating not only human interactions but also human nature. Finally, although the rhetoric always talks of a peaceful world, the reality is very bloody. Hardt and Negri's thesis has some similarity with the Marxist theoretician Karl Kautsky (1914). Writing at the beginning of the First World War he identified a unity of interests between industrialists and financiers, and the dominant powers that would construct an 'ultra-imperialism'. As we shall see, Hardt and Negri point to the importance of additional political actors.

Let us pause for a while and try to make sense of this using the terminology we are familiar with. From our earlier mapping of cycles of imperialism we have already seen that the capitalist world-economy spread to encompass the whole of the globe around the beginning of the twentieth century. This is not new. The 'timeless' rhetoric is not surprising in that it is a useful strategy for marginalizing political alternatives and making the dominant agenda 'common sense'. We have also seen the rhetoric of 'peace' blended with military power. Hegemonic powers preach a 'civilizing mission' while using military force against those who refuse to be 'civilized' in the manner deemed appropriate by the hegemonic power. In addition, the urge to create a particular 'human nature' has been evident in past episodes of imperialism; this was the central theme of the 'civilizing' mission.

But Hardt and Negri (2000) argue that a new form of governance of the capitalist world-economy has come into being; it is not simply states acting as colonial powers that are controlling economic relations and human nature, but a whole range of government agencies, non-governmental organizations (such as Church-based groups), supranational institutions (such as the United Nations), and multinational firms that are regulating social relations and human practices across the globe.

The interesting question to ponder is whether this new form of governance is a moment that can be related to the dynamics of the hegemonic cycle, or whether it is a qualitatively different stage in world history. As in our discussion of the revolutionary heritage, we are again confronted by two conceptualizations of capitalism: one that moves through stages – the classical Marxist approach – and the cyclical framework of world-systems analysis. If the world-systems interpretation is correct, the preponderance of non-state institutions and their role in governing people and economic transactions can be seen as a manifestation of the period of financialization identified by Arrighi. However, we do not make the claim that history will simply repeat itself. Although certain structural constraints exist, the linear trend of increased integration of the capitalist world-economy, plus the specificity of national histories and cultures, will combine to create new political geographies.

Michael Mann (2003) provides a sober analysis of the imperial pretensions of the United States, especially its military capabilities and cultural foundations. Mann (2003: 18) identifies four military resources necessary for the tasks of empire:

➤ Secure defence or deterrence against attack
➤ Offensive strike power
➤ The ability to conquer territories and people
➤ The ability to pacify them afterwards.

Mann (2003) argues that nuclear weapons act as a useful deterrent but an impractical offensive weapon, and although US conventional forces are far superior in numbers and technical capability to any competitor, their form of deployment does not allow for the construction of empire. The forces in Afghanistan and Iraq, for example, are restricted to fortresses isolated from society, and their only presence in indigenous communities is in mobile and heavily armed patrols. Mann contrasts this type of presence to the institutional presence of the European empires in which collaboration with local elites and the establishment of educational and cultural resources that mirrored those of the colonial power embedded social practices that would maintain subservience to the colonial powers throughout society.

What of the culture of the United States? In a provocative passage, Mann argues: 'Lack of an imperial American culture makes for further weaknesses. American kids are not brought up to be as racist, or stoic in combat, as self-denying in crisis, or as obedient to authority, as British kids once were' (Mann 2003: 27). But Mann's framework is very different from that of Hardt and Negri's. Mann still sees the world as defined by separate states. 'Faced with a world of nation states, the US does not have imperial political powers. The Age of Empire has gone' (Mann 2003: 97). The dissonance between Mann and Hardt and Negri is caused by their emphasis upon different logics of capital accumulation: one based on market forces and the other on the power of the state to plunder territory. The world-systems approach has identified just one logic of accumulation, in which states and markets work in tandem in what we have identified as a political economy logic (Chase-Dunn 1982).

One logic of contemporary imperialism

If we follow Hardt and Negri's (2000) logic, then we should be looking at economic and other institutional forms of power in the current age of empire. Imperialism has become a means of global governance, constructing behaviours that promote the accumulation of capital, rather than state projects to control territory and resources. However, it is a stark reality that contemporary identification of 'empire' comes at a time of war: a war waged by the hegemonic power in a region rich with the vital resource of oil. David Harvey's (2005) Marxist approach in *The New Imperialism* relates the role of the United States as an occupying force with the imposition of a neo-liberal agenda. According to Harvey, US power in Iraq was exercised through, and in order to maintain, free-market capitalism. Harvey notes how, in 2003, before the establishment of an Iraqi govern-

ment, the US-run Coalition Provisional Authority made a number of decrees that led to:

➤ privatization of public enterprises;

➤ granting full ownership rights of Iraqi firms to foreign businesses;

➤ complete repatriation of foreign profits;

➤ facilitating foreign ownership of Iraqi banks;

➤ elimination of trade barriers;

➤ outlawing strikes;

➤ restricting the right to unionize.

Such actions violated the Geneva and Hague Conventions regarding the proper conduct of an occupying power (Harvey 2005: 214–15). The decrees were a clear imposition of a neo-liberal agenda and the facilitation of flows of investment we have already discussed as a moment of financialization (Arrighi 1994). In that sense we can interpret the US actions as consistent with a particularly intense form of informal imperialism.

The case of the US invasion of Iraq was an extreme case of contemporary imperialism. However, it is evidence of the one logic of capital accumulation that is a feature of the world-systems approach: 'accumulation through dispossession' (Harvey 2005: 215) facilitates foreign ownership and accumulation through the market. To bring in Hardt and Negri, a combination of governments, firms, and supra-state organizations invoked a neo-liberal rhetoric of markets and put it into practice. This does indeed seem like a new form of imperialism. But it should also be noted that physical military presence was necessary to initiate the process, and the decree of the United States as occupying power made the economic changes happen; and in that there is more than an echo of nineteenth-century style formal imperialism.

Summary

In this section we have:

➤ introduced the contemporary term 'empire';

➤ situated 'empire' within the temporal context of hegemonic cycles;

➤ emphasized the 'one logic' of the world-systems approach.

Chapter summary

In this chapter we have:

➤ discussed the original theories of imperialism; the Hobson–Lenin paradigm;

➤ introduced Marxist theories of the state;

➤ built upon these theories to form a world-systems and political geography explanation of imperialisms;

➤ introduced the concepts formal and informal imperialism and noted the persistence of core–periphery differences;

➤ described the cyclical dynamics of formal and informal imperialism;

➤ situated the politics of imperialism within the cyclical dynamics by discussing trade policy and 'social imperialism';

➤ introduced the concept of 'unequal exchange' to explain the construction and maintenance of core–periphery differences;

➤ introduced the contemporary term 'empire' and situated it within our political economy approach.

In summary, the persistent core–periphery structure of the capitalist world-economy has been maintained by a number of different practices and beliefs that have been referred to as 'imperialism' or 'empire'. We have shown that it is wrong to view these practices as existing only at the global scale as they are a product of household, intra-state and inter-state politics and the decisions and practices of businesses. Only by considering the interplay of different political actors, at different geographical scales, in different geographical locations can imperialisms, past and present, be understood.

Key glossary terms from Chapter 3

administration	'empire'	inter-state system	protectionism
autarky	error of	Kondratieff cycles	protectorate
capitalism	developmentalism	(waves)	Second World War
capitalist world-	feudalism	League of Nations	semi-periphery
economy	First World War	liberal	social imperialism
civil society	formal imperialism	logistic wave	socialism
classes	franchise	Marxism	sovereignty
Cold War	free trade	militarism	space
colonialism	geopolitical code	multinational	state
colony	geopolitics	corporations	structural power
commonwealth	globalization	Napoleonic Wars	territoriality
conservative	government	nation-state	third world
containment	heartland–rimland thesis	neo-conservative	Treaty of Tordesillas
core	heartland theory	neo-liberal	unequal exchange
decolonization	hegemony	non-decision making	United Nations
dependency	households	partition	world-economy
development of	human rights	patriarchy	world-empire
underdevelopment	ideology	periphery	world market
elite	imperialism	power	world-system
empire	informal imperialism	practical politics	world-systems analysis

Suggested reading

Dicken, P. (2003) *Global Shift: Reshaping the Global Economic Map in the 21st Century*, 4th edition. New York: Guilford Press. A textbook that provides a detailed, accessible and stimulating account of how economic activity constructs the contemporary form of the capitalist world-economy.

Gregory, D. (2004) *The Colonial Present: Afghanistan, Palestine, Iraq*. Oxford: Blackwell. A powerful discussion of the way regions are represented to the general public in a manner that mutes public protest towards military action.

Johnson, C. (2004) *The Sorrows of Empire: Militarism, Secrecy, and the End of the Republic*. New York: Metropolitan Books. An accessible account and passionate critique of the militaristic nature of contemporary American foreign policy.

Peet, R. (2003) *Unholy Trinity: The IMF, World Bank and WTO*. London: Zed Books. A compelling history and scathing critique of the key institutions set up by the United States to regulate the world-economy and, especially, the core–periphery relationship.

Smith, N. (2003) *American Empire: Roosevelt's Geographer and the Prelude to Globalization*. Berkeley: University of California Press. An intriguing story of the influence of one of the 'founding fathers' of political geography, Isaiah Bowman, on US foreign policy in the first half of the twentieth century and how this relates to contemporary American 'imperialism'.

Activities

1 Go to the following website and explore the link http://www.saxakali.com/southasia/Voices.htm. Identify at least three concepts we have introduced in Chapters 1 and 3 of this book, for example core and periphery processes, semi-proletarian households, formal and informal imperialism, neo-liberalism. In what way are these concepts engaged by the discussions on these sites, and what forms of resistance are identified? (Hint: they will not use the same terminology as us, but are talking about the same issues.)

2 The concept of fair trade has been used to challenge the core–periphery structure of the capitalist world-economy. Type in 'fair trade' in an internet search engine or try one of the following sites: http://www.fairtraderesource.org/ or http://www.fairtradefederation.org/. Can you identify how the core–periphery structure of the capitalist world-economy is addressed by these sites (Hint: they will not use the same terminology as us, but are talking about the same issues.)

3 Go to the website http://www.globalsecurity.org/. Follow the links to 'military' and 'facilities' and explore the type, size and date of establishment of US military bases, ports, airfields and other facilities across the globe. Plot the location and year of establishment of these bases on a map of the world. Does the historical geography of US bases relate to the cycles of hegemony model, especially the United States' rise to power and the possible need to impose military order during a period of hegemonic decline?

RIOT CONTROL TROOPS, CHUNJU, REPUBLIC OF KOREA, 14 JUNE 1987.
Source: © Nathan Benn/Corbis

Territorial states

What this chapter covers

Cultural politics, state power and the impossibility of grieving

In 2006 reports began to circulate in the US media of demonstrations at the funerals of American service personnel killed during the war in Iraq. But these demonstrations were not a matter of left-wing or pacifist protest against the prosecution of the war. Instead, they were led by Fred Phelps, pastor of the Westboro Baptist Church, Topeka, Kansas. The demonstrators carried banners claiming 'Thank God for Dead Soldiers' and 'Thank God for IEDs'. (IEDs, or improvised explosive devices, were the roadside bombs that targeted US military patrols in Iraq.) By March 2006 the Phelps family had picketed about one hundred military funerals; their cause was a religious fundamentalism that argues the soldiers were fighting for the army of a state (the United States) that accepts homosexuality. And then there were the bikers: calling themselves the Patriot Guard Riders, posses of around four hundred motorcyclists would ride into town and try to stand between the demonstrators and the funeral to allow the grieving family to enact the burial in a

dignified manner. To add to the imagery, the drama was played out in the American West with its symbolism of American frontierism.

The state is at the centre of the funerals, the protests and counter-protests. The funerals are the sad and ultimate manifestation of an overt expression of the power of states: the ability to muster and practise violence. The protestors are resisting the ability of the state to define citizenship in a way that legitimates homosexuality which, according to their fundamentalist reading of the Bible, goes against the word of a higher authority: God. The counter-demonstrators are motivated by patriotism, a love of their country that is focused on the dignity of individual soldiers; but patriotism is an ideology that maintains the state as a legitimate and benevolent actor.

Source: Ed Lavandera, 'Dodge City showdown at funeral'. http://www.cnn.com/2006/US/03/06/btsc.lavandrera.funerals/index.html. 7 March 2006. Accessed 7 March 2006.

Human activities are traditionally divided into three spheres: economic, political and socio-cultural. In turn, these activities define the entities that conventional social scientists study: the economy, the state and civil society. In practice these are interpreted as spatially congruent; for instance, the French economy, the French state and French society are deemed to share identical boundaries, those that limit the sovereign territory of France. But as was emphasized in Chapter 1, world-systems analysis does not take the conventional position on such matters. Rather than multiple national economies there is *one* capitalist world-economy; rather than multiple national societies there is *one* modern world-system. They are two sides of a single logic of social change that has operated for half a millennium. But politics is different: part of the essence of the modern world-system is that there is no single world state (otherwise it would be transformed into a world-empire). Thus, in Chapter 1 we identified states as one of the key institutions of the modern world-system: the latter's logic of social change incorporates *multiple* states

or what Wallerstein (1984a) calls the inter-state system. This divided politics is essentially territorial in nature because sovereignty, legitimate political power, is defined as existing within the boundaries of states. It is territorial states that are the subject of this chapter.

The functions of the state are connected to the need to facilitate economic growth within the state's territory that generates the tax base to support the state itself (Tilly 1990). With money from taxes, the state tends to its territory by providing two types of public goods; these are services available to all its citizens. First, there is the provision of security: a police force and judiciary for internal security, and armed forces for external security. The latter may involve war in the form of invading another state's territory. Second, there is the provision of infrastructure to enable wealth creation: physical infrastructure and/or its regulation (for example, roads, airspace), and social infrastructure (for example, schools, hospitals) to reproduce a population that can serve as a workforce (Mann 1986).

Additional understanding of the state comes from a feminist perspective. Feminist geographers not only discourage separating civil society and economy, but frown upon separating the state as well. As noted above, the state is separated in world-systems analysis for good reason but much can be learned from entwining the political with the economic and social at a practical level. Although analytically distinctive, the state can only operate through and with economic and social processes. Thus, for feminists, 'the market and civil society involve actors and processes that help constitute the state; the procedures and actors of the state similarly influence the market and civil society' (Fincher 2004: 50). The important conclusions are that the state is multifaceted and contested, and that a key outcome is the process of differencing (Kobayashi 1997), defined as 'complex sociospatial processes of empowering and enabling some people and marginalizing and oppressing others on bases of the differences they embody' (Chouinard 2004: 235). The role of the state in creating differences and the political acts of resisting such state practices makes the state the fluid outcome of interaction between the 'private' spaces of the household, the traditional site of unpaid women's work, and the male-dominated 'public' sphere. To understand the state is to identify its pervasiveness or prosaic nature (Painter 2004), and to recognize that the 'public' and the 'private' is a false dualism.

The state is manifest at various geographical scales, from the central to the local government. The exact form of the state is contingent on local circumstances, but in general the local state is seen to be 'closer' to the people than is the central bureaucracy and so helps to legitimate the state. However, the local state has agency too, and can enact policies that challenge the authority of the central state (Kirby 1993). The current form of the state must be understood within the broader context of globalization and the accompanying ideology of neo-liberalism. The result has been the devolution of state power, with greater responsibility for policies and the fostering of economic growth being placed on local governments. Accompanying this movement are new types of local state governance involving the creation of certain business and political behaviours that attempt to attract global investment (Ward 2005).

This introduction only begins to explore the complexity of the political geography of the state. But we have set out some of the tasks that face us in this chapter. First we discuss how the world political map has been created. This was a major question for traditional political geographers but they tended to take a case-by-case approach to the spatial integration of states. Nevertheless, some of their concepts remain salient. In particular, Hartshorne's (1950) idea that state integration is the outcome of two sets of contrary forces remains useful: centrifugal forces (for example, unequal development) pull the state apart whereas centripetal forces (such as a strong 'state idea' or iconography) help bind it together. We begin by exploring the origins of modern states through the use of a simple topological model of the state. This provides the framework for taking topics from traditional political geography and integrating them into a world-systems analysis of the inter-state system. This is important, because we are concerned with a world of multiple states – notice the title of this chapter is in the plural and not the singular.

However, understanding the creation of the world political map is only a preliminary stage for a world-systems analysis of how states operate in the world-economy. In the subsequent section, we look at more fundamental questions concerning the nature of the states themselves. The recent debates within Marxist theories of the state are briefly reviewed as contributions to understanding 'stateness', but we conclude that they are deficient in dealing with 'inter-stateness' – the structural condition of multiple states. Thus we argue for the need for a world-systems theory of the states, that is, of the inter-state system as a whole.

Given that a large element of the idea of contemporary globalization concerns 'de-territorialization', it follows that this literature is often dismissive of the current power of states. It is true that de-territorialization strikes at the very nature of the modern state, but it is not a simple matter of new social forces eliminating old political structures. Those structures are half a millennium old and are, therefore, unlikely to succumb easily to practices with a provenance of just a decade or so. World-systems treatment of inter-stateness points to a direction that does not mean a contemporary demise of the state but its reorientation in new and changing circumstances. The related subjects of citizenship and governance are discussed to illuminate how the state is implicated in the construction of differences as a means to negotiate globalization and maintain power relations of race, gender and class. In our concluding section of this chapter, we use the nature of contemporary conflicts, especially resource wars, to show how our more historically sensitive approach to trans-state processes can put into context over-determined globalization theses on the demise of the state.

Chapter framework

The chapter is framed around four of the seven ideas we introduced at the beginning of Chapter 1:

➤ The construction and maintenance of states is an essential *material* political geographical practice.

➤ The state is an entity and process that requires *conceptualization* in order to be understood.

➤ The growth and form of the state is *contextualized* in time and space.

➤ We identify the state as key political institution within the *wider whole* of the capitalist world-economy.

■ The making of the world political map

Probably the most familiar map of all is the one showing the territories of the states across the world. This world political map is the simple geographical expression of the inter-state system. The minimum requirement for any political geography is to understand this map. And yet the map itself is misleading, since it gives an impression of stability that is completely false. This may account in part for the surprise that many people have felt at the upheavals in the world political map since the east European revolutions of 1989. The addition of new states to the map in the wake of the collapse of the Soviet Union and Yugoslavia is by no means unprecedented in the history of the inter-state system, however. We had just

got too accustomed to our current world map in the Cold War atmosphere before 1989. The simple fact is that any world political map, including today's, provides only a snapshot of states at one point in time; the reality is that this pattern is forever changing. The world political map must be interpreted as a series of patterns that have changed drastically in the past and that will doubtless experience equally major changes in the future. Our concern is to understand how the world political map became the pattern we see today. This is both an empirical–historical question and a conceptual–theoretical question, and our answers will accordingly mix these two approaches.

The origin of territorial states: a topological model

It is sometimes hard for us to imagine a political world that is not organized through states. States are part of our taken-for-granted world, and we hardly ever query their existence. States can even appear to be natural phenomena, which Jackson (1990: 7) blames on the world political map:

> When schoolchildren are repeatedly shown a political map of the world . . . they can easily end up regarding [states] in the same light as the physical features such as rivers or mountain ranges which sometimes delimit their international boundaries. . . . Far from being natural entities, modern sovereign states are entirely historical artifacts the oldest of which have been in existence in their present shape and alignment only for the past three or four centuries.

Jackson's reference to the history of modern states leads the way to counter their ascribed naturalness. Obviously, by describing a recent period when states such as the ones we experience did not exist we undermine their claim to be the only way in which politics can be organized. Furthermore, by investigating the emergence of the modern state we obtain some insights into its essential nature.

Europe in 1500

In 1500, Europe possessed a cultural homogeneity but was politically highly decentralized. As western

Christendom under the leadership of the papacy, Europe consisted of a single civilization. But the Church's secular power was quite limited, so that, politically, Europe constituted a very unusual world-empire. There was a nominal empire that claimed the Roman Empire's heritage, the German 'Holy Roman Empire', but its authority extended to only a small part of Europe, and even there its power was circumscribed. Europe was a complex mixture of hierarchies and territories through which power was organized.

Geographically, this complexity encompassed a variety of scales. First, there were the universal pretensions of the papacy and the Holy Roman Empire, which, although failing to provide a centralized empire, did help to maintain a singular and distinct European political world. Second, and in contrast, there was an excessive localism, with scores of small political authorities scattered across Europe – orders of knights, cities, bishoprics, duchies – all independent for most practical purposes. Third, and loosely tying the localism to the universalism, there were myriad hierarchical linkages as the legacy of feudal Europe. Complexity is hardly an adequate description: Tilly (1975: 24) has estimated that there were 1,500 independent political units in Europe at this time.

How did our world of sovereign territorial states emerge from this situation? Certainly, we should not assume that it was inevitable that power would eventually concentrate at a single geographical scale between universalism and localism. Looking at the world from the vantage point of 1500, Tilly believes that there were five possible alternative futures for Europe: two of localism – continued decentralized feudal arrangements or a new decentralized trading network of cities; two of universalism – a Christian theocratic federation or a politically centralized empire; and the pattern of 'medium-sized' states that finally emerged. With hindsight, we can see that the last was to be most compatible with both the economic changes occurring with the emergence of capitalism and the military revolution that was changing the nature of warfare at that time.

Looking in and looking out

One of the features of the complexity of European politics in 1500 was that territories having allegience to the same sovereign were usually spatially separated. In what Luard (1986) calls the 'age of dynasties' (1400–1559), territories were accumulated by families through a combination of war, marriage and inheritance. This process could lead to successful claims on territory by a family across all parts of Europe. For instance, the most successful family of the period, the Habsburg, accumulated territories in Spain, Austria, Italy and Burgundy to produce a 'realm' that is the geographical antithesis of the modern European state. It is only at the end of this period, according to Luard, that territorial claims began to focus on accumulating land to produce compact and contiguous states. For instance, in 1559 England gave up the French channel port of Calais and claims to other French territory.

What does it mean to produce a world of such compact and contiguous states? At its most elementary, it produces a topology where each state is defined in terms of an 'inside' and an 'outside'. Hence the fundamental nature of the state consists of two relations, what we may term looking inwards and looking outwards. The former case concerns the state's relations with its civil society, the social and economic activities that exist within its territory. The latter case has to do with the state's relations with the rest of the inter-state system of which it is part. In much political analysis, these two relations are treated separately: state–civil society is the domain of political science; and state–state relations are the responsibility of the discipline of international relations. And this mirrors the popular view that divides politics into domestic and foreign policy. But it is the same state that operates in both spheres, looking inwards and outwards simultaneously. For our political geography, based as it is on articulating political relations across different geographical scales, this topological model of the state is the key starting point for understanding the states.

The two stages in creating territorial states

State apparatuses for dealing with domestic and external relations did not evolve at the same time. Strayer (1970) describes a situation where domestic political institutions preceded external ones by about three hundred years. His argument is that the medieval victory of the papacy over the Holy Roman Empire produced a power vacuum that the papacy could not fill. Hence across Europe in the thirteenth century political power accrued to middle-range kingdoms to fill the gap. These kingdoms, some of which have survived (Portugal, France, England) and some of which have not (Navarre, Naples, Burgundy), created only institutions to deal with internal affairs. They were really concerned with large-scale estate management, and the first permanent institutions were the high court and the treasury. Because of this narrow focus, Strayer terms them 'law states'. This form of political organization was to survive the crisis of feudalism after 1350 and was 'available', as we have seen, alongside other political entities as Europe began to construct the modern world after 1500.

It is the existence of these law states before the modern world-system was in existence that allows some modern states to claim a continuity back to the medieval period. But this is misleading since only part of the essential nature of the modern state was created before 1500. Why were there no state institutions for the conduct of foreign affairs? The answer is simple: the concept of foreign affairs had no meaning in the chaotic political geography of the times. In any case, wars and dynastic marriages were family matters requiring the creation of no specialist arm of the state. This situation was slow to change, even during the sixteenth century with the emergence of more compact state territories. For instance, the state with the most advanced apparatus at this time was France, but even here creation of separate institutions for foreign affairs was slow. During the sixteenth century, the need for dealing with foreign matters was recognized, but this was made an added responsibility of the existing state apparatus. The arrangement was that there were four secretaries of state, each responsible for the internal security of a section of France but who in addition dealt with relations with foreign countries bordering on, or closest to, each section (Strayer 1970: 103). By the seventeenth century, France and other countries had evolved a state apparatus that included institutions to deal with external as well as internal relations. Unlike in the medieval period, there now existed an inter-state system, and all states had to compete as territorial entities to survive by looking both inwards and outwards. This was a new world politics premised on territory and sovereignty that we tend to take so much for granted today. In the remainder of this section we look in more detail at the political processes operating within this inter-state system.

Territory and sovereignty

Jean Gottmann (1973: 16) has described the origins of the concept of territory. It derives from the Latin and was originally applied to a city's surrounding district over which it had jurisdiction. Its initial application was to city-states in the classical world, and it reappeared to describe the jurisdictions of medieval Italian cities. It was never applied to the whole Roman Empire or to medieval Christendom, with their universal pretensions. 'Territory' implies a division of political power. In modern usage, its application to cities has become obsolete; it is now applied to modern states. A territory is the land belonging to a ruler of state. This meaning has been traced back to 1494, approximately to the birth of the world-economy.

The modern meaning of territory is closely tied up with the legal concept of sovereignty. In fact, this is a way in which it can be distinguished from the original city-scale definition. Sovereignty implies that there is one final and absolute authority in a political community (Hinsley 1966: 26). This concept was not evolved in the classical Greek world – city territories were not sovereign. Instead, Hinsley traces the concept back to the Roman Empire and the emperor's imperium over the empire. This is a personal political domination with no explicit territorial link given the empire's universal claims. It is this concept that was passed on to medieval Europe in Roman Law and is retained in modern language when a king or queen is referred to as the sovereign of a country. But medieval Europe under feudalism was a hierarchical system of power and authority, not a territorial one. The relations of lord and subject were personal ones of protection and service and were not territorially based. It is the bringing together of territory and sovereignty that provides the legal basis of the modern inter-state system. This emerged in the century after 1494 and was finalized by the Treaty of Westphalia of 1648. This is usually interpreted as the first treaty defining modern international law. It recognized that each state was sovereign in its own territory: that is, interference in the internal affairs of a country was the first offence of international law. The result was a formal recognition of a Europe parcelled up into three hundred sovereign units. This was the original territorial basis of the modern inter-state system – the first 'world political map'.

Territory: security and opportunity

This first mosaic of sovereign territories was a direct outcome of the strife resulting from the religious wars in Europe in the wake of the Reformation and Counter-Reformation. The crucial political issue of the day was order and stability, or rather the lack of it, and the territorial state emerged as the solution to the problem of security (Herz 1957). The legal concept of sovereignty was backed up by a 'hard shell' of defences that made it relatively impenetrable to foreign armies, so that it became the ultimate unit of protection. Herz provides a technical explanation for this development: the gunpowder revolution in warfare, which made individual city ramparts obsolete. The original 'hard shell' of the walled city was replaced by the sovereign state and new defences based upon much larger resources. Such new warfare required a firm territorial basis, not the personal hierarchy of the medieval period.

Herz's explanation is a good one in that it incorporates an important dimension in the origins of modern state formation. But it is only a partial explanation. Tilly (1975) introduces other factors in the 'survival' of these states and the inter-state system. Security provides a stability in which a territory's resources can be mobilized more completely. The territorial state is associated with the rise of absolute monarchs in Europe with their centralized bureaucracies, taxation and large armies. In a world-systems perspective, however, we need to go beyond these 'political' factors. We follow Gottmann (1973) in identifying two basic functions of the territorial state: security and opportunity. The former relates to the origins of the inter-state system, the latter to the emerging world market.

The rise of a world-economy provided different opportunities for entrepreneurs in different locations. In the early world-economy described by Wallerstein (1974a, 1980a), the major groups contesting for advantage in the new world market were the agricultural landed interests on the one hand and the urban merchants on the other. According to Dennis Smith (1978), this conflict is directly related to the rise of the modern state, with the landed aristocracy giving up its medieval rights in return for the sovereign's support against the new rising urban class. But this initial alliance between landed interests and the new managers of the state apparatus soon had to give way to a more flexible 'politics'. In a competitive state system, security requires more than recognition of sovereignty. It requires keeping up with neighbouring states in economic terms. Hence the emergence of mercantilism, which we have discussed briefly in previous chapters. Mercantilism was simply the transfer of the commercial policies of the trading city to the territorial state (Isaacs 1948: 47–8). The scale of territorial restrictions on trade was enhanced to become a major arm of state making.

The rise of a mercantilist world in the seventeenth century relates directly to Dutch hegemony. The Dutch state that emerged from the rebellion against the Habsburgs in the late sixteenth century was a collection of trading cities with a landward territorial buffer against attack. It was an unusual state of its time, therefore, because it was largely run by merchants for merchants. In short, it ruthlessly employed economic measures to enhance wealth accrual in its territory. In contemporary parlance, it pursued policies of economic development. It was the first territorial state to do so. As such, it offered an alternative *raison d'état* focusing on economics rather than the traditional *raison d'état* emphasizing politics, war and the glory of the king (Boogman 1978). The success of the Dutch state meant that the modern world-system was consolidated as a world-economy when other states saw the necessity for an economic policy that was more than large-scale estate management. The result of this retaliation against Dutch hegemony was mercantilism, as Table 3.1 in Chapter 3 shows (Wilson 1958).

The failure of the territorial state?

In the 1990s a new term entered the geopolitical lexicon: 'failed states'. The term was used to denote states that lacked the infrastructural capacity either to constrain and guide social life or to utilize the potential assets within the state. In other words, the state could not maintain social order and economic production. They were identified by the US government and scholars and journalists as a security threat. The 'threat' stemmed from their inability to monopolize violence within their borders and their inability to exploit resources for export. The states were located in the periphery of the world-economy, and so were unable to fulfil their function as zones of cheap production. Instead, there was, according to journalist Robert Kaplan, a 'coming anarchy' (1994). The 'threat' of the states was soon translated into the sphere of national security as the term was morphed into the classification of 'rogue states'. A rogue state was identified as one that did not follow the norms of international security, but the label was attached to those countries that were challenging US power (Klare 1995). After the terrorist attacks of 11 September 2001, failed states were identified as a new form of threat, the 'terrorist haven'.

Historical understandings of sovereignty, territory, and international norms were the foundation for the rhetoric behind 'failed' and 'rogue states'. The Westphalian system was built upon a dominant understanding of security: internal control of social groups and a regulated form of international diplomacy and conflict. 'Failed states' force us to question whether the imposition of the Westphalian system on the periphery of the world-economy is still generally possible. 'Rogue states' force questions about the legitimacy of the norms of behaviour that are promoted by the United States and United Nations.

Mercantilism was based on the premise that each state had to grab as much of the world market as it could by building its industry and commerce at the expense of other states. The power of the state ultimately depended on the success of its mercantilism. The exact nature of different states' policies in the world market reflected the balance of power between the landed and merchant interests. The former succeeded overwhelmingly in eastern Europe to produce its peripheralization, leaving the Dutch to dominate the Baltic trade; in the rest of Europe the balance varied, with anti-Dutch merchants generally most successful in England, although France developed very strong mercantilist policies for a short period under Colbert (Colbertism) after 1661 (see Table 3.1). In all cases, this new concern for economic policy above the scale of cities was the product of the territorial state and the competitive inter-state system. Security and order, opportunity and mercantilism were all premised on the territorial state.

Sovereignty as international capacity

Territory is the platform for engaging in international relations; sovereignty provides the legitimization.

Quite simply: 'Sovereignty is the ground rule of inter-state relations in that it identifies the territorial entities who are eligible to participate in the game' (James 1984: 2). Hence not all territories are sovereign states.

Prior to the twentieth century, when there were still regions outside the world-economy, political entities of the external arena were not recognized as having any political rights. The Iroquois in North America, the Zulus in southern Africa and the Marathas of central India were all equally unrecognized as legitimate actors in the inter-state system. This had the effect of making their territories available for incorporation into expanding sovereign states. Small and Singer (1982) call the resulting wars 'extra-systemic'. Such formal imperialism was therefore a legitimate activity in international law since it violated no recognized sovereignties.

Today, as in the past, it is not possible to become sovereign just by declaring yourself thus. Sovereignty is never a matter for a single state; it is an inter-state arrangement because sovereignty can exist only for 'states who reciprocally recognise each other's legitimate existence within the framework and norms of the inter-state system' (Wallerstein 1984a: 175).

Hence the Bantustans declared independent by South Africa as part of the apartheid policy were recognized by no other states as sovereign and were therefore never part of the inter-state system. Similarly, the republic set up in the northern half of Cyprus following the Turkish invasion of 1974 has obtained no recognition beyond Turkey. Since 1945, recognition of sovereignty has usually been confirmed by acceptance into membership of the United Nations. Hence the very first task of the new post-colonial states of Africa and Asia was to apply to join the UN to prove their entry on to the world stage. This process of recognition has been repeated by the new states formed from the break-up of the Soviet Union and Yugoslavia. In short, sovereignty gives territories an international capacity in the world-economy.

James (1984) points out that territorial sovereignty is a feature of the modern state system that distinguishes it from previous political systems. For Wallerstein (1984a: 33), this is vital:

> It is the fact that states in the capitalist world-economy exist within a framework of our inter-state system that is the *differentia specifica* of the modern state, distinguishing it from other bureaucratic polities.

Hence the historical continuities that are sometimes traced between modern states and medieval polities (Portugal, France and England are the main examples) are misleading at best and confusing at worst: as we have seen, these were originally only 'law states' with just 'internal' sovereignty. In legal terms, fourteenth-century Portugal was not a sovereign state operating in a system of sovereign states, and neither was England or France. They operated a different politics under different rules in a different world-system. Reciprocated sovereignty is found only in the capitalist world-economy.

Conflicting territorial claims

The operation of the twin principles of territory and sovereignty as the basis of international law has an important corollary: states have become the 'collective individuals' around which laws are framed. Hence the 'rights of states' have priority over the interests of other institutions. This is enshrined in Article 2 of the United Nations Charter, which upholds the territorial integrity of member states and outlaws intervention in their domestic affairs (Burghardt 1973: 227). This produces a further corollary: international law is conservative in nature, preserving as it does the status quo in the inter-state system. But we have already noted that stability has not been the norm for the world political map. Changes occur when political claims for territory override legal conservatism. How are they justified?

Burghardt reviews several types of political claim and concludes that just three have had any major influence on the make-up of the world political map. Ranked in order, they are effective control, territorial integrity, and historical and cultural claims.

Effective control as a criterion for accepting a state's right to a territory is used to legitimate armed conquest. In the international sphere as in national courts, 'possession is nine-tenths of the law'. For all the idealism of the United Nations and other world bodies, power politics still lies at the root of international relations. Hence nobody today disputes India's incorporation of Goa into its territory after the successful invasion of the Portuguese colony in 1962. Sovereignty is normally accepted once effective control of a territory is demonstrated. This was the principle that was applied for the partition of Africa among European powers after the Berlin Conference in 1884.

Territorial integrity can be used to challenge the right of a state that has effective control over a territory. Geographical claims can be at any scale. The most famous is the US concept of 'manifest destiny', which justified ocean-to-ocean expansion of the United States (Burghardt 1973: 236). Most claims are much more modest. Currently, the most well-known is the Spanish claim to Gibraltar. Despite British effective control and the wishes of the Gibraltarians to keep their link with Britain, the United Nations voted in 1968 for the transfer of Gibraltar to Spain. The fact that Gibraltar was part of the Iberian peninsular provided the basis for the territorial integrity claim (ibid.).

Historical and cultural claims are much more varied in nature, but they can be summarized as two main types. Historical claims relate to priority or past possession of the land. The former was reduced to nothing more than 'discovery' by European explorers in the partition of Africa. (The priority of occupation by those in the external arena was not recognized, of course – they had no sovereignty.) Murphy (1990) describes in detail the basis for claims based upon restitution of 'lost' lands. Cultural claims have usually been associated with national claims to territory under the heading of 'national self-determination'. We deal with the problems surrounding this concept in Chapter 5. One interesting example of historical priority counteracting cultural patterns is worthy of mention here, however. With few exceptions, today's independent African states have the same boundaries as the colonial territories they superseded, which took little or no account of indigenous African cultural patterns. Despite this, the boundaries of Africa drawn by European powers after 1884 have largely survived intact. This is a good illustration of the conservatism inherent in the inter-state system succeeding in blocking change in the pattern of the world political map. Generally, the new states do not support the division of another state because it would be likely to lead to questioning the integrity of their own inherited territory. Hence when Biafra attempted to secede from Nigeria in the civil war of 1969–71, it obtained little or no political support from other African states. Today, most African boundaries are older than European boundaries.

The use made by politicians of these various claims to territory has varied over time. For example, the Versailles Treaty of 1919 is usually interpreted as the apogee of national self-determination. Murphy (1990: 534) argues that since 1945 there has been what he terms 'the ascendancy of the historical justification'. This is the same principle of restitution as relates to private property. Murphy traces this equation of property and territory back to the earliest formulations of international law in the seventeenth century and argues that it has become particularly relevant since the United Nations outlawed all but 'defensive wars'. The restitution argument can be used to justify war as 'defensive', as Iraq showed in its attacks on both Iran and Kuwait. Murphy agrees that the historical justifications that territory-seeking politicians make are likely to be masks for other motives but show, nonetheless, that the historical

argument does influence the geographical pattern of such claims.

African boundaries and Iraqi aggression both exemplify the basic conclusion of this discussion: the making of the world political map has been ultimately the result of power politics. It is a map of the changing pattern of winners and losers. Territory provides a platform, sovereignty a justification, but neither is an adequate defence for a state against a successful action of power politics by a rival bent on its elimination from the world stage.

Before we leave the issue of competing territorial claims, mention must be made of the important question of sovereignty over the seas. In 1982, the United Nations produced a new Convention on the Law of the Sea, which was signed by 159 countries but not by the United States. For a discussion of the political geography of this law, reference can be made to Glassner (1986) and Blake (1987).

Boundaries and capitals

If territory and sovereignty are the two most important concepts for understanding the world political map, boundary lines and capital cities are the two artifacts that most demand our attention. For most of the history of the world-economy, states were rarely experienced by most of their subjects. Day-to-day activities continued with little or no interference by the state. Boundaries and capitals represent the two major exceptions to this rule. Boundaries and capitals are associated with the growth of two new important forms of behaviour and eventually led to two distinctive types of political landscape.

The two forms of behaviour are smuggling and diplomacy. The political map provided opportunities for entrepreneurs in their continual search for profit. Buying cheap and selling dear could now be achieved by avoiding customs duties when entering territories. Contraband was a major aspect of capitalism from the very beginning of the world-economy and was an extension of the traditional fraudulent practices of avoiding city tolls by earlier traders (Braudel 1984). The new boundaries produced a patchwork of larger markets and varying levels of taxes. It is not surprising, therefore, that smugglers have entered the folklore of border areas throughout the world-economy. Diplomats, on the other hand, are to be found in the capital cities of the new states. The inter-state system was an expensive one to survive in. The medieval European practice of moving government with the personage of the king had to give way to a permanent location of government business. Furthermore, the leading states of the system needed access to information on their rivals' activities and wanted, if possible, to provide input into their decision making. Diplomacy was born as governments despatched permanent representatives to the capital cities of their rivals.

In time, both boundaries and capitals came to be the two locations where the state could be seen to impinge directly on the landscape. Border landscapes with customs houses and associated controls, and varieties of defensive structures, have become distinctive locations in the modern world. Similarly, capital cities have come to represent their states symbolically with a variety of distinctive grand architectures. In boundaries and capitals, we have the two most explicit products of the inter-state system.

We cannot deal with the details of the variety of borders and capitals that have been produced in the making of the world political map. Rather, we describe typologies of this detail as they relate to the workings of the world-economy.

Frontiers and boundaries

Frontiers and boundaries have probably been the most popular topic in political geography. However, as long ago as 1963, Pounds (1963: 93–4) noted a decline of interest in this subject. This reflects the lessening of boundary disputes in the areas where political geography was largely practised – Europe and North America. This contrasts with the first half of the twentieth century in Europe, when boundary issues were central to international politics. Furthermore, many of the early geographers of the twentieth century (for example, Sir Thomas Holdrich) were themselves boundary drawers and surveyors in the imperial division of the periphery. Hence concern for boundaries has waxed and waned with the changing interests of the core countries. Of course, boundary issues continue to be a vital ingredient of politics

beyond Europe and North America. Good reviews of the early work on boundaries are available in Jones (1959), Minghi (1963) and Prescott (1965). Here we reinterpret some of this vast quantity of material in world-systems terms.

The usual starting point in this subject area is to distinguish frontiers from boundaries. This is necessary since the terms are commonly used interchangeably. Kristof (1959) uses the etymology of each term to derive their essential difference. Frontier comes from the notion of 'in front' as the 'spearhead of civilization'. Boundary comes from 'bounds', implying territorial limits. Frontier is therefore outward-oriented, and boundary is inward-oriented. Whereas a boundary is a definite line of separation, a frontier is a zone of contact.

These definitions fit neatly into our world-systems framework. A frontier zone is the area between two social systems or entities. In the case of world-empires, this can be between other world-empires or their juxtaposition with outside mini-systems. Classic cases are the frontiers of China and Rome. Although in each case they built walls between their 'civilization' and barbarians, the walls were part of a wider frontier zone. In Roman Britain, for instance, Hadrian's Wall was just part of the fortifications of the highland military zone or frontier that separated the civilian south and east from the non-Roman north and west. With the rise of the world-economy, a frontier emerged between this system and the systems it was supplanting. The history of imperialism is about pushing forward the frontiers of this new world-system. It produced the 'classic' frontier in the American west, but also other similar frontiers in Australia, southern Africa, north Africa, north-west India and Asiatic Russia. The frontier ended with the closing of the world-system at the beginning of the twentieth century. We now live in a world with one system, so there are no longer any frontiers – they are now phenomena of history.

Frontiers everywhere have been replaced by boundaries, which are a necessary component of the sovereignty of territories. Sovereignty must be bounded: a world of sovereign states is a world divided by boundaries. Boundaries are therefore an essential element of the modern world-economy. But the process of boundary making is very different in the various sections of the world-economy. Jones (1959) identifies five types of boundary concept: natural, national, contractual, geometrical and power-political. These categories are not mutually exclusive; from our perspective, for instance, we would identify all boundaries as reflecting the power politics of their respective producers. Nevertheless, these are useful concepts that, as Jones is able to show, reflect the different ideas of the state in the evolving world-economy. The idea of 'natural' boundaries is a product of the strength of the French state in eighteenth-century Europe and its use of the new rationalist philosophy to claim a larger 'natural' territory (Pounds 1951, 1954). In contrast, the idea of 'national' boundaries is the Germanic reaction to French expansionist ideas. We consider this reaction further in Chapter 5. These two ideas are rationalizations of particular power-political positions in the core and semi-periphery of the world-economy. In the periphery, also, two types of boundary emerged. In non-competitive arenas in the nineteenth century, such as India and Indo-China, the boundaries reflect the expansion of one core state at the expense of weak pre-capitalist social formations. This is where frontiers are extended and then converted to boundaries. The limits are finally achieved when two powers begin to approach one another's peripheral territory. This may lead to the formation of a buffer state in the periphery, as in the cases of Afghanistan between Russia and British India and of Thailand between French Indo-China and British India. In competitive arenas, the boundaries are usually far more arbitrary as they reflect contractual arrangements between competitors. It is in these areas that 'clear' international boundaries are necessary to prevent disputes. Hence such boundaries commonly follow physical features such as rivers or else are simply geometric lines, usually of longitude or latitude. Examples of such 'contractual' boundaries are the United States' western boundaries to north and south along the 49th parallel and the Rio Grande, respectively. The most competitive arena of all, Africa in the late nineteenth century, has the greatest number of 'contractual international boundaries'. Here the concepts of 'natural' or 'national' boundaries had no relevance

as ethnic groups and river basins were divided up in complete contrast to the boundary processes then evolving in the core. Once again, we find contrasting processes in core and periphery, and such contrast is the hallmark of the world-systems approach.

Capital cities as control centres

Many territorial states can trace their origins back to what Pounds and Ball (1964) call their specific 'core area'; for instance, the 'home counties' in the southeast corner of England for Britain and the Isle de France for France. One of the features of core areas is that they usually have the capital city of the state located within them. Paris and London are obvious examples of this. This has led some political geographers to identify 'natural' and 'artificial' capital cities, with the former represented by those in core areas. As Spate (1942) pointed out long ago, any such distinction is to misunderstand the nature of politics in our society. London is no more 'natural' than Canberra: they are both the result of political decisions, albeit over very different time horizons. Spate rightly dismisses this old dichotomy but retreats into case studies as his own approach to capital cities. Whereas case studies are invaluable for understanding the nature of capital cities, they are not sufficient for a full appreciation of the role of these localities in the territorial state. Capital cities are, after all, the control centre of the territory, the focus of political decision making, the symbolic centre of the state and often very much more. As well as the complexity that Spate identifies, which leads him to emphasize their differences, there are important similarities, which we draw upon here.

Many European observers have remarked that Washington, DC, is an unusual capital in that it is not one of the largest cities in its country. Lowenthal (1958) even calls it an 'anti-capital', reflecting as it does initial American 'anti-metropolitan' revolutionary politics. Henrikson (1983) has reviewed this literature and concludes, not surprisingly, that it manifests a Eurocentric bias. Like Spate before him, Henrikson can find no justification for treating Paris and London as 'ideal models' that the rest of the world should follow. Instead, Henrikson identifies two models of the capital city: the European concept of the capital as the opinion-forming centre of the state, dominant in political, cultural and economic spheres; and the American concept of a responsive centre that specializes in politics. Fifer (1981), for instance, refers to Washington, DC, as 'a company town'. Henrikson is more poetic – 'a small, cozy town, global in scope'. However, this is no longer the case. The growth of the US government and its attraction of multiple professional services has resulted in Washington, DC, developing into one of the country's major metropolitan areas, a 'global metropolis' according to Abbott (1999). Nevertheless the city still lags far behind New York, Chicago and Los Angeles in importance and there are no other specialized political capital cities (not even Canberra, Ottawa or Brasilia) that have grown as Washington, DC, has done.

These two concepts do not cover all cases, however. In a famous paper, Mark Jefferson (1939) described a 'law of the primate city' in which he propounded the 'law' that a country's capital city is always 'disproportionately large'. This law 'fits' the European concept but is obviously at variance with the American concept. The reason why this 'law' is still quoted is because it fits so many countries in the periphery. In most Latin American, African and Asian states the capital city is truly primate – Buenos Aires, Lima, Dar es Salaam, Dakar, Jakarta and Manila, to name just two from each continent. This should not be read as meaning that they employ a 'European concept' in the definition of their capitals, since their position is very different.

We can reconcile these problems of definition and add to our analysis of capital cities by employing the world-systems approach. There are three types of capital city: one the result of core processes, one the result of peripheral processes and a third reflecting semi-peripheral political strategies. We describe each type in turn.

The capital cities resulting from core processes are what Henrikson (1983) terms the European concept. The rise of the classic examples of this type is part of the initial economic processes that led to Europe becoming the core of the world-economy. The mercantilist competition of the logistic wave involved the development of vastly increased political involvement centred on the historical capital city. The new bureaucracies were located there, and these cities grew rapidly to dominate their territories. They became the political control centres that were attempting to steer the emerging world-economy in directions beneficial to their particular territory.

In contrast, in the periphery new cities emerged or particular old cities grew where they were useful to the exploitative core–periphery relationship: they were central to Frank's 'development of underdevelopment' outlined in Chapter 1. Most commonly, these were ports directly linked to the core. They were the product of peripheral processes and have often been likened to 'plugholes' sucking out, as it were, the wealth of the periphery. In formal imperialism, political control was associated directly with this process, so these 'parasitic cities' became 'colonial capitals'. Many of them retained their political status following independence and they remain the most extreme examples of Jefferson's primate cities.

But not all colonial administrative centres have remained as capital cities. Some governments have recognized the imperialist basis of their inherited capital city and have relocated their 'control centre'. This is part of a conscious semi-peripheral strategy to break the core–periphery links symbolized and practised through the old capital. Often it is expressed as a 'nationalist' reaction as capitals are moved from the coast inland to the old 'core' of a pre-world-economy social formation. A good example of this can be found in the relocation of the Russian capital. In the original incorporation of Russia into the world-economy, the capital was moved from Moscow to a virgin site on the Baltic Sea, where St Petersburg was built as 'a window on the West'. After the revolution, the capital returned to Moscow as the new Soviet regime retreated from this outward stance in its attempt to break with the peripheralizing processes of the past. Other similar examples of relocation are from Istanbul to Ankara in the centre of Turkey's territory, from Karachi inland to Islamabad in Pakistan, and from Rio de Janeiro inland to Brasilia in central Brazil; currently Nigeria is moving its capital from the colonial port of Lagos inland to a new site at Abuja in central Nigeria. Such policies of capital relocation have been quite common in Africa (Best 1970; Hoyle 1979; Potts 1985). In all of these cases, we can interpret the relocation as part of a semi-peripheral strategy that is attempting to lessen peripheral processes operating in the country.

The semi-peripheral strategy tends to produce what Henrikson (1983) terms the American concept of a capital city, in effect a political 'company town'. The case of Washington, DC, is also the result of an attempt to prevent peripheralization in the creation of the United States. It represents the dominance of 'national politics' over the needs of the world-economy and has been the model for other federal capitals, some of which we have already mentioned. As a compromise between the sectional interests of North and South, its closest parallels are the cases of Ottawa (between French-speaking (Quebec) and English-speaking (Ontario) Canada) and Canberra (between the two largest cities in Australia, Sydney and Melbourne). Washington, DC, Ottawa and Canberra all represent part of a strategy to mobilize a new territory for competition in the world-economy.

In summary, therefore, we can identify three types of capital city reflecting world-economy processes: the initial core processes in Europe and the peripheral processes in Latin America, Africa and Asia, both of which generate 'primate cities'; and capital cities that have developed as part of a conscious semi-peripheral strategy and that tend to be located in past and current semi-peripheral states.

Before we leave the topic of cities and states, we can allude briefly to our future treatment of world cities in Chapter 7. The latter are cities with a major trans-state role under current conditions of globalization. The processes we have described above produce very different potentials for world city status among capital cities. The old imperial primate cities like London and Paris have emerged as leading world cities, whereas semi-peripheral strategies have left other capital cities as secondary in world city terms: the leading world cities of the United States, Canada and Australia are New York, Toronto and Sydney, respectively. Obviously, the latter cities must not be neglected in a world-systems political geography just because they are not formally centres of state power.

Dividing up the state

Capital cities may be the control centres of the world political map, but the territorial states are certainly not the equivalent of the city-states of the past. The territory between capital city and state boundary has never been controlled wholly from the centre. Quite simply, the territories of the inter-state system have been too large for such elementary central organization. Rather, the territories have had to be divided up, with authority delegated to agents of the state in the communities and regions beyond the capital.

In Europe, the states inherited local and regional divisions from their medieval predecessors. In England, for instance, the shires, or counties, were originally the areas controlled by sheriffs (shire = 'sheriffdom'). In France, the accretions of the medieval period produced a wide range of sub-state units with many different degrees of central authority. Political revolution has provided opportunities to eliminate traditional divisions and to reconstruct the structure of the state in the image of the new rulers. Typically, the new boundaries had two purposes: to provide for more rational units; and to undermine traditional loyalties. In 1789, for instance, the Abbé Sieyès drew up a completely new spatial structure for the French state – the current departments – which wiped away all the traditional provincial institutions. A series of regularly shaped spatial units of equal area were created that cut through old social patterns of life. The delineation of these departments was an exercise in spatial–social engineering to break loyalties to the old provinces. To reduce local identification further, the names of the departments avoided any reference to historical, social or economic patterns of life. Instead, the departments were named after 'neutral' physical features such as rivers and mountains. This strategy has become quite common. Poulsen (1971: 228) describes how the Yugoslav government established nine regions in 1931, 'neutrally named after river basins in order to weaken the nationalisms of the major ethnic groups'. Probably the best example of this is King Carol's reorganization of Romania in 1938 into ten completely new districts. These were specifically designed to cut across the traditional ethnic and historical provinces so as not to provide rallying points for sectionalism (Helin 1967: 492–3). Once again, these were named after rivers, mountains and seas to avoid the emergence of new regional identities. This is another example of local government units contributing to the Napoleonic ethos of a 'unified and indivisible nation-state'. Clearly, dividing up the state is not a neutral technical exercise but an essential political policy for all territorial states.

Originally, state territories were divided for administrative and defence purposes. This association continues to obtain. The 'standard regions' of England for instance, originally devised as civil defence regions in the event of invasion, remain the basis for the administrative regions of all British government departments today while retaining their original purpose. In the event of a nuclear attack on Britain that destroys communications to and from London, these regions would have become new sovereign units whose 'capitals' would be underground control centres.

With the increase in activities of the states in the nineteenth and twentieth centuries, divisions of the territories have been required for more than just administration and defence. With the state taking on additional economic and social responsibilities, for example, special policy regions have been designated. Perhaps the two most famous are those associated with the Tennessee Valley Authority in the United States and the regional policy of Britain. The economic regionalization of the state was also integral to the state planning of the former communist regimes of eastern Europe.

One of the original major pressures for the increased state activity in the social and economic spheres is to be found in the extensions to the franchise for electing governments. With the gradual moves towards 'one person, one vote' in Europe and North America, there was a concomitant need to update the electoral divisions or districts. We deal with this topic in Chapter 6. At the same time, democratizing local government produced another tier of divisions that were independent of the state's administration and defence. We deal with this topic in Chapter 7.

Policy regions, electoral districts and local government areas all share one property: they are divisions of the state's territory that do not impinge in any manner on the state's sovereignty. This is not the case with all divisions of the state. Federal divisions of the state and partitions of the state are different in kind from other divisions. The former involves a 'vertical' split in sovereignty, so that it is 'shared' between different geographical scales. The latter is a 'horizontal' or geographical split in sovereignty that produces two or more states where there was one. Federalism and partition are both central processes in the making of the modern world map and were identified as such by traditional political geographers in their concern for state integration. We conclude this section of the chapter, therefore, by concentrating on these two topics.

Federalism and partition

In terms of political integration, we can identify four levels of sovereignty. The first is the unitary state, where sovereignty is undivided. Britain and France are usually considered to be the archetypal unitary states. In Britain, for instance, sovereignty is traditionally held by 'the crown in parliament', providing for the pre-eminence of the latter. With the creation of the European Union, Britain and France are no longer such 'ideal' examples of unitary states. Second, federal states have sovereignty split between two levels of government, as we have seen. Hence in the United States the fifty states and the federal level share sovereignty. In federal states, the constitution is the enabling document that divides power between the two levels. Third, in confederal associations states are legally bound in a much looser arrangement. The key difference with federalism is probably to be found in the lack of opportunity in the latter for states to leave the union. In confederal arrangements, states give up some of their sovereignty to a supranational authority. In the European Union, for instance, the executive Commission routinely takes decisions that are binding on the member states of the Union. Nevertheless, the European Union is far from being the 'United States of Europe' that its founding fathers wished for. The most important limitation of the Union's sovereign powers has been in the field of defence – the constituent states retain the basic function of defending their territories. Finally, partition represents a situation where it is not possible to maintain sovereignty over a territory. The recent break-ups of the Soviet Union and Yugoslavia into multiple new states, in the wake of the post-Cold War geopolitical transition, are classical examples of this process.

Federalism: opportunity and diversity

A federation is, according to K. W. Robinson (1961: 3), 'the most geographically expressive of all political systems.' It is hardly surprising, therefore, that it has attracted much political geography research (Dikshit 1975; Paddison 1983; G. Smith 1995). Generally, federalism is interpreted as the most practical of Hartshorne's centripetal forces in that it has to be consciously designed to fit a particular situation of diversity. A sensitive and carefully designed constitution that is perceived as fair and evenly balanced can contribute to the viability of the state and may even become part of the state-idea, as has been the case for the United States. The problem with this application of the Hartshorne model is that the internal diversity of territories has been emphasized at the expense of the external pressures for producing federations. Historically, it has been the latter pressures that have been the major stimulus to federation. For instance, whereas France and England exemplify the development of unitary states in the early world-economy, Switzerland and the Netherlands represent pioneer experiments in federal structures. Both were defensive combinations of cantons and counties to resist larger neighbours in the era of mercantilist rivalry. Hence we can reasonably argue that we need a more balanced discussion of the internal and external factors behind federalism.

Geographers and political scientists have attempted to specify the conditions under which federalism is the chosen state structure; Paddison (1983: 105) lists four sets of such ideas. Obviously, the reasons are many and various given the many examples of federated states. But one thing does emerge. There must be a powerful group of state builders who are able to convince the members of the territorial sub-units of the benefits of union over separation. The basis of such arguments returns to our original discussion concerning territories in the world-economy. It must be shown that security and opportunity are greater in the larger territory than for separate, smaller territories. This requires an alliance between the state builders and the economic groups that will benefit from the larger territorial arrangement. This is very clearly seen in the American case, as Beard (1914) has argued and Libby (1894) has illustrated. Both Fifer (1976) and Archer and Taylor (1981) use the latter's maps of support for the American federal constitution in 1789 to show distinct differences between different parts of the new colonies. It is not a north–south sectionalism that emerges but a commercial versus frontier cleavage that dominates. In the urban areas and commercial farming areas, which were firmly linked into the world-economy, support for federation was very strong. In the more isolated areas, with their more self-sufficient economies, the advantages of federation were far less obvious. Suspicion of centralization prevailed, and these areas tended to reject the federal constitution. The popular majority who were linked into the world-economy finally prevailed, and the United States was formed on a constitutional basis for its economic groups to challenge the world-economic order using mercantilist policies through the federal government. Although the nature and balance of American federalism has changed over time, the original constitution has contributed to the survival and rise of the state and is now very much part of the American 'state-idea'.

To contrast with the successful example of the United States, we can consider the case of Colombia in the nineteenth century. In Chapter 3, we noted the competition between 'American' parties and 'European' parties in Latin America; in Colombia, this translated into Conservatives and Liberals (Delpar 1981). The latter claimed to represent a cluster of 'modern' and 'rational' ideas. Hence their preference for a federal system of government was accompanied by a secular anti-clerical outlook and support for economic laissez-faire. Centralism was equated with despotism (ibid.: 67). By the 1850s, the Liberals were able to begin to implement federal ideas, and after a civil war they created a federal constitution in 1863. Parallel with these constitutional moves, this 'European' party implemented an economic policy of laissez-faire. But federalism was to become a victim of its association with this economic strategy. As Delpar (ibid.: 71) points out, in Colombia 'private enterprise could not manage without state help.' Quite simply, by the 1880s economic openness had not delivered the goods. After another civil war, a Conservative victory in 1886 produced higher tariffs and a new constitution. A 'unitary republic' was created, with the federal states being converted into administrative departments. The demise of federalism in Colombia, therefore, contrasts with the success of the US federation after the victory of its 'American' (protectionist) party (the Republicans) in the civil war of 1861–65. What these two examples clearly show is that federation can never be considered in isolation from the other political issues that affect the success of the state in the world-economy.

Since 1945, federalism has been associated with the larger countries of the world, such as the United States, the former Soviet Union, India, Nigeria, Brazil, Canada and Australia. (China is the major exception.) It is considered the appropriate constitutional arrangement to cope with the inevitable social and economic differences that large size brings. But this is not the only reason for modern federations. After the Second World War, both Britain and France insisted that the constitution for West Germany should be federal, because this was thought to produce a weaker state that would be less of a threat in the future. This seems to be the centralization–despotism link hypothesis again.

Outside the core of the world-economy, federalism is associated with states that often have acute problems of cultural diversity. In India, for example, 1,652 'mother tongues' have been recorded by the official census. The Indian case shows how federalism has had to be modified to cope with pressures emanating from such cultural complexity. At independence, India was divided into twenty-seven states in a federal constitution that carefully separated powers between the different levels of sovereignty. The new states were combinations of pre-independence units and had no specific relation to the underlying cultural geography of India. As late as 1945, the Congress Party had called for the creation of boundaries based on language, but now that it was in power it changed its policy (Hardgrave 1974). A special commission was set up to investigate the matter and warned that defining states by language would be

a threat to national unity. 'Arbitrary' states were favoured because they provided no constitutional platform for language-based separatists. But this political strategy of the central state builders was a dangerous one, for the states also provided no basis for a popular politics that could aid in the political integration of the country. The constitution was not generally accepted as fair and balanced, since it produced a situation where nearly every cultural group had a grievance against the federal state. Change was not long in coming. In 1955, a States Reorganization Commission conceded the need for the states to match more closely the cultural geography of the country and produced a new federal structure of fourteen language-based states. This structure has been further refined, so that today India is a federation of twenty-two states that broadly reflect the cultural diversity of its territory. The original fears for the national unity of the state have been found to be largely unsubstantiated: Indian federalism has operated as a centripetal force. For a recent assessment, see Corbridge (1997), who places this centripetal force into the wider context of governability to understand why this ethnically diverse state has endured and is likely to survive into the future.

One particular feature of the Indian case is the ease with which the boundaries of the states could be altered: the 1955 reforms required only a majority vote of the federal parliament. This contrasts with federal arrangements in core states, where the constituent units of the federation cannot be so easily changed. This obviously reflects a balance of power in the Indian situation that is biased towards the centre. On some definitions of federalism, this would rule out India as a federal state – Corbridge (1997) refers to India's 'federal mythology'. But this ease of producing constituent units in a federation is shared by other peripheral states. In Nigeria, for instance, it has been developed into a strategy for economic development (Ikporukpo 1986; Dent 1995). The 'peculiarity' of Nigeria is that it inherited three administrative units on independence but is now divided into thirty states (Dent 1995: 129). And this tenfold increase does not represent the whole pressure for new state formation: Ikporukpo (1986) reports twenty-nine states in existence, with another forty-eight outstanding proposals in 1983. The key point is that each state is a 'forum for development' and each new state capital a potential economic growth centre. This seems to be a unique experiment in the use of a federal constitutional framework to try to produce an even spatial pattern of development, with every area having the opportunity to implement its own development plan. This is a new form of state-led economic development. It reminds us once again of the profound differences in circumstances between core and periphery with, in this case, a peripheral state having a history of using its constitutional territorial arrangements in a most innovative fashion.

Today, federalism is as popular as at any time in its history – one observer has even referred to 'a federal revolution sweeping the world' (G. Smith 1995: 1). This is in part the result of political reactions to globalization being expressed as local ethnic mobilizations. In such circumstances, federalism is being used as a means of managing new ethnic conflicts – Smith (ibid.) provides several useful contemporary case studies of this process. However, in the former communist world, notably the Soviet Union and Yugoslavia, federations have been the victim of political reforms that have led to state partition.

Creating new states by partition

Not all federal arrangements have been successes. In the final stages of the dismantling of their empire, the British colonial administrators tried to produce several federations by combining colonies to produce larger and possibly more viable independent states, but by and large this did not work. The West Indian, central African and east African federations collapsed, and Singapore seceded from Malaysia. In traditional political geography terms, there was no state-idea to build upon, so centrifugal forces overwhelmed the new creations. Put another way, if you remove 'rule' from the British imperial tradition of 'divide and rule', all you are left with is 'divide'. This process was to be seen in its most spectacular form in British India, where the partition of 1947 produced Pakistan and India after the loss of 1 million lives and the transfer of 12 million people.

Since 1989, with the collapse of communist rule in eastern Europe, new partitions have taken place.

Beginning with the Baltic states (Latvia, Lithuania and Estonia), the old federation of the Soviet Union has been dismantled into its constituent parts: the world political map has lost one state, and fourteen new states have been added. And this is not the only revisions that cartographers are having to implement. The federation of Yugoslavia has shed all of its units, creating another six sovereign states, and Czechoslovakia is now two states: the Czech Republic and Slovakia. The lesson of these changes is that federation must be based upon consent, not coercion. Without the former, partition will occur as soon as the opportunity arises.

The partitions of the recent past have taken advantage of the political fluidity that is a feature of any geopolitical transition. In more stable periods represented by geopolitical world orders, partitions are generally a much rarer phenomenon. This is because every state partition represents a severe threat to the status quo. It is for this reason that separatist movements usually command very little support in the international community, as we noted in the case of Biafra. In contrast, the separation of Bangladesh from Pakistan in 1971 was quickly accepted by the international community after its creation in the Pakistan civil war by Indian armed intervention.

There had always been doubts about the territorial viability of Pakistan when it consisted of two units, West and East Pakistan, separated by several thousand miles of territory of a hostile neighbour. But the key factor in international acceptance of the partition in 1971 was the failure of the old Pakistani state to accept an election result that put the reins of government in the hands of a political party from the more populous East Pakistan. This brought to a head grievances concerning the way in which the old Pakistani state had favoured West Pakistan at the expense of East Pakistan. Figure 4.1 shows how the state apparatus, both military and civilian, was firmly in the hands of just one part of the country. The state promoted centrifugal forces when it needed to develop very strong centripetal forces to survive as two separated territorial units. When partition came, therefore, it was not interpreted as the result of a typical separatist movement but as a particular and necessary correction of post-colonial boundaries:

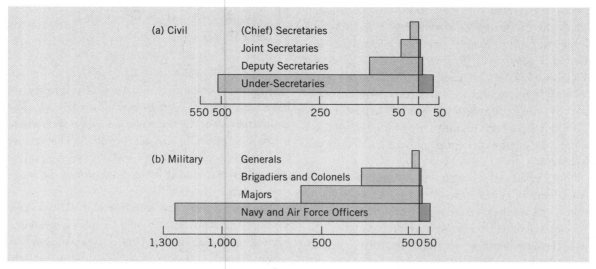

Figure 4.1 The distribution of senior state positions in Pakistan before 1971 between West Pakistan (on the left) and East Pakistan (on the right).

Bangladesh was an exception that would not affect the status quo. Within three years, the new Pakistan (formerly West Pakistan) recognized the new state of Bangladesh (formerly East Pakistan).

In political geography, Waterman (1984, 1987) has considered the processes of partition in some detail. Following Henderson and Lebow (1974), he identifies two very different processes in operation,

Partition as a result of unequal power relations

Israel is close to completing a wall around the occupied territories. The Israelis claimed that the wall was necessary to prevent suicide bomber attacks on Israeli citizens. The Palestinians claimed that the wall was a unilateral delimitation of a potential Palestinian state, and that the Hamas-declared ceasefire was the cause for the near cessation of suicide bombings. The role of the wall in delimiting the territorial limits of a future Palestinian state illustrates the role of power in the practice of partition. Israel, as a recognized sovereign state with greater financial and military resources, was able to impose its vision of the geographical route of the wall; making detours to include some Israeli settlements while literally cutting some Palestinian villages in half. The Palestinians mustered protest, but were unable to prevent the construction of the wall or even to bring about minor adjustments in its course. Politically divided and denied any military force, the Palestinians resorted to a strategy of changing the geographical scope of the conflict by taking the issue to the International Court of Justice. Despite the 2004 Court ruling that the wall was a violation of international law and that the parts already erected should be razed, the construction of the wall continued. The wall is a concrete manifestation that a future Palestinian state will be the product of partition: Palestinians will have little say in the geographical limits of the state, nor will they have control over their own borders and airspace. Partition remains a sign of enduring and complex conflicts in which a geographical 'solution' is imposed by a stronger power. Partition attempts to fix injustices and enmities into a territorial expression that corrals the diversity of an existing political entity into more homogeneous zones. However, in the case of Israel–Palestine the unilateral manner of the partition and the power imbalance are unlikely to catalyse a long-term and peaceful solution.

which are termed divided nations and partitioned states. In the former, the state has a cultural and linguistic unity before partition. Examples are Germany (1949–90), Korea, Mongolia, China and Vietnam (1955–74). These partitions were the result of outside forces and were not considered permanent by their populations. Hence Vietnam and Germany have been reunified, and in the other cases there remains the concept of one nation despite the two states.

Partitioned states, on the other hand, are usually considered to be permanently separated. Here partition is the result of internal pressures. It is a way of solving a destructive diversity within a country. The two classic examples are India–Pakistan partition from the last geopolitical transition and the moves towards a Palestine–Israel partition at the present time. The post-1989 partitions in eastern Europe are clearly of this type. Historically, the main period of success for this type of partition came at the end of the First World War, when national self-determination was accepted as a criterion for state formation, leading to the partition of the old multinational empires. Austria-Hungary, for instance, was partitioned into seven new 'nation-states' (or parts thereof). We deal with this question of nation and state in detail in Chapter 5.

Summary

In this section we have introduced the historical processes and geographical features that have constructed the modern territorial state. In particular, we have emphasized:

➤ the role of territory in grounding the power of the state;

➤ the concept of territory and its relationship to domestic politics and the inter-state system;

➤ the role of boundaries in defining the extent of state power;

➤ the role of capital cities and their relationship to the form of the state;

➤ the political practices of federalism and partition.

■ The nature of the states

Our discussion of the make-up of the world political map has been descriptive in nature. That is to say, we have been more concerned with how the map came about than with why there was a need for such a map in the first place. In this section of the chapter, we shall be more theoretical in orientation as we explore the underlying reasons for the creation of the inter-state system.

The link between this section and the last is made through the topological model of the state. Hence, beyond the usual relations between empirical material and theory, we have a simple model that tells us what a theory to understand states should encompass. In fact, we shall find that from our political geography position most theories of the state are only partial explanations of what states are. There has been a strong tendency in developing theory to concentrate upon internal relations or 'stateness' at the expense of external relations or 'inter-stateness'. In terms of our topological model, the theories look one way only. One of the greatest advantages of world-systems analysis of states is that it looks both ways.

Looking one way only: theories of the state

The notion of sovereignty assumes the existence of the state. But this is a two-way relationship. At the simplest level, the state is defined by its possession of sovereignty. This distinguishes it from all other forms of human organization. As Laski (1935: 21–2) points out, this sovereignty amounts to nothing less than supreme coercive power within a territory – the state 'gives orders to all and receives orders from none' inside its recognized boundaries. Invasion by a foreign power or internal insurgency aiming at creating a new state is a violation of a state's sovereignty. If the invasion or insurgency is not defeated, the state no longer has a monopoly of coercion in its territory and faces extinction. The partition of Poland between Germany and the Soviet Union in 1939 is an example of extinction by external violation of sovereignty.

Since 1945, there has been no such elimination of states, although this would have been Kuwait's fate had Iraqi aggression prevailed in the first Gulf War.

It is important to distinguish between state and government at the outset of this discussion. Again using Laski (1935: 23), government can be interpreted as the major agent of the state and exists to carry out the day-to-day business of the state. Governments are short-term mechanisms for administering the long-term purposes of the state. Hence every state is served by a continuous succession of governments. But governments only represent the state; they cannot replace it. A government is not a sovereign body: opposition to the government is a vital activity at the very heart of liberal democracy; opposition to the state is treason. Governments may try to define themselves as the state and hence condemn their opponents as 'traitors', but this is a very dangerous game. If this strategy fails, the state may find itself challenged within its boundaries by what McColl (1969) termed the insurgent state – one with its own core area, territory and claims to sovereignty. In this case, the fall of the government can precipitate overthrow of the state, as happened, for instance, in South Vietnam in 1975.

It has been suggested that this distinction between state and government is not of practical interest, because all state action must involve some specific government operation acting in its name (Laski 1935: 25). The important point, however, is that this distinction is a theoretical one, and it is at this level that this section is pitched. One of the major problems of much political science and political geography is that they have considered government action without understanding the wider context in which it occurs. That framework can be provided only by developing a theory of the state separate from the particular actions of particular governments. That is the purpose of this section; we return to a consideration of governments in Chapter 6.

Skinner (1978: 352–8) has described the origins of the modern concept of the state, and once again we find that a basic concept in our modern world first appears at the same time as the emergence of the world-economy itself. The word 'state' comes from the Latin *status*, and its medieval usage is related to either the 'state' or condition of a ruler or the 'state' of the realm. The idea of a public power separate from ruler and ruled that is the supreme political authority in a given territory does not occur in medieval or early modern periods. The modern concept develops from this medieval usage in the sixteenth century, first in France and then in England. Skinner argues that this is because these two countries provided early examples of the properties that make up the modern state: a centralized regime based on a bureaucracy operating within well-established boundaries. By the end of the sixteenth century, Skinner claims, the modern concept of the state is well established in these two countries, and modern political analysis focusing on the nature of the state can be said to begin at this time.

We are indebted to Clark and Dear (1984) for their work, which remains the most comprehensive discussion of modern theories of the state in geography. They identify eighteen theories, which they reduce to two modes of analysis: theories of the state in capitalism and theories of the capitalist state. The former mode treats capitalism as a given and concentrates on the functions of the state. These are generally described as liberal or conservative theories of the state. Identification of the 'capitalist state' in the second mode of analysis indicates that the economic relationships of capitalism are brought into the political analysis. These are Marxist theories of the state. Politically, the distinction is that in the former the state is seen as a neutral entity above society, whereas for Marxists the state is a very partisan instrument within the workings of society.

Theories of the neutral state in capitalism

In the core states of the world, we all experience the state as a supplier of public goods (Clark and Dear 1984: 18–19). Tietz (1968) provides a brief catalogue of our uses of the state from birth to death, showing how much our modern way of life is dependent on public goods – schools, hospitals, police, fire prevention, waste disposal, postal services, and so forth. Theories of the state as a supplier of public goods are concerned with the efficiency of this provision. Political geographers, for instance, have been concerned to produce models of the most desirable service areas for the provision of public goods (Massam 1975).

Equally important to our modern way of life is the state as a regulator and facilitator. This is the second theory that Clark and Dear (1984: 19) identify whereby the state operates macroeconomic and other policies to support the economy within its territory. Johnston (1982a: 13) adds the role of the state in producing the physical infrastructure for the smooth running of the economy – roads, railways, power transmission lines, and so on. Political geographers have been concerned with these processes as they relate to spatial integration.

In all of these functions, the neutrality of the state is broadly assumed; in the theory of the state as arbiter this neutrality is explicitly spelt out (Johnston 1982a: 12, 13; Clark and Dear 1984: 21). In this thesis, the state is deemed to be above the endemic conflicts of society and therefore can act as a non-partisan arbiter in adjudicating disputes. This theory underlies most political activity within the core states of the world-economy. Political parties in countries with competitive elections, for instance, normally assume the state to be a neutral institution. After all, the prize of winning control of government would be a rather hollow one if the state hindered or prevented implementation of policies promised in an election campaign. Coates (1975: 142) provides a nice illustration of this assumption for the British Labour Party. Harold Wilson, the Labour prime minister of the 1960s and 1970s, used the analogy of the state as a car. Whoever won an election got the ignition keys and was therefore able to sit in the driving seat and turn the car either left or right. In their most sophisticated form, these ideas form a pluralist theory of the state, which we discuss further in the next section.

One reason why parties of the left have accepted theories of the state in capitalism is because they have been relatively successful in putting many of their policies into practice. Clark and Dear (1984: 20) call this the state as social engineer. This concerns the role of the state in ensuring some degree of distributional justice within its territory. The product is usually termed the welfare state and, as we have seen, the process of its creation can be social imperialism from the top as well as social pressure from below. Welfare

geography, concerned with spatial inequalities, has been a major growth area of modern political geography (Cox 1979).

Johnston (1982a: 12) adds a category of the state as protector, which combines Clark and Dear's public goods and social engineering roles but with more explicit reference to the 'police function'. This is of interest because theories of the state in capitalism generally ignore or underplay the coercive function of the state. However, as we have previously noted, it is precisely this function that distinguishes the state from other social institutions. In the states of the periphery, this particular function is far more obvious as the state is commonly involved in activities of repression. This highlights an important limitation of these theories: their implicit bias towards the activities of states in the core. Furthermore, this is symptomatic of a more fundamental criticism. Theories of the state in capitalism are relatively superficial descriptions of state functions. It is not that these functions are not real but rather that their enumeration does not advance our understanding very far. In Marxist terms, they remain at the level of appearances without engaging with the social reality underlying those appearances (Clark and Dear 1984: 18). All this boils down to the fact that to see the state as merely neutral is naive. The state as a locus of power cannot be outside politics, whether we view it as acting for good or for ill.

Theories of the capitalist state

Developing a more 'political' theory of the state has not been easy. Detailed empirical studies such as we have just presented may undermine the neutrality assumption but they do not immediately provide for its replacement. At worst, they can lead to a myriad of conspiracy theories, as with American 'patriot' groups, which represent social paranoia rather than serious social theory. For the latter, social scientists over the past few decades have turned to Marxist theories, which provide a conflict model of society in which the state is fundamentally implicated. In contradiction to the neutral state and particular conspiracy speculation, this school of thought has provided theories of the capitalist state. However, as we shall see, these are not without their problems.

It is well known that Marx himself never developed a theory of the state. It was a project that he set himself but never completed. Hence Marxist theories of the state are products of his many followers, and this has inevitably led to alternative interpretations of what such a theory should say: there is not a Marxist theory of the state but many Marxist theories of the state. In this discussion, we consider both the differences and similarities of a small number of these theories that have entered political geography. We begin by briefly describing the original orthodox Marxist position before moving on to review three debates surrounding recent attempts by Marxists to update their theory. Finally, we introduce one set of ideas that is particularly related to our world-systems political geography perspective.

There is much raw material in Marx's voluminous political writings for his followers to use to construct theories of the state. Two ideas have dominated Marxist political thinking. The first, from *The Manifesto of the Communist Party* of 1848, dismissed the state as nothing more than 'a committee for managing the common affairs of the whole bourgeoisie'. The second, which can be found in several writings, consists of a 'base–superstructure model' of society where this engineering analogy is used to depict a foundation of economic relationships upon which the ideological and political superstructure is constructed. These two simple ideas are difficult to accommodate to the complexities of the modern state. If they are taken at face value, they lead to a reduction of all politics and the state to a mere reflection of economic forces. Such crude reductionism is termed 'economism'. Hunt (1980: 10) provides a good example of this type of thinking in his criticism of Lenin's contribution to state theory. Lenin proposed a strict relationship between stages of economic development and types of state. Hence parliamentary democracy is the form of state for competitive capitalism, the bureaucratic–military state emerges with monopoly capitalism, and socialism replaces these with 'the dictatorship of the proletariat'. In modern orthodox Marxist theory, economics and politics are fused in another stage known as state monopoly capitalism (Jessop 1982). By directly equating the form of the state with the economic base, Lenin provides a simple explanation, which we now identify as economism. Of course, this model is every bit as developmentalist as 'organic' theories of the state or Rostow's stages of economic growth (Wallerstein 1974b). Most modern Marxist writers distance themselves from such simplistic analyses (Hunt 1980). Nevertheless, this is the heritage for modern Marxist theories of the state; the key issue for contemporary social scientists has been how to challenge economism without losing the essential material basis in Marx's thinking.

In the social science of the English-speaking world, recent concern for Marxist theories of the state can be said to begin with Ralph Miliband's *The State in Capitalist Society* (1969). Miliband's intention was to counter the prevailing pluralist theory of the state in political science, which was the most sophisticated justification for treating the state as neutral. Since modern society consists of many overlapping interests – labour, farmers, business, home owners, consumers, and so on – no one group is ever able to dominate society. In this situation, the role of the state is to act as an umpire adjudicating between competing interests. The balance of interests served will vary as governments change, but the state will remain pluralist in nature and able to respond to a wide range of interests. This is quite obviously the opposite of the Marxist class theory, which in these terms is strictly a non-pluralist account of the state.

The pluralist theory's idea of neutrality is embedded in the practice of electoral politics. As we shall see in Chapter 6, political parties have the role of bringing together coalitions of interests to win an election and thus a term of government. The latter provides for access to the levers of power, the state apparatus. If these were not above the day-to-day politics of the competing parties, then electoral victory would be meaningless. Hence the neutrality assumption has been vital to social democratic strategies in west European states, since it assumes that if a Labour or Social Democratic party wins control of the government in an election, it can use the state to carry out its 'socialist' reforms. The state is 'up for grabs', as it were, in electoral politics. A corollary of this assumption is that the political sphere is treated as separated from the economic sphere. This is the basis of the whole notion of a 'political science', so no problem is envisaged in the relationships of politics and economics. They are separate, although often related, processes that can be adequately studied as autonomous systems. We met this argument in discussing inter-state politics in Chapter 2, and it is repeated at all scales of analysis. From the perspective of this study, pluralist theories of the state represent an important example of the poverty of disciplines.

Economism and pluralist theory can be seen as opposite ends of a scale measuring the autonomy of the political from the economic. At the economism end there is no autonomy. At the political science end there is absolute autonomy. In between, we can identify different degrees of 'relative autonomy' where politics is not determined by economic processes but is not independent of them either. Most modern Marxist analyses have located themselves in the relative autonomy sector of this scale, and it is from just such a position that Miliband (1969) launched his attack on the pluralist theory.

Democratic elections provide the most telling evidence in favour of the pluralist theory of the state. In 1848, the state may have looked very much like a one-class institution, but with franchise reforms the modern liberal democratic state looks to be much closer to the pluralist model. We deal with liberal democracy in some detail in the next chapter, and here we show only how Miliband attempts to uncover the class nature of the apparently pluralist state. Miliband's method is to marshal data on the social, political and economic elites of the modern state and to show that they all come from the same class background. By showing the many interlinkages in terms of family, education and general economic interests, he is able to paint a picture of a single dominant class in which pluralist competition is a myth. This dominant class is able to manipulate the state apparatus irrespective of which party is in government. This has come to be known as the instrumentalist theory of the state. It restates the class basis of the state and completely undermines the neutral pluralist view.

Miliband's thesis stimulated a debate with Poulantzas (1969) upon how to construct a Marxist theory of the state. Poulantzas argued that Miliband's empirical approach, while useful and interesting,

was seriously flawed because it involved analysing the state in terms laid down by non-Marxist theory. It is not that empirical analysis is wrong in itself, but the way in which it is integrated with theory is important. Miliband's study reduces to a description of particular roles within the state apparatus and investigation of links between people carrying out these different roles. The inevitable result is an emphasis on interpersonal relations that is reminiscent of the original pluralist thinking. For Poulantzas, the class nature of the state is not a matter of empirical verification or falsification. A state is not capitalist in nature because a dominant class with capitalist links manipulates the state for capitalist ends. The state is capitalist because it operates within a capitalist mode of production. This sets constraints on the range of action that is possible, so the state has no option but to conform to the needs of capital. The introduction of liberal democracy does not change this fundamental property. This is a logical argument for which empirical proof is unnecessary. Such a structuralist position need not reduce to economism; Poulantzas shares Miliband's relative autonomy assumption, with the state dependent on the economic base only 'in the last instance'.

The Miliband–Poulantzas debate was introduced into geography by Dear and Clark (1978). At about the same time, however, the work of West German Marxists was just becoming known (Holloway and Picciotto 1978), and this completely cut across the arguments that separated Miliband and Poulantzas. This derivation school rejected the notion of relative autonomy and so undermined the arguments of both protagonists. By assuming relative autonomy, any theory implicitly accepts the separation of the economic from the political. By doing this, the proponents of such theories cut politics off from the main motor of change in capitalist society, the accumulation process. For the derivationists, therefore, the key question is not about the degree of separation between the political and economic but why it appears this way in capitalist society. The Marxist political analyses, developed by Miliband and Poulantzas in their different ways, cannot adequately answer this question, because they have forsaken the holism of political economy. For these West German theorists the solution was simple: a Marxist theory of the state must be derived from the mechanisms and concepts described by Marx in his basic theoretical work, *Das Kapital*. To the need for competition in the economic sphere they add the need for cooperation in the political sphere. The state becomes necessary to counter the self-destructive processes of unbridled economic competition. This may even involve political processes that appear to be anti-capitalist. The evolution of the welfare state, for instance, involves ensuring the reproduction of a skilled and healthy labour force for the long-term interests of capital. In this role of coordinating capital and reproducing labour, the state may appear neutral and above the political conflict between capital and labour. This is why non-Marxist theories of the state appear so reasonable and are so widely and easily accepted.

Following Gold et al. (1975), Dear and Clark (1978) introduced a further Marxist approach to the state, which they term 'ideological'. All Marxist theories incorporate some notion of ideology in their formulation, but Gold et al. refer in this context to theories that specifically emphasize the state as a form of mystification whereby class conflicts are hidden behind a national consensus. This is very close to the concept of the state as the scale of ideology that we introduced in Chapter 1. In Marxist literature it is most closely associated with the work of Gramsci and his followers (Jessop 1982). Gramsci is most well known today for his concept of hegemony. This derives from Marx's original argument that the ruling ideas in a society are the ideas of the ruling class. In Gramsci's work, hegemony is the political, intellectual and moral leadership of the dominant class, which results in the dominated class actively consenting to their own domination (Jessop 1982: 17). Hence alongside the coercive state apparatus (police, army, judiciary, and so forth) there is the ideological state apparatus (education, mass media, popular entertainment) through which consent is generated. Notice that these ideological functions need not be carried out by public agencies – in this theory, the state is much more than just the public sector. Historically, the vital battle for state sovereignty involved subjugation of other authority within the state's territory, both local magnates and the

universal ideas of the Church. In the latter case, the issue centred on education and the state's attempts to 'nationalize' its population by converting religious education into a state ideological apparatus. The successful combination of coercion and hegemony will produce an 'integral state'. Here we have a parallel with the territorial integration theory and the concept of state idea and iconography. Gramsci's notion of hegemony, however, is much more pervasive and directly derives from the class basis of the state. In this argument, Marx's original 'committee' assertion of 1848 remains broadly true: the difference between then and now merely relates to the changing relative balance between coercive and ideological means of control.

It is difficult to summarize such a vigorous and expanding field of inquiry as Marxist theories of the state in just a few pages (Short 1982: 109). This is particularly a problem when we are critical of the tradition of thought, as here. We are not able to do justice to the ideas of even the few theorists we have mentioned. Miliband's (1969) original study includes an interesting discussion of dominant class ideology, for instance, and we have only dealt with Poulantzas in his response to Miliband, whereas Jessop (1982) considers his work to be the most impressive of the recent spate of new theory. For key updates on developments in this field the reader is referred to Jessop (2002); for coverage of a range of political geographies of the state, see Brenner et al. (2003). Among the latter political geographies, feminist approaches to the state are important, and we continue this review of state ideas by focusing on this work.

Feminist understandings of the state

The feminist approach to the state began with the identification of the state as the public sphere, a political arena that was dominated by men. The key ideology was not capitalism but patriarchy: the 'system of social structures and practices through which men dominate, oppress and exploit women' (Johnston et al. 2000). The feminist empirical project of 'counting women' in positions of power and noting their absence was soon replaced by more sophisticated theoretical debates. By considering patriarchy an adjective rather than a noun (Butler 1990; Johnston et al.

2000) feminist scholarship identified multiple and overlapping patriachies and, therefore, spaces of women's oppression. In Chapter 8 we shall discuss the household as a site of heteronormative gender relations. In this section we illustrate the manner in which the state is an agent of gender relations through its codifying of difference between groups and how racial and gender relations are exploited by the state to maintain power relationships (economic, political and racial).

Chouinard (2004) describes the changes in feminist approaches to the state. In some ways the debate mirrors that of the previous section, with a challenge to the idea that the state is neutral. Instead, the state was seen as playing a key, even causal, role in perpetuating patriarchal and class relations. Either states were 'embedded within and captives of particular sociospatial orders' (Chouinard 2004: 230), including patriarchy, or states were the institutional means to perpetuate men's violent sexual oppression (MacKinnon 1989; Chouinard 2004). Initial theoretical attempts to connect the patriarchal and capitalist forms and functions of the state were the foundations for contributions that emphasized the conjunctural and geographically contingent nature of the state: class and gender relations came together in different ways in different states (Chouinard 2004).

Contemporary feminist scholarship focuses on the manner in which the 'state regulates, governs, and changes lives' (Chouinard 2004: 231). In some ways, this is a discursive relationship as the state empowers some groups and marginalizes others via its use of language and rhetoric: for example, the manner in which government census categorization inscribes racial identity and heteronormative conceptions of the family and marriage. But the role of the state in empowering some groups and disempowering others is also material. Examples include welfare laws, employment laws, and laws that restrict abortion and contraception. The combination of state laws and discourse of belonging and exclusion have led feminist scholars into analysis of citizenship, a topic to which we return in Chapter 8.

Despite the space limitations above, enough material on theories of the state has been presented to illustrate their deficiency from the point of view of world-systems analysis. We started this discussion with a description of Lenin's economism. The modern theorists have countered this problem in different ways, but they have not tackled the problem of Lenin's developmentalism. Although we have brought the argument a long way forward from simple economism, all the analyses described above remain rooted at the state level. In the next section, we use some of the ideas from Marxist theories but apply them, in terms of our topological model, looking outwards as well as inwards.

Looking both ways: theory of the states

The political sphere that we deal with in this study is not the single state but the whole inter-state system. Hence we need a theory of states – of inter-stateness – where the multiplicity of states is a fundamental property of the theory (Taylor 1995). Clearly, none of the above offers this type of theory. Looking inwards to understand state–civil society relationships is vital, but it is also partial. Hence we can build upon themes elaborated above, but we will need to add the inter-state system to the argument. In fact, we will find that by looking outwards from the state our approach does clarify some of the contentious issues in state theory. In this discussion we begin, in a preliminary way, the bringing together of theories of the state and world-systems analysis. For further discussion see Taylor (1993b, 1994).

One economy, many states

The basic empirical problem confronting the Marxist theories of the state is that the same economic system (capitalism) in a territory was capable of producing very different state forms. Although the United States and Italy are both capitalist states, for instance, they exhibit very different politics. It is this variety of politics that was chosen as the subject matter of modern Marxist political analysis (Miliband 1977; Scase 1980). The break with economism was obviously necessary, and the postulate of relative autonomy of politics from economics seemed to be the way forward for these new analyses. As we have seen, the derivationists have cast severe doubt on the validity of this

Case study

Rape and state terror in Guatemala

For decades the United States has supported the government of Guatemala in its brutal campaign of maintaining a polarized social order. In late 1997 the Guatemalan government signed peace accords with the URNG, an umbrella organization of Guatemalan guerilla groups (Nolin Hanlon and Shankar 2000: 271). Shortly after, reports authored by the United Nations and the Catholic Church detailed the scale of the government abuses. General figures claim that between 1962 and 1996 over 200,000 individuals were killed or 'disappeared'; an all too common euphemism showing the power of the state in some contexts to erase individual life. The population of Guatemala is approximately 50 per cent women and 60 per cent indigenous. Of the 200,000 victims, 83 per cent were indigenous Mayas. The UN report claimed that approximately 25 per cent of the victims were women. The type of state violence suffered by the victims varied by gender. 'Men outnumbered women approximately four to one in the categories of arbitrary executions, torture, forced disappearance, detention, and the category "other". Women and men were found in equal numbers in the statistics of death by forced disappearance, while 99 per cent of sexual violations were experienced by women' (Nolin Hanlon and Shankar 2000: 275).

Nolin Hanlon and Shankar eschew a theoretically informed research and see their work as a form of activism: 'we have chosen to give space and authority to women's voices in the form of *testimonio* [testimony in the

UN and Catholic Church reports] as a way to walk with those who have the courage to speak out' (Nolin Hanlon and Shankar 2000: 265). Such *testimonio* allow us not only to witness the ability of the state to exercise violence, but also to consider its political geography by following Nolin Hanlon and Shankar's discussion of the different scales of state terror. First, a *testimonio*:

> The 15th of September of 1982 we returned with my father from the market of Rabinal . . . we were detained by the soldiers close to the detachment and they locked us up separately . . . they pulled off my clothes, all mounted, the captain first, eight more soldiers . . . the rest touched me, they treated me badly and between themselves they said to the one on top to hurry up, they told me to move and they were hitting me to make me move. Suddenly I saw that they entered with my dad, he was beaten up, they supported him between two [soldiers]. I was naked on the table, and the captain said to my father that if he didn't talk it was going to get bad. So he made the men that he had there start to rape me again . . . I don't believe that my dad was a guerilla, I didn't know what they wanted. Suddenly the captain asked for a machete and cut off my dad's penis and he put it between my legs. My father lost a lot of blood, he suffered a lot, after they took him away. They gave me my

clothing, other clothing, of perhaps another woman and they told me to go. I told my husband what had happened, he answered that the Army had power, that one couldn't protest, that if I hadn't gone to the market nothing would have happened to me. A month later they killed my husband, but deep down I felt relieved. After everything that had happened to me I no longer wanted a man at my side, but they didn't have to die in this way. That is all.

> (CEH 1999, Cap2/Vol. 3, Item 40, quoted in Nolin Hanlon and Shankar 2000: 276)

The Guatemalan Army is one branch of a particular state. The *testimonio* and the preceding statistics illustrate the horrors that result when state power is unleashed and unrestrained. The example shows the ability of the state to forcibly enter different spaces and scales; the community, the home and the body. Rape by soldiers has painfully and dramatically altered the woman's body and mind, and her ability to be a wife or sexual partner. Her family and home have been destroyed. The resignation felt by her husband illustrates that while it is easy for the state to enter other scales and spaces, resistance to the state – through exercise of the body, household practices and community activity – is harder.

Sources: CEH (1999); Nolin Hanlon and Shankar (2000).

position, but they have not replaced it with another means of accounting for the variety of politics under capitalism. World-systems analysis provides an alternative interpretation of this political variety that makes the concept of relative autonomy unnecessary. The crucial step is to return to considering states as a key institution within the world-economy. As an institution the state, any state, is available to be manoeuvred to favour some social groups – classes, peoples – over others within the world-system. In short, we replace relative autonomy by manoeuvrability.

The notion of relative autonomy is based implicitly upon the idea that both state and economy cover the same territory. In Scase (1980), for instance, the issue of the relationship between economics and politics is treated on a state-by-state basis for western Europe. When viewed in this manner, it is easy to see how the problem of relating one 'national economy' to one state polity emerges. But what if there is no such thing as an autonomous 'national economy'? Then the problem simply disappears as a theoretical issue. Instead of a one-to-one relationship there is a one-to-many situation – one world-economy and many states (Figure 4.2). Hence we do not have to appeal to a relative autonomy argument to explain the variety of political forms that states take under capitalism. Instead, there are numerous fragments of the world-economy, each related to its particular sovereign states. Since these 'fragments of capitalism' differ from one another, there is no reason to suppose that the forms that the states take should not differ

from one another. Quite simply, different fragments of capitalism are associated with different state forms. The variety of politics remains to be understood, but there is no need to resort to relative autonomy between economics and politics for explanation.

Dismissal of relative autonomy brings us into line with the state derivation position. But we soon find that there are problems in applying this theory to our framework. In its initial form, we have seen that the state is derived to overcome the anarchic consequences of a 'free' capitalism. If we translate this to a world-economy, then this theory predicts a world government to compensate for global anarchy. This is quite the opposite to the capitalist world-economy as conceived by Wallerstein. As we have seen, multiple states are necessary for manoeuvre by economic actors on the world stage. Production of a world government would therefore signal the end of capitalism as a mode of production. It is manoeuvrability that separates our position from the derivationists and forms the basis of our theory of the states.

State manoeuvrability: deriving a new instrumentalism

Our theory should derive the nature of the state from the nature of the world-economy. Hence we will replace the relative autonomy of the state as a separate entity by the manoeuvrability of states as institutions within the world-economy. These are the particular institutions of the system that wield formal power; they make the rules, and they police them.

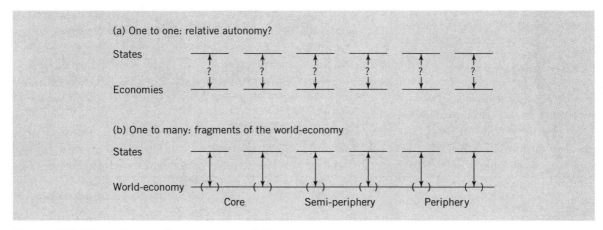

Figure 4.2 Alternative state–economy relations.

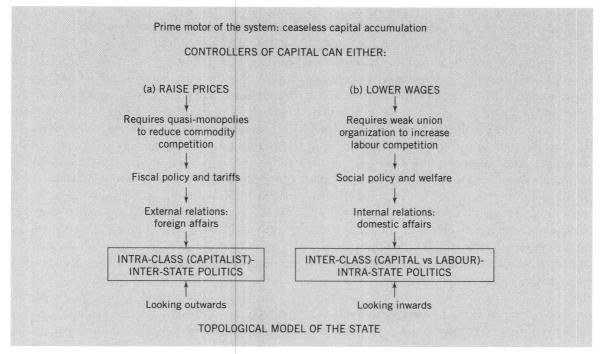

Figure 4.3 Deriving two politics.

States as the repositories of this power are, therefore, instruments of groups who are able to use the power for their own interests. Hence we can derive an instrumental theory of states from the inter-group conflicts within the capitalist world-economy (Figure 4.3).

Let us start with the prime logic of the modern world-system, ceaseless capital accumulation. From Chapter 1, we know that this is organized through class institutions, with the controllers of capital attempting to maximize their particular shares of the world surplus. This surplus is ultimately realized as profits on the world market. Controllers have two basic strategies for increasing their profits: they can either raise prices or lower costs. In political terms, the former requires some degree of pseudo-monopoly to lessen commodity competition, the latter requires the opposite policy with respect to the direct producers, an 'anti-unionism' that enhances competition for jobs by labour. In both strategies, controllers of capital have used states as important instruments in their struggle for the world surplus.

States can be used first and foremost to control flows across their borders. It is through such restrictions on

commodities and finance that pseudo-monopolies can be created and prices manipulated. As we saw in Chapter 3, the limiting cases are autarky (or closed national economy) and pure free trade (or open national economy), but, in reality, states have always pursued policies between these extremes. Whatever policy is adopted, it will favour some controllers of capital both inside and outside the state at the expense of others. For instance, in the late nineteenth century state politics was dominated by the free trade versus protection issue. Today, this conflict continues in many realms of policy, including repatriation of capital and 'hidden protectionism'. A good example of this politics in the 1990s and today is the breakdown of first, the GATT (General Agreement on Tariffs and Trade) talks, and second the WTO (World Trade Organization) negotiations on trade over the degree of freedom in the world market for agricultural commodities. Farmers in the United States, Japan and European Union are using their governments in different strategies to maintain or raise their profits. Generally, this use of the state in external relations is integral to intra-class competition within the strata that control capital.

States are used in a second crucial way to set the legal rules that govern the social relations of production within territories. It is through this means that costs of production can be kept to a minimum. Laws covering wage bargaining, corporation responsibilities for employee welfare and rights to trade union membership are all examples of how state actions can directly affect production costs. Laws restricting or banning trade union activities are common in peripheral states and constitute a key part of the informal imperialism mechanism. A good example of this politics in the 1990s is the British government's refusal to sign the Social Chapter of the 1992 European Community agreement on further economic integration. The motive is that firms producing commodities in Britain will have more power over labour and incur lower labour costs. Generally, this use of the state is integral to inter-class conflict between controllers of capital and direct producers.

Hence we can derive two politics from the two strategies of controllers of capital: raising prices produces an intra-class, inter-state politics looking outwards, and reducing costs produces an inter-class, intra-state politics looking inwards (see Figure 4.3). It is not suggested that these are the only politics operating in the world-economy, but they are the most crucial because they go right to the heart of the system, the capital accumulation process.

These two forms of politics are endemic to the capitalist world-economy, but they become particularly transparent in periods of economic restructuring. It is during Kondratieff B-phases that economic pressures induce different classes to intensify their efforts to use their state to protect their position. We shall illustrate state manoeuvrability with two examples drawn from political reactions to restructuring consequent upon the economic depression of the 1930s, with implications for the economic globalization of the 1990s.

The overthrow of the Weimar Republic in Germany in 1933 and its replacement by a new state, Adolph Hitler's Third Reich, illustrates a very raw case of state manoeuvrability. The Weimar Republic was set up in the aftermath of First World War defeat and survived an immediate revolutionary civil war. The scars of these events were never lost, since the

neutrality of the state was continually challenged. Association with the Versailles peace treaty was seen widely on the right as a betrayal of the German nation, which meant that many reactionary forces, especially in the army, hardly hid their contempt for the constitutional politicians and their parties. One result was the rise of the Nazi Party as an electoral force in the early 1930s. On the left, defeat in the civil war produced a parallel mistrust of the state, leading to the rise in electoral support for the Communist Party in the early 1930s. Tarred simultaneously as national betrayer and class oppressor, the new state needed to build and consolidate a centre politics to be successful, but a period of economic restructuring is not a time for such a middle way. Without the legitimation of neutrality, the Weimar Republic was a political disaster waiting to happen.

The often violent turmoil in Weimar Germany is a reflection of our two basic forms of politics operating in a situation of necessity: the defeated German state required critical manoeuvring within the capitalist world-economy. The intra-class politics of the economic elites centred on disputes between different sectors of the economy and their relations to the wider world. The outward-oriented industries were those engaged in new core-like activities (such as chemicals and electronics), which were competitive on the world market. On the other hand, the inward-looking industries (such as the large Prussian agricultural estates) were seeking protection to stave off competition from the world market. Thus the intra-class struggles within Germany were a classic expression of the quest of semi-peripheral states to increase the presence of core processes and prevent peripheralization within their jurisdiction. However, in this case the situation was crucially complicated by the inter-class politics that derived from the civil war. The failed revolution transmuted into bitter political struggles over wage levels and other social benefits, thus continuing to threaten capital accumulation. In this situation, the intra-class politics intersected with the inter-class politics: the interests of the economic elites were split between core-like industries willing to make concessions towards social policy and wage increases and traditional domestic industries, which wished to maintain the old social order. The result was the rise and fall of coalition governments, which ultimately produced a politically impotent state (Abraham 1986). The Nazi Party exploited the failure to manoeuvre the Weimar state adequately in these difficult times and was thus able to overthrow it. The result was a new state: a reordering of the state apparatus that eliminated the inter-class politics and created a corporate state for internal restructuring through state planning and an aggressive foreign policy to right the 'wrongs' of Versailles. Ultimately, these manoeuverings were to spell disaster for Germany and the rest of the world, but for a period in the 1930s they were popular among Germans, and certainly preferable to the weak Weimar predecessor.

We can conclude that, in general, the actual nature and outcome of the politics of state manoeuvrability are a product of both the structural imperatives of accumulation and restructuring and the manipulation of these conditions by politicians. Like the world political map, the world market is never simply a given constituent of the world-economy. It is forever being fought over and remade to favour some groups over others, and states are central to this process. Contemporary globalization, for all its anti-state biases, is no exception. We are now in a position to generalize about the variety of state forms by treating them as instruments in the struggle for the world's surplus.

State manoeuvrability and contemporary economic restructuring

The Kondratieff model of economic restructuring we presented in Chapter 1 defines economic restructuring as a recurring set of processes within the *longue durée* of the capitalist world-economy. The different periods of restructuring are given different labels to illustrate their uniqueness. The current period has attracted the label 'neo-liberalism'. We shall discuss neo-liberalism as a particular moment of the capitalist world-economy, a moment in which the role of the state has come under intense scrutiny. Neo-liberalist polices illustrate how state manoeuvrability is partially constrained by the broader dynamics of the capitalist world-economy, but that states still have the ability and desire to adopt policies that they think will increase economic growth within their borders.

The central belief of neoliberalism is that 'open, competitive, and unregulated markets, liberated from all forms of state interference, represent the optimal mechanism for economic development' (Brenner and Theodore 2002: 350). The philosophical roots of neo-liberalism can be found in the essays of Friedrich Hayek and Milton Friedman, but they gained political currency in the 1970s and 1980s, most visibly in the programmes of Thatcherism and Reaganism. However, given the world-systems identification of a global period of economic restructuring, Kondratieff IVB, it is not surprising that neo-liberalism became a global phenomenon. The World Trade Organization (WTO), International Monetary Fund (IMF) and World Bank were institutional agents aimed at disseminating neo-liberal policies across the globe. Global economic restructuring created a global context for neo-liberal policies, but the policies were enacted differently by different states (Brenner and Theodore 2002).

Marxist scholars have approached neo-liberalism as a form of economic restructuring, but without placing it in the *longue durée* of the capitalist world-economy. Moreover, Marxists will remain interested in the role of the state, and the manner in which neo-liberal policies are both the product of state policies and constraints upon the ability of the state to regulate, or intervene in, economic activity. The state is seen as regulating economic activity within its boundaries in six ways:

➤ Regulating wages – of course not totally, but by enacting employment laws and setting the institutional context within which workers and owners can negotiate.

➤ The form of intercapitalist competition – how firms interact.

➤ Forms of monetary and financial regulation – how capital investment is regulated.

➤ The state and other forms of governance – the institutional arrangements.

➤ The international dimension – defining state relations with the world-economy.

➤ The regulation of uneven development – how regional inequalities are managed.

Table 4.1 Dimensions of capitalism in Fordism and neo-liberalism

Dimension	Fordism	Neo-liberalism
Wage relations	Organized labour Collective bargaining	Local and atomized negotiation of wages Lower wages and new gender division of labour
Intercapitalist competition	State support for 'nationalized' industries National control of foreign investment	Selective state support for industries Free trade zones Global capital markets
Financial and monetary regulation	Bretton Woods agreement State control of exchange rates	Speculation-driven currency markets Offshore financial centres and tax havens International banking
Role of the state	State regulation Keynesian demand management Welfare provision Public sector employment Democratic oversight	Monetarist supply-side economics Public–private partnerships Privatization Reduced welfare support Less public accountability
International configuration and uneven development	National regulation State economic aid for struggling regions and localities	Capital mobility within trade blocs Market-mediated competition between states, regions and localities

Source: Brenner and Theodore (2002), table 1, pp. 364–6.

These dimensions are an additional step in the Marxist theory of the state, they recognize the international arena or 'the mechanisms through which national and subnational economic relations are articulated with worldwide processes of capital accumulation' (Brenner and Theodore 2002: 354). However, the form of the 'articulation' and the nature of the global economy are under-theorized. The world-systems approach interprets neo-liberalism as a general ideology within a period of global economic restructuring, but one that is adopted and utilized by different states in different ways.

In the period we identified as Kondratieff IVA, or the post-Second World War economic boom, the dominant form of state–market interaction was labelled Fordism. Summarizing a fuller table in Brenner and Theodore (2002) we can compare and contrast the form of the six dimensions of capitalism under Fordism and neo-liberalism (Table 4.1). The dominant theme is the erosion of the state's willingness and ability to intervene in economic activity. States have enacted laws weakening the power of trade unions, have withdrawn financial support of

some industries through denationalization, lost the power to define the value of currency to global financial markets, and promoted policies to make poorer regions attractive locations for the investment of globally mobile capital rather than the destination of government aid. Table 4.1 illustrates the dramatic change in the role of the state. The state has reduced its involvement in the market in some ways (such as welfare provision) while maintaining a role in others (underwriting private investment in large projects, such as the Channel Tunnel, for example).

These changes have also changed the relationships between different geographical scales of government. We discuss this in greater detail in Chapter 7, but suffice to say here that as state involvement has decreased then so has the concentration upon the national scale. In its place, businesses and non-governmental organizations have become involved in regulating and defining economic and political behaviour (known as governance) and have directed their attention to localities and the global economy. Cities have become a key scale in this new form of governance (see Chapter 7).

State intervention and non-intervention in the economy

Two stories sat next to each other in the 11 March 2006 edition of *The Economist*. One article was entitled 'Soft landing at Longbridge' and detailed the fortunes of the British car industry. Longbridge, near Birmingham, is the site for MG Rover's car factory that was closed in April 2005 'after half a century of bungled government interventions and bail-outs'. Given *The Economist*'s pro-market editorial stance it is not surprising that this story is critical of the Fordist policies of government ownership and support of the British car industry and supports its privatization. The economic restructuring that has stemmed from the plant's closure is seen as a success of neo-liberal policy: 'Only 6,271 workers lost their jobs . . . more than three-fifths have found new jobs elsewhere.' The tone of the article indicates the dynamic nature of economic restructuring and that new jobs are likely to replace the old car industry. However, the state is not invisible. Government assistance during the long decline of the factory gave the region and workers time to adjust and find new opportunities and the British government 'is spending some £250m (US$436m) between 2000 and 2008 to support suppliers and workers'. Government intervention in the car industry has waned to the point that it no longer plays the role of employer, but is there to cushion the blow of unemployment and to try to retrain workers in order to attract new investment.

The second story, 'Pick and mix' discussed a new British government system to control immigration. It is a points-system to attract immigrants with skills related to shortfalls in the UK workforce, and so exclude those who do not meet this criteria. For example, having a job offer places the candidate in tier 2 where they must amass 50 points to gain entry. Points are calculated on the basis of education (a PhD is worth 15 points) and salary (the higher the greater number of points). The article points out that this move is not a big change in policy; it has more to do with political moves to appease public sentiment about refugees and immigration.

Together the two stories illustrate the British government's role in a period of neo-liberalism. It has withdrawn its Fordist-type support of industry, and has let the British car industry fend for itself. It is actively seeking flows of foreign investment into the country. But it is very much concerned with managing another type of flow: immigration. In a neo-liberal context the state is concerned about managing movement of people across its borders, and must balance the construction of a workforce that is attractive to global investment strategies while appeasing public opinion over immigration. Here we witness the tension of the one economy/ many states structure of the capitalist world-economy. A period of economic restructuring requires the 'creative destruction' (Brenner and Theodore 2002) of dismantling existing state practices in order to attract investment that creates new forms of employment.

Sources: 'Soft landing at Longbridge', *The Economist*, 11 March 2006, p. 53. 'Pick and mix', ibid., pp. 52–3.

Paradoxically, states have attempted to increase their abilities to maximize opportunities for state manoeuvrability in the current period of economic restructuring by reducing some of their roles and functions. Although this is a global phenomenon, we should also expect a geography of actions and implications. It is to the question of the differential form of states in the capitalist world-economy that we now turn.

The variety of state forms: a space–time introduction

The form that states take depends upon the particular combination of economic, social and political forces in their territory in the past and at the present time. Hence, strictly speaking, every state form will be unique. But we can generalize in terms of the space–time structures previously discussed. Particular combinations can be aggregated into a simple 3 × 2 matrix, with time represented by A- and B-phases of growth and stagnation, and space apportioned among core, semi-periphery and periphery. We should then be able to discuss the form that states take within each of these six positions, allowing for further variation due to different histories or past positions. In all cases, the states as institutions will have to carry out two basic tasks: (1) provide the conditions for accumulation of capital; and (2) maintain legitimation of the system. O'Connor (1973) identifies these two basic functions of the state in his study of the fiscal crisis with specific references to the United States. Here we extend his ideas, based as they are on German derivationists, to states generally within our space–time categories of the world-economy. In particular, we consider the legitimation function as a varying balance between forces of coercion and consensus.

Core states are the most stable, and consensus has been far more important than coercion in maintaining control. This is because the controllers of capital in core states are strongly placed in the world market and can pass on some of their surplus to core labour. This is the process of social imperialism described in the previous chapter. It is much more than simple bribery, however, and involves the incorporation of labour into the system. These are the truly 'hegemonized' states. But the strength of this hegemony varies between A- and B-phases. In periods of growth, more resources are available to keep the system stable. These are periods of building hegemony in the wake of union successes, social security advances and finally the welfare state. With the onset of stagnation, pressures to maintain conditions for accumulation lead to cutbacks in public expenditure with resulting dangers in the loss of legitimacy. In core states, however, the class hegemony has coped remarkably well with the current B-phase.

The opposite situation obtains in peripheral states, where instability dominates. Here there is no surplus to buy off labour, which is largely left to fend for itself while being coerced into submission. The result is an 'overdeveloped superstructure' relative to the economic base, according to Alavi (1979). But this does not represent strength; it reflects the weakness of

peripheral states in the world-economy. As before in our arguments, overtly 'strong' states deceive by their appearance. Alavi argues that in post-colonial societies a military–bureaucratic group emerges to oversee the interests of three exploiting classes: (1) the metropolitan core interest, (2) the local urban industrial interest, and (3) the landowning interest. These three 'capitals' have a common interest in maintaining order as the basic condition for accumulation, but they compete in terms of the relationship of the state to the world-economy. Generally speaking, metropolitan core interests and local landowners favour an open economy, whereas local industrial interests favour protection. This is the peripheral/semi-peripheral strategy dichotomy discussed in Chapter 3 and illustrated using Frank's discussion of 'American' and 'European' parties in nineteenth-century Latin America. To the extent that urban industrial groups are able to have their interests promoted by the military–bureaucratic regime, the state becomes more like a semi-peripheral state. But such promotion can fail, and the state will sink even lower into the periphery. The case of Ghana and the failure of Nkrumah's 'African socialism' is an example of this (Osei-Kwame and Taylor 1984). It is a matter of the pattern of opportunity existing for advance in the world-economy, which varies with A- and B-phases. We can see from the current B-phase that in periods of restricted opportunities it is the periphery and its people who suffer the most. In these circumstances, the very word 'globalization' seems to be a misnomer. It is hardly global, for example, if economically much of Africa has been largely bypassed by contemporary globalization. Holm and Sorensen's (1995) phrase 'uneven globalization' neatly captures this circumstance.

Finally, we come to the most interesting examples: the semi-peripheral states. As we argued in Chapter 1, this is the dynamic sector of the world-economy where political actions by states can affect the future structure of the system. According to Chase-Dunn (1982), this is where class struggle is greatest, where the balance between coercion and consensus is most critical. State governments in this zone specialize in strategies that emphasize accumulation, as we have previously indicated. Semi-peripheral economic policy is all about 'catching up'; it is the zone of protection in particular and mercantilism in general. This will make legitimation difficult, so much of the semi-periphery is associated with dictatorial regimes. But coercion itself is an expensive form of control and will stretch resources to the extent of hindering the 'catching up'. Hence the semi-periphery is also associated with powerful consensus forces, specifically fascism and communism, and generally nationalism. More recently religion has been used as an integrative force. These are strategies for mobilizing the state's population behind the dominant classes without the greater material expenses of the social imperialism of the core. The political pressure in the semi-periphery has been dramatically illustrated in the current B-phase. In recent years, repressive regimes have collapsed throughout the semi-periphery – military dictatorships in Latin America and communist governments in eastern Europe. The subsequent 'rise of democracy' is discussed further in Chapter 6. In terms of globalization, the rise of Pacific Asia beyond Japan as a new 'globalizing arena' (Beaverstock et al. 1999) based on the rise of world cities, notably Hong Kong and Singapore, is both an example of a general semi-periphery case taking advantage of restructuring and a specific glimpse of a new possible city-led world. However, with Hong Kong reunited with its territorial state, China, only Singapore is left as a 'city-state' in the inter-state system. Hence for any contemporary assessment of world political processes the territorial state must remain the focus of attention.

Case study

Theocracy

A theocracy is a state in which the government itself is a religious authority, or is subject to religious authority or doctrines. The term has become central to the lexicon of geopolitics, as it is identified as the mirror image of western liberal democracy. As a component of the War on Terrorism, the United States has partially justified its military presence in the Middle East through the rhetoric of democratization. The invasion of Afghanistan after the terrorist attacks of 11 September 2001 was initially heralded as an attempt to capture or kill Osama bin Laden. However, it was soon portrayed as a successful liberation of the Afghani people from a strict fundamentalist theocracy imposed by the ruling Taliban. Theocracy has become synonymous with Islamic fundamentalism – though of course the definition applies to any religion.

In March 2006, President George W. Bush released the National Security Strategy of the United States, an annual revision of the country's geopolitical code. Iran was identified as the most prominent threat to the United States; an Islamic theocracy that was striving to possess the atomic bomb. However, Iranian politics is more complex than the label theocracy would imply. The Islamic Republic of Iran came into being in 1979 after the pro-western Shah was overthrown. Although the Ayatollah Ruhollah Khomeini became the dominant figure of the revolution, the opposition to the Shah consisted of a 'diverse coalition of secularists, liberals, and fundamentalists uneasily cooperating in the overthrow of the monarchy' (Takeyh 2003: 42). Since then Iranian politics has involved competition between different centres of power; the theocratic bases of the Spiritual Leader and the Guardian Council on the one hand, and the elected president, parliament and municipal councils on the other. Since the death of Ayatollah Khomeini in 1989 there has been an ebb and flow of attempted democratic secular reforms and resistance by the Islamists (Takeyh 2003; Gresh and Vidal 2004: 135–9).

Iran is illustrative of the politics of theocracies in general and the world-systems interpretation in particular. Far from being the monolithic structures portrayed in certain geopolitical representations, theocracies contain political tensions and dynamics representing the particular cultural and institutional make-up of the society.

Sources: Takeyh (2003); Gresh and Vidal (2004).

Summary

In this section we have discussed the theoretical conceptualization of the state. In particular we have noted:

➤ theories that view the state as a neutral institution;

➤ Marxist theories that view the state as an instrument of capitalism;

➤ feminist views of the state which emphasize the construction of difference;

➤ the world-systems approach and its situation of the state within the dynamics and structure of the capitalist world-economy;

➤ the role of the state within contemporary economic restructuring, commonly referred to as neo-liberalism.

■ Territorial states under conditions of globalization

The idea that the authority of states is being eroded is much older than current debates about globalization. For Deutsch (1981), the entry of the world into the 'nuclear age' in 1945 meant that states could no longer even perform their most basic function, the defence of their people. Brown, in his provocatively titled book *World without Borders* (1973), provided an 'inventory' of problems for humanity that transcend the territorial state, such as the environmental crisis, the population problem and the widening of the rich–poor gap. He highlighted the growing economic interdependence of the world and claimed that 'national sovereignty is being gradually but steadily sacrificed for affluence' (Brown 1973: 187). Like globalization today, these writers asked whether the end of the territorial state and the inter-state system was in sight? Was the demise of the state nigh?

In fact, the message of this chapter is that claims about the demise of the nation-state are overstated. If you still need convincing (and assuming you are not a corporation), try to avoid paying your taxes! But the power of states is being renegotiated. States have real-

ized that their sovereignty is under attack from the global flows of capital within the world-economy and have elected to transfer some of their power to other institutions. Thus the dilution of state sovereignty has often been the result of processes initiated by the states themselves (Sassen 1996).

One such example of this is the construction of the European Union (EU). There have been two separate views of the role of the EU. Those who wish to retain the maximum amount of state sovereignty have proposed an intergovernmentalist perspective. They hold a vision of a 'Europe of Nations'. Intergovernmentalism sees the EU as being no more than the sum of its parts, with decisions requiring the consensus of all the member states. On the other hand, the supranationalist stance envisions a decline in state sovereignty, with more decisions being made within the institutional framework of the EU. They hold a vision of a 'United States of Europe'. Initially, the member states retained an intergovernmentalist perspective, but recently more tasks and functions have been passed to the EU. The major battleground between the intergovernmentalists (such as the British position) and the supranationalists (such as the Germans) has been in the voting methods adopted at meetings of ministers from the constituent states. Intergovernmentalist states have argued for the larger countries in the EU to retain veto power over key decisions, whereas the supranationalists are more amenable to decisions being carried by a majority vote.

In practice, these problems have been solved geographically by allowing differential application of EU policies. Hence, while the single economic market covers all EU states, the single currency began in just eleven of the fifteen EU countries (Britain being the major absentee). Similarly, the Schengen Agreement, which allows the free flow of people across borders, does not cover all of the EU (again Britain has opted out). However, the very existence of these trans-state arrangements over the majority of EU states does suggest that the political momentum lies with the supranationalists. Nevertheless, policies are the product of meetings and agreements between the leaders of individual states. Sovereignty has been released willingly to the EU. To explain this apparent paradox, it must be acknowledged that the countries of Europe had already lost sovereignty to economic processes. As Nugent (1991) argues, the sovereignty of European states was already challenged by flows of capital within multinational companies and in the international financial markets. Also, superpower competition within Europe had undermined military and political independence. Jacques Delors, then president of the Commission of the European Communities, summed it up nicely when he said: 'our Community is the fruit not only of history and necessity but also of political will'. The EU was seen as a means by which states could combine their power to retain influence over global economic processes. In particular, the aim is to develop the single currency, the euro, to rival the US dollar as an international currency, thus levering some economic advantage from the United States (Martin and Schumann 1997). Similar to the viability and utility of federal states discussed earlier, the states of western Europe have acted to deepen and widen the reach of the EU in order to facilitate their interaction with the world-economy.

The contemporary world has sometimes been portrayed as consisting of a great competition in which large private corporations are in the process of undermining the traditional territorial states. One common way of expressing this competition is to rank countries and corporations together in terms of their gross domestic product (GDP) and total sales, respectively. Many years ago, for instance, Brown found that General Motors was economically larger than most states, ranking twenty-third in his combined league table. He concluded enthusiastically that 'today the sun does set on the British Empire, but not on the scores of global corporate empires' (Brown 1973: 215–16). If the territorial states are facing demise, therefore, the consensus would seem to be that large corporations will be their replacement.

The argument for the corporations winning this competition is based upon their greater geographical manoeuvrability compared with that of territorial states. With the end of formal imperialism, the state's economic location policies are internal ones. These can be regional policies to maintain the territorial integrity of the state for legitimation reasons or policies of free-trade enclaves to promote capital accumulation in the state's territory. In either case, state strategy is limited to operating its economic policies within its own boundaries. Corporations, on the other hand, can develop economic policies across several territories. In their investment decisions, they can play one country off against another. Once production starts, they can control their overall tax bills by the method of transfer pricing. This involves manipulating prices for components being transferred between plants in different countries but within the corporation: the purpose is to ensure that large profits are declared for production in low-tax states and small profits or losses declared in high-tax states. For instance, Martin and Schumann (1997: 198) report that in 1994–95 the German electronics company Siemens made a worldwide profit of DM 2.1 billion (€1 billion) and paid no German taxes. It would seem that corporate taxes are becoming a thing of the past as global capitalism becomes the greatest freeloader in history. In fact, we can take the argument even further: it is corporations that are taxing governments (sometimes called a subsidy) for the pleasure of their company within a state's territory – in 1993, the German car giant BMW was able to report a loss on its German activities and thus qualified for DM 32 million (€16 million) tax refund. This is a new economic version of the strategy of divide and rule. The ultimate victory of the corporations seems inevitable.

Of course, it is not as simple as this. There is one important characteristic that corporations do not possess and that is formal power, the right to make laws. Hence when a corporate executive suggests that eventually all financial corporations will be 'head-quartered on a ship floating mid-ocean' (Martin and Schumann 1997: 84) he is missing one vital ingredient: ultimately wealth is material, not virtual. The properties of all corporations are guaranteed in the last instance by the property laws of the states in whose territories their property is located. Thus the notion of competition between state and corporation covers only part of the relationship between them. More generally, state and corporation exist in a sort of symbiotic relationship, with each needing the other. Every state requires capital accumulation within its territory to provide the material basis of its power. Every corporation requires the legal conditions for accumulation that the state provides.

If the corporation is not replacing the state, what does the rise of the large trans-state corporation since 1945 represent? For many Marxists, it represents a new stage of capitalism: we have reached a new age of global capitalism where production transcends the barriers of the state. In world-systems analysis, it represents a secular trend of increasing concentration of capital, but it does not mark any fundamental new structure. In the capitalist world-economy, production has always transcended state boundaries: remember that the world-economy is defined in terms of a system-wide division of labour. The system-wide organization of capital has been structured in different ways at different times – charter companies in the Dutch hegemonic cycle, investment portfolios in the British cycle and corporations during US hegemony – but these are each merely alternative means to the same end: accumulation on a world scale. Thus the world-economy continues to develop in the same cyclical manner as before. Despite all the claims of new powers that the corporations are deemed to possess (see, for example, Barnett and Muller 1974), they were powerless to prevent the Kondratieff downturn of the world-economy in the 1970s. Like all other actors in the system, they had to react and adapt to the new circumstances to survive. The fact that many did so successfully to produce a new series of even larger corporations (Taylor and Thrift 1982) represents a further deepening of the same mechanisms of capital concentration. But it is not at all certain that the balance between state and capital has tilted towards the latter. For example, none of today's enterprises would seem to be powerful enough to bankrupt the two leading states, as happened to Spain and France in 1557.

The problem with the globalization thesis on the demise of the state is that it confuses state adaptations to new circumstances with the erosion of the state (Taylor 1994, 1995). The modern state in its multiplicity is not eternal and will one day disappear when the modern world-system reaches its demise. But in the meantime, the inter-state system is integral to the operation of the world-economy. Without multiple states, the economic enterprises would not have their windows of opportunity from state control that have allowed them to expand and prosper. Hence the ambiguous relationship between the territorial states and capital. To use Deutsch's (1981: 331) phrase, the states are 'both indispensable and inadequate' today and throughout the history of the world-economy. The bottom line is that without the territorial states there would be no capitalist system (Chase-Dunn 1989).

Summary

In this section we have:

➤ discussed the role and form of the state under conditions of globalization;

➤ considered the way that the state continues to play an essential role in the capitalist world-economy;

➤ noted challenges to state sovereignty and the manner in which states have responded.

Chapter summary

In this chapter we have:

➤ illustrated the historical political geography of the formation of states;

➤ highlighted the key concepts of territory and sovereignty;

➤ introduced key features of territorial states; boundaries, capital cities and frontiers;

➤ discussed theories of the state, noting the relationship between capitalism and the state;

➤ exemplified the feminist geographical understanding of the state, and the emphasis upon public and private space and the inscription of difference;

➤ introduced the world-systems approach to territorial states by using the term manoeuvrability to situate states within the context of the capitalist world-economy;

➤ situated the political geography of states within the processes of globalization to illustrate continuity and change in the function and form of the state.

The state is a key geographical scale and actor in our political geography framework, but it can be understood only with reference to its role in the capitalist world-economy and its role (as arena and tool) in political struggles over access to resources and power. The feminist approach to political geography considers how the state apparatus facilitates the maintenance of power by some groups and the marginalization of others. Within the context of globalization the essential roles of the state have been maintained, though the concept of manoeuvrability has increased its saliency.

Key glossary terms from Chapter 4

administration
apartheid
aristocracy
autarky
balance of power
boundary
bourgeoisie
capital city
capitalism
capitalist world-
   economy
centralization
centrifugal forces
centripetal forces
civil society
classes
Cold War
colony
communism
conservative
constitution

core
core area of states
democracy
derivationists
development of
   underdevelopment
diplomacy
economism
elite
empire
European Union (EU)
fascism
federation
First World War
franchise
free trade
frontier
geopolitical code
geopolitical transition
geopolitical world order
geopolitics

globalization
government
hegemony
home
households
iconography
idealism
ideology
imperialism
informal imperialism
instrumental theory of
   the state
inter-state system
Islam
judiciary
Kondratieff cycles
   (waves)
left-wing
liberal
liberal democracy
liberation movement

local government
local state
logistic wave
longue durée
mini-systems
mode of production
multi-party system
nation
national self-
   determination
nation-state
neo-liberal
opposition
partition
peoples
periphery
place
pluralist theories of the
   state
political parties
power

protectionism	semi-periphery	superpower	world-economy
relative autonomy of the state	social imperialism	Treaty of Versailles	world-empire
scope of conflict	socialism	Treaty of Westphalia	world market
Second World War	sovereignty	unitary state	world-system
sectionalism	space	United Nations	world-systems analysis
	state	world cities	

Suggested reading

Brenner, N., Jessop, B., Jones, M., and Macleod, G. (2003) *State/Space: A Reader.* **Oxford: Blackwell.** A collection of classic and contemporary articles addressing theories of the state, intra-state geographies and globalization.

Jessop, B. (2002) *The Future of the Capitalist State.* **Cambridge: Polity Press.** This book facilitates an explora-

tion of current changes in governance and the welfare state within the context of globalization.

Storey, D. (2001) *Territory: The Claiming of Space.* **Harlow: Pearson Education.** This text provides an introduction to the state, citizenship and intra-state political divisions.

Activities

1 Make a list of all the ways you and members of your household or family have experienced the influence of the state over the past week or two. Sort your experiences by identifying them with the three forms of power we identified in Chapter 1.

2 Consider the notion of state manoeuvrability we introduced in this chapter by reading a policy statement by a state leader or other politician of your choice. In what ways did the speech refer to opportunities offered or constraints imposed by the 'global economy'? What were the implications for people within the state?

3 States are grounded on their sovereign territory. Use a historical atlas to identify the changing boundaries of your own state (or another of your choosing) and delineate the power politics behind the initial modern boundaries and all subsequent changes.

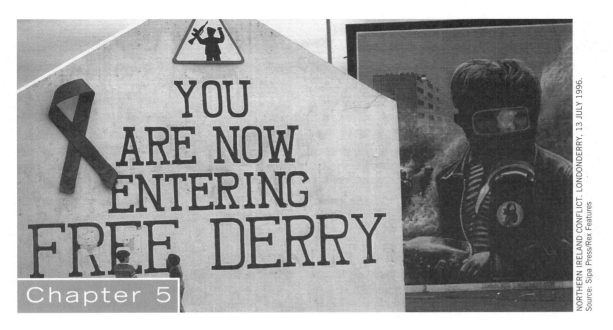

NORTHERN IRELAND CONFLICT. LONDONDERRY, 13 JULY 1996.
Source: Sipa Press/Rex Features

Chapter 5

Nation and nationalism

Nationalism versus Europeanism

The European Union is a political project driven by the economic imperative of a single European market, but one with key political implications. The implications are focused upon the sovereignty of individual states, or their ability to define and enact their own policies. In March 2006, *The Economist* reported on a summit of European leaders and the tendency to invoke 'economic nationalism' rather than pursue the European project of integration. But why the term 'nationalism'? What does it mean in this context, and is it an appropriate term?

The Economist juxtaposes two different geographies but wrestles with two very different politics. On the one hand is the belief in greater economic efficiency through the continued development of a single European economic space. The other politics is a resistance to that project by the individual states who are still calculating economic and political benefits and risks through the lens of a narrow 'domestic' agenda. The tension faced by politicians between the European and the domestic demands is enhanced by the fact that their constituents still look at the world predominately through a national lens: the belief that their primary identity is German or French, for example. These national identities have much greater resonance than an attachment to a European identity. Although economic processes are challenging established notions of state sovereignty and, in the process, established understandings of national identity, the strength of national identity is still strong and is an important political factor.

Anyone who doubts the power of national identity should stand among the fans watching the World Cup finals, or, more tragically, consider the mass graves of victims of the ethno-nationalist conflict in the former Yugoslavia. Anyone who doubts that established notions of national identity are not being challenged should consider the road-signs in southern Arizona near the Mexican border, where the distances are reported in kilometres rather than miles, as is the norm. Nationalism is a persistent but dynamic political ideology.

Source: 'Charlemagne: Union blues', *The Economist*, 18 March 2006, p. 56.

There has been an interesting paradox in the treatment of nationalism in political geography, which David Knight (1982) identified. Nationalism is generally considered to be the most geographical of all political movements, and yet it was neglected as a research topic. Traditional political geography was largely organized around the trilogy of territory–state–nation, so that behind every successful territorial state there was a vibrant nation. Hence territory becomes national 'homeland', sometimes even 'fatherland' or 'motherland', imbued with the symbolic significance of nationalism, and the state becomes the 'nation-state' as the ideal expression of the political will of nationalism. It is surprising, therefore, to have to report a dearth of studies of nationalism per se in political geography until quite recently: there is no equivalent of Mackinder's contribution in geopolitics or Hartshorne's on the territorial state.

The paradox of research neglect of such a central concept is initially quite mystifying. It has not been ignored, as has imperialism; quite the opposite. Political geography has been imbued with nationalism. The idea of 'nation' permeates political geography studies to such a degree that the concept was seen as largely unproblematical. Nation and nationalism were a 'given'; they were part of the assumptions of analysis that were not investigated. And we have not been alone in perpetuating this peculiar form of neglect. Until recently, it was commonplace for social scientists in general to bemoan the paucity of research on such an important topic (A. D. Smith 1979). All this has changed in the past two decades with the development of new theories of nationalism, including important contributions from political geographers. The idea of the nation and associated nationalism is no longer considered unproblematical in contemporary social science, including political geography.

The basic reason why nationalism could elude serious study for so long was because its practices were deeply embedded within modern society. Nationalism became particularly noticeable in periods of political crisis such as war, but it has always been more than mere politics. Recently, Michael Billig (1995) has illustrated this with his concept of 'banal nationalism'. He shows that nationalism is part of everyday life in all societies. The nation and the nation-state are naturalized as obvious and unquestioned necessities that organize our lives and frame our outlooks. According to Billig, the nation is 'flagged' for us every day as we recognize, use and seek comfort from flags, coinage and other national symbols that have commonplace functions. Similarly, 'we' and 'us' are commonly used in daily newspapers to constantly remind us that we are part of a nation and different from others. All news, in newspapers, on radio or TV, is habitually divided into 'home news' and 'foreign news'. Advertisements often signal their product as associated with, or even part of, the nation of the targeted consumers. Many millions of holidaymakers are now thoroughly familiar with queuing at passport controls as part of their holiday experience. This is a concrete experience of boundaries far more indicative than abstract maps of boundaries hanging on a classroom wall. The continual presence of these reminders in our contemporary lives creates an unquestioned acceptance of the 'naturalness' of both nations in general and ours in particular.

It is in these circumstances of latent national identity that politicians in times of war or other crises are able to activate national support for the geopolitical actions of nation-states. In Billig's words:

> One might think that people today go about their daily lives, carrying with them a piece of psychological machinery called 'a national identity'. Like a mobile telephone, this piece of psychological equipment lies quiet for most of the time. Then, the crisis occurs; the president calls; bells ring; the citizens answer; and the patriotic identity is connected.
>
> (Billig 1995: 7)

The continual flagging of nationhood is necessary to enable the call to arms whenever it is required to legitimate geopolitical actions. Our task in this chapter is to utilize political geography to help to 'disembed' nationalism intellectually so as to understand its construction and to evaluate its contemporary salience under conditions of globalization.

Nation and nationalism continue to be particularly important to our understanding in political geography for one basic reason: they are both explicitly territorial in nature. As Anderson (1986: 117) points out, nations do not simply occupy space as other social institutions or organizations do; they claim association with a particular geographical location. They share this property with the modern sovereign state, and this shared territoriality is expressed in the concept of the nation-state. This is the pivot of our political geography framework of ideology separating experience from reality, described in Chapter 1. The territorial state as nation-state represents our scale of ideology. In the first section of this chapter, we expand on this notion by investigating the ideological heritage that lies behind our studies of nationalism. We conclude that nationalism is not an eternal expression of nations through the ages but consists of a family of political practices that are barely two hundred years old.

Having disposed of this mystifying heritage, in the remainder of the chapter we draw upon critical historical studies and modern theories of nationalism to aid an understanding of this ideology and practice, which imbues so much of our thinking. We begin our discussion by concentrating upon nationalism in practice, drawing upon historical studies to understand its meaning in contemporary politics. We then consider how this political practice has been interpreted in recent debates on the theory of nationalism. Many past theorists thought that nationalism was a passing phase of modernization that had either disappeared or was soon to meet such a fate. That this is not true has been clear for some decades now but has been highlighted in the 1990s with the rise of new national conflicts on all continents. Rather than homogenizing the world, globalization elicits political reactions producing hybrid mixtures of local and global (Featherstone et al. 1995). Under conditions of globalization, therefore, nationalism is being renegotiated, which we briefly introduce as both theory and practice. We conclude with a synthesis that attempts to link together the politics surrounding the four key institutions: nation, state, class and household. The crucial, indeed pivotal, role of the hybrid institution, the nation-state, is highlighted once again.

Chapter framework

This chapter engages the following components of a political geography framework we introduced in Chapter 1:

➤ The ideology of nationalism justifies the *material* division of the world into a mosaic of nation-states.

➤ The content of national ideology is a *representation* of the collective identity of a 'people' and *their place in the wider world.*

➤ Analysis of the construction of national identity has until recently been *silent* regarding the role of the representation of *gender.*

➤ We address how the ideology of nationalism reinforces the dominant representation of the *scale* of the nation-state as the dominant legitimate political institution.

■ The ideological heritage

The vast majority of writings on nationalism, prior to the recent critical turn, were undertaken by self-conscious nationalists. The result was a body of literature extolling the virtues of this nation or that, with little or no comparative perspective. According to A. D. Smith (1986: 191), the intellectuals who have produced this material failed to adhere to the canons of scientific research; indeed, 'objectivity was not their main concern'. They produced a tradition of thinking in which nationalism was interpreted as an autonomous and natural force operating throughout history. This is, as it were, a view of nationalism through the lens of nationalism. Although challenged by more recent work, the lens of nationalism still assumes a dominant position in interpretations of nationalism, both popular and academic (Agnew 1987). It is the purpose of this section to reveal the

basic limitations of this heritage to help to free political geography from its deadening grasp.

A world of nations

We shall try to be a little more objective in our treatment of nations. Why is the existence of nations taken for granted? Why do nations appear to be as natural as families and kinship groups? We must start with these questions in order to begin to understand why we live in 'a world of nations'.

Primordial versus modernist interpretations

We all belong to one nation or another. We do not choose which nation to join; it is ascribed to us: we are born into a nation. This is the natural basis of nationalism. The word 'nation' comes from the Latin *nasci*, meaning to be born. Nations would seem to be historical communities, therefore, that share a common ancestry. Hence the origins of today's 'nations' are to be found in yesterday's 'tribes'. This produces an evolutionary view (see, for example, De Blij 1967) that culminates in a world of nations, each with its own particular and unique genealogy.

A. D. Smith (1986) describes this view as 'primordialist' because of its insistence on the primordial ties of ethnicity and language. For primordialists, ethnic communities emerged out of prehistoric times and entered history as the basic units of human experience. For instance, the Germanic people first enter history as the German-speaking tribes fighting the Roman armies along the Rhine frontier. In this view, nations are natural and perennial. The human species is genetically divided into a limited number of kin-related groups of individuals. These groups have always existed, although they may not have always expressed themselves as forcibly as in the very recent past. Hence all periods of history will contain nations, and some of these will have survived migrations, assimilations and conquests to form the origins of modern nations: the Babylonians and Assyrians have not survived, the Germans and Chinese have.

Nobody doubts that ethnic communities of one sort or another have existed before the modern era. The important question is the relationship between these communities and the nations we see around us today. Smith (1982) has proposed that we view nations as a particular form of ethnic community that merges cultural identity with political demands. It is this link that is very recent. Breuilly (1993: 3) gives a very revealing example to make this point. In the Middle Ages, Dante identified an Italian 'nation' in terms of language. He wanted this language to be the literary language of all Italian poets. He also wrote about politics and defined his ideal form of government. But there was no cross-referencing between these works: the notion of Italian political unification was totally absent. Nation was a cultural phenomenon, government a political one, and Dante saw no reason why they should be related one to another.

The modernist view of nationalism, to use Smith's (1986) terminology, treats it as a historically recent phenomenon that has provided a unique and powerful link between politics and culture. Nationalism as a term emerged only in the late nineteenth century. As an idea, though, it can be traced back to the earlier concept known as the 'principle of nationality' (Hobsbawm 1987: 142–3). This principle was a very simple and powerful one: every nation has the right to its own state. This idea emerged in the eighteenth century, became a major force in world politics in the nineteenth century and dominated the politics of the twentieth century. Hence as Hobsbawm (1990: 14) states: 'The basic characteristic of the modern nation and everything connected with it is its modernism.' It is this recognition expressed in a series of key books over the past two decades that has produced the critical literature that has made this chapter possible.

Hence we shall treat nation and nationalism from a 'modernist' rather than 'primordialist' position. But although the primordialist heritage is concerned more with justification than with explanation, it cannot be ignored. Like the power-political heritage of geopolitics, it is necessary to know this heritage in order to develop an understanding of the political geography of our modern world. Following Breuilly (1993), we shall emphasize nationalism as a political strategy, so the activities of nationalist politicians and intellectuals are our object of study. If these are indeed modern political practices, then we must start by investigating their origins.

The French Revolution as crucible

The modern idea of nation and nationalism is different from all previous expressions of loyalty and identity because of its appeal to 'the people'. This new politics was originally associated with the rationalism of the eighteenth century. The key step was the questioning of the personal authority of the monarch as the sole source of a state's sovereignty. The alternative formula to be devised was that sovereignty lay with the people. This notion is given its most famous political expression in the opening sentence of the American Constitution of 1787: 'We the people . . .'. But the new United States did not originally call itself a nation (Billington 1980: 57). In contrast, the idea that the people constitute the nation was central to the French Revolution:

> The concept of *la nation* gave tangible definition as well as higher definition to the revolution. The revolution acquired spatial dimensions and was henceforth embodied not in complex republican institutions but in simple concentric circles. The borders of France were an outer ideological mote, Paris the inner citadel and the National Assembly the 'perfect point' of authority within Paris itself.
> (Billington 1980: 57–8)

This new nation was associated with unitary and centralizing characteristics, 'one and indivisible' as the French termed it, which accounts for the politicians of the federal United States generally avoiding the word (Hobsbawm 1990: 18). Nevertheless, with the French Revolution the idea of nation as a popular political entity possessing a state entered world politics.

Billington (1980: 526) tells a very instructive story illustrating the novelty of the nation. In April 1790, a band of revolutionary peasants, suspicious of a well-dressed man passing through their district, forced him to cry 'Vive la nation!' Then rather sheepishly they begged him: 'Explain to us what is the nation.' Clearly, the French political elites of the revolution were constructing a new politics in the setting up of a National Assembly in 1789 and proclaiming France 'la grande nation' in 1790 (Rustow 1967: 21, 27).

This construction included two crucial principles that have subsequently been at the heart of

161

nationalist political ideas. First, language was used to define the nation. A new 'post-aristocratic form of French' was invented by literary elites and proclaimed *la langue universelle de la République* (Billington 1980: 36). This was imposed on the dialect-rich provinces as a means of breaking down local loyalties. We know from Weber (1976) that this language uniformity was not achieved before 1914, but the centralizing policy begins in 1792. In that year, central government documents were no longer translated into the various dialects, and Billington (1980: 36) suggests that 'the body of the language replaced the body of the king as the symbol of French unity'. Here was the first critical use of a cultural attribute as a political tool; nationalism had arrived.

Second, in this French Revolution crucible we find the explicit link between nation and territory forged. Transferring the imagery of the body of the king to the body of the nation produced medical analogies promoting territoriality. Loss of territory was equated with 'amputation' and the possible 'death' of the nation. Hence the unprecedented provision in the 1793 Constitution preventing ever making peace with a foreign power while it occupied any French territory (Billington 1980: 66). War was no longer about defending the territory of the state; it was now national territory that was at stake.

We are heirs to this politics linking people, culture and territory. Although it has changed many times and in many ways since these origins two hundred years ago, there is a general continuity, a doctrine of nationalism, that dominates politics to this day.

The doctrine of nationalism

We should never underestimate the ideology of nationalism. As A. D. Smith (1979: 1) so rightly says: 'No other vision has set its stamp so thoroughly on the map of the world and on our sense of identity.' In fact, the idea of 'nation' is so embedded in our consciousness that it is even reflected in the use of terms that describe counter-national arrangements: 'supranational', 'multinational' and 'transnational' all assume the prior reality of nations (Tivey 1981: 6). Similarly, the two great liberal organizations of states in the twentieth century have found it unnecessary to men-

tion the states themselves – we have had successively the League of Nations and the United Nations. Of course, 'nations' that do not possess states, such as the Kurds, cannot join these organizations, but the many states with a melange of many ethnicities, such as most post-colonial African states, do join immediately on independence to reinforce their sovereign status. As Seton-Watson (1977: 2) has humorously put it: 'The United Nations in fact has proved to be little more than the meeting place for representatives of disunited states.' Both the League and the UN are, or were, special clubs for states, not nations. Their misleading titles merely reflect the strength of the doctrine of nationalism in the twentieth century. The fact that the practice and study of inter-state relations is called 'international relations' just shows how embedded nations are in both popular and academic language.

But what is this 'doctrine', this ideal that all nationalists subscribe to? It is more than a simple theory linking an individual to the nation of his or her birth. It provides a national identity to an individual, but it is premised on a wider acceptance of a world of nations. It provides for much more than a simple 'them and us' dichotomy such as between a civilized 'us' and a barbarian 'them' in the former world-empires. In the world made by nationalism there are multiple 'thems'. Numerous other nations and nationalisms are recognized as equal to and the equivalent of 'our' nation and 'our' nationalism. This has been expressed most clearly by a member of the Kach vigilantes, a group of Israeli nationalists who are usually considered to be at the extreme end of Israel's political spectrum:

> I personally have nothing against the Arabs. We in Kach do not hate Arabs. We love the Jews. The Arabs are a danger to us. If they were Chinamen I would fight them too. If I was in the place of the Arabs I would do the same as them. It's only natural that they support the PLO, it's their liberation organization. I understand it. But I won't let them kill Jews.

This statement of mutual respect from such an unlikely source tells us a lot about the world according to nationalism. We can describe it in terms of the following propositions drawn from Tivey (1981: 5–6) and Smith (1982: 150), and related successively to our three scales of analysis.

A1: The world consists of a mosaic of nations.

A2: World order and harmony depend upon expressing this mosaic in a system of free nation-states.

B1: Nations are the natural units of society.

B2: Nations have a cultural homogeneity based upon common ancestry and/or history.

B3: Every nation requires its own sovereign state for the true expression of its culture.

B4: All nations (rather than states) have an inalienable right to a territory or homeland.

C1: Every individual must belong to a nation.

C2: A person's primary loyalty is to the nation.

C3: Only through the nation can a person find true freedom.

We can term this list the common doctrine of nationalism. It has justified the following scale effects: the world is politically divided rather than unified; the state as nation-state is the basic arena of politics; the local scale is bypassed as experiences are transcended by 'higher' and more remote ideals.

The effects on social relations are no less profound. Other political ideologies have had to adapt to or be crushed by nationalism. According to Smith (1979: 8), nationalism split with liberalism at the time of the 1848 revolutions. The simple individualism of classical liberalism had to give way to a 'national' liberalism in order to survive. At the other end of the scale, the internationalism of socialism was fundamentally defeated at the outbreak of the First World War in 1914. Worker fought worker under their different national banners. Like liberalism, the various brands of socialism have had to adapt to the political reality of nationalism in order to survive. Liberal individualism and socialist internationalism counted for little against the doctrine of nationalism. The result is our three-tier scale structure of the world-economy pivoting around the nation-state.

This common doctrine is general to all nationalisms. But every nationalism is based on particularism. Each has its own character. These are what Smith

(1982: 150) terms special secondary theories. Nairn (1977) calls them national*ism*, emphasizing the common and ubiquitous ideology; and *national*ism emphasizing the uniqueness of each nation, the qualities that make one confident in identifying with a particular set of flags, symbols and histories. Since every nation is different, a brief description of just a single example of *national*ism will have to suffice.

Watson (1970) quotes approvingly from a letter written in 1782, which lists eight features of American society that were to become American 'national characteristics'. These are 'a love of newness', 'nearness to nature', 'freedom to move', 'the mixing of peoples', 'individualism', 'a sense of destiny', 'violence' and 'man as a whole'. The first four reflect American history, especially frontier history. The next two features are key ideological props of 'the American way of life'. Individual competition to achieve personal success – log cabin to White House – is the basis of American liberal ideology. Manifest destiny has set national goals originally continental in scope and latterly global. The final two features are contradictory as they contrast conflicts in US society with the idealism of belief in rational accommodation. Together, these eight features add up to what Watson calls 'the myth of America'; in our terms they are the special secondary theory of American nationalism. It is quite remarkable how these descriptions of a very different American world of two centuries ago can have an immediate and obvious contemporary salience. This is *national*ism embedded in society.

Clearly, it is at this level of theory that we can identify the various elements that political geographers have studied in the past – *raison d'être*, iconography, state-idea, and so on. It has been the concentration of effort at this level of symbolism that has prevented political geographers breaking through to the general doctrine in understanding nation and nationalism. This is because nationalism is based upon a series of myths embodied in the secondary theories. These myths consist of distorted histories concerning society/ ethnic origins, past heroic ages and betrayals, and the 'special' place of the nation in world history. There may well be only one deity, but he or she has certainly been generous in designating 'chosen people'!

Nationalist uses of history: the 'modern Janus'

It will have been noticed that special secondary theories are essentially historical in nature. History is as important as culture and territory in the make-up of nations. But it not just any history that is our subject matter here; these are histories with a nationalist purpose, where myth and fact are entangled in complex ways. One authority has even argued that 'Getting its history wrong is part of being a nation' (Hobsbawm 1990: 12). Certainly, creating nations has always involved creating new histories.

The entry of the 'people' on to the political stage produced a change in the historical requirements of the state. In the absolutist states, legitimation was vested in the sovereign, whose right to rule rested on his or her personal lineage. Like the world-empires of the past, the most important history was the monarch's family tree. This was hardly appropriate for the new world of nations. The people, and not just a symbolic head, have to have a link with the past.

One of the inventions of the new national histories was the ascribing of special significance to particular dates. Centenaries became celebrated for the first time, for instance (Hobsbawm 1987: 13). The centennials of both the American and the French Revolutions were celebrated with international expositions in 1876 and 1889, respectively. Other nations proclaimed anniversaries of their foundations in a dim and distant past. Hungarian patriots found their nation to have had its origins in the invasion of Magyars in 896 just in time to celebrate the millennium in 1896. Similarly, the Swiss chose 1291 as the date of the foundation of Switzerland for appropriate celebrations in 1891 (Anderson 1983: 117). More recently in Poland, in the early 1960s, other patriots traced the origins of the Polish nation to its conversion to Christianity in 966, again just in time for a millennial celebration. We can be sure there were no celebrations in Hungary, Switzerland or Poland in 1796, 1791 or 1766, respectively, for the simple reason that this was before the national need for new histories. Quite simply, there was nothing to celebrate.

The essence of celebrations of such anniversaries is their appeal across the whole national community. The past is shared by everybody in a display of national unity. The meaning for the nation is profound: 'history is the precondition of destiny, the guarantee of our immortality, the lesson for posterity' (Smith 1986: 208). Without history, there can be no nation.

Poetic landscapes and golden ages

A. D. Smith (1986: 178) identifies three forms of national history. Where the new nation has had a formal political existence over a long period, there will be more than a sufficiency of historical material. In this case, the history is produced by rediscovery, selecting a new amalgam of facts for the new history. Where the new nation is less well endowed with material, the history has to be created by conjecture in a process of reconstruction. In rare cases, the history may be produced by simple fabrication. All such histories consist of different balances between what Anderson (1986: 130) calls the 'rich amalgam of fact, folklore and fiction'.

Each historical drama has two major components: a story of continuity that fills in the gap from the origins of the nation to the present; and a small number of symbolic tableaux depicting key events that the whole nation can identify with. The two most famous are probably William Tell shooting the apple off his son's head and Joan of Arc being burnt at the stake (Smith 1986: 180). Both stories evoke the innocence of a patriot fighting devious and foreign threats to the nation.

National histories both define and direct by providing the space–time coordinates of the nation. In the romantic tradition, they create what Smith (ibid.: 182) terms a poetic space and a golden age. The former defines the place of the nation, its landscape, its sacred sites and historical monuments. This can eulogize both the capital city with its monuments and the ordinary landscape that represents the true habitat of the people – for England both the historical sites of London and the thatched cottage on the village green. We should not doubt the contemporary importance of this process. In recent years, this history has been designated 'heritage', and museums have become 'heritage centres'. According to Hewison (1987: 9), one such centre opens every week in Britain, threatening to turn the country into 'one vast museum'.

This geography of 'special places' is interwoven with the national history. Typically, all such histories include what we may term the Sleeping Beauty complex. That is to say, every nation has its golden age of heroes, since when it has suffered decline; but now, like Sleeping Beauty after the kiss, the nation is reawakening to reclaim its former glories. This is the central drama in the series of eight myths that Smith (1986: 192) describes as typical of any national mythology. The whole process is undertaken to generate a vision of the past to shape a direction for the future. Hence Nairn's (1977) particularly apt phrase describing nationalism as the 'modern Janus'. Janus was the classical god who faced both forwards and backwards. All nationalisms do the same.

Case study

Inventing traditions: Dutch tartans and English kilts

We have noted that complete fabrications in the creation of national histories are rare. Nevertheless, such exercises do illustrate the lengths to which nationalists may go in order to provide a suitable antiquity for a modern nation. This dubious activity by forgers and confidence tricksters can be amazingly successful when taken up by other innocent nationalists.

The national costume of Scotland is perhaps the most widely known of all European nations. All readers will associate tartan kilts with Scottish clans. The general assumption is that they represent the ancient clothing of Scottish kinship groups. It will come as a surprise to many readers, therefore, to find out that the tartan originates from the Netherlands, that kilts come from England, and that there were no 'clan tartans' before 1844. This is a tradition that has been invented as part of a fabricated Scottish history.

Trevor-Roper (1983) has carefully documented this invention. He traces three stages in the making of the new history. In the first stage, a highly distinctive culture for Celtic Scotland was created. This required reversing the cultural dominance of Ireland over Highland Scotland in the medieval period. This was achieved in the eighteenth century by forging an 'ancient epic poem', the *Ossian*, which portrayed great Scottish cultural achievements in which the true core of Celtic culture, Ireland, was relegated to a cultural backwater.

The second stage was the creation of new Highland traditions. This centred on the tartan kilt. Tartan designs were imported into England and Scotland from the Netherlands in the sixteenth century. Quite separately, in 1727, a Lancashire weaver invented new clothes designed for tree felling based on the old Saxon smock. At the time of the Jacobite Rebellion of 1745 therefore, kilts were a very recent invention for labourers, and clan tartans did not exist in the Highlands. After the defeat of the rebellion, the Highlands became a major source of men for the British army, and tartan kilts gradually became the uniforms differentiating the new regiments. Their transfer from lower-class and soldier wear to upper-class clothes was a subsequent product of the Romantic movement of the late eighteenth century. In 1778, the Highland Society was established in London, and in 1820 the Celtic Society was founded in Edinburgh. In 1819, demand for tartans was so great that the first pattern book was published. By the time George IV paid a state visit to Edinburgh in 1822, the new Highland traditions were an accepted part of Scottish life and the ceremony was dominated by tartans and kilts.

The third stage saw the adoption of the Highland tradition by the lowland region, where the vast majority of the Scottish population lived. This was achieved by another forgery. Two brothers claiming to be descended from the royal Stuart family produced a document, *Vestiarum Scotium*, which purported to show that each medieval clan had its own tartan pattern. This document was published as a historical source in 1842, and two years later the brothers reported its contents in their book *Costume of the Clans*, which was suspiciously similar to the earlier commercial pattern book. This allowed all Scots people to find their clan tartan from their surname. Highland tradition soon became popular in the lowlands, and a new large clothing business was created. Since this time, Scots people across the world have celebrated their Scottishness in the tartan kilts of the Highland tradition.

Any selection of historical facts will tend to favour one particular future action over another. For instance, national histories of Greece have two alternative golden ages on which the national mythology may hinge. One looks back to the glories of Byzantium, the preservation of Greek culture in the Orthodox Church and Greece as a major eastern Mediterranean power centred on Constantinople (today's Istanbul in Turkey) as the 'second Rome'. The second possible golden age is that of classical Greece, the rationality of the city-states and Greece as a nation-state in the territory of the peninsula. Promoting the former mythology commits the Greek nation-state to a policy of intervention in Turkey. The second mythology treats the Byzantium episode as an alien Roman interlude and promotes a policy of nation building within a much smaller Greek territory. After the defeat by Turkey in 1923, the latter mythology has tended to prevail. However, the point is that different histories suit different presents and imply alternative futures. The modern Janus is a slippery customer: whatever else national histories are primarily about, it is not the past.

Summary

In this section we have introduced the nation as a political geographic concept and noted:

➤ the competing views of nationalism as a primordial tendency versus a modern construction;

➤ the manner in which national identity has been constructed;

➤ the 'doctrine of nationalism' as the dominant manner in which the political geography of the modern world is conceptualized;

➤ how nationalism is used in political projects.

■ Nationalism in practice

Nationalism is above all a political practice. The ideological heritage is crucial but it does not constitute nationalism as many nationalists have supposed. Furthermore, we should always remember that nationalism has not always been the dominant ideo-

logy of the modern world-system. As we have shown, this political use of the idea of nation dates from the French Revolution and no earlier. In the two or three centuries before 1800, the world-economy evolved without a politics based on nationalism. There were states to be sure but, as Wallerstein (1974a: 102) points out, the political practice was anti-national where ethnic regional powers stood in the way of the absolute state's centralization. There was 'statism' or mercantilism as we have seen, but until the nineteenth century no nationalism. Since that time, however, we have experienced a very wide variety of political practices that are nationally based. The first task of our discussion below is to put some order into this diversity by presenting a typology of nationalism. We then consider contrasting interpretations of these particular political practices.

Variety: the good, the bad and the ugly

We will begin with a standard typology of nationalisms, loosely based on the work of Orridge (1981a). Our listing below is largely drawn from his discussion of the sequence of various nationalisms. We identify five basic types: proto-nationalism, unification nationalism, separation nationalism, liberation nationalism and renewal nationalism. We consider each in turn.

Proto-nationalism

This is the nationalism of the original core states, the medium-sized states of western Europe. This nationalism is the source of much dispute over the timing of the emergence of nation and nationalism. Gottmann (1973: 33–6), for instance, is able to trace the idea of *pro patria mori* – dying for one's country – back to about 1300, so that by 1430 Joan of Arc could employ such sentiments freely to mobilize the French knights against English encroachment on French soil. The statements of English 'nationalism' to be found in Shakespeare's plays of the late sixteenth and early seventeenth century are even more familiar examples of this early 'patriotism'. But these are both examples of loyalty to monarch or state or even country, not to the collective idea of a people as a nation incorporating all sections and classes. Nevertheless, the centralizing tendencies of these states within relatively stable boundaries did lead to a degree of cultural homogeneity by 1800 that was not found in other areas of comparable size. England and France are the key examples here, but similar nation-states were emerging in Portugal, Sweden, the Netherlands and, to a lesser extent, Spain. In all of these cases, state preceded nation, and it can even be said that state produced nation. The result was what Orridge (1981a) terms proto-nation-states. The 'people' were entering politics, but nationalism as an ideology was not fully developed until later in the nineteenth century. Hence we can say that nation preceded nationalism.

Unification nationalism

For the full development of the ideology of nationalism we have to look elsewhere. In central Europe, such medium-sized states had been prevented from evolving under the contradictory pressures of small (city-scale) states and large multi-ethnic empires. In particular, Germany and Italy were a mosaic of small independent states mixed with provinces of larger empires. After 1800, the Napoleonic Wars disrupted this pattern, which had been imposed a century and half earlier at the Treaty of Westphalia (1648). Although the Congress of Vienna in 1815 attempted to reconstitute the old Europe, new forces had been unleashed that were to dominate the rest of the century. Nationalism was the justification for uniting most of the German cultural area under Prussian leadership into a new German nation-state and transforming Italy from a mere 'geographical expression' to an Italian nation-state. These are the prime examples of unification nationalism and are generally considered to be the heartlands of the ideology.

Separation nationalism

Most successful nationalisms have involved the disintegration of existing sovereign states. In the nineteenth and early twentieth centuries, this nationalism lay behind the creation of a large number of new states out of the Austro-Hungarian, Ottoman and Russian Empires. Starting with Greece in 1821, a whole tier of new states was created in eastern Europe from Bulgaria through the Balkans to Scandinavia.

Surviving examples are Norway, Finland, Poland, Hungary, Romania, Bulgaria, Albania and Greece. Ireland also comes into this category. This type of nationalism was, until recently, considered to be a phenomenon of the past, at least in core countries. However, in the past two decades further 'autonomous' nationalisms have been appearing in many states. Some of the most well known are found in Scotland, Wales, the Basque country, Corsica, Quebec and Wallonia. None of these has been successful in establishing its own nation-state, but all have been granted political concessions within the framework of their existing states. We discuss this 'new' nationalism further below.

Liberation nationalism

The break-up of European overseas empires was described in Chapter 4 but is of interest here as representing probably the most common form of nationalism. Nearly all such movements for independence have been 'national liberation movements'. The earliest was the American colonists, whose War of Independence in 1776 finally led to a constitution giving sovereignty to 'the people'. The Latin American revolutions after the Napoleonic Wars were more explicitly 'nationalist' in character. These can be considered liberal nationalist movements. In the twentieth century, such movements have invariably been socialist nationalist movements, varying in their socialism from India's mild version to Vietnam's revolutionary version. Another way of dividing up liberation nationalism is between those based upon European settler groups and those based upon indigenous peoples. In the former case, we have the United States and Latin America plus the original 'white' Commonwealth states of South Africa, Canada, Australia and New Zealand, which negotiated independence without liberation movements. In the latter case, there are the states of Africa and Asia that have become independent since 1945.

Renewal nationalism

In some regions beyond the core, 'ancient' cultures withstood European political capture for a variety of reasons and were able to emulate the proto-nationalism of the core, often using a politics similar to unification nationalism. These countries had a long history as ethnic communities upon which they could easily build their new nationalism. National renewal to former greatness became the basic cry. Hence Iran could rediscover first its Persian heritage and then, after the 1978 Islamic revolution, its Shia Muslim origins. Turkey, after losing its Ottoman Empire in the First World War, could concentrate on its Turkish ethnicity. The classic cases of this type of nationalism are to be found in twentieth-century Japan and China, two former world-empires that were incorporated into the modern world-system in the nineteenth century largely intact. Israel is a very distinctive case of renewal nationalism, based as it is on reversing a diaspora, which meant 'renewing' (conquering) its territory as well as its people.

This form of nationalism can also occur as part of a process of creating a new state identity that attempts to redefine the relations of the state to the world-economy. As such, this renewal is associated with modern revolutions. Stalin's 'socialism in one country' had many of the trappings of a renewal of the Russian nation, for instance. Other such radical renewals have occurred in Mexico, Egypt, Vietnam and China.

These five types and their various sub-types seem to provide a reasonable cover of the variety of nationalisms that have existed. The sequence from the early types to their emulation in the later ones illustrates, according to Orridge (1981a), the basic process of copying and adaptation that had lain behind the strength of the ideology. Political leaders in a wide range of contexts have been able to appeal to the nationalist doctrine to justify their actions. And this has led to a curious ambivalence in how nationalism is perceived. Hence, although our standard typology does its job as a descriptive tool in outlining the range of nationalisms, it is of less help in interpreting the politics of this ubiquitous though varied practice.

We are used to thinking politically in right–left terms, contrasting conservative with liberal positions or revolutionary activities with reactionary order. Nationalism has been available for use to both sides of the political spectrum. Hence one of the fundamentally confusing things about nationalism is that it

has been both a liberating force and a tool of repression. On the one hand, it is a good thing, a positive force in world history, as when it is associated with liberation movements freeing themselves from foreign rule. But it also has an ugly side, the negative force associated with Nazism and fascism in Europe and military dictatorship throughout the semi-periphery and periphery. The resulting ambivalence towards nationalism is most clearly seen in attitudes after the two world wars in the twentieth century. In 1919, the First World War was blamed on the suppression of nationalism; in 1945, the Second World War was blamed on the expression of nationalism (Rustow 1967: 21). We have to look to political practices prior to the twentieth century to see how this unusual situation came about.

The transformation of nationalism in the nineteenth century

The defeat of the French in the Revolutionary and Napoleonic Wars did not stem the new ideology and practice of nationalism. The French monarchy was restored, and the Concert of Europe was set up to control political change. But the nationalist genie was out of the bottle. Nationalist revolutions were successful in creating new states where the great powers were not affected (Greece in 1821, Belgium in 1830 and Romania in 1859), but the great nationalist cause for Polish independence failed until 1919. But these few examples do not do justice to the nationalist movement that engulfed Europe in the mid-nineteenth century.

The leader of this movement was the Italian nationalist Giuseppe Mazzini. In 1831, he set up the Young Italy society, and three years later Young Europe was organized as a federation of nationalist movements. Within a year, it had hundreds of branches across Europe; Billington (1980) calls it the Mazzinian International. This phrase hints at Mazzini's theory of nations. As proper 'organic units', every nation was distinct and separate and therefore there would be no reason for conflict between them. Force would be required to sweep away the old order, but once the political world of true nations existed, wars would be unnecessary in the new harmonious conditions. This idealism is a long way from the aggressive nationalism that we are familiar with in our century, but it represents early faith in what we have described as the doctrine of nationalism.

The irony is that when Italian unification was achieved it was not the result of a national revolution but traditional realpolitik converting the king of Sardinia into the new king of Italy. This was symptomatic of the changes that were happening to nationalism in the second half of the nineteenth century. First, nationalism became a tool of the great state builders of the era, notably Bismark and the unification of Germany; and second, it became integral to the imperialist politics at the end of the century. The politics of nation was shifting to the right, and the term 'nationalism' was coined at this time to describe the new aggressive politics (Hobsbawm 1990: 102). It could hardly be further from Mazzini's peaceful idealism. From a position where nationalism facilitated political self-determination and included diverse social, political and religious attitudes, it began to demand uniformity and primary allegiance as it became a 'civic religion', or integral nationalism, that 'determined how people saw the world and their place in it' (Mosse 1993: 1). In order to become a civic religion that was coherent, uniform and worthy of loyalty, the nation created 'a fully-worked out liturgy that, with its symbols and mass actions, would come to direct people's thought and needs' (ibid.: 2). The result was that the nation 'tended to deprive the individual of any space he could call his own' (ibid.: 3). National anthems, flags, monuments, ceremonies and the worship of a mythical national history came together to form a structure that defined the individual's space of identity. No wonder Billington (1980: 324) calls this change from revolutionary to reactionary nationalism 'a dramatic metamorphosis'.

And so the twentieth century inherited two very different nationalisms, one revolutionary, the other a tool of the state, which we shall term national self-determination and national determinism, respectively. In the former case, the emphasis is on choice – people say which nation, and hence which state, they wish to belong to. This idea has destroyed empires in the twentieth century. The latter forces people into the nation as defined by the state, for instance by outlawing minority languages. These two processes were at the forefront of the politics of the construction of new states after the First World War. This period is illustrative for two reasons. First, it represents the apogee of the recognition of nationalism as a legitimate force in world politics. Second, nationalism is easier to define as a doctrine than it is to define on the ground. The negotiators at Paris in 1919 had the task of redrawing the map of central and eastern Europe. The difficulties of their task and different means used to achieve their ends illustrate the two sides of nationalism in a very concrete manner.

During the First World War, national self-determination gradually came to be regarded, especially after the entry of United States, as the principal war aim of the Allied forces. By the time the Paris Peace Conference was convened, new states had already been created out of the territories of the defeated states of Germany and Austria-Hungary (Cobban 1969: 556). Hence the peace negotiators spent most of their time putting boundaries around existing political creations. National self-determination should have provided a simple guide to this process, but the simplicity proved to be deceptive.

State and nation since 1945

The reorganization of the world political map after the First World War is generally considered to be the apogee of nationalist politics because it represents the culmination of the nineteenth-century European nationalist movements. But this standard interpretation does not look so clear-cut from the perspective of the early twenty-first century. Today, we see the fruits of nationalist movements beyond Europe, which have produced far more states since 1945 than were created after the First World War. Although we have previously considered these as the collapse of European imperialism, this was engineered by new national liberation movements throughout the world. In addition, with the collapse of the Yugoslav and Soviet federations, more states are being created in Europe itself. The result is that by far the majority of states that appear on today's world political map have been created since 1945. It is clearly time to take a fresh look at state and nation from a global rather than a purely European perspective.

Nationalism as a force for challenging existing states derives from another important change in the nature of this politics that happened in the late nineteenth century. Hobsbawm (1990: 31) calls this the abandonment of the 'threshold principle of nationality'. Mazzini's nationalism recognized the existence only of large nations that would make economically viable states. His 1857 map of European nationalities, for instance, included just twelve nations. With the demise of such a political threshold, the number of nations that can claim a right to statehood increased dramatically. At the Paris Peace Conference in 1919, a Europe of twenty-seven 'nation-states' was constructed. Today, the number is even higher, but the implications of opening the possibility of statehood to all ethnic groupings that may be 'nations' has global implications for the stability of the world political map.

Gunnar Nielsson (1985) has investigated the question of the relation between today's states and ethnic groups – 'nations' and 'national minorities'. He produced data for 164 states and 589 ethnic groups covering the whole world. The fact that there are three times as many ethnic groups as states does not augur well for finding an ideal world of nation-states. In fact, Nielsson produced a typology in which a range of relationships between state and nation are identified.

According to Mikesell (1983), Iceland is the only authentic example of a nation-state in the sense of one people, one state. All other states have a degree of mixed population, which makes their credentials as nation-states doubtful. Nevertheless, lack of 'cultural purity' has not prevented most states in the world claiming to be nation-states. Let us investigate the nature of such claims. Nielsson (1985) defines nation-states as those states where over 60 per cent of the population are from one ethnic group: 107 of the 164 states qualify on this very loose definition. These can be divided into two main types. First, there are those where the ethnic group is dispersed across several states. The best example is the Arab nation, which dominates 17 nation-states. We shall term these part-nation-states, and there are 52 such cases. Egypt and Syria are typical examples, along with the 'divided nations' that we described in Chapter 4 – the

2 'Koreas' and still 2 'Germanies' in Nielsson's data. Second, there are the single nation-states, where an ethnic group dominates only one state. On a few occasions, the ethnic group contributes more than 95 per cent of a state's population. These are closest to the nation-state ideal, and there are 23 such cases. Iceland, Japan and Somalia are typical examples. It is more common where one ethnic group dominates the state, but not to the degree of the 'ideal' category of single nation-states. There are 32 such cases. Britain and the United States are examples of this category, along with Nicaragua, Sri Lanka and Zimbabwe.

Nielsson identifies 57 non-nation-states, where no single ethnic group has 60 per cent of the state's population. These can be divided into three types. An intermediate non-nation-state type occurs where there is a single dominant ethnic group in a state, but it constitutes only about half the population (40–60 per cent in Nielsson's definition). There are 17 such states, examples being the former Soviet Union, the Philippines and Sudan. Bi-nation-states are identified by Nielsson where two ethnic groups provide a combined percentage of a state's population of over 65 per cent. There are 21 cases, and Belgium, Peru and Fiji are typical examples. Finally, there are 19 cases of what Nielsson terms multinational states with a high degree of ethnic fragmentation, so they do not fall into any of the above groups. India, Malaysia and Nigeria are typical examples.

In Table 5.1, we show how these types of state are distributed across five continents. The geographical pattern is as we would expect. The two continents with the oldest modern states, Europe and the Americas, have the most single nation-states of both categories. Part-nation-states are most common in Asia, which reflects a fragmentation of their large nations. Multinational states and bi-nation-states are most common in Africa, reflecting the arbitrary boundaries imposed on the continent in the colonial division. But we can also note that most types are found on all continents. The major exception is the dearth of multinational states in Europe, which reflects their dismantling after the First World War. We can conclude, therefore, that nation-states of one sort or another form the majority of the world's states

Table 5.1 Geographical distribution of nation-states and non-nation-states

	Nation-states			Non-nation-states		
	Part-nation-states	Single nation-states Over 95% one ethnic group	60–94% one ethnic group	Intermediate non-nation-states	Bi-nation-states	Multinational states
Africa	7	4	3	9	9	14
Americas	7	6	11	1	5	3
Asia	22	2	6	6	3	2
Europe	12	9	9	1	2	0
Oceania	4	2	3	0	2	0
Total	**52**	**23**	**32**	**17**	**21**	**19**

but that the nation-state ideal as reflected in single nation-states has not managed to dominate even in its 'heartland' of Europe. The non-national states are obviously susceptible to nationalist challenges from within their boundaries. Governments in these states have to be very sensitive to the needs of the range of ethnic groups in their territory. In Chapter 4, for instance, we showed how the Indian government conceded a language-based federal structure to contain ethnic discontent. But we should not think that ethnic challenges are limited to the non-nation-states. Our loose definition of the nation-state allows for important minority ethnic groups to reside in the single nation-state category, and it is in these states that some of the main examples of 'ethnic resurgency' have been found in recent years.

Nation against state

The resurgence of minority nationalisms within western European states from the 1960s onwards was a surprise to most political scientists. Their developmentalist models predicted a gradual decline in territorially based loyalties as communications within the state brought all the population into a single community (Deutsch 1961). This would be accompanied by modern functional cleavages such as class replacing traditional core–periphery ethnic cleavages (Rokkan 1980). And, in any case, this was Europe: the continent whose boundaries had been redrawn after two world wars to produce nation-states. Nevertheless, separatist and autonomist nationalisms have grown and survived just as political scientists were writing their obituaries. In the 1980s, minority national

resurgence spread to eastern Europe, especially the Soviet Union, a state that had long seemed to have safely accommodated its ethnic diversity (G. E. Smith 1985). With the collapse of communism in eastern Europe, the forces of nationalism have been released to create new states. In terms of the new nationalism, the East has politically overtaken the West in recent years. Clearly, yet another round of nationalist politics has occurred throughout Europe in the 1990s.

A further surprising feature of the new nationalism is that the oldest states of Europe – Spain, Britain and France – are not immune, in fact quite the opposite. We can use these three states to illustrate the range of political practices to be found in the new nationalism (Williams 1986). The most threatening form of politics is violence against the state and its agents. This political practice dominates the Basque struggle in Spain, the Ulster conflict in Britain and the separatist movement of Corsica in France. Various forms of non-violent resistance dominate the nationalist politics of Catalonia in Spain, of Wales in Britain and of Brittany in France. Finally, party political opposition is the major strategy of the nationalists of Galicia in Spain, of Scotland in Britain and of Alsace in France. The new nationalism is clearly a complex phenomenon expressed in different ways in different places. But they all have in common a challenge to the contemporary 'nation-state'.

What is the future for these various nationalist groups? Recently, both Irish and Basque nationalists have declared ceasefires. They have put faith in the political process to achieve their goals, and at the time of writing there is hope for peaceful solutions.

However, a return to violence is possible in both cases. The nationalist movements are asking for national self-determination, the principle employed to draw new state boundaries after the First World War. Can this new round of nationalism produce a new round of boundary drawing and a new political map? The answer may well be yes, but not in the same manner as in the early twentieth century. The difference is the existence of the European Union and its provision of a higher level of political administration that provides a political filter between smaller political entities and the capitalist world-economy. In other words, the scale of the state is itself multi-scalar, and the construction of a multi-state entity (the EU) has facilitated the belief that nations such as Scotland or the Basque region may, for example, construct their own state apparatus but within the administrative apparatus of another layer of government.

The rights of indigenous populations

In Nielsson's (1985: tables 2.2 and 2.3) analysis of ethnic groups, nearly half (289) reside in just one state and constitute less than 10 per cent of that state's population. Some of the new nationalisms of Europe are included in this category. The indigenous populations of the European-settled states of America and Oceania make up another important example of such 'minorities'. These peoples occupied the territory before the invasion of Europeans. After defeat, they were generally ignored, neglected and cheated by the new colonial states. The major source of conflict with the settlers was land, and this continues to be the major practical grievance of indigenous peoples today.

In recent years, indigenous peoples have found a new voice in their struggle for survival. The most striking examples are the World Council of Indigenous Peoples and the activities of the United Nations on their behalf. In 1985, for instance, over two hundred representatives of indigenous peoples participated in a meeting of the UN Commission on Human Rights Working Group on Indigenous Populations (Knight 1988: 128). This provides an indication of the size and scope of this political struggle. It is difficult to find good estimates of the numbers of people involved, but Burger (1987: 37) provides the following figures: Canada 800,000;

United States 22 million; Latin America 30 million; Australia 200,000; and New Zealand nearly 400,000, plus many millions more in Asia and Africa. In all cases, the indigenous peoples are demanding a similar package of rights. First, they require the right to preserve their cultural identity. This requires, second, a right to their territory, to the land, resources and water of their homeland. Third, they need the right to have responsibility over the fate of their people and their environment. Finally, this leads on to the right to control their own land and people, or what is commonly called, in other circumstances, national self-determination.

What does all this mean for the indigenous groups? First, they require recognition as 'nations' rather than as mere ethnic 'minorities'. But as nations are they asking for state sovereignty? There has been some confusion over terminology in this area (Knight 1982), and it seems that most demands envisage autonomy rather than separate sovereignty. Without the demand for secession, some states may grant indigenous peoples many of the rights they are striving for now that they have the United Nations as a forum for their challenge. But past treatment of these peoples does not leave much room for optimism.

Recent court cases and legislation in Australia provide a good example of the difficulties faced by indigenous peoples. Torres Strait Islanders and Aboriginal peoples have made some gains in reclaiming their heritage. For example, it is now government policy to replace the Anglo-Saxon names of mountains and other natural features with Aboriginal nomenclature – Ayers Rock is now Uluru (Mercer 1993). However, the more important fight for Australia's indigenous peoples is for access and title to land that was once theirs. The Australian legal system denied such title through the principle of *terra nullius*, the notion of which was based upon the 1889 court case of *Cooper* v. *Stuart*, which claimed that the colony of New South Wales was unoccupied at the time of its peaceful annexation by Britain in the eighteenth century (ibid.: 305). This case rests upon two views of history: first that the land was empty prior to European settlement; and second, that there was no conflict between Aboriginal peoples and European settlers. Recent histories reject both of these propositions. It is now estimated that the continent of Australia was home to 750,000 persons prior to colonization. A current estimate of the Aboriginal population of what is now New South Wales and Victoria is 250,000, four times the estimate made in the 1930s (ibid.). There has also been ample evidence of wars of resistance by Australia's indigenous peoples. Nonetheless, subsequent court cases upheld *Cooper* v. *Stuart* on the basis that it was the law of the land rather than historical fact.

In 1967, a national referendum supported a change in the constitution that allowed the states and the Commonwealth of Australia to enact special laws regarding Aborigines, extended the franchise to indigenous people and included them in the national census (Mercer 1993). Further progress was made in 1974 when Neville Bonner, the country's only Aboriginal senator at the time, successfully moved a motion that Australia's Upper House

> accepts the fact that the indigenous people of Australia, now known as Aborigines and Torres Strait Islanders, were in possession of this entire nation prior to the 1788 First Fleet landings . . . urges the Australian government to admit prior ownership by the said indigenous people and introduce legislation to compensate the people known as Aborigines and Torres Strait Islanders for dispossession of their land.
>
> (Quoted in Mercer 1993: 307–8)

However, in 1988 (European Australia's bicentennial year) both the Upper and Lower Houses of parliament acknowledged that Australia was occupied by Aborigines and Torres Strait Islanders and that they had suffered dispossession and discrimination, while avoiding the issue of land ownership. The statements of 1967 and 1988 did challenge the notion of *terra nullius*, and in 1992 the case of *Mabo* v. *Queensland and the Commonwealth of Australia* finally put it to rest. Eddie Koiki Mabo was a Torres Strait Islander who used the courts to stop the Queensland government taking property and fishing rights. The success of this case depended upon proof of an uninterrupted connection with the land via the maintenance of gardens and trespass disputes between Aborigines. The necessary weight of evidence required to show

Aboriginal claim to the land was unlikely to lead to a mass of other successful land claims (Mercer 1993). The need to show uninterrupted connection with the land was especially problematical given the past policies of forced Aboriginal relocation (Mercer 1997).

The reaction to the *Mabo* case within white and corporate Australia was strong. Within the broad context of globalization, farming and mining interests mounted a campaign that argued that Aboriginal land claims would be obstacles to Australian economic success (Mercer 1997). The outcome of the *Mabo* case was the Native Title Acts of 1993 and 1998, which have been negotiated within a political context of increasing racism and xenophobia among white Australians. Although these Acts reflect the opinion of the *Mabo* court case, and others, that Aborigines and Torres Strait Islanders do have rights to the land and the sea, they appear to offer an opportunity for further political and legal wrangling rather than a satisfactory solution. Uncertainties remain about the amount of compensation that Aboriginal peoples will be able to claim. Most significantly, the principle of uninterrupted use, rather than sacred attachment to the land, is likely to hinder any major shift in the balance of Australian land ownership.

Generally, we can conclude with respect to all forms of nationalist struggles today that the relative stability of the world political map in the late twentieth century does not mean that the ideal of the nation-state has been realized. Rather, any stability reflects the power of the status quo in the inter-state system. It is the realism of power politics and not the idealism of national self-determination that matters.

Summary

In this section we have:

➤ identified different forms of nationalism;

➤ shown the major historical shifts in the politics of nationalism;

➤ emphasized the continuing mismatch between the geography of nations and the geography of state borders;

➤ noted the violence inherent in a geography of multinational states.

Nation building: empire constructing nations?

In the wake of the Cold War, the US armed forces lacked a clearly defined mission. Their established enemy, the Soviet Union, had ceased to exist and what had been constructed as an immediate threat of war was no longer plausible. In response, the US forces devised the term MOOTWA, an acronym for Military Operations Other Than War. The armed forces were portrayed as an organization that could solve environmental and social problems. MOOTWA was associated with the concept of nation building – or that belief that an outside power could play a constructive role in creating the institutions and practices of society, to be precise a western democratic nation-state.

In the United States the value and possibility of nation building was a matter of political debate. While campaigning to become president in 2000, George W. Bush dismissed the practice. However, in the wake of the terrorist attacks of 11 September 2001 nation building was adopted as part a component of the War on Terrorism. It was related to two other geopolitical constructs: 'failed states' and 'terrorist havens', the idea being that constructing democratic nations was a form of counter-terrorism: 'vibrant' nations, in the western model, would not promote extremist

political views or harbour terrorists. In the words of then Secretary of State Colin Powell:

> But the war on terrorism starts within each of our respective sovereign borders. It will be fought with increased support for democracy programs, judicial reform, conflict resolution, poverty alleviation, economic reform and health and education programs. All of these together deny the reason for terrorists to exist or to find safe havens within those borders.
>
> (Remarks to UN Security Council, Secretary Colin L. Powell, 12 November 2001, http://www.state.gov/, quoted in Flint 2004: 375)

There are, however, two critical problems with this strategy. The first is that the nations were to be 'built' within existing state borders. As we have seen from our earlier discussion of 'nation against state', state borders, especially those imposed by colonial powers, do not follow patterns of national identity. In Iraq, the desire to create a united Iraqi nation has been violently disrupted by the primacy of other collective identities (Sunni and Shia Muslims) on the one hand, as well as the presence of Kurds, who view themselves as a

separate nation, on the other. 'Building' an Iraqi nation is not just a matter of providing certain institutions. It also assumes a homogeneous nation that is coincidental with the state borders.

The second problem is that US practices of nation building can be critiqued as the imposition of empire. In Iraq, for example, the post-Hussein constitution was written by US scholars and politicians and the United States, as an occupying force, enacted harsh anti-labour laws. Nation building is far from being a neutral agenda. It is the attempted construction of a particular form of state. Hence, the nation building of the twenty-first century is a very different form of politics from either national self-determination or national determinism. Nation building is the attempt of a hegemonic power to assert norms of political behaviour that are calculated to maintain a particular geopolitical world order. The Iraqi insurgency and the persistence of warlordism in Afghanistan show that attempts to build collective identity by an external power using political boundaries imposed by earlier colonial powers is fraught with difficulty, at best, and is a hubristic exercise that is likely to provoke violent reaction at worst.

Modern theories of nationalism

The ideological doctrine of nationalism presented in the first section of the chapter is what we may term the nationalist's theory of nationalism. It serves as the conceptual basis for the practice of nationalism. But it is not of any use as a framework for understanding

nationalism. For a critical perspective on nation and nationalism we have to turn to modern theories that have distanced themselves from the doctrine. This allows us to look at nationalism dispassionately from outside the ideology.

The first step in separating our analysis from the doctrine is to recognize what Anderson (1983: 14) terms the three paradoxes of nationalism. First, there is an objective modernity that is built upon a subjective antiquity. This is Nairn's modern Janus analogy. Second, there is the universality of concept that covers the particularity of applications. In the world of nations, every nation is unique. Third, there is the political power of the idea, which contrasts spectacularly with its philosophical poverty. All theories have to come to terms with these paradoxes and explain their existence.

Modern theories of nationalism can be said to begin with the attempt by Tom Nairn (1977) to develop a new Marxist theory of nationalism. This was a highly controversial project and was severely criticized by Marxists and non-Marxists alike (Hobsbawm 1977; Blaut 1980, 1987; Orridge 1981b; Gellner 1983). In some ways, he has acted as a catalyst provoking critical but constructive reactions that have provided alternative theories. It is appropriate therefore that we begin our discussion with Nairn's theory of nationalism and uneven development. We categorize this theory as one of 'nationalism from above' because it emphasizes the role of the bourgeoisie in nationalist movements. The most cogent criticism of Nairn has come from Blaut (1987), who argues for an alternative 'nationalism from below' perspective. The remainder of this section deals with these two positions in turn. In the following section we turn to some current views on the theory and practice of nationalisms, where 'age-old' ideas are being renegotiated.

Nationalism from above

Traditionally, Marxist theories of nationalism have treated this form of politics as characteristic of a particular phase of capitalism: the rise of the bourgeoisie in Europe in the nineteenth century. Once this class had consolidated its hold on the state, the need for new nationalisms would lessen. With the stage of 'rising capitalism' over, therefore, any subsequent nationalisms would be irrelevant to mainstream political struggle (Blaut 1987: 26–7). Consequently, Marxist theorists were ill-prepared for the outbreak of new nationalisms in Europe in the second half of the twentieth century, and hence Nairn's oft-quoted remark that 'the theory of nation-

alism represents Marxism's great historical failure' (Nairn 1977: 329). Nairn sets himself the task of filling this theoretical void.

Nationalism and uneven development

Nationalism did not occur in isolation. As well as this new political phenomenon, the nineteenth century witnessed many other economic and social changes, often summarized by the terms 'industrialization' and 'urbanization'. Of course, these parallel developments are not a coincidence, but adequate linkages between the various strands have been difficult to find. Nairn (1977), drawing on Gellner (1964), has provided a basic model to integrate these great processes of change. Nationalism is the result of the 'tidal wave of modernization' that swept Europe in the nineteenth century. This modernization, or more precisely economic development, was not evenly spread, so that western Europe was more advanced than central and eastern Europe. Nationalism is, then, a compensatory reaction to this uneven development. Beginning in the area bordering the advanced zone, the new rising urban industrial interests were unable to compete with the more efficient core producers. They had to evolve strategies to survive, to prevent their peripheralization. We have already discussed one part of their reaction – List's new economic theories, which led to the Zollverein (customs union) in 1834. But economic policy alone is not enough. How can workers be persuaded that dearer food caused by tariffs is in their interests? The answer is to appeal to 'higher' values than mere material needs. The only resource available to these local capitalist interests not available to the metropolitan interests was their cultural affinity to their 'people'. By emphasizing differences between ethnic communities, local interests could form broad 'national' alliances with which to challenge the core. Hence the regions where nationalism blossomed as a major force on the European stage were central Europe and the unification nationalisms of Germany and Italy. From this beginning, nationalism spread like a tidal wave following the uneven development unleashed on the world by 'modernization'. Next came the separation nationalisms of eastern Europe, to be followed by the liberation nationalisms outside Europe in the twentieth century. But nationalism became more than a reaction of semi-periphery and periphery. With the demise of British hegemony, the new vigorous nationalism spread to the core as a major force in the subsequent political competition, resulting, as we have seen, in the second phase of formal imperialism. By the twentieth century, nationalism was to become the dominant ideology throughout all zones of the world-economy.

Where do the two sides of nationalism fit into this theory? So far the use of 'tradition' has been made clear, but what of the 'progress' side of the ideology? Although the urban industrial interests of the semi-periphery and periphery used images of an idyllic past to mobilize the people, they could not be true conservatives and prevent progress. The whole purpose of the strategy was to close ranks with a view to catching up with the core. This could be done only by borrowing the salient features of the modernization against which the movement was reacting. A 'medieval' Germany or Italy would be no match for a modern Britain or France. Hence the rhetoric of nationalism, with its glorification of the past, was ultimately only a cover for rapid modernization. The classic case is late-nineteenth-century Germany, where massive industrial growth went hand in hand with a popular German cultural 'revival'.

The intelligentsia have played an important role in the rise of nationalist movements everywhere (A. D. Smith 1981: 81), not least by giving them a respectability that they hardly deserved. The new class of intellectuals of nineteenth-century Europe, often from lower-middle-class backgrounds, was able to provide the historical, philosophical, ethnographic and even geographical basis of the new nationalisms. As Nairn (1977: 100) puts it, the dilemma of underdevelopment only becomes nationalism when it is 'refracted' into a society in a certain way. The intelligentsia were the agents of refraction, the most 'advanced' part of the new national middle class. Nairn (ibid.: 117) postulates a social diffusion process starting with a small intelligentsia, initially reacting to the French Revolution, which he terms phase A. This is followed by phase B, from 1815 to 1848 in Europe, where the ideology spreads through the middle classes. But it is still only a minority movement,

and this accounts for its ultimate failure in the 1848 revolutions. Phase C occurs in the second half of the nineteenth century in Europe, when it diffuses to the lower classes and modern popular nationalism is born. The intelligentsia were therefore the initial purveyors of ideas to be used for the general interests of their class. But they continued to be of use in sustaining and developing the special theories underlying each nationalism.

Finally, Gellner (1964) and Nairn (1977) point out that there is an important psychological aspect to nationalism. As Giddens (1981) observes, the dominant elites did not have to force feed the masses their nationalism; there was a receptive public out there waiting to be mobilized. The strong sentiments that were aroused indicate a particular need for identity. As well as threatening the dominant interests outside the core, the tidal wave of modernization was undermining the everyday life of ordinary people. Many were migrating to industrial zones, where their existence was routinized in ways never previously experienced. Here we have the alienation of mass society based upon the breakdown of tradition. Nationalism gave the people back their tradition and provided an identity in an alien world. It is for this reason that nationalism is particularly associated with periods of radical disruption such as wars and, as Giddens emphasizes, strong charismatic leadership. In short, nationalism acts as compensation for the alienation of mass society by producing what Anderson (1983) has termed 'imagined communities'.

Nations as imagined communities

Benedict Anderson (1983) argues that nationalism should not be linked automatically to self-consciously held political ideologies such as liberalism and socialism. Rather, nationalism has more in common with the larger cultural systems that preceded it. Nationalism is a political ideology to be sure, but it is much more than that. This is the reason why it is constantly being underestimated by theoreticians.

Societies have always been held together by more than physical coercion. The population of the former world-empires consisted of 'imagined communities' integrated by religious ideology. In these religious imagined communities, a sacred language operated as a medium for creating a whole cosmology in which the people could play their part. From the Christendom of feudal Europe, the early modern world-system inherited a religion that was fragmenting. In seventeenth-century Europe, tens of thousands of people fought and died for their religion. In twentieth-century Europe, millions of people fought and died for their country. Nation had replaced religion as the cultural system within which people could find their identity. The measure of the change was that it had become 'normal' for German Protestants to fight English Protestants and German Catholics to fight French Catholics. Religion was no longer the basis of the people's imagined communities; it had been superseded in its cultural role by nation.

Anderson (1983) argues that this new force of imagined community was made possible by the convergence of capitalism with the impact of print technology on Europe's diversity of languages. The latter ensured that the communities would be inherently limited in scope. In the seventeenth century, Latin was still widely used as the medium in intellectual circles, but as more books, newspapers and pamphlets came to be written and printed in vernacular languages, so Europe became divided into separate language-based print markets. This made it possible to conceive of a 'nation' consisting of people existing together in space and time. The key concept is simultaneity: the new technology enabled people to appreciate that their lives ran parallel with the lives of other people like themselves. Newspapers, in particular, could provide a feeling of community in time and space. Since no one person can ever meet all other persons in this community, it is an 'imagined' community like its more metaphysical religious precursors.

The geography of Anderson's theory of nationalism is distinctively different from Nairn's (1977). This is of particular interest because one of the main criticisms of Nairn's theory made by Blaut (1980) is its Eurocentric bias. Blaut finds the earliest examples of national liberation to have occurred outside Europe – in the United States in 1776 and Haiti in 1804. Anderson accommodates Blaut's point in his identification of creole nationalism as the first of three types of nationalism. This first phase of nation-

alism, from 1770 to 1830, resulted in the creation of new American states by national independence struggles. These were not mass movements but served to promote the interests of the local-born European settler groups (creoles) against imperial rule. The grievances of the creoles were expressed in the new print technology to produce 'creole communities'. The break-up of the Spanish Empire into several new states was a result of the limited geographical reach of the new technology, which created creole communities at the scale of imperial administrative divisions. Notice this first 'separatist nationalism' is not based on language – the creoles used the languages of their imperial masters – but on physical separation, with print technology providing the means of creating new communities across the Atlantic.

It is in the second phase of nationalism, from 1870 to 1920 in Europe, that languages become important in defining new communities. In the complex cultural mixture of peoples in Europe, this pluralism was being reduced by the printing press in its creation of literary languages or print languages. Rising state bureaucracies further produced a need for literacy in the 'state languages'. The end result was the creation of a whole new literate class to run affairs in both the private and public sectors. As Anderson (1983: 74) observes:

> An illiterate bourgeoisie is scarcely imaginable. Thus in world-historical terms bourgeoisies were the first classes to achieve solidarity on an essentially imagined basis.

This literate bourgeoisie had 'by the second decade of the nineteenth century, if not earlier, a model of the independent national state available for pirating' (ibid.: 78), and they proceeded to turn the nineteenth century into an 'age of nationalism'. This is the 'classical nationalism' at the heart of Nairn's theory and provides examples of the first truly popular nationalism.

Overlapping with this popular nationalism, but continuing beyond 1920 to the present day, Anderson identifies an official nationalism where the state attempts, usually successfully, to harness the popularity of nationalism to bolster its own legitimacy. This is the change in nationalism we recorded as a 'shift to the right' previously. In the established states, this involved the nationalization of the old European dynasties. This is 'the age of the primary school' (Hobsbawm 1987: 150), where the technical need for a literate and numerate workforce was supplemented by secular state and national propaganda. Geography and history became favoured vehicles for transmitting the new secular religion (Grano 1981). In its extreme form, this type of nationalism involved forced language policies, as in the Tsarist Russification of non-Russian peoples, Germanification of Danes and Poles within the German Empire and even Macaulay's famous Anglicization in British India. Hence, 'official nationalisms were conservative, not to say reactionary, policies adopted from the model of the largely spontaneous popular nationalisms that preceded them' (Anderson 1983: 102). In this destruction of plurality, dialects, or what Hobsbawm (1987: 156) provocatively calls 'languages without an army or police force', are gradually eroded in a new uniformity that is the modern nation-state. The fluidity of the language–dialect distinction is illustrated by dialects being reconstituted as languages. The example we are thinking of here is the recent break-up of Yugoslavia, which has sounded the death knell of its Serbo-Croat language. One of the first tasks of both Serbs and Croats has been to create new dictionaries that purge either Serb or Croat adulterations from new 'pure' Serbian and Croatian languages fit for their new states.

Language is only one of many symbolic nationalist tools. Official nationalisms have wallowed in monuments to their own glory. German nationalism is illustrative in its identification of national monuments that 'by revealing a universe of symbol and myth determine the secret music of our soul' (Mosse 1975: 47). In accordance with Anderson's religion analogy of nationalism, at the beginning of the twentieth century, the German architect Theodor Fischer argued that national monuments were a new type of church that facilitated the worship of the nation: 'We must create buildings . . . through which men can once more be formed into a higher, cosmic community' (Mosse 1975: 67). Note the gender of those who are to be transformed! Worship of the nation

within these buildings required that man 'removes his hat and woman restrains her tongue' (ibid.: 67). In this ideology, men were the chief architects, messengers and protectors of the nation, while women were the passive mothers of the nation, who needed protection. (We discuss such gendered differences in national identity later in this chapter and in Chapter 8.) The Nazi regime continued the construction of buildings and landscapes intended to glorify both the German nation and Nazism. In addition, the plazas and stadia constructed during the Nazi regime were designed to hold masses of people to endorse feelings of nationalism and facilitate mass participation. Of course, the role of national monuments in cementing national identity is not restricted to Germany in the first half of the twentieth century. All capital cities have their national monuments. The Millennium Dome in London was a hybrid of science and memorialization of British achievements. It served to illustrate the importance of past achievements and future challenges to national identity. In such projects we find the modern Janus alive and kicking us into the twenty-first century: the promotion of past glories and conquests acts as a rallying call for future success in breaking new technological, and subsequently trade, frontiers.

In general, national monuments are the clearest representations of nationalism from above – they depict national histories in an attempt to cover the divisions and alternative images that will exist in all nations.

Finally, it should be noted that Anderson's identification of official nationalism meets another of Blaut's (1987) important criticisms of the original Nairn theory. Nairn's study, like most treatments of nineteenth-century nationalism, neglects 'large nation' nationalism in favour of 'small nation' nationalism. Secessionist nationalism was not opposed by some 'anti-nationalist' movement but by an alternative larger nationalism. For instance, Irish nationalists were not just fighting the British as an alien force; they were confronting a larger nationalist project: British imperialism. The latter type of official nationalism should not be underplayed; big nationalisms were the generators of the popularity of the new imperialism in the late nineteenth century and thus legitimated, often democratically in elections, the oppression of other cultures worldwide. There are two sides to every nationalist conflict, and both are nationalist.

Since 1945, according to Anderson (1983), the many new states that have been created contain a mixture of creole, popular and official nationalisms. National liberation movements have been dominated by their 'youth' wings, which represent the first new 'schooled' generation. When they have successfully taken control of their new states, official nationalism comes to dominate the imagined community resulting in new reactionary policies. Anderson (ibid.: 11) considers the conflicts between communist China, communist Vietnam and communist Kampuchea in 1979 to be 'wars of world-historical importance', since they show the strength of official nationalism even in self-consciously revolutionary socialist regimes. For Anderson, nationalism remains essentially bourgeois in nature, and he remains pessimistic about the achievements of recent national liberation movements. In this, he is directly opposed to the position that Blaut (1987) has developed.

Nationalism as resistance

The major stimulus to Blaut's (1987) work has been to correct the serious neglect of anti-colonial struggles by mainstream writers on nationalism, both Marxist and non-Marxist. He states his position as two propositions:

1 You really cannot understand the modern world as a whole if you do not understand the dynamic of that part of it which has endured and struggled against colonialism.
2 There can be no adequate theory of development, of imperialism, of accumulation on a world scale, and of much more beside, if there is not an adequate theory of national liberation, of national struggle in its anti-colonial form.

(Blaut 1987: 9)

But he finds that current theories of nationalism are of little use in this task because most of them are theories of nationalism from above, whereas he inter-

prets the anti-colonial struggle as very much a case of nationalism from below. In particular, he is determined to refute what he calls the 'all-nationalism-is-bourgeois theory', first, empirically by documenting the numerous examples of national liberation movements dominated by workers and peasants, and second, by developing an alternative theory of nationalism in which full weight is given to anti-colonial struggles. Blaut is interested in nationalism as resistance, as a form of class struggle.

National struggle as class struggle

For Nairn (1977), nationalism is separate from class struggle and provides an alternative mechanism of change in the modern world. Blaut (1980) is at pains to show that nationalism is not an autonomous force but is a particular sort of class struggle. Such struggle will always be focused on the state because that is where political power lies. Class struggle then takes a particular national form when control of the state is in the hands of a foreign dominant class. A national struggle against external exploitation is just as much a class struggle as resistance to an internal dominant class. Hence nationalism is not an autonomous force.

Nationalism is not necessarily 'progressive', however. Blaut quotes approvingly Horace Davis's (1978) argument that nationalism is a neutral tool that is available to all classes or alliances of classes. Hence Nairn's (1977) identification of Nazism and fascism as the 'archetype' of nationalism is to overemphasize one extreme reactionary use of nationalism from among the wide range of uses in the modern world. The credibility of national liberation movements in the third world is thereby severely undermined by an unfair link to a politics we all abhor.

Blaut (1987) recovers what he sees as Lenin's theory of nationalist struggle as a part of the wider theory of imperialism. As the international conflict between great power nationalisms heightens, the exploitation of colonies is intensified. This produces a new phenomenon, since members of all classes are liable to suffer within the severely exploited territory. Hence there is an essentially 'multi-class struggle' directed against imperialism (Blaut 1987: 129). This differs from the earlier nineteenth-century national struggles, which were basically bourgeois battles against traditional forces to produce bourgeois states. Hence the products of this new nationalism need not be a reactionary bourgeois state but can be a revolutionary socialist state such as Cuba.

In this way, Blaut recognizes three forms of nationalism in the world today: the original bourgeois variety, which is lessening in importance; an intensified bourgeois nationalism of the large capitalist states; and the national liberation struggles of the periphery. Hence any modern theory of nationalism must confront the current opposition between great power nationalism and the nationalism of resistance in the periphery. This is precisely what Blaut's 'Leninist' theory of 'nationalism within imperialism' achieves.

Any evaluation of this theory must be centred on the interpretation of the new states created by this latest nationalism. If we follow Anderson in his disillusionment with the behaviour of revolutionary socialist states in 1979, then we would express doubts concerning the progressiveness that Blaut finds in the nationalisms of the periphery. Frank (1984: 187) identifies the south-east Asian wars of 1979 in the same light as Anderson and talks of 'nationalist cats in socialist disguise'. Samir Amin (1987), in particular, has voiced his disappointment that radical national movements have, after all, produced national bourgeois states that seem to operate in the same way as the bourgeois states produced by previous phases of nationalist movements. Hence much doubt can be cast on Blaut's assertion of a new form of nationalism in the second half of the twentieth century. However, this debate is becoming less relevant. With the demise of communist states in Europe, radical nationalisms in the old third world have come under increasing pressure to conform in the 1990s. The most important example is Cuba. Without the backing of the Soviet Union, Cuba's nationalist resistance is being fought in conditions of poverty under a continuing US trade embargo. The country is no longer able to project itself as a beacon of equality, and the fate of Cuba suggests that the old-style radical nationalism, so important to the politics of the periphery in the twentieth century, is now a phenomenon of history.

Radicalism has not disappeared but is now expressed in other forms of resistance – new social movements – which we deal with in Chapter 8.

National struggle as an anti-systemic movement

In world-systems analysis, all political struggle reflects a basic ambiguity of the capitalist world-economy: political activities have focused on the state, whereas production is based on a trans-state division of labour. The conflict and tension resulting from this structural situation has to be understood and incorporated into any theory of nationalism.

For Wallerstein (1984a), national struggle is part of what he terms the anti-systemic movements. These movements originated in the ideas and political activity of the French Revolution and developed during the nineteenth century to challenge the system. Initially, they took two separate forms. The social movement grew to articulate the demands of exploited classes, especially the urban proletariat. The national movement grew to articulate the demands of exploited peoples, especially in the semi-periphery. Both movements provided a forum for those dissatisfied with the status quo, but they identified different causes of the system's pernicious outcomes and therefore targeted different enemies. But they both agreed on the same political strategy: to control state structures within the system. Before 1917, the two arms of the anti-systemic movement were generally seen as opposing one another. This generated a large theoretical discussion on class versus nation as alternative bases for popular mobilization. After 1917, however, this dichotomy became far less important, and after 1945 all liberation movements claimed to be both national and socialist. In the recent national struggles that Blaut (1987) emphasizes, therefore, the two nineteenth-century forms of anti-systemic politics have coalesced (Phillips and Wallerstein 1986). Hence Wallerstein agrees with Blaut on the revolutionary credentials of this politics of the periphery. But he disagrees in the interpretation of the meaning of the national liberation successes.

Although Wallerstein (1984a) is supportive of national struggles as elements of anti-systemic politics, he counsels that we must understand the limitations of this politics. And here we return to the antinomy of multiple politics and one economy. Winning control of a state structure, whether as a socialist, a nationalist or a modern national liberation front, provides the opportunity to operate within the rules and constraints of the world-economy and inter-state system. This does produce genuine possibilities for tackling urgent problems within a territory: all states have some leeway within the system, some manoeuvrability, as we have argued in Chapter 4. But this politics is a very constrained politics, as all revolutionary regimes have found to the cost of their ideals – hence the disappointments expressed at the behaviour of modern socialist states as in the south-east Asian wars of 1979. It is not that 'socialists have become nationalists' but rather that by taking the strategic decision to pursue a politics leading to state power in the late nineteenth century, the social movement inevitably converged with the national movement (Taylor 1987, 1991b, 1992c). Today, there is little doubt which of these two is the more powerful politics: the vision of a world of nations remains central to world politics. As Michael Billig (1995: 4) argues:

> In our age, it seems as if an aura attends the very idea of nationhood. The rape of a motherland is far worse than the rape of actual mothers; the death of a nation is the ultimate tragedy, beyond the death of flesh and blood.

We seem not to have advanced at all from the French revolutionary analogy of defending national territory as if it were a human body.

Summary

In this section we have outlined different theoretical interpretations of nationalism and:

➤ noted the role of state elites in formulating nationalism as an antidote to class divisions;

➤ identified the role of nationalism in resisting imperialisms;

➤ discussed the ambiguous nature of nationalism as a form of resistance within the capitalist world-economy.

Renegotiating the nation?

Despite their symbiotic relationship, today there is far less talk of the end of nationalism than there is of the demise of the state. The experience of ethnic resurgence in the 1990s meant that the typical position is that 'the nation is here to stay; nationalism has proved enduring, surviving murderous wars as well as the forty-five years of postwar Bolshevik rule in Eastern Europe' (Mosse 1993: 10). However, Mosse's argument is more complex than this suggests, since he is very sensitive to the changing nature of nationalism over time. His analysis of how the ideological nature of the nation must constantly change in order to survive leads us to an investigation of the changing relationship of the nation with other scales and territorial units. Despite the horrors of ethnic cleansing, there are signs that the monolithic official nationalisms that have dominated the twentieth century are losing their powers. We argued that nationalism underwent a massive metamorphosis in the late nineteenth century. Is there evidence of a second metamorphosis occurring today?

A renegotiation of nation can only mean a departure from nationalism's monolithic tendencies, from national determinism. Quite simply, nationalism from above is under attack. Challenging the unchallengeable can come in many unexpected places. Whereas it could be reasonably anticipated that Australia might want to end the tradition of having the queen of England as its head of state, the sudden unpopularity of the monarchy in Britain in the wake of the death of Diana, Princess of Wales was surprising. Earlier queries about the cost of the monarchy have extended to querying the flag and the national anthem – how 'national' is 'God Save the Queen'? Englishness is being renegotiated. Another example of such a challenge comes from the United States, where southern state flags bearing the old Confederate standard are challenged by African-Americans as symbols of past brutalities (Leib 1995). People are no longer willing to accept that the nation and its trimmings are immutable.

In keeping with our world-systems approach, we focus upon the rescaling of national ideas to explain the renegotiation of nationalism (see also Paasi 1997).

Is the scale of ideology rupturing? This question is asked in the context of the globalization of economic activity, which has resulted in a decrease in the ability of nation-states to manage their internal affairs. As Wallace (1991: 66–7) argues:

> Inward and outward investment, multinational production, migration, mass travel, mass communications, all erode the boundaries that 19th century governments built between the national and the foreign.

Quite simply, through the creation of the nation-state, the challenge to state sovereignty has automatically resulted in a questioning of national identity. But the new identities being constructed are more complex than either a new reactive national identity or a naive embracing of a new cosmopolitan culture. According to Booth (1991: 542):

> Sovereignty is disintegrating. States are less able to perform their traditional functions. Global factors increasingly impinge on all decisions made by governments. Identity patterns are becoming more complex, as people assert local loyalties but want to share in global values and lifestyles.

But how does globalization result in a renegotiated nationalism, and what form does that nationalism take? Appadurai (1991) argues that globalization has produced a deterritorialization of identity as ethnic and national groups display interactions that transcend territorial boundaries and also engage other identities. Global media networks and migration patterns have ignited new 'imaginations' (Appadurai 1991), which are more complex than Anderson's (1983) national 'imagined community'. The imagination of the nation was a grand narrative in the sense that it denied differences within the nation as well as the validity of other identities. The complex global–local relations that are a feature of globalization have undermined the grand narrative of nationalism and facilitated more complex multiple identities.

Rodriguez (1995) defines three dimensions of change in national and ethnic identity related to the globalization of economic relations. One dimension is the global diasporas created by massive international migration, especially from the periphery to the core. Second is the growing 'global–urban context of racial and ethnic intergroup relations' (ibid.: 213). Inter-group relations within urban centres occur within the context of the role played by the city in the global economy. For example, falling world oil prices in the mid-1980s resulted in an increase in unemployment among white office workers in Houston, Texas. The related collapse of the real estate market allowed Hispanics and blacks to move into previously all-white neighbourhoods and resulted in inter-group tensions (ibid.: 214). The third dimension of global change is the growth of bi-national communities as rapid communication and transportation allowed for the reproduction of households in two different countries. For example, bus and courier services criss-cross the US–Mexican border, and Mexican radio stations have increased the power of their transmitters to reach American cities (ibid.: 215). Appadurai (1991) tells an anecdote of his family's travels from the United States to India to visit a Hindu temple, only to find that the priest they had come to see was in Houston setting up a new temple for the Indian community there.

It is within the context of global flows and connections and the way in which people experience them in different places that 'internally fractured and externally multiple' (Bondi 1993: 97) identities are constructed. These new identities constitute a politics of difference that challenges identities imposed by colonialism and nationalism (Bhabha 1990). Thus collective identities are currently in a state of flux as people try to find shared histories that have meaning and resonance but also facilitate participation in a globalized economy. As the role of the state is being renegotiated then so too must national identity. The tension between national identity and loyalty to the state is further explored in Chapter 8.

Is a new European identity emerging?

The obvious first place to look for such rescaling of identities is the European Union, initially mentioned at the beginning of the chapter. The idea of Europe can be traced back to classical Greece and its 'three-continent' model of the world, but there has never been a modern sense of Europeanness to rival its constituent nationalisms. The political project of the

EU has definite state-building goals, as we discussed in the previous chapter; here we consider its more ambitious attempts to capture political and cultural identities.

The question of a contemporary European identity has been investigated thoroughly by A. D. Smith (1995), who defines two competing models for the creation of collective identities (ibid.: 126–7). First, identities may be seen as 'socially constructed artifacts, which can be brought into being and shaped by active intervention and planning'. Thus, proactive policies by European elites can create a supranational European identity, just as these elites created a supranational institutional framework. The second model sees cultural identities as 'collective memory' or 'the precipitate of generations of shared memories and experiences'. Thus, according to this model, a European identity would evolve in a random and unplanned manner as a variety of symbols, myths and traditions from across Europe coalesced into a supranational identity that included all the peoples of Europe.

Current evidence suggests that the possibility of the first model is problematical. For instance, previous popular reactions to the imposition of the EU's bureaucracy in terms of national referenda aimed at endorsing European policies have often displayed antipathy, or just apathy, towards an increase in the power of the EU and a related decline in national sovereignty. Popular responses to an imposed loss of national identity as a supranational identity is imposed make the task of elites in the first of Smith's models quite problematic.

Evidence for the processes underlying Smith's second model is equally dubious. The construction of a transnational identity through the recognition of relevant European myths, symbols and values is difficult precisely because of the content of national identities across the continent. European history is pregnant with internal hatreds, wars, massacres and genocides. As Schlesinger (1992) points out, an important component of memory is forgetting, or collective amnesia. But can Europeans afford the luxury of forgetting the Holocaust, for instance, in light of the rise of neo-Nazism and attacks on immigrants? Because of these bitter memories and their continued

salience, a European identity would have to be abstract while simultaneously promoting an embedded solidarity (Smith 1995: 133).

After listing a number of possible symbols and values that may be deep enough to act as the scaffolding for a European identity, Smith (1995: 133–41) finds each of them lacking in sufficient geographical breadth. First, there is the possibility of Judeo-Christian values. However, there are strong and persistent divisions within the Christian Churches, let alone, as Schlesinger points out, the legacy of the Holocaust and contemporary anti-Semitism. Second, there is the Indo-European heritage of language and origin, but some European languages (Basque, Finnish, Estonian and Hungarian) have different roots. Third, there is an imperialist tradition that offers an image of white European superiority, but this is a negative image unable to integrate non-white immigrants or Bosnians and Turks. Alternatively, there are European heroes, humanists and scientists, but these are often divisive characters (such as Napoleon) or practitioners of universal, rather than exclusively European, arts and sciences (such as Shakespeare and Einstein).

In addition to the geographical variation in the salience of these European symbols is the fact that most of them are embedded in a national rather than a transnational framework. Language, religion, and past imperial and military glories have been captured by national ideologies. Any attempt to use these same myths in a transnational identity is likely to reinforce rather than reduce national identity. Smith (1995: 139) dramatically concludes:

> Without shared memories and meanings, without common symbols and myths, without shrines and ceremonies and monuments, except the bitter reminders of recent holocausts and wars, who will feel European in the depths of their being, and who will sacrifice themselves for so abstract an ideal? In short, who will die for Europe?

Clearly, Smith is very pessimistic about a new European identity in the foreseeable future. The evidence he marshalls produces a powerful argument, but there are historical precedents for the creation of successful supranational identities. Ironically, the best example is the creation of British identity after the union of England and Scotland in 1707. Although historical enemies, they both invested in the new state as 'north British' and 'south British' to create a new Protestant imperial identity against the Catholic French Other (Colley 1992). The economic successes of the new state helped to sustain the British identity above English and Scottish nationalisms into the second half of the twentieth century (Nairn 1977). It is by no means beyond the realms of possibility that renewed economic growth equivalent to that in the first decade of the European Economic Community, plus popular concern for threats from a Muslim Other in North Africa and the Middle East, could change the prospects for a European identity dramatically. Of course, the British example came at a time when nationalism was just beginning to emerge, and the EU has to deal with deeply embedded 'mature' nationalisms. But the future of the nation is there to be negotiated.

Cascadia?

A more modest attempt at renegotiating collective identity is currently under construction on the Pacific coast of North America. In contrast to a European identity, this identity is a contemporary invention.

'Cascadia: A Region Without Borders' is a transnational region bridging British Columbia and the Pacific north-western states of Washington and Oregon (Sparke 2005). This region is under the process of construction by local business elites, who are following Kenichi Ohmae's (1995) mantra, taken from his book of the same title, that we are witnessing 'the end of the nation-state' and 'the rise of the regional economies'. For Ohmae, regional economies are essential in a globalized economy in their role as conduits for cross-border trade and as the sources for 'new networks of local cross-border interdependency' (ibid.). However, as noted by Sparke, it is often hard to distinguish whether such notions are descriptions of what is actually occurring or just part of the language of local boosterism. In other words, are these interrelated networks of global and local investment actually happening and so creating Cascadia? Or is the image of Cascadia being created to attract such investment?

Our particular interest in Cascadia is not one of what came first, the image or the investment. Instead, we contrast the problem of spatial referents in European identity with the centrality of spatial themes in the embryonic Cascadian identity, or, as Sparke puts it, 'the eclipse of the *eco-logic* by the *eco-nomic* face of Cascadia' (italics in the original; also see Henkel 1993). The region of Cascadia is based upon a bioregion of temperate rainforest and the Cascades mountain range. Thus, it is expected that a regional identity will refer to the natural landscape as well as a mythology of frontier conquest. However, the purpose of this region is to act as a functional economic unit, and so such imagery must be meshed with job creation and the attraction of investment.

The following quote from the boosterist magazine *The New Pacific* does exactly as expected. Using images of natural bounty to symbolize economic prosperity and the carving out of new economic opportunities, it says:

> Across the Pacific Northwest, from Burnaby to Boise, from Corvallis to Calgary, high-tech companies have sprouted up like mushrooms in a rain forest, emerging from the lush soils of the region and attracting an inflow of technical talent from across the continent. Cascadia is not yet the heart of the technology world. But as the glow of Silicon Valley fades, it's right where the high-tech sun is rising. And it has what many regions wish they could replicate: a natural environment where entrepreneurs thrive and techies long to live.
>
> (Yang 1992)

However, such spatial referents are inadequate on their own. What is also necessary is the attraction of investors and consumers as well as the construction of a regional political identity to cement the image of Cascadia. Investors and consumers are being sought for transnational transport infrastructure, and via commercials celebrating the attractions of 'the two-nation vacation'. In addition, commentators are simultaneously portraying Cascadian interests as being dependent upon global free trade and, therefore, free of intervention from Canadian and US politics. For example, the Cascadian booster Alan Artibise (1996) argues that Cascadians possess a 'certain bemused antipathy toward the two national capitals'. Thus, common to the people of Cascadia is a western alienation from the centres of federal governance in Ottawa and Washington, DC (Sparke 2005). Such an image suggests that the Cascadian population is predisposed towards policies that promote decentralization, deregulation and free trade (ibid.). Thus, from biogeographical spatial referents to frontier mythology, the Cascadian population is given a collective identity that facilitates the future economic function of the region: a modern Janus in the making.

Competing collective commitments? Religion and nationalism

We began the chapter with one paradox. While nationalism was the dominant political ideology through the late nineteenth century, most of the twentieth-century political geographers ignored it (Knight 1982). Another paradox can be identified. As geographers and other social scientists were theorizing nationalism at the end of the twentieth century the concept was being challenged by other forms of collective identity. One of these identities is religion, which we discuss in relation to nationalism.

Religion was not a form of 'peoples' that warranted as much attention in the 1980s. This has clearly changed. The airwaves and the current affairs sections of bookshops are replete with discussions of religion as a key geopolitical trope. Initiated by Samuel Huntington's article 'The clash of civilizations' in 1993, the trend has been to identify religion as a form of identity replacing the saliency of national identity. Huntington exemplifies the tendency to essentialize religious affiliation, in the sense that all adherents to a particular religion manifest particular behavioural tendencies. Huntington's article gained notoriety for labelling Islam as a violent religion. Other writers, notably Bernard Lewis (2002) in his book *What Went Wrong?*, focused upon Islam as a religion that was 'backward' or 'undeveloped', making these claims in comparison to the Christian West. The term 'jihad' perhaps best illustrates the tendency to essentialize Islam and construct it as a geopolitical other. Jihad or to '"strive or struggle" in the way of

God' (Esposito 1998: 93) is a central component of the Islamic faith, and refers to the actions of individuals as well as Muslim communities to realize the will of God. However, the dominant representation in the western media, facilitated by the underlying tones of a 'clash of civilizations', is as holy war – with emphasis upon war.

Religion and national identity are entwined rather than separate or competing identities. Nationalist ideologies are often reinforced by religious identity and the institutions of the state may include a national religion, such as the Church of England. As a contrast, the communist regimes of the Cold War period defined themselves through their secularity. The tension between a secular communist state and the freedom of religious expression continues in China. The example we will use to illustrate the connection between religious identity and nationalism is one of extreme politics: the Ku Klux Klan in the United States. However, it is important to stress that the most common ties are 'banal' in Billig's (1995) sense of the term. They are everyday and taken for granted, such as the acceptance of the right of the Archbishop of Canterbury to offer social commentary.

At times though, the role of religion in mobilizing particular ingredients of national identity becomes prominent. The panel at the beginning of the previous chapter illustrated a religious challenge to the legitimacy of the US government. World-systems analysis offers one avenue into understanding the timing of the political usage of religious identity. The Ku Klux Klan (KKK) has been a presence in the United States since its inception in 1865 in the wake of the Civil War. However, its visibility and influence has waxed and waned (Flint 2001). In addition to its violent campaigns against African-Americans, other racial minorities and homosexuals, the KKK has been a vehicle for promoting Protestantism as a vital component of American national identity. In the 1920s it identified the Roman Catholic Church as a threat precisely because Catholics had allegiance to a Church that was aterritorial. This was seen to threaten the security and territorially based identity of an 'American' identity. Enmity was also directed towards Jews for the same political geographic reasons. The anti-Semitism of the KKK, and other extreme right-wing groups in the United States and other countries, revolves around an imagined geography of international Jewish conspiracy that threatens the taken-for-granted geography of territorial national identity.

The religious component of the KKK's ideology produces a very spatial political agenda in which threats to their particular view of what society should be emanate from particular places and require the defence of others (Flint 2001). Two periods in the history of the state of Pennsylvania illustrate this political geography of nation, religion and hate, and place it within the dynamics of the capitalist world-economy. In the 1930s the KKK in Pennsylvania showed a clear base in the rural northern counties (Figure 5.1). This is unsurprising given its mobilization against the 'sin' of the modern cities epitomized by the spread of jazz and flapper girls (Leuchtenburg 1958; Shideler 1973). The United States was becoming an urban country. For the first time in US history the 1920 census found the rural population to be a minority. As the economy changed the KKK mobilized a threat to established identities and ways of life, and the identification of 'sin' was a key vehicle (Annan 1967).

The pattern of hate activity in Pennsylvania in the 1990s is very different. Although the KKK was the dominant hate group in the 1920s, a number of hate groups existed in the 1990s; hence, the pattern of hate crimes is a better representation of the geographical manifestation of hate in the latter period. The location of hate crimes has shifted to the suburban counties in the east of the state (Figure 5.2). Over the course of the twentieth century the idealized notion of what constituted Americanism had shifted from a rural setting to suburbia (Fishman 1987; Stilgoe 1982). This change is related to the role of the United States, as hegemonic power, in defining and

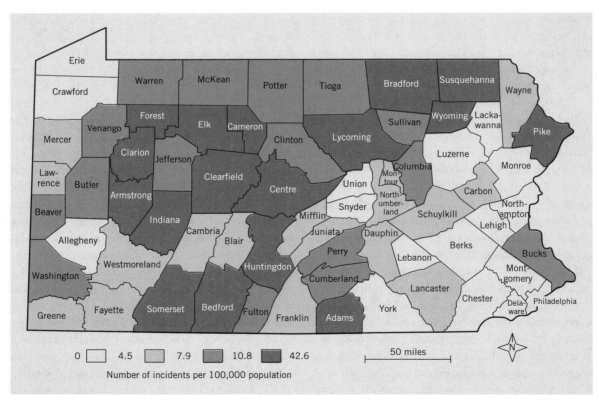

Figure 5.1 Ku Klux Klan Klaverns (cells) per 100,000 people in Pennsylvania, 1920.
Source: Flint (2001) p. 774.

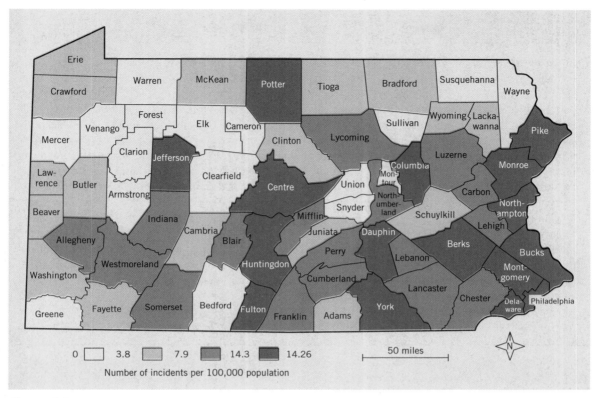

Figure 5.2 Hate incidents per 100,000 people in Pennsylvania, 1984–99.
Source: Flint (2001) p. 781.

disseminating a prime modernity around an idealized image of the suburban lifestyle as modern (see page 292). Suburbia was a racialized landscape, it was the destination of mainly white families leaving inner-city areas, the process of 'white flight' (Danielson 1976). However, in the 1990s the relative racial homogeneity of suburbia was changing as other racial groups began to move to the suburbs. An essential component of the US society, the self-segregation of races through the operation of the housing market, was being partially, though not dramatically, eroded (Crump 2004). Hence, it is not surprising that hate crimes (not necessarily committed by members of the KKK) predominated in suburban areas; that was where races mixed and it was also a challenge to what had become viewed as the landscape of the twentieth-century American way of life.

Religious and national identities are integrated to produce the particularity of national identities.

Such integration facilitates the territorialization of religious identity: in the KKK's ideology America is a Protestant nation. However, that is only one form of the interaction between religious and national identity. Religious national identity may also be a challenge to the ideology of nationhood, in the sense of a territorial and homogeneous collective identity. Tensions over the wearing of traditional Islamic clothing and over customs, especially the practices of veiling and covering hair, in schools in Britain and France are a contemporary case in point. State schools in all countries have played a vital role in establishing a sense of national identity through disseminating a common sense of national history and, hence, common identity. In some cases, British or French citizens wearing Islamic dress are seen by some as a threat to an understanding of commonality in cultural behaviour that defines the particularity of a nation.

Building nations and killing Christians

In March 2006, as the war in Iraq was becoming increasingly unpopular amongst US voters, President George W. Bush's administration faced a particularly embarrassing issue. In Afghanistan, the country that had been 'liberated' from the Islamic fundamentalism laws of the Taliban in the first military action of the War on Terrorism, an Afghan man, Abdul Rahman, was facing the death penalty for converting to Christianity. Renouncing the Muslim faith, and converting to another religion, is illegal under Muslim law. Although the pro-western President of Afghanistan, Hamid Karzai, did not favour this punishment, the Islamic clerics were pushing for the traditional punishment. Even if Rahman was released it was understood that his life was in danger if he remained in Afghanistan. He was released and given asylum abroad.

This incident illustrates the difficulty, if not

impossibility, of creating a sense of nationhood that is founded on the imposition of western style democracy rather than on a sense of collective identity. Statements by the United States are informative by what they did not say. The US embassy in Kabul was reported as saying: 'We have a profound commitment to a person's fundamental freedom to choose how he or she worships and we must be true to that commitment', and Secretary of State Condoleezza Rice stated: 'We've been very clear with the Afghan government that it has to understand the vital importance of religious freedom to democracy.' There is no mention of an Afghan nation in these statements. The scales addressed are the individual and confession.

Source: 'Afghan Christian convert is released', CNN.com, 28 March 2006. http://www.cnn.com/2006/WORLD/asiapcf/03/28/christian.convert/index.html. Accessed 28 March 2006.

The gendered nation: feminist understandings of the nation

According to Mayer (2004) consideration of the nation from a feminist perspective was initiated by non-western writers. Identification of the role of women in anti-colonialist movements was linked to theoretical developments (under the rubric of post-structuralism) that investigated the sources and nature of power relations. In combination, the differential role of women in constructing nations and nationalism as well as the sexist nature of the nation-state was brought into focus. The feminist critique identified the relative exclusion of women from the public sphere of government and business operations (see Table 5.2). In addition, the way in which women and men are portrayed in nationalist ideology was highlighted. We stress that this is a matter of the banal and the everyday, just as in our discussion of religion and nationalism. However, it is the extreme right that brings such processes into the clearest focus.

Table 5.2 Regional averages of the percentage of women in parliament, April 2006

Region	Mean percentage of women representatives in upper and lower houses
Nordic countries	40%
Americas	21%
Europe (non-Nordic)	18%
Sub-Saharan Africa	17%
Asia	16%
Pacific	14%
Arab states	8%

Source: Inter-Parliamentary Union, 'Women in parliaments', 30 April 2006. http://www.ipu.org/english/home.htm. Accessed 3 May 2006.

Nationalist struggles often include calls for women to fulfil the biological function of sexual reproduction as part of the political project of reproducing the nation (Fluri and Dowler 2004). Such calls could be found, for example, in Nazi Germany, as well as in the statements of the Irish Republican Army and

Palestinian nationalist groups. The flip side is programmes of reproductive control restricting the reproduction of racial and ethnic groups that are deemed to threaten the racial purity of the nation. Again Nazi Germany provides the strongest historical example. Within many historical and contemporary wars rape is a weapon, partially undertaken to destroy any sense of national purity (Mayer 2004).

The sexism of the ideology of nationalism revolves around the sentiment that women's roles should be relegated to the private sphere of the household – to give birth to and raise subsequent generations of the nation – while it is the duty of men to manage and defend the nation (Enloe 1989). It will not be hard to think of a war movie that includes the scene of men marching off to war as the tearful women remain at home. Such practice requires an ideological understanding of the household as well as of the nation: the household or home as a haven from external threats and the source of national reproduction.

Fluri and Dowler's (2004) analysis of white separatist groups in the United States is a vivid illustration of the combination of sexist and racist practices and ideologies in the mutual construction of households and nations. One female white separatist activist, Jane, is quoted as saying:

My children understand the importance of not mixing the races and have agreed to not date other races. It's not about hating; it's about the survival of our people. For example, my older sister married a half Mexican, half Japanese man. She had two children. My brother married a Mexican woman and I'm hoping they don't have children. Now, out of my siblings, there are my three white children when there could be five and possibly more. Do you see how we are a dying race in just this instance? I do not want our people to become extinct. We are the only ones on the planet with blue and green eyes. All the other races are colored basically the same, brown skin, brown eyes, and black hair. Only the Europeans have the genetic traits we do.

(Quoted in Fluri and Dowler 2004: 75)

This quote reveals a number of key points: a belief in biologically pure and distinct races, the political need to racialize nations (Mexico and Japan as non-white), and the political role of a woman to bear

racially pure children and socialize them in a manner that ensures the future of a racially pure nation. Jane sees biological reproduction as part of the national project. But the sexism extends beyond the reproductive role. It also requires defining gender roles in which women and men are identified as playing specific roles in the socialization of the next generation. Another white racialist activist is quoted as saying:

Children are an integral part of my culture within the local white racialist community I belong to, and we all help to teach and nurture those children. For example, I am helping and encouraging one young man to build a website for our church. An elder of our church recently took another young man out and successfully taught him to hunt and prepare a deer for meat. We all take great joy in seeing and sharing the accomplishments of our children as they grow and learn from newborn stages into adulthood.

(Quoted in Fluri and Dowler 2004: 80)

The beliefs and practices of white racialists are a stark, and extreme, exemplification of how sexist gender roles are implicated in the mutual construction of a patriarchal household and nation. Another point can be made too, and that is the tenuous relationship between nation and state. White racialists in the United States view the federal government as illegitimate because of its promotion of a diverse multicultural society. One white racialist group, the National Alliance, claims:

With the growth of mass democracy (the abolition of poll taxes and other qualifications for voters, the enfranchisement of women and of non-whites), the rise in the influence of the mass media on public opinion, and the insinuation of the Jews into a position of control over the media, the U.S. government was gradually transformed into the malignant monster it is today: the single most dangerous and destructive enemy our race has ever known.

(Quoted in Fluri and Dowler 2004: 74)

In summary, the construction of the household, the nation, the state and race are intertwined in different and particular political projects. The example of the contemporary white racialist movement in the United States seen through a feminist perspective highlights the connections between these institutions. Although we have discussed an extreme case, the connections between gender roles and nationalism, and the mutual constructions of household, state and nation are pervasive. Feminist scholarship plays an essential role in identifying not only the multitude of power relations and their interconnections but also the ways in which established notions of the nation and gender relations are being challenged. However, although these relations are dynamic, and their complexity now identified, it does not necessarily mean that the nation and the nation-state are being challenged to such a degree that their demise is in sight. We conclude the chapter with a discussion of the persistence of the nation and its role in the capitalist world-economy.

Summary

In this section we have discussed contemporary changes in the form of nationalism and noted:

➤ the manner in which globalization and flows across borders have changed national identity;

➤ the existence of competing collective identities, and especially the relevance of religious communities;

➤ how, from a feminist perspective, divisions and inequities within nations may challenge the legitimacy of those nations.

■ Synthesis: the power of nationalism

Earlier in the chapter we identified the mismatch between the power of nationalism and its philosophical poverty (Anderson 1983). Hence it is essential that we come to terms with this power in our conclusion to this chapter. The first point to make is that we do not see nationalism and the nation-state becoming irrelevant in the short or medium term. We agree with A. D. Smith (1995: 153–9), who gives

three reasons why the nation is here to stay. First, nationalism is politically necessary to anchor the multi-state system upon the principle of popular sovereignty. Second, national myths provide social cohesion and the basis of political action. Third, the nation is historically embedded by being the heir to premodern ethnic identity. For Smith, the nation is a functional institution that is unlikely to be replaced, because of its relationship to primordial identities. We would emphasize the territorial link, however, as the crucial hinge in the power of the modern nation-state. Here we attempt a synthesis of theories of the nation and nationalism by returning to the institutional vortex that we introduced in Chapter 1 wherein we located both states and nations. Also, by returning to this framework we provide a chance to take stock of where our argument has led us in terms of the four key institutions introduced in Chapter 1.

Beyond the institutional vortex

According to Wallerstein (1984a), the four key institutions of state, class, nation and household exist in an institutional vortex, each supporting and sustaining the other. The image we are presented with is a kaleidoscope of interlocking institutions that define our social world and its politics. This metaphor captures the complexity of the situation but not the concentration of power that has occurred in the last century. As well as the variety of relationships within and between these institutions, there has been a real accretion of power centred on the state as nation-state. In effect, two of the institutions have coalesced in a pooling of power potentials that has come to dominate our contemporary world. That is the real message of the interchangeability of the terms 'state' and 'nation' in everyday language.

We have seen that the crucial period when the state began to harness the political power of nationalism was the later decades of the nineteenth century. This period is also important for the strategic decisions made by non-nationalist movements. Basically, the state began to be seen as an instrument that could be used to achieve radical ends. That is to say that the instrumental theory of the state we presented above was extended beyond the manoeuvrings of the political elite and their economic allies. With extensions to the franchise, what had been a dominating state came to look more and more like an enabling state (Taylor 1991b). The women's movement, with its great campaign for female suffrage, and the socialists, with their organization into political parties to contest state elections, both focused their politics on the state. The state has been at the centre of the vortex sucking in power with the indispensable help of the nation. Once subjects became converted into citizens with rights, then states could be equated with the collective citizenry, the nation. States as nation-states have become our imaginary communities for whom millions of individuals have laid down their lives in the twentieth century. Such power is awesome.

Nation-state: the territorial link

The concentration of power into the dual institution of nation-state has been a very complex and contradictory process that is ongoing. One crucial feature that brings much of our previous discussion together is territory. Both institutions, state and nation, are distinctive in their relationship to space. Whereas all institutions, major and minor, use space and operate within space, only state and nation have a relationship to a particular segment of space, a place. According to Mann (1986), the power of all states has been territory-based; in the modern world-system, the state is defined by possession of its sovereign territory. According to Anderson (1986), nationalism is formally a territorial ideology; a nation without its homeland (fatherland or motherland) is unthinkable. It is the equating of these two territorial essences – sovereign territory equals national homeland – that has made possible the dual institution that is the nation-state.

Several geographers have developed political theories of regional class coalitions based upon shared commitment to place. For instance, Harvey (1985) argues that segments of capital that are place-bound, such as local banks and property developers, can make common cause with local labour interests to produce a place politics that transcends traditional political differences. City boosterism in the United States and regional development agencies in Europe are often expressions of such politics, and we consider them in Chapter 7. State politics is a place politics related to the above in that its economic policies consist of large-scale boosterism and development based upon implicit class alliance. But this class alliance is qualitatively different; it is a nation. The place that is the state is also an imagined community, with all that that entails for individual identity.

At its very core, the nation-state provides individuals – its citizens, its nationals – with their fundamental space–time identities. This is our final conclusion. We have arrived at what we may term a 'double-Janus' model of the state. From the previous chapter's topological model we know that spatially the state looks inwards to its civil society and outwards to the inter-state system. Individuals are quite literally located in the world-system in terms of who and where they are and who and where they are not. With the addition of nation to the state, Nairn's Janus model is added looking backwards to past national struggles and forwards to a secure national future. Individuals are given identity in terms of where they have come from and where they are going to. In summary, nation-states define the space–time dimensions of the imagined communities that we all belong to. Herein lies the power of the nation-state, the pivot around which the politics of the modern world-system is conducted.

Chapter summary

In this chapter we have:

➤ noted the theoretical heritage that is the foundation and foil for contemporary political geographies of the nation;

➤ emphasized the historical and geographical components of nationalism via the term 'double Janus';

➤ highlighted the different forms that the politics of nationalism takes;

➤ exemplified the contentious and often violent politics between nation and state;

➤ noted the role that the politics of nationalism plays within the politics of economic and racial difference in the capitalist world-economy;

➤ discussed the changing form of national identity within the context of globalization;

➤ illustrated the relationship between the nation and the politics of religion;

➤ shown the gendered nature of nationalism;

➤ discussed the role of nationalism and the nation-state within the institutions of the capitalist world-economy.

The nation is the dominant form of political identity in the capitalist world-economy, ensuring widespread loyalty to the state and a general acceptance that politics is organized within a mosaic of sovereign states. The politics of nationalism is intertwined with the politics of gender, race and class in what we have introduced as the institutional vortex. The context of war is likely to promote nationalist allegiance, whereas, on the other hand, the economic links of globalization and the universalism of religious beliefs may disrupt and transcend national identity. The politics of nationalism is also susceptible to the politics of marginalization within national groups.

Key glossary terms from Chapter 5

administration	conservative	fundamentalism	League of Nations
Arab League	constitution	geopolitical world order	liberal
autonomy	containment	geopolitics	liberation movement
boundary	core	globalization	Marxism
bourgeoisie	decentralization	government	militarism
capital city	democracy	hegemony	minorities
capitalism	elite	home	Napoleonic Wars
capitalist world-	empire	homeland	nation
economy	'empire'	households	national determinism
centralization	European Union (EU)	iconography	national self-
civil society	fascism	idealism	determination
classes	federation	ideology	nationalism
Cold War	First World War	imperialism	nation-state
colonialism	formal imperialism	instrumental theory of	opposition
colony	franchise	the state	peoples
commonwealth	free trade	inter-state system	periphery
communism	frontier	Islam	place

plebiscite	right-wing	space	United Nations
pluralism	secession	state	world-economy
political parties	Second World War	suffrage	world-empire
power	semi-periphery	territoriality	world-system
racism	socialism	third world	world-systems analysis
realism	sovereignty	Treaty of Westphalia	

Suggested reading

Billig, M. (1995) *Banal Nationalism*. Thousand Oaks, CA: Sage Publications. A short and accessible monograph outlining how nationalism is constructed by everyday behaviour.

Hutchinson, J. and Smith, A. D. (1994) *Nationalism*. Oxford: Oxford University Press. This reader brings together the essential writings on the theory and practice of nationalism in an abridged format.

Smith, A. D. (2001) *Nationalism: Theory, Ideology, History*. Cambridge: Polity Press. An accessible introduction to the history and theory of nationalism, with an extended discussion of nationalism and globalization.

Yuval-Davis, N. (1997) *Gender and Nation*. Thousand Oaks, CA: Sage Publications. This essential guide to the feminist discussion of nationalism addresses the role of gender relations in the representation of the nation and the connections to citizenship, militarism and ethnicity.

Activities

1 Consider the national myths of your own country, or any other of your choosing. What are the popular historical characters and legends? What are seen as key historical events? Then, taking a feminist perspective, consider the gender roles that are exemplified by the common understanding of these legends and characters. In what way do the male characters occupy and define public spaces and the female characters private spaces? Do contemporary representations of these gendered roles differ from traditional ones?

2 Write down three or four attributes of what could be classified as the 'national values' of your country. Then ask family members, especially older generations, to do the same. In what way do the values, and attachment to them, differ from generation to generation? If you are in a class, do you see the values identified vary by gender and cultural identity?

3 Consider your national homeland (or another of your choosing). What are the revered landscapes? Are there places that have especial national meaning? Does the capital city 'represent the nation' in its architecture and monuments? What groups are missing or underrepresented in the way the national homeland is depicted? (Hint: a good place to start are official national tourist websites and commercial guidebooks and their websites.) Note how official and commercial information might differ in emphasis.

VILLAGERS VOTE FOR TOWN CHIEF IN CHONGQING.
Source: China Photos/Getty images

Chapter 6

Rethinking electoral geography

What this chapter covers

If success can be measured by quantity of production, then electoral geography has been the success story of modern political geography. In the past three decades there have been hundreds of studies on the geography of elections, so much so that some have argued that the growth has been 'disproportionate in relation to the general needs of political geography' (Muir 1981: 204). In fact, as Muir (ibid.: 203) pointed out, ideas on the role of electoral geography in political geography have ranged from those who imply that it is 'the very core and substance of political geography' to those who feel that it does not belong in political geography at all. Obviously, the position taken in such a debate depends ultimately on the definition of political geography being employed. In our political geography, elections play a key role at the scale of ideology in channelling conflicts safely into constitutional arenas. Hence we need to consider electoral geography, but not necessarily in its usual form.

There is a second debate on the nature of electoral geography that is much more important. Given the rapid increase in these studies, what are they adding to our knowledge of political geography? It is not at all clear where electoral geography has been leading. The goal of most studies seems to be nothing more than understanding the particular situation under consideration. The result has been a general failure to link geographies of elections together into a coherent body of knowledge. In short, we have a bitty and uncoordinated pattern of researches, which has produced a large number of isolated findings but few generalizations. There are some exceptions, and these are described below, but on the whole quality has lagged behind quantity of production, and few would now claim electoral geography as a success story. This is no longer acceptable at a time when the spread of democratic practices across the world provides some hope for humanizing globalization. To be credible, contemporary political geography must contribute to debates on democratization and to act as an empirical vehicle to make 'theoretical connections between the actions of voters within localities and global flows and structures' (Flint 2002: 395). Hence the need for some serious rethinking of electoral geography.

Although electoral geography has not been explicit in its theorizing, its implicit theory is easy to identify.

In the main, electoral geographers have simply accepted the political assumptions of the core countries in which they study. These assumptions can be summarized by the term 'liberal democracy', and in the first section of this chapter we review electoral studies in geography as a 'liberal heritage'. Unfortunately, the narrowness of these past studies has led to major omissions in coverage. In particular, comparison between elections in different countries is only conspicuous by its absence in electoral geography. Hence we have to 'bolster' this liberal heritage by drawing on some work in comparative politics that enables us to appreciate more fully the global implications of the liberal assumptions. The critique of this work leads on to our world-systems interpretation of elections, which forms the framework for the remainder of the chapter.

The most conspicuous omission from traditional electoral geography has been elections in countries of the semi-periphery and periphery. Our world-systems interpretation helps us to overcome this deficiency by providing a framework for studying elections in both core and periphery. Hence the two substantive sections after the heritage section deal with elections in the core and 'beyond the core', respectively. The thesis presented argues that these represent two very different sets of political processes. The implications of this for the current success of democratization across the world are very serious. Here we intersect with current US foreign policy that treats 'democracy' as a political panacea for reforming former enemy states. This emerged as part of post-Cold War American triumphalism that argued liberal democracy to be the 'end of history' – this political practice had shown itself to be superior and therefore the final stage of political change (Fukuyama 1992). But this is western universalism at its worst. We treat liberal democracy as a particular political process that developed through conflicts specifically in core countries from the nineteenth century to the present. It is arrogant to think this process can be taken out of this context and simply transplanted into other, very different contexts and made to 'work' immediately. Democratic institutions can be legislated into being in any state but this does not translate into diffusion of liberal democracy. Rather there are multiple 'hybrid politics' that may or may not work for the state in question. We pose the question about what elections do in these circumstances, specifically in terms of their geographical outcomes.

The first testbed of post-Cold War diffusion of democracy was the former Soviet bloc countries of eastern Europe. We ask, in particular, where do the former 'second world' countries emerging from communist rule fit into this core–periphery dichotomy of political process? The traditional emphasis on liberal assumptions in electoral geography is hardly surprising given that during the Cold War competitive elections were a key differentiating characteristic between East and West. With the end of the Cold War, western economic and political processes began to be transferred to the former communist countries, extending the geographical range of electoral studies. New electoral geographies are being described, for instance the process of establishing democracy and competitive party politics in Ukraine (O'Loughlin and Bell 1999; O'Loughlin 2001), to give fascinating insights into this new world of elections. Our world-systems analysis poses the key question, however: are the countries of eastern Europe that have newly experienced competitive elections going to develop a politics that is core-like or one similar to that of the third world? It is too early to say, but our theoretical approach provides the basis for understanding which is more likely. We conclude this chapter by considering the critical issue of whether 'democracy' really is 'on the move' across the world in the twenty-first century.

Chapter framework

The chapter relates the empirical analysis of elections to the following components of political geography that we introduced in Chapter 1:

➤ Voting is *conceptualized* within a model of liberal social democracy.

➤ The geography of liberal social democracy is understood within the *persistent* and *spatial differences* of the capitalist world-economy.

➤ The individual act of voting is understood within the related *scales* of local context, nation-state political systems, and the capitalist world-economy.

■ The liberal heritage

In some ways, electoral geography is like geopolitics in being represented in the work of some of the founding fathers of modern geography. In the case of electoral geography, we can cite André Siegfried of the French regional school, whose 1913 study of western France under the Third Republic has long been regarded as a classic of its genre. Siegfried is usually considered the father of electoral geography because of his method of mapping election results and comparing them with maps of possible explanatory factors. At about the same time, Carl Sauer (1918) was contributing to the perennial American debate on how to define congressional districts. As the founder of America's cultural–regional school of geography, it is perhaps not surprising that his solution involved representation by geographical region. Other studies could be cited, but, until the 1960s, electoral studies in geography were sporadic with no sustained effort, except perhaps in France.

All this changed with the so-called quantitative revolution in geography. This affected human geography in particular and resulted in the decline of qualitative regional studies and the rise of quantitative systematic studies, especially in economic and urban geography. As has often been noted, most of political geography was passed by in these intellectual upheavals, but this was not so for electoral geography. The regular publication of volumes of electoral data neatly organized by electoral areas provided a wealth of material for the new quantitatively oriented geographers (Taylor 1978). Hence the massive rise of electoral studies in geography and the 'disproportionate effort' that electoral geography attracted within political geography. The first part of this section describes this quantitative electoral geography; the second part describes an attempt to coordinate this effort in a systems framework; and the final part looks beyond these electoral studies to a critique of a global model of liberal democracy, which geographers never achieved.

Quantitative geographies of voting

There were three aspects of the new quantitative approach that were applied to electoral geography: geographies of voting; geographical influences upon voting; and geographical analyses of electoral districts. This trilogy of electoral geography studies was first identified by McPhail (1971) and subsequently used by Busteed (1975) and Taylor and Johnston (1979). The most common type of electoral geography study was the first, in which standard statistical analyses were widely applied to geographical patterns of voting. This is what we focus upon in this section. Within geography generally, there has been a growing interest in the role of spatial factors in human behaviour since the 1960s, and this is what has been reflected in geographical influences in elections. This has led to concern about the 'neighbourhood effect' in election results, which we deal with briefly in Chapter 7. In terms of the geographical study of electoral districts, probability modelling of spatial distributions has been employed in studies of the geography of representation. This geography of representation is treated in Chapter 7.

The geography of voting follows in the Siegfried tradition in that its purpose is the explanation of particular voting maps. In modern geography of voting studies, however, cartographic comparisons have given way to statistical analysis. It is these studies that have borne the brunt of the criticism of electoral geography. Generally speaking, the explanation of a particular voting pattern has become an end in itself, with the result that myriad sound quantitative analyses added up to very little in terms of understanding elections. Taylor and Johnston (1979) attempted to overcome this fundamental deficiency by introducing the work of Stein Rokkan (1970) into electoral geography to provide a framework within which geographies of voting could be interpreted. This model is particularly salient to geographers because it incorporates a spatial dimension into its analysis.

Rokkan's model of party cleavages

Rokkan (1970) argued that there have been four major conflicts in modern Europe that resulted from the two fundamental processes of modernization: the national revolution emanating from France and the Industrial Revolution emanating from Britain. Each of these processes produces two potential conflicts. These are subject versus dominant culture and Church versus state from the national revolution, and agriculture versus industry and capital versus labour for the Industrial Revolution. Each conflict may produce a social cleavage within any one country, but each country has a unique history in which these conflicts are played out. Rokkan argued that the particular mixtures of cleavages that occur in European states deriving from these conflicts are reflected today in the variety of political party systems in Europe.

Rokkan terms this a model of alternative alliances and oppositions. For each European state, the nation-building group made alliances with one side or other in these various conflicts, forcing the opposition to forge a counter-alliance. Before the full effects of franchise extensions were felt (c.1900) nation builders in the dominant culture had a choice of secular or religious alliances and agricultural or industrial alliances. According to Rokkan, the choices made in these conflicts have largely determined the great variety of political parties on the centre and right of European politics. After 1900, as franchise reforms became effective, the final capital–labour cleavage came into operation to produce much more uniformity on the left in European politics. For example, in Britain, the nation-building group allied with a national Church and landed interests (the Tories and later the Conservatives) against an alliance of dissenters, industrial interests and minority cultures (the Whigs, Radicals and later the Liberals). After 1900, the rise of the Labour Party produced a particular right–centre–left cleavage represented by three political parties whose pattern of support in terms of votes still reflects these historical cleavages. Although the Conservatives have eclipsed the Liberals and taken their 'industrial interest' vote, the latter party remained strong in non-conformist and particularly peripheral zones of Britain. Throughout their decline, Liberals maintained MPs from Celtic Britain (Cornwall, central Wales and northern Scotland) when everywhere else rejected their candidates.

Case study

The persistence of geography: residual analysis of UK voting patterns

It is national core–periphery cleavages that have attracted most geographical research attention, as might be expected. Even in the longest established nation-states, political mobilization may not be complete, as we have seen in the discussion of the resurgence of national separatism in Chapter 5. Models incorporating a geographical (core–periphery) dimension, such as Rokkan's, imply a lessening of the relevance of location as socio-economic criteria come to dominate modern politics. Hence to many social scientists recent expressions of separation nationalism were unexpected. But they did not appear out of nothing. In Britain, for instance, Scottish and Welsh nationalism may have become a significant political force only in the late

1960s and 1970s, but that does not mean that before then Scottish and Welsh voters all behaved as typical British citizens. In a classic study, Hechter (1975) showed that sectionalism persisted in British politics throughout both the two-party systems of Conservative–Liberal and Conservative–Labour. Figure 6.1 shows a small part of his results. Eight elections from 1885 to 1966 were analysed by counties in terms of predicting the Conservative vote from seven socio-economic variables. The residuals from each analysis indicate how the vote percentage in a particular county deviates from that expected due to its socio-economic structure. Positive residuals indicate a pro-Conservative bias and negative residuals an anti-Conservative bias.

The average residuals over all eight elections are distributed in Figure 6.1(a), where two distinct 'tails' are prominent. On the pro-Conservative side there are three overwhelmingly strong counties – the three most Protestant Ulster counties, which consistently voted Unionist and hence Conservative in this period. The anti-Conservative tail is larger and includes fourteen counties, all in Celtic Wales and Scotland. The geography of these extreme residuals is shown in Figure 6.1(b), which defines three contiguous sections that have not conformed to normal British voting behaviour. Quite simply, British elections were never completely nationalized, and sectionalism persisted to be reactivated in part as new 'nationalisms' after 1966.

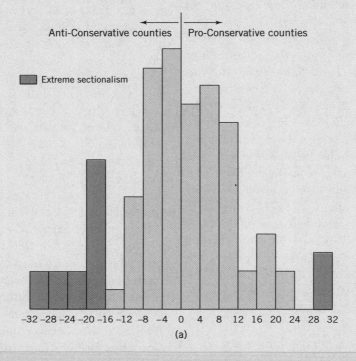

Figure 6.1 Sectionalism in British elections, 1885–1966: (a) distribution of residuals.

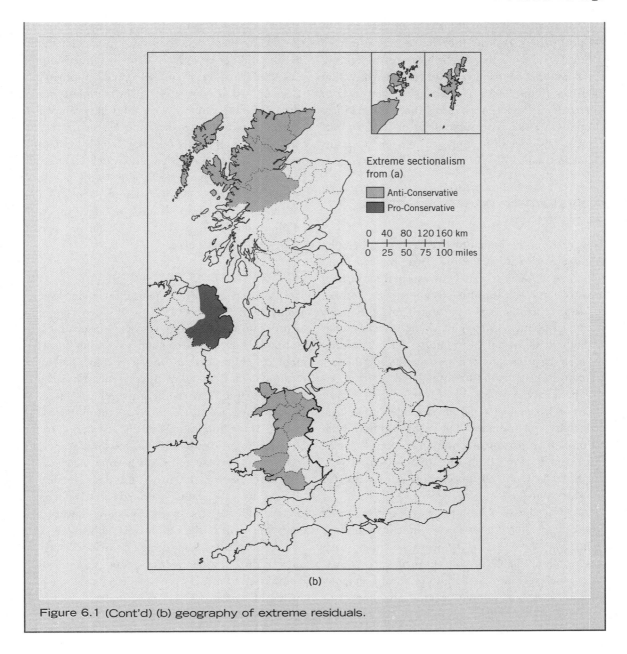

(b)

Figure 6.1 (Cont'd) (b) geography of extreme residuals.

Rokkan identifies eight different patterns of nation-building alliances and opposition counter-alliances, all of which are reflected in at least one western European state today. This particular model summarizes European history and is not, therefore, generally applicable outside Europe. Nevertheless, the process of modernization that Rokkan describes has spread beyond Europe, and similar cleavages can be identified in non-European countries. This is further explored in Taylor and Johnston (1979: 196–206).

For a recent evaluation of the relevance of Rokkan's model to electoral geography, see Johnston et al. (1990). The use of Rokkan's model exemplified in the case study of the UK is illustrative of the way in which electoral geographers can contribute to theoretical discussions in human geography. National elections

199

are not treated as a single pattern of responses to political parties. Instead, they are seen as a form of political action, one that can be measured in terms of votes cast, that contributes to the social construction of places (Flint 1998). This change in the scale of interest is part of a general recognition that 'national elections' are not geographically as national in outcome as has often been supposed. We follow up this theme in Chapter 8.

A systems model of electoral geography

The organization of quantitative electoral geography is in many ways unsatisfactory. Two problems stand out. First, the subject matter consists of three distinct areas of interest, with little cross-reference. Second, the subject matter is not integrated into the mainstream of political geography. To be sure, examples of both linkages can be found – the relationship between the geography of representation and the segregation in geographies of voting, and the relationship between the geography of voting in peripheries and more general core–periphery models in political geography, for example – but such links have been no more than perfunctory and undeveloped. For all the research effort from the 1970s, electoral geography had become an uncoordinated and isolated sub-discipline.

The way out of this impasse was not for more empirical research but for a new synthetic framework to integrate these various themes. Such an approach was readily available in conventional systems analysis, which was introduced into political geography in the 1970s but without making any important

impact. Although textbooks (for example, Bergman 1975; Muir 1981) paid lip service to systems thinking, this did not go as far as actually influencing the presentation of most of their political geography (Burnett and Taylor 1981). The simplest model employed was Easton's (1965) political system which could be reduced to just four elements: input, throughput, output and feedback. Such a framework seems to fit electoral geography, which is, after all, commonly thought of as dealing with 'electoral systems'. Johnston (1979) attempted to use this approach to integrate studies of elections with administrative systems, but no satisfactory overall framework was produced.

Input, throughput and output

The most explicit use of systems thinking in electoral geography is Taylor's (1978) use of systems concepts to order his discussion in a review of electoral geography (Figure 6.2). Geography of voting and geographical influences on voting become the input to the system. Geography of representation becomes the throughput, leaving the geographical effects of the resulting legislature/executive as the output of the system. Three implications of this reordering of the literature were identified. First, electoral geography was given a purpose and focus beyond the elections themselves. Second, the current emphasis on one part of the system – the input – seemed excessive in comparison with the other parts. Third, in contrast, the massive neglect of the output was highlighted. This was the reason electoral geography seemed to have no clear direction. By treating elections as an end in themselves, the actual purpose of elections – the producing of legislatures and governments – had been

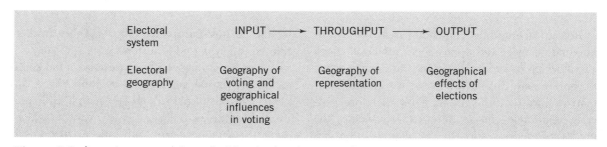

Figure 6.2 A systems model applied to electoral geography.

Power and the systems model of electoral geography

The US presidential election of 2000 was a historical event. Republican candidate George W. Bush trailed Democrat Al Gore in the popular vote (the actual number of votes cast across the country). However, the presidency is gained through a system of weighted representation based on the population of each state, the Electoral College (Shelley 2002). Although several states, including Florida, were decided by less than 2.5 per cent of popular votes (Archer 2002), the result in the Electoral College came down to who had received the most votes in one pivotal state, Florida. The Republican candidate George W. Bush ultimately gained the presidency at the expense of his Democrat rival Al Gore on the basis of decisions made by the Federal Supreme Court (Webster 2002). The details of the cases rest upon the scale of government, namely whether the Federal Supreme Court had the right to overrule decisions made by Florida's Supreme Court and impose its own decisions. In the end, the Federal Supreme Court did assert its authority and made rulings about the ability and manner of the recount that led Al Gore to concede defeat.

On the one hand, the case could be made that the US presidential election illustrates the efficacy of the liberal assumptions of the systems model (Figure 6.2). The institutions of the electoral system, including the courts, facilitated the translation of votes (input) into throughput (representation). However, the decisions made in the Federal Supreme Court were highly political (Webster 2002). In essence, conservative (read pro-Republican) judges went against previous decision-making logic and ruled that the federal state could and should intervene within states and not defer to the decisions of the state Supreme Court, in this case Florida. In this pivotal decision that put President George W. Bush in the White House the conservative judges reversed their traditional support of 'state's rights'. The furore around the decision of the federal Supreme Court provides us with one lesson: the systems model of elections is naive in its exclusion of conflicts and the social bases of power.

largely forgotten. Hence the systems approach pointed towards a redirection of electoral geography beyond input and throughput.

Johnston (1980) has taken the argument one stage further and located electoral geography as part of a systems-oriented political geography. Feedback loops between input and output round off the system and integrate electoral and political processes. This is clearly illustrated by American pork-barrel politics, whereby politicians ensure that the area they represent is well treated by government, and they expect, in return, the gratitude of citizens to be expressed in their votes. In the US Congress, for instance, representatives and senators try to obtain places on spending committees that affect policies most relevant to their own constituents' interests. If they can achieve such committee positions, especially chairmanships, then they can usually benefit the district or state they represent. Although this process is difficult to

substantiate statistically (Johnston 1982a), there is no doubt that it is an important element of American politics. Although less recognized, the same process will occur in other countries such as Britain, where there are other explicitly geographical government effects such as in regional and inner city policy.

Critique: liberal assumptions

While providing a solution to the problems of electoral geography enumerated at the beginning of this discussion, the systems approach has proved to be more important for what it reveals than for what it solves. Quite simply, the systems approach has opened up a real Pandora's box. The assumptions upon which electoral geography has been built are laid bare (see panel above). They turn out to be the classic liberal assumptions of the twentieth-century core states: a receptive government responds to an electorate that articulates its demands through its representatives. All of a sudden, conflicts have disappeared, history is forgotten and political parties are nothing more than vehicles for transmitting candidate and voter preferences. But things are never as simple as they seem. Sometimes there is a complete mismatch between input and output. When this occurs, it is difficult to see how the simple liberal model can cope with the evidence. We give just one example here to illustrate the argument, and we consider such mismatches in more detail later in the chapter.

We can term our example Hobson's paradox. In a contemporary study of the British general election of January 1910, Hobson (1968) noted that the country could be divided between north and south in terms of economic interests. The industrial north he termed 'Producer's England', and the residential south became 'Consumer's England'. This division was reflected heavily in the voting returns, with the Liberals polling strongly in the north and the Conservatives likewise in the south. The interesting point, however, is that this voting pattern is inconsistent with each region's economic interests. Both parties inherited nineteenth-century traditions based upon the old urban–rural cleavage, so Liberals maintained their free-trade stance and Conservatives campaigned on a policy of tariff reform (protectionism). The paradox is that Consumer's England voted for protection and hence increased prices, whereas Producer's England voted for free trade, exposing its industries to American and German competition. We will not resolve this paradox here – indeed, it is difficult to see how the systems approach can cope with it beyond designating it a historical anomaly. We shall provide other paradoxes below to show that such 'anomalies' are indeed quite common.

The reason why electoral geography can get into such a muddle is that its liberal assumptions are interpreted as self-evident and innately proper. The liberal democracy model upon which it is based is a normative argument that regards itself as the rational and correct culmination of the western political tradition. But if it is so good, why has the West had so much trouble transplanting its parliamentary ideals to the periphery? The popular will as expressed in elections may produce changes in government in core countries, but elsewhere several other procedures can be found. For example, in Figure 6.3 all 'irregular executive transfers' as identified by Taylor and Hudson (1971: 150–3) for the period 1948–67 are plotted. Taylor and Hudson define these transfers as a change of government accomplished outside the conventional legal procedures in effect at the time of the change, and with actual or threatened violence. Their period of study covers the post-Second World War economic boom in the world-economy, but this did not prevent 147 of these irregular transfers occurring. Only one occurred in the core – De Gaulle's

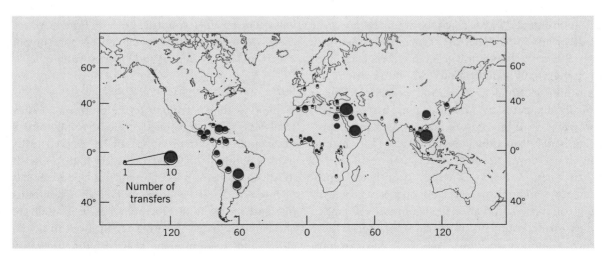

Figure 6.3 Irregular executive transfers, 1948–67.
Source: based on data in C. Taylor and Hudson (1971).

return to power in France in 1958 – and there are only two other cases in Europe, in Czechoslovakia in 1948 and Greece in 1967. The remaining 144 cases are in Latin America, Africa and Asia. This is a classic case of a peripheral political process that is almost entirely missing in the core. Electoral geography should no longer treat elections as an ideal; they should be seen realistically as just one of several means of choosing governments, and one with an extreme geographical bias.

The only political geographer to comment on the relative lack of liberal democracy in the periphery is Prescott (1969: 378), but he treats it merely as a data problem that prevents 'profitable geographical analysis'. Clearly, the geography of elections as the global pattern of a particular government-creating institution is a topic sorely missing from quantitative electoral geography. Fortunately, it is not a theme neglected by political scientists. Given our world-systems approach, it is important that we consider global-scale analyses of liberal democracy as a prelude to our own interpretation. We present one such study, whose methodology is entirely consistent with quantitative approaches in modern geography.

Summary

In the previous sections we have:

➤ noted the tradition of quantitative electoral geography;

➤ described Rokkan's cleavage model as the standard explanation of electoral divisions in European politics;

➤ introduced and critiqued a systems model of liberal democracy.

We now build upon the critique of standard explanations of elections and electoral geography using the world-systems approach.

Coulter's global model of liberal democracy

A major contribution of political science to the rise of modern quantitative social science was the produc-tion of large-scale data sets covering almost all countries in the world (for example, Banks and Textor 1963; Russett et al. 1963). This allowed researchers to carry out comparative political studies on a scale never previously achieved. The comparative politics study most well known to political geographers is Russett (1967), but in many ways Coulter's (1975) study of liberal democracy is much more relevant to political geography. Russett's study appealed to geographers because of its use of the regional concept and its relation to political integration, but such integration between states is not a major issue in most parts of the world. Coulter (1975), on the other hand, attempted to test Deutsch's (1961) model of social mobilization at a global scale. It remained the most comprehensive study at this scale until the recent concern for global democratization, which we deal with below. Since Deutsch's work has been widely used in political geography, it follows that Coulter's study remains of particular interest in linking Deutsch's model to the global scale.

Liberal democracy and social mobilization

Coulter uses a classic quantitative geography research design. The first step is to define the 'problem map' to be explained. This involves measuring degrees of liberal democracy across eighty-five states. He identifies three aspects of liberal democracy – political competitiveness, political participation and public liberties – and combines them into a single index (Coulter 1975: 1–3). Multi-party elections, voter participation and freedom of group oppositions are all elements of this index, so it effectively measures the variations in the degree of importance of elections in determining governments. Variations in liberal democracy for the period 1946–66 are shown in Figure 6.4, which can be interpreted as the obverse of Figure 6.3.

In defining the variables to explain this map, he keeps very close to Deutsch's ideas on social mobilization and democracy. Deutsch (1961) postulates the mobilization of people out of traditional patterns of life and into new values and behaviours. This occurs to the extent that a population is urbanized, is literate, is exposed to mass media, is employed in non-primary occupations and is relatively affluent.

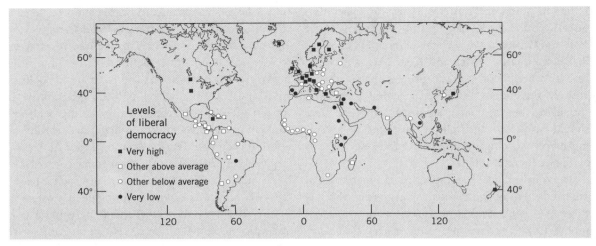

Figure 6.4 World map of liberal democracy, 1946–66.
Source: based on data in Coulter (1975).

Coulter therefore defines five sets of variables to produce indices of urbanization, education, communication, industrialization and economic development. These are measured both as levels for 1960 and as rates of change for 1946–66.

The global model of liberal democracy is a multiple regression analysis with liberal democracy as dependent and the five aspects of social mobilization as independent variables. The results are reasonably good, with economic development being the best explanation of liberal democracy followed by the communication index. However, these are not really separate explanations, since they are very highly correlated ($r = +0.94$). Nevertheless, Coulter has shown that liberal democracy can be statistically accounted for, in large measure, by the indices of social mobilization.

Two interpretations of a relationship

Coulter's results are summarized in Figure 6.5(a), which shows the basic trend line whereby an increase in social mobilization is associated with an increase in liberal democracy. Also shown on this diagram is Coulter's interpretation of his results. All countries that lie within one standard error of the trend line are termed optimally democratized. By this he means that the level of liberal democracy in these countries

is about as high as would be expected on the basis of their social mobilization. By using the term 'optimal', he implies that politics in these countries is correctly adjusted to their social situation. Among this group we find all the western European states, as we might expect, but Haiti and South Africa are also designated optimally democratized. Countries that lie below the optimally democratized band in Figure 6.5(a) are designated under-democratized, indicating a lower level of liberal democracy than would be expected on the basis of social mobilization. Such countries include Spain and Portugal, so we might be tempted to argue that the democratic revolutions in these two countries after 1966 represent a move to conform with Deutsch's model of political development. Countries lying above the middle band are designated over-democratized, since they have 'more' liberal democracy than their social mobilization would warrant. These include Greece, Uganda and Chile, and we may interpret moves against liberal democracy after 1966 by the Greek colonels, Idi Amin and General Augusto Pinochet as similarly contributing to their countries' conforming to Deutsch's model.

Perhaps the most surprising result of this analysis is that Coulter (1975) finds – in 1966 remember, in the middle of the Cold War – the Soviet Union to be optimally democratized and the United States to

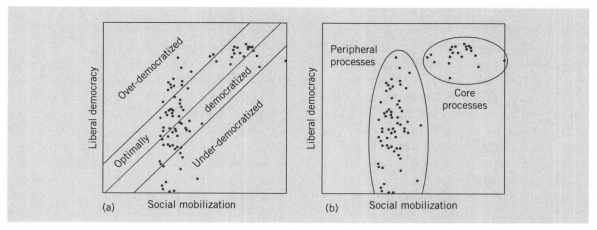

Figure 6.5 Liberal democracy and social mobilization: (a) as a trend line; (b) as two clusters.

be under-democratized. This is counter to our expectations, although it does not mean that the Soviet Union was more liberal than the United States but simply that, relative to their respective levels of social mobilization, the Soviet Union scored higher on liberal democracy. However, this result must make us wonder about the model. Either the measurements are unsatisfactory or the structure of the model is incorrect. We argue here that both are at fault. In Figure 6.5(b), the same scatter of points is presented in a completely different way. Instead of concentrating on a trend line, we identify two clusters of points, one defined by a vertical oval and the other by a horizontal oval. They represent separate, non-overlapping levels of social mobilization. But as we have seen, the most important component of social mobilization is economic development. We shall, therefore, interpret these two distinct levels of 'social mobilization' as representing economic core and peripheral processes. Now the scatter of points begins to make some sense. All core countries experience liberal democracy. In peripheral countries there is a wide range of political systems, showing many different levels of liberal democracy. This will depend upon the nature of the peripheral state, as discussed in Chapter 4.

This interpretation makes much more sense than Coulter's global model. It is entirely consistent with our world-systems framework in its emphasis upon two different sets of processes operating in the world-economy. Once again, a world-systems interpretation has been found to be superior to a developmental one, in this case one that sees countries on an 'optimal path to political development'. Quite simply, politics do not develop separately country by country but are all part of a larger unfolding system of political economy.

A world-systems interpretation of elections

Of all modern social institutions, elections would seem to be a set of activities that have to be understood at the scale of the individual state. Elections occur separately on a country-by-country basis and are strictly organized within one state at a time. The geography of elections therefore presents a particular challenge within political geography to the world-systems approach and its one-society assumption. Surely, here we have a case where we need a multiple-society view of the world to make sense of national elections.

In fact, it is relatively easy to show that the activities surrounding elections are in no way insulated from the world-economy. The remainder of this chapter is an illustration of this point. For the moment, we can use the example of party labels to show how trans-state processes are directly implicated in national elections. From the period when elections moved beyond the stage of confirming local elites in power, political parties have come to dominate electoral activities. Most parties represent a set of ideas that will be linked to a political ideology, however loosely conceived. Hence the plethora of Labour parties and Liberal parties, Christian Democrats and Conservatives, Communist and Social Democratic parties. Every one of these parties adheres to a set of ideas that are in no sense unique to that party. Particular interpretations of these general ideas within different countries will inevitably vary, but no parties are independent of the political world beyond their country's border. Perhaps an extreme example will help to fix ideas. The power of English liberalism at the time of British hegemony is reflected in this statement by a nineteenth-century Brazilian liberal politician:

> When I enter the Chamber [of deputies] I am entirely under the influence of English liberalism, as if I were working under orders of Gladstone. . . . I am an English liberal . . . in the Brazilian Parliament.
> (T. Smith 1981: 34)

As usual, hegemonic processes give us a limiting case, but in general we can conclude that all electoral politics occurs within the overall political processes of the world-economy.

A world-systems approach to electoral geography has two tasks to fulfil. First, there is the need to understand the variations in the use and meaning of elections in different zones of the world-economy. We consider this topic in the remainder of this section. Second, electoral activities within states remain the prime concern of our electoral geography and make up the remainder of the chapter.

Liberal democracy and social democracy

The idea of liberal democracy is an even more recent phenomenon in the world-economy than nationalism. For most of the nineteenth century, for instance, liberals faced what they saw as the dilemma of democracy. A fully democratized state represented 'the great fear' that the lower classes would take control of the state and use it to attack property and privilege (Arblaster 1984). This 'liberals versus democracy' phase is the very antithesis of liberal

democracy and is often forgotten in simple developmental theories about democracy. These evolutionary arguments were a product of the optimistic era of social science in the post-1945 period, when liberal democracy was viewed as the natural result of political progress. But as late as 1939, about half of the European liberal democracies of the 1950s had been under authoritarian rule. The 1930s were a time when pessimism about democracy reigned. We have to transcend these phases of pessimism and optimism about liberal democracy. What they tell us in world-systems terms is that liberal democracy is concentrated in time as well as place – in the core zone after 1945. But to understand this world-systems location, we have to return to the nineteenth century.

In Chapter 5 we concentrated on one particular problem confronting politicians in the nineteenth century: the national question. This was in reality just one of many new political questions that were competing to join the political agenda at this time. Three important questions relate to the emergence of liberal democracy (Figure 6.6). First, there was the constitutional problem posed by the liberals. They advocated the replacement of arbitrary power by constitutional checks and balances. Second, there was the political question posed by the democrats. They argued that the people as a whole should wield power in the new liberal constitutions. Third, there was the social question posed by the socialists. They asked how the new elected governments were going to deal with the new urban poverty. The answer to the first two questions was the liberal democratic state; the answer to the latter two questions was the social democratic state. We shall consider each in turn before we describe their crucial historical coalescence after 1945 (Figure 6.6).

We shall interpret liberal democracy as much more than a label for a party or policy; it is a type of state. Liberal democratic states have three basic properties. First, pluralistic elections, in which there is competition between two or more parties to form the government, are held regularly. Second, all adult citizens are entitled to vote in these elections. Third, there are political freedoms that allow all citizens to associate freely and express their political opinions. These properties are found with only minor blemishes in all core countries at the present time. In addition,

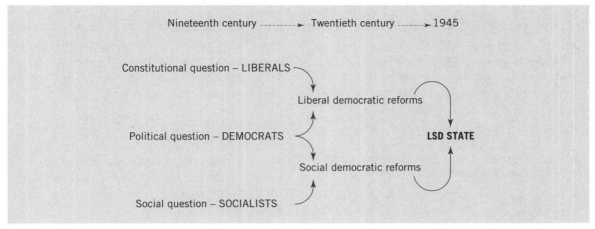

Figure 6.6 Three questions and the liberal–social democratic (LSD) state.

these states exhibit a further important property: political stability. Since 1945, countries in the core have experienced continuous liberal democracy. Hence they are liberal democratic states. They can be distinguished from states that have had liberal democratic interludes alternating with illiberal regimes. These more unstable states are typical of much of the world beyond the core. Analyses such as that by Coulter (1975), reported above, confuse these two categories. It is central to any world-systems analysis to distinguish between the liberal democratic state and liberal democratic interludes in other states.

In order to understand the space and time concentration of the liberal democratic state we need to consider social democracy. Again, we shall interpret this politics to represent more than a party or a policy; it is a type of state. Social democratic states have three basic properties. First, the state takes responsibility for the basic welfare of its citizens, so a wide range of social services and supports are provided. Second, there is a political consensus among all major party competitors for government that this historically large welfare expenditure is both necessary and proper. Third, the welfare is paid for through progressive taxation, which involves some degree of redistribution of income by the state. These properties are found in all core countries from 1945 to different degrees through to the present time, ranging from New Deal and Great Society-type programmes in the United States to the more redistributive socialism of Sweden. Some of the origins of this type of state are to be found in the social imperialism processes discussed in Chapter 3. Whatever the means, however, by the 1940s 'welfare states' were being created throughout the core, and they have remained a typical characteristic of core political processes despite some recent cutbacks.

The three political problems from the nineteenth century have generated two forms of state in the mid-twentieth century. And it is not a coincidence that these forms of state have coalesced: all liberal democracies today are social democracies (see Figure 6.6). We might say that we are looking at the same state from two different angles.

From a world-systems perspective, this 'liberal–social democratic' state is the result of two processes, one economic and one political. First, the world-economy location in the core during the fourth Kondratieff cycle enabled this small number of countries to develop a politics of redistribution that was not possible in states at other times and in other places. This means that these states were rich enough to have meaningful competition between parties on the distribution of the national 'cake', where all citizens could potentially benefit. Elections matter; the issue of 'who gets what' encompasses all strata. Second, in the emerging Cold War geopolitical world order, the liberal–social democratic form of state is easily the most preferable for providing an alternative 'social progressive' politics to communism. Hence the new politics of redistribution was encouraged by the United States because it provided a bulwark against communism, especially in western Europe. But notice that the ideological concept of 'free world', first coined to describe non-communist Europe, has not transferred easily to areas beyond the core.

Theoretical corollary: the paradox of democratization and globalization

The above argument leads on to an important theoretical corollary. Since the world-economy is inherently polarized, the political benefits of liberal and social democracy can never be wholly transferred to the periphery. Hence the ideal of the liberal–social democratic state that is offered to these countries is beyond their reach. But this is the only modern state form in which liberal democracy has prospered. Why should the vast majority of a state's people participate in an election if there is little or no politics of redistribution? This means that the simple call for 'a return to democracy' in poor countries, which we have heard since the 1960s, is simply not a sustainable goal. Elections have too often turned out to be mini civil wars, with campaign deaths counted as well as votes. The implications of this US promotion of a 'free world' based upon democracy during the Cold War were debilitating; the subsequent implications for the 1990s post-Cold War spread of democracy are even more worrying.

The contemporary paradox is a simple one. In a world where economic polarization over more than two decades has reversed social democratic tendencies, more and more countries have adopted democratic means of forming governments. From our analysis these two trends are contradictory, implying one of two things. Either the new democracies, for instance in Russia, South Africa and Brazil, will turn out to be very fragile and current 'democratic gains' will be soon reversed, or the new democracies, or at least some of them, will invent new social arrangements that may make continuing elections and increasing polarization compatible. This would be a different sort of democracy from that described above. Perhaps half a century of sustaining Indian

democracy in conditions of mass poverty may be more of a pointer to the future of the new democracies than past North American and western European democratic experiences. These are issues we deal with later in this chapter, after we have fleshed out our world-systems model more fully.

Electoral geography contrasts between core and periphery

Our theory implies that elections will be substantially different in different zones of the world-economy. This will be reflected in contrasting electoral geographies. Consider the situation in the core, for instance. A viable politics of redistribution allows parties to build stable support bases as they implement policies that favour their supporters. The process that Rokkan (1970) has described for Europe has produced a stable pattern of votes based on social cleavages, for instance. This is translated into a stable geography of voting, because social groups are geographically segregated. In Britain, for instance, Labour does better in working-class districts, and the Conservatives obtain more support from middle-class districts. In contrast, in non-core areas without a viable politics of redistribution, the basic mechanism of keeping voter loyalty is missing. A party is far less able to reward the mass of its supporters and sustain its votes. Hence we expect less stable bases for party support, which will be reflected in unstable geographies of voting. We can test this basic hypothesis with a simple empirical analysis of contrasting geographies of election.

The degree of geographical stability of a party's vote can be measured by factor analysing the geographical pattern of the vote over a series of elections. If the pattern is exactly the same in every election, the first factor in the analysis will account for 100 per cent of the variance. The less geographically stable the vote over time the further the first factor will be from the 100 per cent limit. Hence the 'importance' of the first factor provides a sort of percentage geographical stability score. Such measures are reported from the major parties of ten countries in Table 6.1, covering elections between 1950 and 1980.

The countries in Table 6.1 are divided into seven core states and three peripheral states to highlight the differences in the electoral politics of the two zones.

Living in the core and voting in the periphery

Perhaps more than any topic in political geography, electoral geography has followed the multiple society assumption of social science. The voter has been identified as the citizen of a particular state, and the geographical extent of the political system has been assumed to be delimited by the borders of the state. However, these assumptions do not hold. Indeed, it is estimated that 175 million people live outside their country of citizenship, about 3 per cent of world's population (Collyer 2006). Initial investigation into these topics promises an interesting geography of electoral behaviour that is not constrained by state borders. Such analysis would integrate electoral geography into the broader project of analysing and understanding globalization, while retaining the interest in the contextual influence of geographical location upon the voter.

Theoretically, two categories of voters are of interest: residents who are able to vote in the elections of their host state but are not citizens, and citizens of a state who have migrated abroad but may still vote in their 'home' elections. For example, the first round of elections in Iraq after the overthrow of Saddam Hussein was conducted across the globe as the Iraqi diaspora was deemed eligible to vote. Different countries have different attitudes to and laws governing the voting rights of non-resident citizens (Collyer 2006). For example, in the Republic of Ireland non-residents are prevented from voting, but Italian non-resident citizens may return to vote (and, if they meet the financial criteria, receive monetary aid to do so), and the United States and the United Kingdom allow non-resident citizens to vote from abroad. In an interesting case, the Cook Islands residents and non-residents vote on separate slates of representatives.

Why the current interest in non-resident voting? One of the arguments rests on the importance of financial remittances sent by migrants – an increasingly important source of income for peripheral countries as foreign direct investment has declined (Itzigsohn 2000). However, Collyer (2006) concludes that the reasons behind states' facilitating non-resident citizen voting vary by region: the explanations found for Latin America were different from those for the Maghreb region of North Africa. Morocco is a good example, in which those advocating allowing non-resident citizens to vote to maintain the flow of remittances are faced by a counter-argument that such voting would weaken existing and well-established bases of power (Collyer 2006).

Table 6.1 Geographical stability of voting patterns c.1950–80 by major parties in selected countries

Core countries		Peripheral countries	
Italy	95	Jamaica	59
Belgium	94	Ghana	35
Netherlands	94	India	33
Britain	93		
West Germany	88		
Denmark	86		
France	83		

Source: derived from data in Johnston et al. (1987). For all countries except Ghana and India the scores for stability are the average for the two main parties in the country. For Ghana the scores are for the pro-Nkrumah party at each election and for India the Congress Party.

In all the European core countries the geographical stability in the post-1945 period is very high. In Rokkan's (1970) terms, the major parties in these countries are able to successfully 'renew their clienteles' over time. In contrast, in the three peripheral states the degree of clientele renewal is very low. Jamaica seems to come closest to a level geographical stability consistent with a viable politics of redistribution, but even in this case the percentage levels are well below European states, with their more fully developed politics of redistribution. What this means is that in these peripheral states the geographical pattern of support changes appreciably from one election to the next. Parties are unable to maintain the support of those who have voted for them in the past.

In short, there seems to be a very different political process going on. Even though the Jamaican, Ghanaian and Indian elections upon which this analysis is based may be as fair and as open as the European elections, this fundamentally different electoral geography is indicative of something other than a 'liberal–social democratic' state.

We must conclude, therefore, that in order to study elections worldwide we will need to bear in mind the very different politics resulting from huge differences in material well-being between countries in the core and in the periphery. In the next section, we deal with electoral geography in the core, and in the final section, we consider the role of elections in the far harsher politics beyond the core.

Summary

In the previous sections we have introduced a world-systems interpretation of the geography of elections. In doing so we have:

➤ introduced the concept liberal social democracy;

➤ built upon this concept to argue that we will see different politics, including different electoral politics, in the core and periphery of the world-economy;

➤ considered the discussion of the geography of elections within the context of globalization.

■ Liberal democracy in the core

In traditional electoral geography, political parties have been viewed as either reflections of social cleavages (Taylor and Johnston 1979) or simple vote-buying mechanisms (Johnston 1979). Parties may carry out either or both of these roles, but they are much more than this. The concept conspicuous by its absence in such analysis is power. The main purpose of political parties is to gain power – to take control of a state apparatus. By adding power to our analysis we can go beyond the linear systems model described above and create an alternative and more critical

model. This is described in the first part of this section. We then apply the model to the making of liberal democracy in the core. We concentrate on the specific relationships between parties and governments and the dynamic nature of the capitalist world-economy.

The dialectics of electoral geography

A dialectical process is one that unfolds through history from two opposites to a resolution of the opposition. These are termed thesis, antithesis and synthesis, respectively. It will be argued here that this process has occurred in the electoral politics of states. This model of political change is laid out in Figure 6.7. In the discussions that follow, we draw on some ideas introduced previously and link them with new concepts for dealing with elections.

The initial basic opposition is between the relentless pursuit of accumulation and the need to legitimate this pursuit. Since accumulation concentrates capital in the hands of the few, it acts against its own legitimation in the eyes of the many. But both thesis and antithesis are necessary if the system is going to advance beyond simple coercion of the many by the few. Political parties have been key actors in the movement from this situation of opposition to today's resolution in the liberal–social democratic state. Let us briefly trace the steps:

1 The first need of capital is order to counter the full implications of the anarchy of production. Parties have provided what we shall call the 'great act of organization' – from the myriad of possible packages of policies, parties simplify the choices down to just a few, often just two, options (for example Republican versus Democrat policies in the United States).

2 But parties also carry out what we shall call the 'great act of mobilization' whereby the population is brought into the political process. This is to legitimate the politics.

3 Different types of party were associated originally with these two processes – we define them as cadre and mass parties below – but their coming

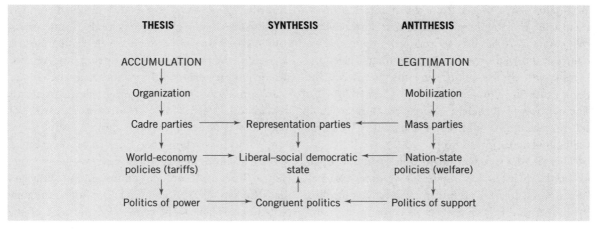

Figure 6.7 The dialectics of electoral geography.

together after 1945 as a new form of representation party provides the key step in synthesizing the needs of accumulation and legitimation in the political system.

4 The cadre party's emphasis on external policies (such as trade) and the mass party's emphasis on internal policies (such as welfare) comes together in the liberal–social democratic state.

5 From this, we define alternative politics concerned with power and support, which since 1945 define a congruent politics that is at the heart of contemporary politics in the core.

6 It is the combination of representation party, liberal–social democratic state and congruent politics that forms the synthesis that has resolved the initial oppositions.

Let us now consider the ideas and concepts contained in this model in more detail.

Organization and mobilization

Political parties carry out two basic tasks. First, they set, or at least influence, the political agenda. Second, they appeal for support among the population. These two activities are closely related, since success or failure in one is likely to produce success or failure in the other. For instance, the demise of the Liberal Party in Britain in the first half of the twentieth century was the result of its failure to continue to dominate the political agenda as it had in the nineteenth century, with a resulting loss of popular support. It was quite literally being seen as irrelevant to the needs of many newly enfranchised voters. The Labour Party, on the other hand, was much more successful in producing a new agenda with which it was able to capture much former Liberal support as well as new voters. Eventually, Labour replaced the Liberal Party as a party of government. A significant transfer of power had taken place.

From the point of view of the state, these two activities by political parties are highly functional. Parties first organize the politics of the state and then they mobilize the population behind that politics. But parties cannot do this singly; such a useful outcome depends on the creation of a competitive party system.

A party system depends upon opposition groups being perceived as alternative governments rather than as threats to the state. From the nineteenth century onwards, state-building groups in core countries and some peripheral countries have come to accept this position, so that elections become the means of selecting governments. In the United States, this fundamental position was reached in the second party system of Democrats versus Whigs, which developed in the 1830s. In the first party system a generation earlier, each party fought elections with the intent of eliminating their rivals from the political scene – eventually, the Democrat–Republicans succeeded in reducing the Federalists to political impotence. In contrast, the Democrats and Whigs fought elections merely with a view to securing the presidency for their candidate (Archer and Taylor 1981: 54–61). The election of President George W. Bush in 2000 was the twenty-second transfer of executive power in the United States by election.

This transfer of government office is not as open as the above discussion implies. Government formation is not a free-for-all but is a carefully controlled process. And this is where parties come in. In many countries, there is a duopoly of power to form governments. In the United States, for instance, all presidents since the Civil War have been the nominees of the Republican or Democratic Party. In Britain, the Conservative and Liberal Parties until the 1920s, and then the Conservative and Labour parties, have had a similar duopoly of power. Even in multi-party systems, there remain severe constraints on voter choice, with relatively few effective votes available. But this is the whole point of a party system. From the vast range of positions on a large number of topics, voters are asked to support just one of a limited number of 'manifestos' or 'platforms'. This is what Schattschneider (1960: 59) terms 'the great act of organization', with political alternatives reduced 'to the extreme limit of simplification'. The power of parties is simply that electors can vote for or against particular party candidates, but they cannot vote for or against a party system (Jahnige 1971: 473).

Electoral politics as constrained by political parties is, therefore, an important control mechanism in liberal democracies. The actual organization operating at any particular election, however, is not normally designed for that election. As we have seen, parties and party systems are the product of the specific histories of their countries. The manipulation of the political agenda is not a conspiracy of ruling elites but rather reflects the relative power of different interests in the evolution of a party system.

For Schattschneider (1960), this power over choice enables parties to define the politics of a country. There are an infinite number of potential conflicts in any complex modern society. By controlling alternatives offered to voters, parties decide which conflicts are organized into a country's politics and which conflicts are organized out. Hence electoral politics is defined by the party system, producing massive constraints on the nature of the political agenda.

In some ways, this integrating role of parties is paradoxical. 'Party' comes from the same root as 'part' and indicates a division within a political system. Political parties, therefore, have the second role of accommodating differences within a state. Hence the social conflicts and resulting cleavages that Rokkan (1970) identifies do not ultimately pull the state apart but rather become part of the state. Parties can therefore convert even potentially rebellious subjects into mere voters. The rise of Christian Democratic parties throughout Europe, but especially in Italy, represents a victory of the state over the transnational pretensions of the Catholic Church. Devout Catholics became mobilized into state politics via their Church parties. More generally, we may term this process the party system's 'great act of mobilization'. Today throughout the core it is difficult to conceive of state politics without political parties.

The development of political parties

Liberal democracies may have been created by political parties, but how did they come to be so important? It is a complex story in every country, but in general terms it can be simplified by reference to mobilization and organization. These two tasks have their origins in the development of two very different sorts of party in the nineteenth century.

The acceptance of legitimate oppositions within states was reflected initially in the formal organization of parliamentary factions, or 'parties'. These loose groupings of politicians represented different special interests within the dominant class. In the mid-nineteenth century, a distinction began to be drawn between factions serving particular interests and parties organized by principles and claiming to represent the public interest from different perspectives. In Britain, Whigs and Tories were replaced by Liberals and Conservatives, for instance. These parties were originally organized only in parliament and did not constitute modern political parties, according to

Duverger (1954). They became proper parties only when the parliamentary organizations were forced to mobilize support in the country in the wake of suffrage extensions and competition from new parties. The formation of 'electoral committees' in electoral districts to organize campaigns converted these 'traditional parties' into fully fledged modern parties. They became what Blondel (1978) terms cadre parties, since their organization is merely to find supporters, and power continues to reside at the centre.

As suffrage reforms began to reach down to the direct producers, another very different sort of political party was created outside parliament. These extra-parliamentary parties had only one source of resources: their members. Hence they were forced to mobilize voters and potential voters into mass parties. The most successful were the socialist parties, which in 1889 created the Second International as an alliance of socialist parties from numerous countries. Populist agrarian parties and some Christian parties also developed as mass parties in selected countries or regions.

Hence at the beginning of the twentieth century there were two very different types of party in most of today's liberal democracies: mass parties emphasizing mobilization and cadre parties emphasizing organization. But between them they have defined, for the most part, the party systems that exist today. According to Rokkan (1970), European party systems were 'frozen' into their current structure in the first two decades of the twentieth century. Hence elections today take place between political parties, most of which were competing with one another before the First World War. But we should not let the similarity in party labels lead us to assume that electoral politics has not changed fundamentally between these two periods. Blondel (1978) has pointed out that the party systems in Europe before the First World War were much less stable than they seem in hindsight. The mix of unresponsive cadre parties and mobilizing mass parties was a recipe for conflict rather than consensus. The latter in particular were potentially divisive, since they developed political ideologies that were inclusive. Socialist ideology, for instance, looked forward to mobilizing all the working class in a coun-

try, which would eventually provide the party with a permanent parliamentary majority. There was little room for pluralism here. The most extreme case was in Germany, where the Social Democrats seem to have created an alternative 'class nation'. The question is, therefore, how did this unstable and potent mix of dissimilar political parties become converted into the stable liberal democracies of today?

We begin to answer this question by considering the US case, which was the major exception to the party development described above, since both agrarian (Populist) and socialist mobilizations failed to produce major political parties to compete for government. According to Burnham (1967) this is the decisive step where the American party system diverges from the European experience. Hence US elections remained competitions between cadre parties well into the twentieth century. But they were forced to shed the unresponsiveness typical of cadre parties by the economic collapse after 1929. With the coming of the New Deal in the 1930s, we find the generation of a new form of party, which Blondel (1978) terms a representation party. This new type of party developed as a synthesis of elements of the cadre and mass types. Representation parties are more responsive than traditional cadre parties in that they make direct concerted appeals to the electorate beyond narrow party channels, but they are not mass parties, since they are not primarily concerned with mobilizing voters to accept a special political cause. Representation parties are pragmatic and eschew ideology. In the age of new mass communications, first radio and then TV, political leaders can appeal to the electorate directly, and we enter the world of policy packaging, image making and the 'selling' of the candidate. Elections are about which party and its leader can best represent the public mood of the time. In the new age of nationwide radio, Roosevelt's New Deal Democrats (1932–45) can claim to be the first representation party.

In Europe between the two world wars, the cadre and mass parties continued to exist side by side in an uneasy politics with strained party systems. After the Second World War, both cadre and mass parties have metamorphosed into representation parties. For the cadre parties, this was a relatively easy transition as they extended their electoral campaigning to incorporate new techniques. At the same time, the mass parties changed fundamentally. The old socialist parties have come to rely on pollsters and advertising agencies just as much as their right-wing rivals. Socialist parties now claim to represent public opinion rather than guide it. Rokkan (1970) terms this process the 'domestication of socialist parties', and during the Cold War this distinguished socialist parties from communist parties in western Europe. But since Rokkan's analysis, the 'Eurocommunist' tendency produced new 'respectable' western European communist parties even before the end of the Cold War. This was the final act in the conversion of mobilizing mass parties into representation parties. The elimination of mass mobilizing parties can be interpreted as a sort of 'Americanization' of European party systems. The key point is that representation parties are a necessary step in the political resolution of the accumulation/legitimation dialectic in the core of the world-economy (Figure 6.7).

Two politics and two geographies in every election

In Chapter 4 we derived two politics in an instrumental theory of the state. Since elections are about competition for formal control over the state apparatus, it follows that political parties should reflect these two politics: inter-state, intra-class and intra-state, inter-class. In fact, this is the case. Generally, the distinguishing feature of cadre parties was in terms of different policies towards the rest of the world-economy. Each party represented economic interests within the dominant class within a state favouring either free trade or protection. For instance, in the United States the Republicans were protectionist and the Democrats were the free traders, whereas in Britain these roles were taken by the Conservatives (tariff reform) and the Liberals, respectively. In contrast, the new mass parties based their mobilizations on domestic distribution politics: more for small farmers in the case of the US Populist Party and agrarian peasant parties in Europe; more for working people in the case of socialist parties.

At the beginning of the twentieth century these two politics, promoted by their respective parties, operated side by side in elections. This is the source of the potential instability that Blondel observes and accounts for the confusion of the 1910 British election. Basically, a cadre politics (free trade versus tariff reform) was mixed up with a new mobilization of class politics (urban/north versus rural/south), leading to a mismatch between interests and voting. We can extrapolate from this example to argue that in every election two distinct processes will be operating. First, there is the politics of power, which can be traced back to accumulation. It is about winning elections to promote policies favouring particular interests in the pursuit of capital accumulation. All governing parties of whatever political hue must promote accumulation of some form within their state's territory. But equally, a party cannot govern until it wins an election. Hence there is a politics of support that parties develop and nurture. These two politics operate together: every policy is advantageous to some interest group, which may fund the party that introduces it, while the overall package of policies that a party presents is designed to appeal generally to voters. But they remain separate processes and can produce quite paradoxical election results, as the 1910 example illustrates. Representation parties since 1945 have blurred the distinction in their claims to reflect the public good, and it is our job in electoral studies to unravel this plausible appearance.

The lesson we can draw is that these two processes will have two electoral geographies associated with them. There are the familiar studies in the 'geography of voting' that we described above and that here become the geography of support. Second, there is the much less familiar geography of power, of interest-group funding and policy outcomes. With two electoral geographies to deal with, we are now in a position to introduce a revised model (Figure 6.8), which should be compared with the simple 'linear' model presented in Figure 6.2.

Most of electoral geography has been concerned with the geography of support, partly as a simple result of data availability. Elections must be very public exercises to function as legitimizing forces, so voting returns are readily available to produce geographies of support. As we pointed out at the beginning of the chapter, this has been the basis of the

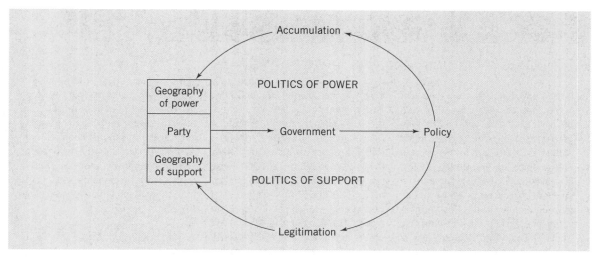

Figure 6.8 A revised model of electoral geography.

spectacular rise in electoral geography in recent years. But data availability for the politics of power is much more likely to be limited. Where the politics is based upon covert actions we may never know about it. CIA funding of 'friendly' foreign parties or destabilization of 'unfriendly' foreign governments is part of the geography of power in some states that we are only just beginning to learn about. Of course, there is no inventory of foreign funding of political parties across the world for us to consult on such matters. Certainly, we cannot expect simple official tabulations to produce geographies of power in the way that we can produce geographies of support. In addition, the study of the geography of policy outputs has not been systematically developed. In short, the geography of power in elections has been relatively neglected. And yet the politics of both support and power are equally essential to the smooth operation of liberal democracy. We should not neglect one merely because it is more difficult to investigate. The major lesson of our reformed model is to redirect electoral geography towards its neglected half, the geography of power.

Three types of electoral politics

This new model for electoral geography enables us to identify different types of electoral politics. We can define three basic relationships between the two geographies: an inverse relationship, no relationship and a positive relationship. These are termed contradictory, disconnected and congruent politics, respectively, in Figure 6.9. This diagram illustrates these politics in simple abstract terms. For the geography of support, we identify just a simple cleavage dividing the electorate into two halves. For the geography of power, the sum of all interested groups is divided into two clusters of policy interests. Shading indicates compatibility between policy interests and voting segment. The three forms of electoral politics are then clear to see.

The confusion of the 1910 British election can now be identified in our model as a case of contradictory politics: Producers' England votes for free trade, Consumers' England votes for tariffs. Disconnected politics can be illustrated by countries using a proportional representation (PR) system of voting. This is because the voting and the government formation are distinct processes. Normally, no one party is able to form a government, so negotiations between parties over policy are necessary in order to construct a coalition government. This highlights the two political processes as separate mechanisms allowing for the illustration of disconnected politics. We can show this by employing the results of two surveys of European liberal democracies carried out by Arend Lijphart.

Case study

US intervention and the dirty politics of power

Hugo Chávez was elected President of Venezuela in February 1999 with 59 per cent of the vote. Since then he has been targeted by the United States as a 'threat' to the region. The essence of the United States' displeasure is Chávez's domestic policies of economic redistribution that run counter to the Bush Administration's agenda to integrate the whole of South and Central America into a free-market zone dominated by the United States. In addition, Venezuela is an important oil exporter, and the United States is wary of Chávez's control of this vital resource. Since Chávez came to power, as a result of a politics of support, he has faced a particular politics of power: aggressive external interference by a powerful state.

Back in 1983, President Ronald Reagan established the Office of Public Diplomacy for Latin America and the Caribbean (LPD). The purpose was 'Psyops' (Psychological Operations), or promoting stories within the US media that would justify US actions in Central America. In 1987, the US General Accounting Office, the government's own watchdog, claimed that the LPD had conducted illegal and unethical practices in violation of government regulations. However, the institutional basis for monitoring, and interfering in, the internal politics of Latin American and Caribbean countries had been established.

The most blatant example of contemporary interference is Venezuela. The US campaign against Chávez began, in a form that is evident in other countries across the globe, by the funding of opposition parties. In other words, the players in the politics of power were partially created by, and

mainly funded by, the United States. Two government agencies are involved: the US Agency for International Development (USAID) and the National Endowment for Democracy (NED) (James 2006). In 2000 these two agencies were allocated a total of US$200,000 for activities in Venezuela. The amount leapt to around US$4 million in 2002. One of the institutions receiving money from NED was the International Republican Institute (IRI). It was given money for Venezuela to 'train national and/or local branches of existing and/or newly created political parties on such topics as party structure, management, and organization; internal and external party communications; and coalition building' (www.venezuelafoia.info, quoted in James 2006): in other words, to build a landscape of the politics of power that would embolden pro-US parties. However, this form of the politics of power was not deemed successful enough by the financial backers.

In April 2002, Chávez was briefly deposed in a *coup d'état*. Government documents, obtained through the US Freedom of Information Act, show the close connections between the coup leaders and the United States. CIA briefs show knowledge of the impending coup just five days before it happened. Pedro Carmona was established as the new president. He had recently visited the United States and met with government officials. Carmona immediately dissolved the national assembly and other branches of government, with the exception of the executive branch. It was claimed that Chávez had 'resigned', but when it became clear to lower-ranking military officers and the general population that he had been removed by

force he was reinstated on a wave of public outrage. Public statements by the US government claiming ignorance of the impending coup, as well as involvement with the instigators, are challenged by the US government's own documents and communications immediately preceding the coup.

Despite this failure to remove Chávez as democratically elected president, the US campaign to oust him continued. After the attempted coup USAID and NED pumped even more money into oppositional politics. Tactics have changed annually, ranging from the support of strikes in 2003, funding a referendum opposing Chávez in 2004 and rigging exit polls to claim he had been defeated (monitoring groups classified the referendum 'free and fair' and support for Chávez remained at 59 per cent), and, in 2006, the new tactic of diplomatic isolation. As the presidential elections of December 2006 approached, the United States continued using taxpayers' money to mould the politics of power within Venezuela to the liking of President George W. Bush's Administration.

Sources: Deborah James, 'US intervention in Venezuela: a clear and present danger', January 2006. http://www.globalexchange.org/countries/americas/venezuela/USVZrelations.pdf. Accessed 13 April 2006. Eva Golinger, 'US aggression towards Venezuela: the rise of black propaganda and dirty war tactics (again)', 30 March 2006. http://www.venezuelanalysis.com/articles.php?artno=1409. Accessed 13 April 2006. National Endowment for Democracy, *Summary of Projects Approved 2000–2004, Venezuela*, n.d. Available at www.venezuelafoia.info. Accessed 13 April 2006.

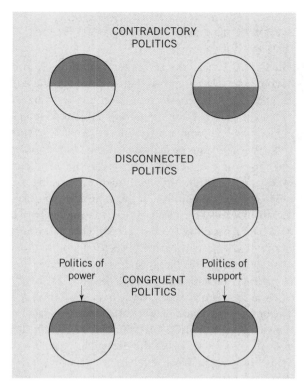

CONTRADICTORY
POLITICS

DISCONNECTED
POLITICS

Politics of
power CONGRUENT
POLITICS Politics of
support

Figure 6.9 Three types of electoral geography.

Table 6.2 Contrasting politics of support and power in seven west European countries

	Politics of support		Politics of power	
	Electoral cleavages[a]		Coalition formation[b]	
	Religious	Economic	Religious	Economic
France	59	15	25	75
Italy	51	19	0	75
Germany	40	27	12	70
Netherlands	73	26	0	100
Belgium	72	25	17	70
Austria	54	31	5	95
Switzerland	59	26	0	49

[a] Percentage bias towards religious and socialist parties by church attenders and manual workers, respectively.
[b] Percentage of 'coalition years' for the period 1919–79.

In the first study (Lijphart 1971), the electoral cleavage is investigated by finding the percentage bias of different groups towards their 'natural' party. In the first two columns of results in Table 6.2, we report on just the religious and economic cleavages for seven European countries that have employed a PR system. The figures represent the percentage bias of church attenders towards Christian Democrat parties and the percentage bias of manual workers towards socialist/communist parties. For the European liberal democracies reported here, the most outstanding feature of these results is the predominance of the religious cleavage. In the final two columns of Table 6.2, Lijphart's (1982) results for a study of coalition formation for these same countries are shown. Using information on all governments for the period 1919–79, Lijphart defines each coalition as either religious or economic on the basis of party membership. Table 6.2 shows the total years of government for each category as a percentage of all years of coalition government. The most notable feature of these results is their contrast with the electoral cleavage findings. Roughly speaking, religion is three times as important as the economic cleavage in voting patterns, whereas economic criteria are eight times as important as religion in the process of forming governments. This is the classic case of mobilizing voters on cultural grounds while forming governments around economic coalitions to produce a disconnected politics.

Finally, we come to congruent politics. Here parties pursue policies that broadly reflect the interests of the people who support them. These are the typical elections fought by representation parties since 1945. Generally speaking, parties of the right produce policies that favour the upper–middle end of the class spectrum and base their support on this group, whereas parties of the left produce policies favouring the lower–middle end of the spectrum and gain their support accordingly. To be sure, there are many complications country by country, but in essence this is the congruent politics that we experience today in the core. Such a politics is a crucial element of the political synthesis that we have explored in this discussion (see Figure 6.7). It produces the politics of distribution, which as we have seen is the hallmark of liberal–social democratic states. This trio of party, politics and state complemented one another to resolve the accumulation/legitimation dialectic in the core of the world-economy after 1945 (see Figure 6.7).

The making of liberal democracies

With three types of electoral politics, we are in a far better position to understand the making of liberal democracies than we were with the one electoral politics underlying the linear systems model. In the latter case, an evolutionary history is typically produced whereby liberal democracy is a 'natural' outcome of democratizing trends over the past century or so. But from our previous introduction to a world-systems analysis of elections, we know that there was no such smooth transition to democracy. Therborn (1977) in particular has recorded the extent of the political opposition to democracy in all countries that are now liberal democracies and promote democracy worldwide. Our simple typology of three types of electoral politics enables us to address both the 'dilemma of democracy' in the nineteenth century and the 'triumph of democracy' in the twentieth century.

The purpose of this section is to illustrate the model that we have developed in terms of concrete examples of geographies of elections. We begin with the cadre parties of the United States before the New Deal, since they provide the best continuous example of non-congruent politics to have existed. We then consider the nature of the congruent politics that developed after 1945 before finally looking in some detail at the British case.

Non-congruent electoral politics: section and party in the United States before the New Deal

The United States has the longest continuous record of competitive elections based upon a broad franchise. As early as 1828, presidential elections were based upon white male suffrage involving millions of voters. Hence, while politicians in Europe were worrying about the dilemma of democracy, the United States was practising a politics that incorporated direct producers but without their influence intruding too much on government. It is at this time that the designation of somebody as a 'politician' comes to have a derogatory meaning (Ceaser 1979). This is a direct result of the non-congruent electoral politics that was developed. This highly successful experiment in democracy warrants further investigation.

Here we report briefly on a part of Archer and Taylor's (1981) factor analyses of American presidential elections from 1828 to 1980, updated by Shelley and Archer (1994). Using the percentage vote for all Democratic candidates, Archer and Taylor derive different patterns of party support across states. When several elections show the same pattern, these are defined as 'normal votes' in the sense that a tradition of voting has been established. For the analysis of eastern states from 1828 to 1920, two such patterns dominate the analysis, and we report on this finding and its relation to our previous discussion.

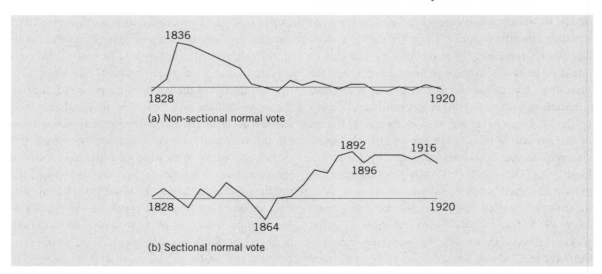

Figure 6.10 Normal vote profiles: United States, 1828–1920: (a) non-sectional normal vote; (b) sectional normal vote.

In Figure 6.10, we show the strength of these two patterns over time. They clearly represent pre-Civil War and post-Civil War normal vote patterns, the first being important from 1836 to 1852 and the second from 1876 onwards. In Figure 6.11, their distributions across states are shown, and from this we have derived their names – the non-sectional normal vote and the sectional normal vote, respectively. The latter represents what was the usual pattern of voting in the United States for much of the twentieth century: solid Democratic South, less solid Democratic 'border states' and a slightly pro-Republican North, with Vermont and Michigan particularly so. This is the sectional voting for which America was famous. It contrasts strikingly with what went before. In the non-sectional normal vote pattern, strong Democrat states are found both North and South. The distribution seems haphazard, even random. Northerners and Southerners supported both Whigs and Democrats with relatively little sectional favour. This is best illustrated by identifying the most pro-Whig and

pro-Democrat states of this era. Interestingly enough, they turn out to be contiguous – Vermont and New Hampshire, which are as similar a pair of states in social, economic and ethnic terms as you could expect to find in the period prior to the US Civil War. This is non-sectional voting *par excellence*.

There are several important aspects of this finding. First, political developmental models that assume a decline in the territorial basis of voting over time are exposed here. In the United States the extreme sectional voting follows non-sectional voting and not vice versa. It seems that the United States was highly 'integrated' before the Civil War. This interpretation is consistent with an emphasis on the voting pattern but must be laid aside as soon as we consider the party system. What the non-sectional voting pattern represents is merely a conglomeration of local alliances that come together on the national stage to support a selected presidential candidate. The forums for this national activity are the two political parties: Democrats and Whigs. This local agglomeration

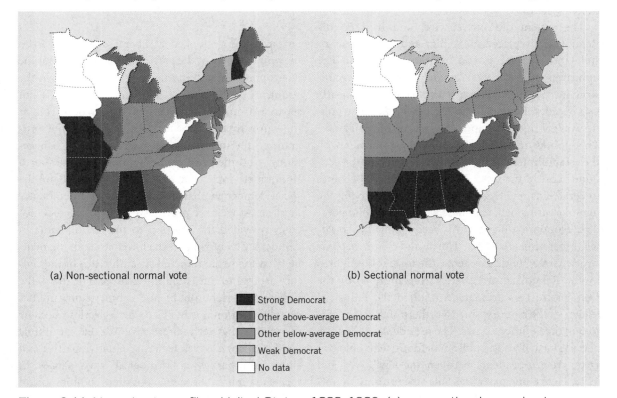

(a) Non-sectional normal vote

(b) Sectional normal vote

■ Strong Democrat
■ Other above-average Democrat
■ Other below-average Democrat
■ Weak Democrat
□ No data

Figure 6.11 Normal vote profiles: United States, 1828–1920: (a) non-sectional normal vote; (b) sectional normal vote.

produced 'national' parties for the only time in American history (McCormick 1967: 109).

There is a paradox here. Just as the country was undergoing the strains of sectional competition, which was to erupt in the Civil War, elections show no sectional bias. This means that the North versus South cleavage was being kept off the political agenda. Quite simply, there was no 'North' party or 'South' party to vote for until the 1850s. Hence the tensions developing in the country were organized off the political agenda by a non-sectional party system. The parties acted as an integrating force in a politics of sectional compromise. Martin Van Buren, president from 1837 to 1841, is usually credited as the architect of this highly successful control of a political agenda (Archer and Taylor 1981: 81–4). This was party versus section and, for a generation, party won. It was, in Ceaser's (1979: 138) words, 'a complete antidote for sectional prejudices'. This remains a classic example of electoral politics being diverted from a major issue by political parties in a disconnected politics.

The system did not survive, and in 1856 the Republicans replaced the Whigs and the Democrats split into Northern and Southern factions. The upheavals of the Civil War and subsequent reconstruction produced no normal vote pattern for nearly two decades. By the late 1870s, however, a new politics was arising based upon the sectional normal vote. This politics of sectional dominance coincides with the establishment of the Northern section as the economic core of the state. The state's territory becomes integrated as one large functional region serving the manufacturing belt. And here we have a second paradox: economic integration is accompanied by political separation. In effect, North and South become two separate political systems, the former dominated by the Republicans, the latter even more so by the Democrats. This arrangement suited the industrial leaders of the country, since Northern states were able to outvote Southern states. The presidency, therefore, became virtually a Republican fiefdom, with only two Democratic presidents between the Civil War and 1932. Hence one party was able to control the political agenda by relegating its opponents to a peripheral region.

What about the electors? What were they voting for while this control was going on? Fortunately, historians have attempted to answer this question by using correlation and regression methods on voting and census data. Their findings are fascinating. The major determinants of voting in this period are always cultural variables (McCormick 1974; Kleppner 1979). Hence voters were expressing religious and ethnic identification in elections. And yet in this whole period there are only two 'cultural' policies that reach the political agenda: slavery as a moral issue at the beginning of the period, and prohibition at the end. Through the whole of this period, the major dividing line between the parties was protection (Republicans) versus free trade (Democrats). Hence while voters were expressing their culture in elections, party elites were competing for economic stakes in terms of US relations to the world-economy. There can be no clearer example of the separation of voters from government by parties in a disconnected politics.

Congruent electoral politics in the Cold War era, and after?

So far, we have looked at liberal democracy from the point of view of the parties that created it. But the development of a domestic congruent politics in the countries of the core of the world-economy is not separate from the international events occurring at the same time. The New Deal provided the programme for internal social peace during US hegemony. And the great economic boom of the fourth Kondratieff cycle enabled the construction of the social democracy that maintained liberal democracy in place. As we have previously argued, social democracy provided the solution to the old dilemma of democracy by buying off the lower strata in the manner that the social imperialists at the turn of the century aspired to but could not achieve. For them, the time was not right. But by mid-century a new politics of redistribution was in place where elections became the choice between alternative baskets of public goods. This was a task for representative parties, not aloof cadre parties or 'ideological' mass parties. In this situation, a broadly congruent politics could emerge around the question of more or less public goods. This competitive domestic politics was carried

Immigration and political realignment

In the first few months of 2006 cities across the United States were filled with massive peaceful protests regarding the question of immigration legislation. The US Congress was debating a bill that would make it easier for the 11 million illegal immigrants in the country to gain citizenship. At the time of writing, April 2006, Democrats and Republicans had failed to agree on the legislation, while large demonstrations by immigrants continued. The connection between the legislative wangling, demonstrations and electoral politics was soon made. CNN stated that, 'politicians may face an angry Hispanic electorate in which Republicans would be the biggest losers'. One activist, Elias Bermudez, head of advocacy group Immigrants Without Borders, was quoted as saying, 'we will see this [the demonstrations] transfer into political power. If we cannot change their minds, we will change them [politicians].' The politicians were keeping a close eye on the political situation, noting the tendency of Hispanic voters to support the Democrats as well as the geographical concentration of voters in key Electoral College states such as California, Texas and Florida. 'It's a real high risk situation for Republicans, and it's almost all down side,' Republican political consultant Bill Miller was quoted as saying. The demographic changes wrought by immigration were changing the politics of support, and political parties were reacting to the changing social-political geography.

Source: CNN, 'Immigration marches could mark rise of Hispanic power'. http://www.cnn.com/2006/POLITICS/04/11/immigration. power.reut/html. 11 April 2006. Accessed 12 April 2006.

on within a foreign policy consensus. This was important for the development of the Cold War geopolitical world order. In 1945, some circles in the US government were unsure of the nature of the socialist parties that were enjoying electoral success in Europe. They soon realized, however, that they represented the best defence against Soviet-backed communist parties. Hence socialist parties such as the British Labour Party were the first to be 'domesticated' and provide a 'safe' politics from the point of view of US hegemony. The British Labour government, for instance, played a leading role in forming NATO in 1949 to bring the United States into western European defence.

The social democratic consensus that produced the congruent politics was always more than just a welfare state. In western Europe in particular, a corporate state was developed where government brought representatives of both capital and labour into economic decision making. The liberal–social democratic state had something for everybody – while the boom lasted. With the onset of the B-phase of the fourth Kondratieff after 1970, the slowdown in economic growth led to intense pressures on the public expenditure that sustained the corporate state.

In the 1980s, this led to a major political attack on many of the programmes that constituted the social democratic consensus and the inclusion of labour in the corporate state. These new policies have been associated with the right-wing leaders of the United States and Britain at this time – President Reagan and Reaganomics, Mrs Thatcher and Thatcherism – but we should not conclude that the change was limited to these two old hegemonic states. The new emphasis on market forces was a corollary to the necessity to cut government expenditure in times of economic difficulty. This was a general policy across all core countries irrespective of the colour of the party in power. The new politics was closer in content to the previous right-wing position in the social democratic consensus (less government) and so generally favoured right-wing parties to 'represent' the public's 'new realism', but this was not the case everywhere. In fact, some of the most severe policy changes occurred where there were left-wing governments. In New Zealand, a Labour government probably made the most severe cuts in public expenditure in the 1980s, and in Spain it was a socialist government that provided possibly the biggest attack on the trade union movement in the 1980s. In both cases, these are classic representation parties appealing to a general public that there is no alternative, the slogan of all such parties in recession. But what of the victims of this process?

In these affluent societies of the core created in the post-1945 boom there has developed what Galbraith (1992) calls the new politics of contentment. This is a situation where for the majority of the population the good life continues and they no longer need the state social provisions that were part of the social democratic consensus. This is the ultimate world of representation parties as they compete to serve the contented. With society becoming more polarized economically and with no one representing the lower end of the class spectrum, the parties are gradually losing their legitimation function. They are no longer accommodating differences but are exacerbating them. We can see the synthesis unravelling in the inner cities of the core states today. Another dialectic is in the making.

In the core of the world-economy, the dominant political discourse promotes elections as the major, if not the only, legitimate means of conducting politics. The effectiveness and appropriateness of demonstrations and strikes, for example, are commonly questioned in terms of their democratic credentials. The basic political rhetoric employed in liberal democracies promotes elections as the legitimate means of addressing issues emanating at the global scale but experienced at the local scale. Furthermore, these experiences lead to demands for ameliorative actions by the state. However, following Schattschneider (1960), we note that electoral choices reflect the agenda-setting practices of political elites. The electoral cycle is, then, driven by elites, and some of their concerns will include the relationship of their particular state with the rest of the world-economy. On the other hand, elections are the opportunity for voters, or non-elites, to express their opinions over how well the state and the national economy have met their needs. Again, such concerns will be partially determined by the state's relative success in the world-economy.

In order to make theoretical sense of the linkages between elections, the state and the world-economy, we will use Jürgen Habermas's (1975) typology of crises. Habermas argues that capitalism is always

prone to crisis. Two crises are of particular interest to us since they relate to the two electoral politics we have identified. First, in the politics of power, a rationality crisis occurs when the state does not succeed in managing the economy to the satisfaction and needs of the owners of capital and the business elite. In other words, in the opinion of the business elite, the state fails to manage the economy correctly either by being too involved in what are perceived as market decisions or by making poor decisions over taxes and trade, for example. Second, in the politics of support, a legitimation crisis occurs when the state does not manage to meet the social and economic needs of the masses while simultaneously meeting the imperatives set by the business elite. In other words, the economic or social conditions of the majority of the population become intolerable, even though the economy continues to produce at levels acceptable to business owners and managers.

The construction of liberal social democracy in the core of the world-economy illustrates the dual role of rationality and legitimation crises. The liberal imperative – the perceived imperative to maintain control of the political agenda and apparatus – reflects the dominance of the elites in creating the political system. The purpose of cadre parties was to manage the economy to their own ends. On the other hand, the extension of the franchise and the redistribution of economic benefits reflect the need to legitimate both the economic and political systems, hence representation parties.

Our world-systems framework should also allow us to predict the timing of the two types of crisis. In the A-phase of a Kondratieff cycle we would not expect a legitimation crisis, since we expect governments to be able to afford to meet their supporters' wishes. However, this will mean more government activity in the economy, which is likely to create a rationality crisis for business as voter demands grow and the economic boom begins to falter. Once the B-phase begins, governments will no longer be able to satisfy voter demands and a legitimation crisis is likely. This may result in a dangerous period of double crisis, rationality and legitimation. It was in just such a period that the democratic Weimar Republic in Germany collapsed in the early 1930s. If democracy

Seediness, power and the *Rolling Stone*

The geography of the politics of support is easier to identify and analyse than that of the politics of power. In territorially based electoral systems, such as the United Kingdom and the United States, the spatial pattern of voting tendencies is matched, with varying degrees of sophistication, to the geography of socio-economic attributes and conditions (refer back to Figure 6.1). The geography of power is harder to map. It is a matter of interconnection between political and economic elites, much of which takes place in informal settings rather than public spaces.

Rolling Stone magazine cast a critical eye over the power-plays that drive the US political scene. The conviction of Jack Abramoff for fraud and charges of illegally pocketing lobbyist fees focused attention upon the side of Washington politics that is not as visible as the party political jockeying for votes. Lobbyists in Washington are paid to represent interests to politicians and 'encourage' them to vote along particular issues central to the operation of those interests. Interests include political pressure groups across the political spectrum, businesses and trade unions, and foreign countries. It is regulated to operate within certain limits of how much money politicians can receive, from whom they can receive it, and what they can and cannot spend it on. Even legal lobbying is viewed by many as a corruption of the political system, and the case of Jack Abramoff only heightens the invective. For example,

> To most Americans, Jack Abramoff is the bloodsucking bogeyman with a wad of bills in his teeth who came through the window in the middle of the night and stole their voice in government.
>
> (Taibbi 2006a: 38)

In our theoretical, and more sober, language, Abramoff represents the secrecy and unjustness of the politics of power. Deals are cut between politicians and other elites with no democratic oversight, quite often illegally, and with no concern for the opinions of voters that politicians are deemed to represent.

To illustrate the process further, *Rolling Stone* ran an exposé of lobbyists at work (Taibbi 2006b). Their journalist, Matt Taibbi, invited himself to a birthday party for Montana Senator Conrad Burns, costing US$1,000 for organizations and US$500 for individuals to enter and schmooze with the senator. Despite the presence of professional lobbyists at the party, Taibbi was able to talk with one of the senator's staffers and pitch himself as a lobbyist for a bogus Russian energy company (given a fake Russian-sounding name that more or less translated to FartOilGas) planning to drill for oil in one of the United States' most highly regarded National Parks, the Grand Canyon. Despite the absurdity of the pitch, the magic words 'regulatory relief' were understood by the staffer and the promise of a meeting and Senator Burns' interest in 'talking' was established. The voters of Arizona, the location of the Grand Canyon, and Montana, the state that elected Senator Burns, were not at the party or in the conversation.

---

does survive, forced cutbacks in government expenditure are likely to help to resolve the rationality crisis. It is just this sort of pattern that we can see in the changing nature of British electoral politics.

World-economy and 'new politics': the case of Britain

Our discussion has brought the concept of power to centre stage in electoral geography, but we must not overemphasize the power of political parties. For all their dominance of government in liberal democracies, they are still constrained by the operation of the world-economy, as we have just seen. Although parties may be powerful within their state's boundary, there is no guarantee of power beyond the borders. In this final section on core elections, we link the politics we have been discussing with the broader issues that have concerned us in earlier chapters.

The problem for all political parties is that whereas the politics of support is an internal matter within

Table 6.3 'New' politics in Britain, 1918–97

Period	World-economy	'New' politics	Major events
1918–31	Stagnation B (i)	Politics of crisis I	Rise of Labour Party, General Strike
1931–40	Stagnation B (ii)	Politics of national interest I	Dominance of National Coalition
1940–60	Growth A (i)	Social democratic consensus	Establishment of welfare state
1960–72	Growth A (ii)	Technocratic politics	Application to join EEC
1972–82	Stagnation B (i)	Politics of crisis II	Conflict/accommodation with unions
1982–97	Stagnation B (ii)	Politics of national interest II	Falklands War

their country, the politics of power inevitably extends beyond the boundary of the state. In the medium term, the crucial matter is the cyclical nature of the world-economy. Every time the world-economy moves from an A-phase to a B-phase, or vice versa, the constraints and opportunities facing every individual state change fundamentally. Political parties operating within these states have to tailor their policies accordingly. The result is a series of 'new politics' within each state corresponding to the particular reactions of the political parties to the new world circumstances. Reaganomics and Thatcherism are recent examples of such new politics in their respective countries.

The most interesting feature of these 'new politics' is that they do not necessarily arise out of elections. Usually, there is not an election where voters are asked to choose between 'old' and 'new' politics: this is not a matter for the politics of support, it is an issue for the politics of power. For each new politics, old assumptions are swept away and new items appear on the political agenda, but all major parties accept the new politics. A new party competition arises but takes place within the new politics, being limited to matters of emphasis and degree. Hence the mobilizing powers of the parties are usually able to bring the voters into line with new economic circumstances. It is for this reason that the stability of voting patterns commonly found in electoral geography in core states is not a good index of the changing politics of the state.

Britain's long-term economic decline has elicited a variety of political responses. Part of the political reorganization in post-1945 Britain can be seen as a response to a rationality crisis as the British state tried to become more competitive within the world-

economy. Other electoral issues reflect the desire for legitimation, such as the rise of nationalist movements in Scotland and Wales. If we consider the period from the First World War, we can identify six phases of 'new politics', each developing as consecutive pairs of responses to the A- and B-phases of the world-economy. These are shown in Table 6.3, and we consider each new politics in turn below.

We begin this sequence of politics with the depression following the First World War. The initial reaction was a politics of general crisis. The old party system of Liberals versus Conservatives crumbled and the new Labour versus Conservative system emerged. But two-party politics did not arrive straight away. Since no party was seen to have the answer to the economic problems, every election led to the defeat of the governing party – a classic effect of a legitimation crisis. This process ceased in 1931. The culmination of the political crisis was the fall of the Labour government and its replacement by a coalition National government confirmed by general election, which solved the legitimation crisis. Labour retreated to the political wilderness as the National government maintained support through the 1935 election. This replaced two-party competition as the electors were mobilized to reduce their economic expectations.

The Second World War swept away the assumptions of the 1930s. A new social democratic consensus emerged, and with the return of a Labour government in 1945 a welfare state was created. The revision of the domestic agenda did not extend to foreign affairs. The combination of funding the welfare state and maintaining a global military presence led to severe economic cycles: the rationality crisis of the famous 'stop–go' sequence in British economic performance. The problem became acute when the relative performance of Britain became a political issue. The 'reappraisal of 1960' and the return of a Labour government in 1964 promising 'a white hot technological revolution' ushered in a new technocratic politics. This involved widespread reform of state institutions to make Britain as competitive as its rivals. Reorganization became the key word as local government, the welfare state and other central state departments were 'streamlined' for 'efficiency'. In foreign affairs, the retreat from global power to European power was sealed by membership of the European Economic Community – the ultimate technocratic solution to Britain's rationality crisis.

With the onset of the current recession, it soon became clear that tinkering with the administration of the state was not working. Once again we enter a politics of general crisis. Haseler (1976) identifies thirty-five events between 1966 and 1975 that he interprets as signifying the breakdown of the political system. In this new legitimation crisis, again no party was seen as having the answer, and we return to electoral defeats for governing parties and even a period of minority government. The party system was under stress with the rise of nationalist parties in Scotland and Wales, increased support for the Liberal Party and finally a split in the Labour Party. All this changed with the rise of a new politics of national interest in the 1980s, which resolved the legitimation crisis. The Conservative government was returned to power with a large increased majority as voters learnt to mellow their demands and expectations. In fact, the Conservative electoral victory in the difficult economic circumstances of 1992 has been widely acclaimed as the first clear example of the politics of contentment. But a new rationality crisis soon emerged with the debacle in the currency market in 1993. Although the Major government limped on through its full term, in effect a new politics began in 1993 with the administration consistently recording the lowest poll ratings in history, culminating in the massive Labour landslide victory of 1997. 'Blairism' is the current new politics, which proclaims a modernization revolution not unlike that of 1945. However, this is a very mixed politics continuing Thatcherism

in some respects while also pursuing a social democratic agenda in exclusion and poverty elimination based upon full employment and economic growth. Currently mired in a failed foreign policy adventure, Blair's legacy may well mirror that of US President Johnson in the 1960s who is remembered for the Vietnam War and not domestic successes. Overall, Blairism does look like the latest growth-based new politics to add to Table 6.3.

Further details of these new politics can be found in Taylor (1991c). For our purpose here, there are just two main points requiring further emphasis. First, the various 'new politics' have not emerged at elections. In 1931, the new political battle lines were drawn before the election. The key example, however, is probably the practice of identifying the new politics of the social democratic consensus with the 1945 Labour victory. This is incorrect. Although the Labour government was responsible for setting up the welfare state, the major policy decisions had been agreed by the Conservative-led wartime coalition government. Hence we can date the emergence of this new politics to 1940 with the creation of the new coalition government. Similarly, technocratic politics were not the result of any one election – the appraisal of 1960 came just after a massive government election victory, which presumably endorsed past policies. We have dated the emergence of the new politics of crisis to 1972, when the Conservative government gave in to miners' union demands and became tarnished by its 'U-turn' image. Finally, the Falklands War of 1982 produced a nationalistic reaction among voters, represented by a turnaround in Conservative government popularity that was then consolidated in the 1983 election. The year 1982 obviously marks the beginning of the latest politics of national interest. It can be noted that the 'New Labour' politics of Blair were forged before the 1997 election victory. Hence in all cases new politics have emerged independently of elections. The politics of power precedes the politics of support.

The second main point is that throughout this period of various new politics there was only one major geography of support. Generally speaking, the areas voting Labour and Conservative in the first politics of crisis have continued their political biases right through to the present day: Labour maintains a northern, urban pattern of support, while Conservatives are the rural and suburban party. Despite electoral swings back and forth between the parties, this electoral geography has remained remarkably stable. Even the rise of a centre political alliance in 1983 could not break this geographical mould. The geography of Blair's 'New' Labour support is similar to 'old' Labour support in terms of locations of relative strength and weakness. The two major political parties may not always have been able to counteract Britain's decline in the world-economy, but they have continued successfully to maintain their support pattern through thick and thin. This fascinating mixture of impotence and strength is the hallmark of political parties of all countries, although the balance will differ greatly between core states and peripheral states.

Summary

In this section we have discussed electoral politics in the core and noted the importance of understanding the role of political parties. Thus we have:

➤ defined different types of political parties: cadre and mass parties;

➤ identified two types of politics: the politics of power and the politics of support;

➤ illustrated the congruent and non-congruent relationship between the politics of power and support.

■ Elections beyond the core

In January 1988, the Philippines under President Corazon Aquino carried out nationwide local elections. This involved deploying 156,000 troops in 930 so-called election hot-spots. In the event, the elections had to be suspended in ten of the seventy-three provinces, and in nineteen other provinces they had to be placed under the control of special commissions. The final death toll was 103, including 39 candidates. Mrs Aquino was able to claim a 'substantial reduction in blood letting' – in the previous local elections in 1971 under President Ferdinand Marcos, 905 people lost their lives.

What are we to make of such events? The violence of the Philippines poll was by no means atypical. Two months earlier in Haiti, 150 people were killed in an election, including 14 voters at a polling station. There is no need to produce a fuller catalogue of electoral violence to make the point that elections beyond the core are qualitatively different political processes from elections in liberal democracies.

This conclusion is hardly surprising given the different historical backgrounds and material circumstances that exist between the liberal democracies and the remainder of the world. Perhaps what is surprising is how often elections are held in such unpromising situations. Even the communist states felt the need to legitimate their government with periodic elections, albeit with choice limited to one party. Nevertheless, this does show the power that the electoral process possesses. Here we concentrate on only genuinely competitive elections beyond the core. In these third world countries, there have been two very different routes to competitive elections. In most countries, elections were a transplanted political process written into constitutions at the time of independence after 1945. Throughout Africa and Asia, these constitutions failed to protect this politics, and military coups have replaced elections as the most common means of changing government. Where elections do survive they are often traumatic and dangerous events, as we have seen.

In Latin America, however, with its much longer period of independence, the history of elections is very different. For instance, as Wesson (1982: 15) has pointed out:

> In 1929 every major Latin American government was civilian with some reason to claim that it was democratic; it seemed a reasonable assumption that this was the way of ever-improving civilization.

But now we know that this was not to be. Most Latin American countries have experienced military coups that have abruptly stopped this trend towards democracy. The most striking feature, therefore, is that despite their contrasting histories both sections of the third world have generated a similar outcome of fragile democracy and generals commonly becoming politicians. This provides strong evidence for materialist explanations of the relative failure of democracy, since mass poverty is the one feature shared by all countries in the third world. What happened in Latin America after 1929 and in other third world countries soon after their independence was that they were unable to provide the resources to sustain a viable politics of redistribution. Hence liberal democratic political processes failed because it was not possible to link them with a social democracy to produce a viable liberal–social democratic state. Without the consensus of the latter, the state reverts to a coercive mode of control.

How do parties operate in these circumstances? The first point to make is that party competition in elections is just one of several routes to power. Election campaigns and military campaigns can sometimes merge into a single process. Second, party victory in an election provides access to a state apparatus, which gives two important capabilities: opponents can be persecuted and prosecuted through the legitimate agencies of the state; and allies can be given access to the spoils of office. This has tended to produce a clientistic type of politics, with parties controlled by 'strongmen' in the battles for the spoils. Such parties consist of narrow groupings of people who support the 'boss' in return for favours. In some countries, the power of these extreme cadre parties has been undermined by new mass parties of a populist variety (Canovan 1981; Mouzelis 1986). Populist parties have been successful in a few countries in bringing the rural and urban masses into a state's political system. But this mobilization could not overcome the impossibility of a large-scale politics of redistribution. It is often the ignominious failure of populist policies, such as Peronism in Argentina, that have led to military intervention and coercion of the popular forces. In short, the rural and urban masses are removed from politics once again. In recent years, and especially since the end of the Cold War, multiparty democracy has spread widely in the third world. The remainder of this chapter explains why we think that what these political changes represent are new

'liberal democratic interludes' rather than sustained liberal democracy. We conclude this chapter with a further consideration of the paradox of increasing democratization under conditions of increased inequalities with globalization.

The politics of failure

We can describe the processes outlined above as the politics of failure. Given the world-system location of these countries, they are unable to develop the luxury of a congruent politics. In this situation, all governments in the eyes of most of their population turn out to be failures. This produces the instability of government for which the third world is notorious. The extreme case of the politics of failure is Bolivia, which has now experienced more than two hundred governments in its less than two hundred years of independence. More generally, where elections continue to be used to produce governments, the politics of failure is reflected in a continuous turnaround in party fortunes.

Democratic musical chairs

Suppose that in the material circumstances beyond the core there is a country that is able to sustain competitive elections over a decade or more. What sort of political system would we expect? Whatever the particular reasons that enable elections to continue, we would predict that given that the material situation produces government 'failure', then every party government would have severe difficulties in being re-elected. In short, this is an electoral situation made for opposition parties. Our expectations are, therefore, that one party would rule for one term of office, to be immediately replaced by the opposition party, and so on. This process is the opposite of what we see in the United States and Britain today; in the third world, it is the incumbents who have the disadvantage and who are voted out of office.

We can observe this process operating in Latin American states in the post-1945 era. Dix (1984) has investigated what he terms 'electoral turnover' in nine states, and his figures have been updated by Werz (1987). The one country with a continuous record of competitive elections is Costa Rica. In ten

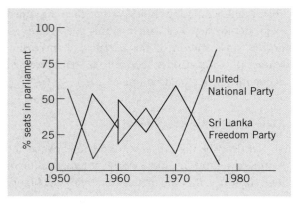

Figure 6.12 Democratic musical chairs in Sri Lankan elections.

elections between 1948 and 1986, the government has been swept from office on eight occasions. Chile, Venezuela and Ecuador have each had five elections, for just one returned government in each case. Dix and Werz find only eleven successful government re-elections out of a total of forty-three elections.

The best example of this democratic musical chairs comes from outside Latin America, however. Sri Lanka has been governed by elected politicians since independence in 1948. The two-party politics that evolved in Sri Lanka was remarkable for its extreme changes in party fortune (Figure 6.12). The seven parliamentary elections since 1952 have resulted in six changes of government. Subsequently, parliament has agreed a new constitution with a presidential electoral system, but the country has drifted into civil war.

The geographies of a politics of failure: the case of Ghana

What is the electoral geography of this political instability? This question has been answered in some detail in a case study of Ghana (Osei-Kwame and Taylor 1984). We used a finding from this research in Table 6.1 to show how Ghana as a peripheral country had a low level of geographical stability in its voting patterns. This was based on the analysis of eight elections between 1954 and 1979, which pitted one group of politicians who supported the first president, Kwame Nkrumah, against his opponents centred on

his great rival Busia. The party names changed over time, but these two political groupings can be easily identified. The pro-Nkrumah group are the centralists, who favour a 'modernizing' strategy of using cocoa-based export earnings from the central Akan region to develop modern industry. Their early emphasis on planning and protectionism meant that they were sometimes identified as the 'socialists' in the party system. The opposition were originally federalists who did not favour exploiting agricultural areas for the benefit of the coastal ports and towns. They have been the free traders, the 'liberals' in the party system. The analysis of voting returns concentrated on the pattern of votes for the centralists.

Whereas in the core countries of Table 6.1 there has been one major pattern of votes since the Second World War, for the peripheral countries low geographical stability means several patterns of votes. In the case of Ghana, the eight elections reveal four distinct patterns. In Figure 6.13, the changing geographical bases of support for the centralists are shown. Starting with a pre-independence pattern along the coast and in the south, by 1960 the centralists had spread their support inland around the Akan region, but not as far as either the northern or eastern boundary. They do reach these boundaries in subsequent elections, but at the expense of their original support bases. In 1969, the centralists have an extreme south-eastern bias to their voting map; a decade later, this disappears to be replaced by an extreme northern and south-western bias. Clearly, there is little of Rokkan's (1970) 'renewing of clienteles' here. What is the politics behind this geographical fluidity?

The politics of support in Ghana has been dominated by ethnicity. Nkrumah's Convention People's Party led the drive to independence but never produced a national movement across the country. Its original support was among the 'modernizing' elites of the coastal area and Nkrumah's home region in the south-west. Elsewhere, the party was overwhelmingly rejected. In most of the country, traditional power elites were successful as either independents or candidates for small regional/ethnic parties. After independence in 1957, Nkrumah's party was able to extend its support further inland but was still firmly resisted in

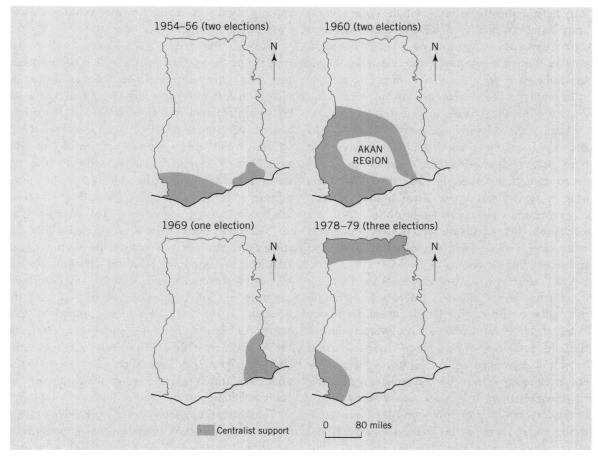

1954–56 (two elections)

1960 (two elections)

AKAN REGION

1969 (one election)

1978–79 (three elections)

◼ Centralist support

0   80 miles

Figure 6.13 The changing geography of voting in Ghana.

the Akan region (Busia's home area) and the north and east. When Nkrumah was toppled by a coup in 1966, therefore, he had not managed to become a national leader transcending ethnic rivalries.

With the departure of Nkrumah, the subsequent pattern of centralist support is wholly ethnic in nature. In 1969, the only time the centralists lost an election, they were pushed back to a south-east core, which was the region of origin of their new leader. His demise produced other new leaders in 1978–79 whose northern and south-western origins became yet further new support bases.

We can therefore conclude that Ghana presents a classic case of a disconnected politics. Ethnic geographies of support exist alongside a politics of power that is concerned with alternative ways of managing

an economy dependent on cocoa export revenues. Only the cocoa-growing Akan region is consistent in resisting the centralists. Otherwise, different regions will support centralist policies, depending on the ethnic origin of the party leader. Hence the Ghanaian political process consists of a cultural geography base that is transformed into different political geographies that provide a capability to produce alternative economic geographies.

India as 'the largest liberal democracy in the world'

The association between political instability and elections is not found in all countries of the third world. In Mexico, for instance, regular elections over more than half a century have resulted in the same political party being returned to office. We argue here that these are special cases that warrant specific investigation. In Mexico, for instance, the association of the governing party with the revolution is fundamental to its ability to keep the opposition small and fragmented. Similarly, in India the Congress Party was at the head of the independence movement and has by and large been able to dominate Indian government since 1947. We describe such successful governing parties as aggregative parties, since they display an ability to aggregate electoral support across a wide spectrum of the electorate. The discussion below focuses on the mechanisms behind this electoral strategy as operated by the Congress Party, first in their politics of support and second in their politics of power.

The reason why we concentrate on the Congress Party is because of Blondel's (1978: 55) contention that India 'constitutes the largest liberal democracy in the world.' In our world-systems interpretation of elections, India cannot be a liberal democratic state because it does not have the resources for a viable politics of redistribution. But it does have regular free and open elections. If Blondel is correct, this would be the exception to prove our world-systems analysis wrong. Hence Indian politics represents a crucial test of our political geography. By concentrating on the electoral geographies of the Congress Party, we will show that India has a very different politics from that found in the liberal democracies of the core.

Congress geographies of support

The geographical scale, material inequalities and cultural diversity of India provide an immense range of potential social cleavages for political parties to try to organize. It is therefore not surprising that there have been literally hundreds of parties competing in Indian elections. Most have disappeared, but Congress survives and prospers. Above all the complexity that is Indian politics there is the simple fact of one-party dominance. Since 1947, Congress has been in power for all but two short periods. By any criterion it is a supremely successful political party. How has the Congress Party achieved this?

The Congress Party was formed in 1885 and until the end of the First World War it behaved as a typical cadre party for the small Anglicized Indian middle class. After 1919, this role was taken over by the breakaway Liberal Party, and Congress developed into the mass mobilizing party of the independence movement. After independence, however, it was not able to maintain its mass mobilizing function or to convert into a modern representative party. Congress never developed a single doctrine or ideology behind which it could build a national constituency of support. Hence Congress, for all its dominance, has not had the electoral integrative role that parties have performed in liberal democracies. Rather, Congress has become an aggregative party. This means that at any particular election the party's support base is a mere aggregation of different groups. Such aggregation may be temporary, since the creation of these support bases does not depend on any consistent doctrine or policy position, what Park and de Mesquita (1979: 113) refer to as Congress's 'purposive ambiguity'. The most important feature of the aggregation is that it constitutes a majority. In short, Congress has been successfully opportunist and pragmatic in its handling of the complexity of Indian politics.

The electoral geography of the aggregative strategy has been illustrated by analysis of voting patterns in the state of Punjab. In Table 6.1, the low level of geographical stability recorded for India was based on a study of Congress voting patterns in Punjab (Dikshit and Sharma 1982). We can see the details of this geographical instability in votes for Congress in another analysis by Brass (1975). Elections from 1952 to 1972 for the three main parties competing in Punjab are depicted in Figure 6.14. For each election, a party's vote pattern is correlated against the geographical distribution of the Hindu population. For instance, the Jana Singh party campaigns as a Hindu party, and its pattern of votes is always positively correlated with the distribution of Hindu people, as we would expect. In contrast, the Akali Dal is the party of the Sikhs, and this party's votes are always negatively correlated with the Hindu distribution. But Congress has no such simple correlation. In 1952, the correlation indicates largely Hindu support. At the next election, however, the correlation suggests largely non-Hindu support before reverting to Hindu support, and so on back

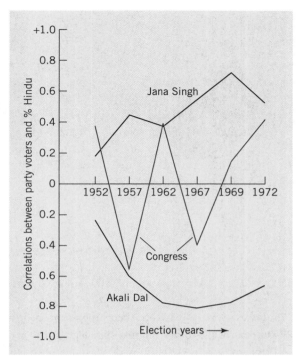

Figure 6.14 The changing ethnic bases of party voting in Punjab.

and forth between the two communities. This reflects a series of politics of failure that have been accommodated by changing the party's support base. Put simply, Congress lets down in turn each community that votes for it. Its rejection at the next election is countered, however, by creation of a new base of support. In this way an aggregative party can maintain control even though it is being continually rejected by its previous supporters in a politics of failure. Congress won every election in Punjab to 1967. Hence here we have a different expression of the politics of failure from the turnover elections described above for other third world countries.

There are no electoral geography studies of the whole of India for us to see how this aggregation strategy works out across the country, but it is clear that the politics of failure is reflected at this scale despite Congress domination of government. The first cracks in Congress's control came in 1967, when it lost several state elections. We shall consider Congress's response to being in opposition in our

discussion of the politics of power below. In 1971, the old 'official' Congress was defeated by the new 'Indira' Congress of Mrs Gandhi in a personalized and populist campaign. In 1975, a state of emergency was declared and 676 political opponents were jailed. In all, 110,000 people were arrested and detained without trial. Congress had clearly lost control of India's politics of failure. When the next election eventually came, Mrs Gandhi was swept from power by an opposition united by their persecution. The successful Janata Party was another aggregative party with no clear doctrine or policy except opposition to Congress and the emergency. Unlike Congress, however, Janata soon disintegrated, allowing Mrs Gandhi to win the 1980 election and return to power.

The geographies of the two elections of 1971 and 1977 show an extreme geographical instability (Figure 6.15). In 1971, the pattern of support was concentrated in the Hindu heartland of the north. In 1977, the north was lost, and most Congress votes came from the non-Hindu southern states, producing a completely new pattern of votes: these were turnover elections in terms of geography as well as politics. This is as we would expect for a third world country, where stable congruent politics is not a possibility.

Congress geographies of power

Congress has not relied solely on its politics of support to maintain itself in power. Its 'purposive ambiguity' in policy positions allows it to forge links with a wide range of special interests. For instance, Congress has always claimed to be a democratic socialist party and can rely on the formal support of many trade unions and peasants' organizations, outflanking India's two main socialist parties. At the same time, it has received six times the amount of contributions that the 'free enterprise' party, Swatandra, obtains from big business and rich landlords (Sadasivan 1977: 307–8). Clearly, a successful aggregate party depends upon a politics of power just as much as a politics of support.

Congress's politics of power is best illustrated by the 1967 state elections and the subsequent events. Kashyap (1969: 3) calls this the first watershed in post-independence Indian politics because it 'exposed

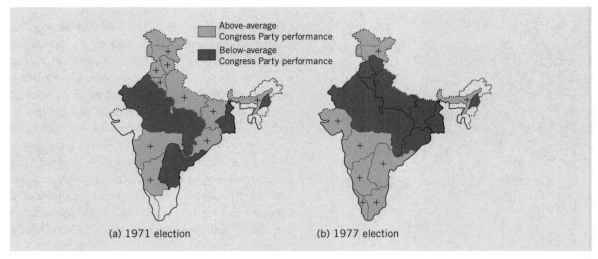

Above-average
Congress Party performance

Below-average
Congress Party performance

(a) 1971 election   (b) 1977 election

Figure 6.15 Changing geographies of Indian elections.

the artificial level of political stability'. Prior to the election, there were no anti-Congress governments in any state, but in the election, Congress lost control of seven out of twenty-one states. But this was by no means the end of the story. Losing an election in a state does not necessarily mean losing power. There are two ways in which this can be prevented. First, opponents can be bribed to leave the party for whom they were elected and join Congress. Second, if political instability occurs, the central Congress government can deem the state ungovernable and declare 'president's rule', which effectively means control by Congress from the centre. Both strategies have been widely used throughout post-independence Indian politics but became more widespread as Congress's support base began to crumble in 1967. We shall consider each in turn.

Sadasivan (1977) reports 542 defections of legislators between 1957 and 1967. In the year after the 1967 elections there were 438 defections. Bihar state was the foremost casualty, with 85 defections among just 318 legislative seats. This was one of the states that Congress lost in the election, but by January 1968 a pro-Congress coalition government had been installed. Defections were a two-way affair, as Congress legislators deprived of office could keep a share of the spoils by defecting. In this period, Hartmann (1980: 182) reports 139 gains for Congress against

175 losses, which between them represent nearly 20 per cent of the total Congress state legislators.

President's rule was imposed on states only ten times between 1951 and 1967 but seventeen times in the four-year term after the 1967 election (Dua 1979). In the year following the 1967 elections, president's rule was imposed on five of the seven states that Congress had failed to hold (Kashyap 1969). Often this strategy was combined with defections. For instance, Hartmann (1980) presents a case study of Punjab, where Congress's defeat in 1967 produced at first a non-Congress government. This was initially undermined by defections, producing a pro-Congress coalition by November 1967 and culminating in president's rule in August 1968.

The resulting geography of the 1967 election results is thus quite complicated. In the two years following the election, and without any further reference to the electors, there were thirteen changes of government, and of the original non-Congress governments only Kerala remained untouched. The detailed geography of this politics of power is illustrated by Kashyap (1969).

Since this 'first watershed' in post-independence Indian politics, these politics of power have continued to operate. Dua (1979: 19), for instance, reports a further fifteen examples of presidential rule between 1971 and 1974. After the 'second watershed', the election of a non-Congress central government in 1977, it was defections that brought down the Janata government in 1979. Kamath (1985) reports numerous additional defections in the 1980s, culminating in the outlawing of the practice in 1985. But he thinks it unlikely that defections will be eradicated, since politicians will be able to get around the law (ibid.: 1053). Quite simply, legislative seats are too valuable a commodity to be removed from the market. Securing election under any party label becomes a personal resource. It provides access to the state and its resources. Kamath (ibid.: 1048) reports prices for defectors in the 1980s ranging from US$100,000 to US$250,000, plus additional inducements such as ministerial office or the dropping of criminal charges.

After only this brief exploration into Indian politics of support and power we can refute Blondel's (1978) claim for an Indian liberal democracy. The political processes operating in India are far closer to other third world countries with their different political institutions than they are to the liberal democratic states of the core (Taylor 1986). This has become even clearer in the 1990s as the Congress Party has lost its grip on Indian politics, resulting in weak government as the electorate gives no one party a majority. Musical chairs democracy is in operation. But these times have witnessed a dramatic turn in Indian politics eschewing Nehru's moderate socialism for free market reforms. Successful in terms of stimulating economic growth, the non-Congress coalition that implemented the reforms has been cast aside in the latest election; Congress, now a free market party, rules again.

We conclude that since liberal democracy is a particular product of the history of one part of the world-economy where recent relatively abundant material circumstances have been crucial to its establishment, the attempt to diffuse this political product has explicitly failed in most countries. In some countries, however, institutions exist that look superficially like liberal democracy but that form a distinctive form of polity. India is such a case. It has invented a political system that marries a democratic politics with mass poverty, a remarkable, and to the present, unique achievement.

Democratization and globalization

There is no doubt that the end of the Cold War has given a world-political stimulus to 'democracy' in regions beyond the core. This has been both 'bottom-up' and 'top-down': there have been genuine uprisings of peoples in the third world demanding 'people power' as well as core governments imposing the condition of multi-party democracy before dispensing economic aid. Both of these factors have led to a spread of competitive electoral politics, especially in Africa. From our analysis above, it will be understood that we expect these to be liberal democratic interludes only, another round in the politics of failure.

Empirical analysis tends to confirm our theoretical scepticism. O'Loughlin et al. (1998) conducted a broad study of the diffusion of democracy from 1946 to 1994 and found that about 60 per cent of countries can now be classified as democracies, compared with 28 per cent in 1950. However, these aggregate statistics do not represent a smooth and uniform trend towards the democratization of the globe. There is a distinct regionalization of democracies and autocracies, with similar political systems clustering next to each other. Also, there have been spurts of democratization followed by periods of reversal as some of the newly democratic countries reverted to autocracy (Huntington 1991; O'Loughlin et al. 1998). The clustering of democratization in time and space is consistent with our material explanation of the geography of democracy. The structure and dynamics of the world-system provide limited opportunity for the expansion of democracy. However, it would be churlish to ignore the slow spread of democratic practices.

Figure 6.16 maps the distribution of democracies for three snapshots since 1946 (O'Loughlin et al. 1998). These maps illustrate the instances of decolonization during Kondratieff IVA, spurred on by the hegemonic ideology of the United States, and the increase in the number of countries classified as strongly democratic. However, the existence of reversals towards autocracy should also be noted: India and Venezuela between 1972 and 1994; Egypt, Turkey and Brazil between 1950 and 1972; and Indonesia between 1950 and 1972, for example. The recent trend towards an increase in the level of democracy in

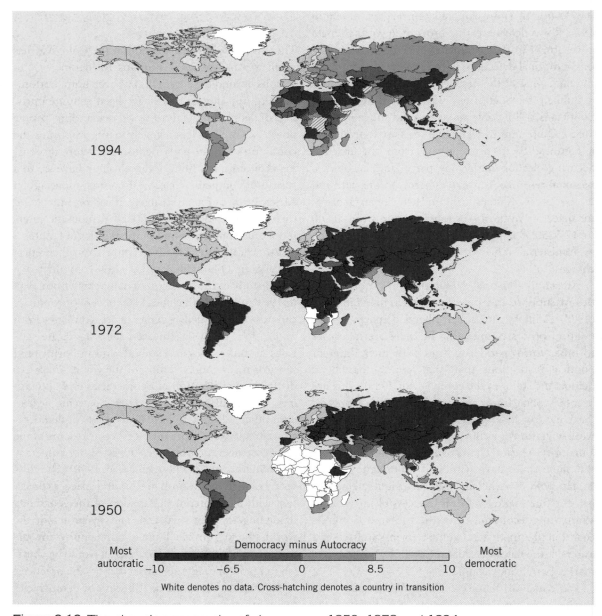

Figure 6.16 The changing geography of democracy, 1950, 1972 and 1994.

Source: O'Loughlin et al. (1998).

the world-system is further illustrated by calculating the mean democracy score for all the years since 1946 (Table 6.4). Although the number of countries changes for each year (O'Loughlin et al. did not include colonies in their calculations), it is clear that the level of democracy fell to a low of −2.40 in 1977 and then increased to a maximum of 2.98 in 1994.

The fall in the 1960s is a function of the inclusion of newly independent African countries and their turn towards autocracy after independence – or the politics of failure.

Despite the overarching picture of a trend towards democracy, further analysis raises doubts about the sustainability of some democratic countries.

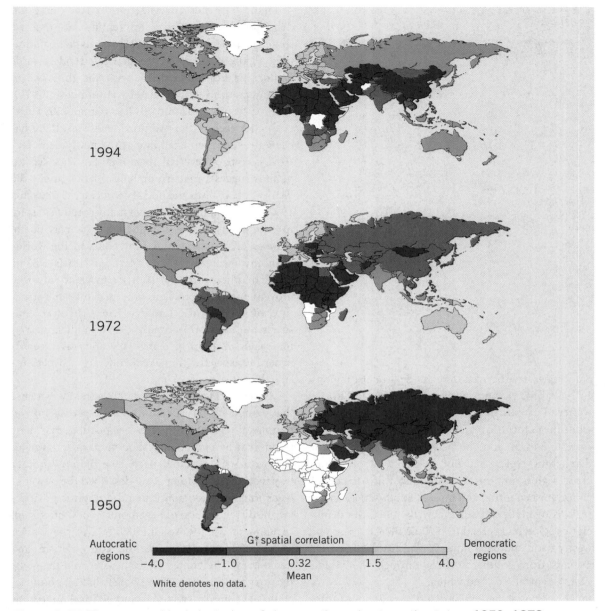

Figure 6.17 The geographical clustering of democratic and autocratic states, 1950, 1972 and 1994.

Source: O'Loughlin et al. (1998).

Figure 6.17 maps the regionalization of democracies and autocracies. For each country, a statistic is calculated that measures the extent to which it is surrounded by countries with similar democracy scores. For example, a high positive score is obtained if a democratic country has other democracies as neigh-bours, while a country obtains a high negative score if it is an autocracy surrounded by other autocracies. Low scores are given to countries if they are democracies surrounded by autocracies or autocracies surrounded by democracies. Figure 6.17 clearly shows the extent to which democracies and

Table 6.4 Mean global democracy scores, 1946–94

Year	Number of countries	Mean democracy score
1946	76	1.11
1948	81	0.38
1949	86	0.09
1955	92	0.12
1960	109	−0.25
1965	128	−1.04
1970	135	−1.50
1975	141	−2.04
1977	142	−2.40
1980	142	−1.72
1985	142	−1.16
1989	142	−0.50
1990	141	0.89
1991	155	2.18
1992	156	2.30
1993	157	2.71
1994	157	2.98

Source: O'Loughlin et al. (1998).

autocracies have been clustered into particular regions. In 1950, only North America, Australasia and north-western Europe can be treated as democratic regions, while the autocratic region was centred in communist eastern Europe and the Middle East. In 1972, the democratic region had not changed, but the autocratic region now encompassed most of Africa. The communist countries and most of South America were regions of moderate autocracy. By 1994, the democratic region had spread to include the Americas and western and southern Europe. The autocratic region extends from southern Africa through the Middle East to central Asia and China.

The regional pattern of democracy and autocracy is consistent with our material framework; democracy is an option only for those countries that are able to extract enough of the global surplus to distribute to their populations. The spread of democracy represents political changes stemming from changes in the capitalist world-economy. But does the increase in democracy challenge the essential core–periphery structure of the world-economy? Without the benefit of foresight, we cannot answer that question. However, our material framework suggests two responses.

First, a further investigation of the temporal dynamics of democratization reminds us that we should not infer a one-way street towards democratization. Figure 6.18 shows the general trend towards democratization since 1815, with the three waves of democratization identified by Huntington (1991): 1828–1926, 1943–62 and 1974 to the present. However, after each of the two previous waves, there has been a reverse trend as some of the newly democratized countries reverted to autocracy. This cyclical pattern suggests that triumphalist claims about the victory of liberalism and liberal democracy (Fukuyama 1992) are premature. Although many countries aspire to core status, and receive benefits such as liberal democracy, the structural constraints of the world-economy mean that some of those efforts will be futile. In other words, the short-term agency of social movements and politicians is impeded by the structure of the world-economy. The structural constraints of the world-economy are also illustrated by the regionalization of democracy and autocracy. It is hard for societies to make democracy prosper outside the core.

However, O'Loughlin et al. (1998) do show evidence that democracy has spread into semi-peripheral and peripheral regions. Although it is possible that these trends may be a function of short-term material gains, the world-systems perspective provides another explanation. In Chapter 2, we described the power of hegemons to shape political and economic practices via the dissemination of hegemonic codes. Important components of the hegemonic code of the United States were self-determination, consumerism and democracy. The United States has ordered the globe through the espousal of the belief that all countries could have similar economic and political opportunities to those in the United States. The spread of democracy has been stimulated by the imperatives of current US hegemonic practices: were the east European revolutions of 1989 spurred on by the thought of democracy or the promise of consumerism? The answer is both, but we should not underestimate the latter. In fact, our materialist perspective suggests that the diffusion of such core-like political practices is unsustainable without some widespread achievement of the 'good life', which is

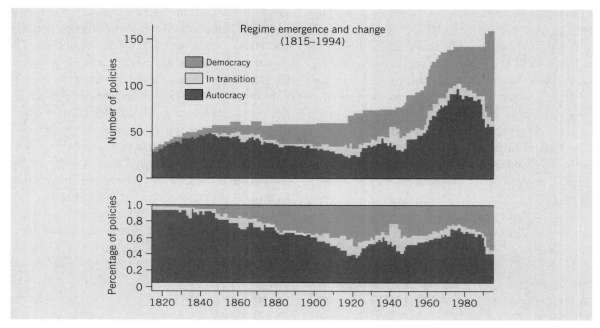

Figure 6.18 Regime emergence and change, 1815–1994.

Source: O'Loughlin et al. (1998).

ultimately impossible with increased polarization. The pressing question, therefore, is whether reverse trends towards autocracy will lead to greater social unrest now that US ideology has let the democratic genie out of the bottle.

The new democracies of eastern Europe are good examples of the problems in establishing liberal social democracy. Two key elections took place in the first half of 1989 that illustrated the lack of support for communist regimes in eastern Europe and set the stage for the revolutions later that year. In the Soviet Union in March, the first competitive elections for the Congress of People's Deputies produced notable defeats for party nominees, especially in Moscow (which overwhelmingly elected Boris Yeltsin) and the Baltic republics (where the nationalists were success-ful). 'People power' had arrived, but the opposition was not organized so the effect was muted. This was not the case with the Polish election in June. Here the united opposition, Solidarity, was allowed to contest only a minority of the seats (163) but defeated the government's candidates in every one. The shock

to the system of this massive endorsement for the opposition led to the formation of a non-communist government in August 1989. The remaining Soviet-allied communist states disappeared with the revolutions of late 1989.

In repudiating communism, with its non-competitive elections, one of the first tasks of the new regimes was the organization of pluralistic elections. Hence 1990 saw a series of what are generally called 'foundation elections' (Bogdanor 1992) across eastern Europe. This terminology is itself interesting, since it implies the future continuity of liberal democracy. The many new elections in the third world are not called 'foundation elections' because commentators are far less confident about the prospects for demo-cracy there. In fact, the elections in eastern Europe founded little beyond the principle of holding free elections and repudiating the Communist Party. Even the latter was not universal: in Bulgaria, the Socialist Party (former Communist Party) won easily with 48 per cent of the vote; in Romania, the success-ful National Salvation Front contained many recent

Communist Party members in its leadership; and the Serbian communists retained their administration within the federation of Yugoslavia. These three cases reflect a lack of opposition organization after the rapid political changes. In the other former communist states, oppositions were organized and reduced the communists to between 10 and 20 per cent of the vote in Hungary, Czechoslovakia and East Germany. Non-communists also won in the Yugoslavian republics of Slovenia and Croatia. A new East–West division seemed to be forming in the old eastern Europe.

Organizing opposition to the communist state is not the same as generating a multi-party democracy. The existence of umbrella groups such as Solidarity in Poland and Civic Forum in Czechoslovakia may win elections, but they are essentially anti-pluralist in practice. There is no simple route from broad alliances covering a wide range of opinion to a functioning party system that produces alternative governments via the electoral process. In Poland, the splintering of Solidarity resulted in twenty-seven parties winning seats after the 1991 election, making the formation of a government a very difficult task. According to Bogdanor (1992), we will not be able to assess where these political systems are going until there have been at least two more elections. The question is, will these elections really be foundations for the spread of liberal democracy to eastern Europe? The jury is still out, but the intriguing thing has been how the new non-communist governments, for all the voter goodwill they began with, soon faced their own legitimation crises in country after country as new market policies destroyed public services. Hence, as a group, former communist parties remain the largest electoral force in eastern Europe.

We have argued a materialist theory of liberal democracy above. Clearly, the economic difficulties of the ex-communist states put a clear constraint on the possibilities of democratic continuity. This is countered temporarily by the popular anti-communism that forced the political changes and allowed free elections. But this goodwill is exhaustible; certainly, the current economic austerity programmes are eating it up rapidly. The people realize only too well that elections alone will not solve their problems – state-

wide local elections in Poland and Hungary attracted well under half the electorate to the polling booths in 1991 (White 1992: 286). If the new liberal democracies come up with goods, and this depends at least as much on there being a strong upturn in the world-economy as on anything governments themselves do, then they can become more than liberal democratic interludes and develop representation parties in a congruent politics creating liberal–social democratic states. In short, they have to join the core of the world-economy in its next growth phase. It is unlikely that all of eastern Europe will be successful. The lesson of the 'foundation elections' is that a new East–West division is developing in Europe. It seems to us that a new central Europe (including the Czech Republic, Hungary and probably Poland and Slovenia) closely integrated into the German economy may well become core-like and establish stable liberal democracies, leaving a smaller eastern Europe including the former Soviet Union in the semi-periphery and unable to sustain liberal democracy. This does not necessarily mean third world-type instability in the new eastern Europe. This region may well construct a new politics accommodating its political elites and people compatible with its strong semi-peripheral position, like Mexico or India. But we do not know; what we do know is that it will turn out to be a good test of the materialist thesis proposed in this chapter.

The economic difficulties of establishing liberal social democracy in the semi-periphery and periphery can be further complicated by geographies of ethnic and religious difference. The geopolitics of democratization, as part of a global hegemonic project, is enacted by a diverse set of political groups within nation-states, each with its own agenda. The hegemonic power's attempt to catalyse the diffusion of democracy was most evident in the US-sponsored War on Terrorism. A new region was targeted as being ripe for democratic change, the Middle East and central Asia. Elections were held in Afghanistan after the US-led invasion had overthrown the Islamic fundamentalist Taliban regime. Iraq was next on the agenda and dominated the news. However, the call for elections in the first Palestinian parliamentary elections in January 2006 caused a diplomatic problem. The winners were Hamas, whose political

Democratization and the War on Terrorism

The spread of democracy to the Middle East was identified as a key component of the War on Terrorism. After the overthrow of Saddam Hussein, US scholars were involved in writing the constitution for post-Hussein Iraq (Feldman 2003). Much political, and military, effort was put into a referendum on the nature of the constitution and subsequent elections. However, the situation in the first months of 2006 was not so rosy. Violence between the Sunni and Shia communities was a daily occurrence of bombings and execution-style shootings. Debate bubbled over whether Iraq was in a state of civil war or not. What of the project of democratizing the Middle East?

The political commentator Robert Kaplan, identified as a 'leading neo-conservative' by the Egyptian newspaper *Al-Ahram*, was having second thoughts about the geopolitical value of extending democracy to the Middle East. In an article in the *Washington Post* (2 March 2006), he voiced the desirability of a despotic state rather than a democratic one: 'For the average person who just wants to walk the streets without being brutalized or blown up by criminal gangs, a despotic state that can protect him is more moral and far more useful than a democratic one that cannot.' Clearly, the ability to spread western-style electoral democracy into Iraq was floundering and an imperative of

'stability' under a non-democratic state was taking precedence. Interestingly, Kaplan sub-titled his article 'Creating normality is the real Mideast challenge': an acknowledgment that (1) the War on Iraq had been a massive disruption and (2) non-democratic politics should be expected in this region.

The questioning of the democratization project was not restricted to Iraq. *Al-Ahram* reported that Egyptian president Hosni Mubarak believed that conversations with US Secretary of State Condoleezza Rice in March 2006 conveyed a sense that the US was satisfied with the process of democratization in Egypt. *Al-Ahram* seized on this statement to argue that growing US disgruntlement with the democratization in the Middle East had facilitated the Egyptian government's repression and persecution of political dissidents, notably the arrest of members of the Muslim Brotherhood. The changing attitude towards democratization of the Middle East facilitated state repression of political opposition without US diplomatic censure.

Sources: Robert D. Kaplan, 'We can't force democracy', *Washington Post*, 2 March 2006. www.washingtonpost.com/wp-dyn/content/article/2006/03/01/AR2006030101937_pf.html. Accessed 13 April 2006. Amira Howeidy, 'Democracy's backlash', *Al-Ahram Weekly Online*, 9–15 March 2006. http://weekly.ahram.org.eg/2006/785/eg3.html. Accessed 16 March 2006.

agenda included outright hostility to Israel as well as the promotion of religious law; this was not the sort of party that the United States was hoping would emerge from a bout of democratization. Hence, there is an interesting paradox emerging in the geopolitics of democratization. Democracy is being diffused by the United States in its role as hegemonic power in an attempt to secure political influence in the Middle East and other regions, but the process is bringing parties to power that are antagonistic to US goals. Perhaps, a reverse-wave of democratic failure in the Middle East will leave anti-US parties even more emboldened within their nation-states?

Summary

In this section we have discussed electoral politics outside of the core of the world-economy by:

➤ introducing the concept the politics of failure;

➤ describing the geography of democratization;

➤ identifying the likely limits to democratization;

➤ discussing the relationship between democratization and 'empire'.

Chapter summary

In this chapter we have used the world-systems approach to explore the geography of elections, especially the persistent spatial pattern of democratic elections being restricted to the core of the world-economy. The key points we have addressed are:

➤ the quantitative heritage of electoral geography;

➤ the liberal assumptions that underlie mainstream electoral geography analysis;

➤ the conceptualization of liberal social democracy;

➤ using a world-systems approach in conjunction with the concept of liberal social democracy to explain the general pattern of elections being limited to the core of the world-economy;

➤ focusing upon political parties to explain electoral democracy in the core;

➤ noting the operation of two politics in the core: the politics of power and the politics of support;

➤ introducing the concept 'the politics of failure' to explain the instability of electoral support in elections held in the periphery;

➤ describing the diffusion of electoral democracy, or democratization;

➤ arguing that democratization will be limited spatially and temporally.

In sum, our focus on the geography of elections has exposed the assumptions of liberal understandings of electoral democracy. Instead, we have situated electoral politics within the structural power relations of the capitalist world-economy. With this approach we have noted and explained the core–periphery pattern of democratic and non-democratic politics. Looking within the core, we have emphasized the separate but related processes of the politics of power and those of support, while within the periphery we have highlighted the politics of failure. We argue that as long as the core–periphery structure of the world-economy remains, then democratization will be stilted.

Key glossary terms from Chapter 6

absolutism	conservative	geopolitics	liberal democratic
administration	constitution	gerrymander	interlude
anarchy of production	contradictory politics	globalization	local government
authoritarian	core	government	malapportionment
bloc	decolonization	hegemony	multi-party system
boundary	democracy	ideology	nationalism
capitalism	disconnected politics	instrumental theory of	nation-state
capitalist world-	elite	the state	neighbourhood effect
economy	executive	Islam	neo-conservative
centre/centrist	faction	isolation	opposition
Christian democracy	federation	Kondratieff cycles	peoples
civil liberties	First World War	(waves)	periphery
classes	franchise	left-wing	place
Cold War	free trade	legislature	pluralism
communism	geopolitical world	liberal	pluralist theories of the
congruent politics	order	liberal democracy	state

political parties	proportional	semi-periphery	state
politics of failure	representation	social democracy	suffrage
politics of power	protectionism	social imperialism	third world
politics of support	right-wing	socialism	world-economy
power	Second World War	space	world-systems analysis
	sectionalism		

Suggested reading

Agnew, J. (2002) *Place and Politics in Modern Italy.* Chicago: Chicago University Press. Outlines a theoretical framework showing the mutual construction of place and politics, with in-depth empirical analysis.

Flint, C. (2001) 'A TimeSpace for electoral geography: economic restructuring, political agency, and the rise of the Nazi party', *Political Geography* 20: 301–29. This article connects the geographical pattern of electoral support for the Nazi party in inter-war Germany to the economic and political dynamics of the capitalist world-economy.

Johnston, R. J., Shelley, F. M. and Taylor, P. J. (1990). *Developments in Electoral Geography.* London: Routledge. Accessible discussion of trends and issues in contemporary electoral geography.

Activities

1 At what scale does the act of voting take place? Consider the manner in which institutions at different geographical scales enable a person to vote. In turn, consider a recent election that you either participated in or are aware of. Were particular issues identified with particular scales? In what way was the designation of issues to particular scales a denial of the connections between scales?

2 Identify a political party of your choice and browse its website. In what way does the party's manifesto address the three properties of liberal social democracy we identified on page 207?

3 Select two countries, one from the core and one from outside the core. Use a search engine to find election stories by typing in the country name with the word 'election' (for example German election and then Philippine election). List similarities and differences between the two elections. Do your findings fit our model of different 'liberal democracies' across the world?

DERELICT DWELLINGS IN ROME, ITALY.
Source: © Rolf W. Hapke/zefa/Corbis

Locality politics

What this chapter covers

Burmese capital city moved

The population of Burma saw their new capital city for the first time as the backdrop to General Than Shwe's televised speech to 12,000 troops extolling the national need for a large and modern army. Burma is ruled by a military junta, with no possibility of democracy on the immediate political horizon. The new capital city of Pyinmana has been quickly and dramatically transformed from a quiet and peripheral logging town. On 6 November 2005 the junta made a surprise announcement that the capital was being moved 250 miles from the old site of Rangoon. Immediately, surprised civil servants were moved by bus, and just five months later Pyinmana was revealed as the new capital.

What are the political processes driving the change in physical location of the Burmese capital and the transformation of a logging town into a capital city? On the one hand, the military junta, in power since 1962, has been facing increased domestic protest from the democracy movement led by Aung San Suu Kyi. Pro-democracy demonstrations in Rangoon may have provoked the movement of government functions to a new location away from the geographical site of the protest movement. On the other hand, it is also suspected that the junta has become nervous about possible US invasion within the global scope of the War on Terrorism. Hence, no surprise that it was a phalanx of troops that the TV showed populating the capital city and not health service workers.

What does this story suggest about the politics of the locality? Politics occurs within places. The movement to change the regime and make Burma a democratic country is organized and active within particular places. One part of the junta's strategy to negate the movement was to withdraw to a new locality where the democracy movement would have to build strength from scratch. Second, the politics of localities must be understood within the context of the capitalist world-economy. Specifically, in this case, the perceived intent of the hegemonic state has been partially responsible for Pyinmana's new function and form. Finally, locations as venues offer possibilities and constraints for political actors. The junta hopes that moving the central government to Pyinmana constrains the pro-democracy movement and facilitates its own hold on power.

The case of Pyinmana shows the importance of integrating a variety of politics when considering the geography of capital cities (as discussed in Chapter 4).

Source: 'Defiant Burmese junta unveils new capital', *Guardian Unlimited*, 27 March 2006. www.guardian.co.uk/print/ 0,,329443970-111494,00.html. Accessed 27 March 2006.

We have reached the scale of experience in our framework for political geography. The range of this scale is defined by the day-to-day activities of people in the ordinary business of their lives. We may start with Hagerstrand's time geography concept, which treats the strict space–time constraints on an individual's behaviour. Starting from his or her 'home base', every person's day-to-day life consists of a series of regular paths that must be organized to enable return to the home base every night. The range of these paths depends upon the transport available to the individual. This 'physicalist' view of geography places a 'space–time' prism of constraint around every individual that is unique to that individual. Hence everybody's direct experience of the world is distinctive to that person.

Being distinct does not mean being unrelated. Society does not consist of an aggregation of random 'prisms' but is a highly organized and structured phenomenon. Space–time prisms are clustered, as anybody who has spent time in a city's rush hour knows only too well. In much of the modern world, these clusters are the 'daily urban systems' of activities around our major cities. Although usually defined just in terms of commuting, these urban systems incorporate a wide range of people's needs in modern society, including shopping, education and recreation as well as employment. In many parts of the world, daily urban systems are the limit of routine behaviour and therefore represent in concrete terms the scale of experience.

Before we consider the nature of this scale, we must emphasize once again that our identification of scales is a pedagogical device and does not indicate any separateness or different systems at different scales. Every individual in his or her prism that is part of a daily urban system is no less a member of a nation-state as a citizen and participates in the world-economy as a producer and consumer. Although the dependency heritage of the world-systems approach emphasizes the global scale, the second heritage that we identified in Chapter 1, the French Annales school of history, equally emphasizes the everyday life of ordinary people. For Braudel (1973), it is these routine behaviour patterns that provide the long-term structures in which 'world events' occur. The scale of experience is just as integral to the world-economy as the scale of reality.

As we have already indicated, contemporary studies of the scale of experience will normally be urban in nature. Urban studies in social science were the largest growth area in the twentieth century. Cities have been a major concern of sociologists, political scientists and geographers for several generations. Political geography, however, apart from its treatment of capital cities, has been largely immune to this academic endeavour, and urban political geography can only be said to start from about 1970. Since then, studies at the urban scale have been a major growth point in political geography. Why, therefore, has this chapter not been entitled 'urban political geography'? The reason is simply that we do not adhere to the idea that there can be a coherent, conceptual framework called 'urban theory'. Given that the vast majority of the world's people, especially in core zones, live in daily urban systems, the idea of an 'urban way of life' is becoming a ubiquitous condition. Hence we replace the distinctively 'urban' by the more general idea of locality and search for the nature of the relationship between experience defined by locality and the nation-state and world-economy that encompasses it.

This is a very difficult task. The most sustained attempt to research this question was the British 'Changing Urban and Regional System' research programme (Cooke 1989), which focused on seven localities and attempted to relate them to economic restructuring at the state and global scales. This work has generated a large debate, much of it methodological argument, and in practice the studies got bogged down in local detail at the expense of broader structures. We shall not rehearse this debate here but are mindful of the difficulties that the British project

encountered. We approach localities in two distinct ways. In this chapter, we describe the politics that occurs at the local scale. This locality politics draws on several themes from previous chapters and shows how they work out in practice in localities. In this argument, there is nothing special about the localities, they simply constitute a new scale at which politics unfolds.

We emphasize two points in this chapter. The first is that localities are settings for political action. They create contexts that provide opportunities for political actors, but also limit what can be done. This does not mean that the nature of localities fully determines what people and groups can and cannot do. Rather, the first parts of the chapter explore how the physical nature of localities and the meanings attached to them are created by political activity and act as a context for ongoing politics. Second, we illustrate that localities can only be fully understood by placing them within a hierarchy of geographical scales and the interconnections of localities across the spatial extent of the capitalist world-economy.

Locality politics follows a long tradition of social science research that tends to abstract social relations from their particular context. The limiting case of this can be found in the spatial school of geography, which imposed regular geometric patterns on messy real-world patterns. We begin this chapter, therefore, with this spatial heritage, which eschews politics for 'administration'. The remainder of the chapter can be interpreted as a retreat from this apolitical beginning. In the second section we bring together a series of examples that show that localities do matter in politics. This is where we revisit some earlier themes and treat them at this new scale of activity. This is followed by the presentation, derived from several sources, of a new theory of the politics of localities. This third section goes beyond new illustrations of old politics to build a coherent model of locational politics based upon the distinction between everyday practical politics and the formal politics of the state, represented here by the local state. We then illustrate the material and rhetorical tranformations of localities within the context of the War on Terrorism. Finally, we choose as our concluding topic the current debate on relations between the local and the global. With calls to 'reassert the power of the local' (Cox 1997) and new concepts such as global–local nexus and 'glocalization' (Swyngedouw 1997), we reaffirm our consideration of localities as integral to globalization through a brief introduction to world cities, very special localities that are becoming more and more powerful in the operation of the world-economy. The concept of world cities brings together the two ends of the geographical scale dimension around which this book has been organized. In addition, the contrasting locality politics of world cities and strategic sites in the War on Terrorism pose questions of what new political geographies are in the process of being created.

Chapter framework

This chapter focuses upon six of the seven components of a political geography framework we introduced in Chapter 1:

➤ The *material* form of localities is identified as both a product of politics and a mediator of political possibilities and expectations.

➤ We *conceptualize* localities as the *scale* of experience. Localities are the venues for political action that can only be fully understood within the *wider whole* of the capitalist world-economy.

➤ The role of key localities, conceptualized as world cities, is *contextualized* within the temporal dynamics of the capitalist world-economy.

➤ Focus upon localities, as venues of the politics of the everyday, *challenges the traditional silences* regarding the politics of gender and statism.

■ Theoretical foundations

Three parallel debates in the social sciences may be identified as the basis for the contemporary political geography of localities. The first is the challenge to the quantitatively driven spatial analyses of the 1960s and 1970s. The discovery of quantitative techniques and emerging computer power led to a series of analyses on what were identified as 'local' issues. Two

such bodies of work were the provision of services (such as health care and education) and the pattern of the political boundaries of voting districts or constituencies. Paradoxically, such work was initially seen as apolitical. The discovery of data and the development of mathematical models resulted in a conceptualization of the provision of services by the government as a matter of rational cost–benefit analysis rather than politics. For example, Gould and Leinbach's (1966) analysis of the optimal location of hospitals in Guatemala gave no consideration to the consequences and manner of military rule. Increasingly, the political context of such decisions was seen as something that could not be avoided. Civic unrest in American cities, and the politics of the anti-Vietnam War movement, radicalized human geography (Flint 1999). In an academic epiphany, David Harvey (1973) declared that radical politics, based on the writings of Marx, was necessary to understand why services are located in particular places. Power relations, which Harvey identified from a Marxist perspective, determined differential social and geographical access to services – and not a purely spatial equation based on the distribution of population. In other words, the 'population' used to calculate optimal location of services is differentiated: there are haves and have-nots. Consideration of politics is necessary to understand the construction and maintenance of such inequities.

The other two debates are parallel but quite distinct from the rediscovery of local politics; they are about the rediscovery of localities. First, the argument began with a debate in urban studies revolving around the question of what is and what is not 'urban'. There was a debate on the 'spatial' nature of the urban, pitting the idea that the urban is essentially spatial against the view that it has no spatial basis at all. Second, there was a debate between the idea that urban is not a meaningful category for study and a reaffirmation of the importance of urban space as locality in a rediscovery of this neglected theme in social science.

How 'spatial' is the 'urban'?

R. E. Pahl (1970) set about the task of redefining urban sociology and in the process emphasized constraints of behaviour rather than choice, so introducing power into his project. Pahl's schema is famous for introducing the notion of social gatekeepers – planners, social workers, estate agents, developers, and so forth – who organize and control the scarce urban resources. The result is a socio-spatial system that reflects power in society as mediated by these managers. Pahl therefore proposes a spatial sociology where the organization and control of facilities, access to such resources and their effects on the life-chances of individuals replace traditional urban sociology's emphasis on the study of a particular type of place identified as 'urban'.

Pahl's approach has come to be known as the managerial thesis and has initiated a large debate. While generally praised for emphasizing the constraints that exist in the urban system over individual choices, the nature of the constraints identified by Pahl is controversial. In short, he is criticized for dealing with the 'middle dogs' at the expense of

Case study

The racial political geography of housing

'Cities in the United States are the most racially segregated urban areas in the world' (Crump 2004: 227). Most striking is the existence of African-American urban ghettoes and the suburbs, increasingly gated, of white Americans. The construction and maintenance of this geography is the product of decision makers identified by Pahl (1970) as social gatekeepers: bankers, real estate agents, developers, and national and local government officials.

The current pattern of urban segregation in the United States has its roots in the mid-twentieth century and the interrelated politics of home ownership and racial identity. A sense of 'whiteness' was constructed in opposition to the label of 'black'. Owning a home was not merely a sign of economic position but was seen as a means of defining racial differences. Especially for working-class white Americans facing economic insecurity, the symbolic role of home ownership was important. Simply put, 'owning a home in an all-white neighborhood was an important symbol of whiteness' (Crump 2004: 229). The politics of identity, a sense of whiteness, involved the construction of the

racially differentiated physical form of the American city.

What of Pahl's social gatekeepers? After the Second World War the Federal Housing Authority (FHA) was blatantly racist. Its underwriting manuals for loan officers 'clearly stated that blacks were a threat to the maintenance of property values and explicitly forbid the financial support of integrated neighborhoods' (Crump 2004: 232). Racist ideology was coupled with the politics of economic development and the financial need to maintain property values. Home owners, state officials, bankers and real estate agents all had a motivation, within an overarching ideology of racism, to produce segregated cities.

However, there is more to the story. What political actions ensure that social gatekeepers will act in this way? One factor is economic self-interest. But another is the expression of political will by white home owners. Urban race riots were the vehicle for displaying the hopes and fears of the white working class. The clearest case is in Chicago, though similar processes and events were evident in all American cities at this time. Between 1940 and 1950, Chicago's

African-American population grew immensely and led to severe overcrowding in the black ghetto. In addition, the job market was strong and African-Americans found themselves in a position to buy houses outside of the ghetto. From another perspective, blacks were 'encroaching' into white neighbourhoods. There were numerous incidents of violence towards African-Americans with the express intent of intimidating blacks and preventing desegregation. For example, Chicago's 1951 Cicero housing riot, sparked by blacks moving into an apartment building, provoked vicious rioting by between 2,000 and 5,000 whites (Crump 2004).

The response by the Chicago Housing Authority (CHA) was to use federal housing funds to construct large public housing projects for African-Americans. The goal was to create new housing for blacks separate from the white neighbourhoods. The decisions by the social gatekeepers who controlled the expenditure of government housing money were a product of the political violence of white rioters. Simply, 'CHA policy was made in the streets' (Hirsch 1983: 232, quoted in Crump 2004: 236).

the top dogs of the system. By this it is meant that by concentrating on the immediate managers of urban resources, such as public housing managers, the ultimate constraints on their actions are neglected. Middle managers may control 'who gets what' to some degree, but they have little or no say in how much there is to distribute in the first place.

Pahl's programme of 'spatial sociology' in Britain was paralleled by a similar argument for a spatial urban political science in the United States by Williams (1971), who attempted to move American urban studies away from an emphasis on individual choice to one in which space is socially controlled in order to manipulate accessibility to resources. Spatially defined interest groups rather than managers are the focus of attention, but the result is similar to that of Pahl. Analysis of the urban field consists of studying the spatial organization of urban resources and facilities and the conflicts that generates. Not surprisingly, these ideas appealed to geographers, and it is at this time that urban political geography can be said to begin, the first substantial works being David Harvey's (1973) *Social Justice and the City* and Kevin Cox's (1973) *Conflict, Power and Politics in the City: Geographic View*. But no sooner had geography found a spatial framework to borrow than this framework came under severe attack.

Although urban studies have largely been neglected in Marxist thought (Tabb and Sawers 1978), the most influential urban study of the 1970s was undoubtedly Manuel Castells' (1977) *The Urban Question*, which derived directly from the French school of structuralist Marxism. Castells approached the concept of the urban as an 'everyday notion'. It is, in his words, the 'domain of experience' where people live their day-to-day lives. Thus the urban unit is 'the everyday space of a delimited fraction of the labour force' (ibid.: 445). Hence the 'domain' corresponds to the daily urban systems that we identified in the introduction to this chapter. So far, therefore, Castells' schema is entirely consistent with our scale of experience. But Castells narrows his concern to only certain activities within daily urban systems. In advanced capitalist societies, he argues, the processes of production and reproduction can be interpreted as relating to essen-

tially different scales of operation. Whereas production is organized at a national and international scale, the reproduction of labour remains an urban phenomenon. This reproduction involves consumption processes in which the state has become increasingly involved. This is termed collective consumption and includes urban planning, public transport, public housing, education, public health care, and so on. Hence the city becomes a 'unit of collective consumption' (Castells 1978: 148), which explains why 'urban problems' usually reduce to local conflicts over public management of consumption in transport, planning, housing, and so on. The result of this argument is what Dunleavy (1980: 50) terms a 'content' definition of the urban field. On this basis, Dunleavy has defined an urban political analysis in which the spatial basis of 'urban' or 'city' is entirely banished. All we are left with is a set of processes concerning the reproduction of labour that are deemed to be 'urban' in nature.

Disposing of 'urban' and discovering 'locality'

The final demise of the urban as a distinct object for study can be found in the work of Philip Abrams (1978). He castigated the whole idea of urban studies as an exercise in reifying the city as if it were a social object. He argued that in reality it is 'nothing more than a phenomenon' (ibid.: 10). Hence the reason that we have no adequate theory of the city is that it is impossible. The city is not an entity that we can develop theory around because it is not an operating system. Cities and towns are merely arenas for the unfolding of social relations; they do not themselves represent social relations. In Abrams' words, towns should be seen 'as battles rather than monuments' (ibid.: 31).

A major advantage of the idea of collective consumption is that it links 'urban problems' directly to increased state activities as described by, among many others, O'Connor's (1973) work on the functions of the capitalist state. It is true that the fiscal crisis of state expenditure is acutely seen in our inner city areas owing to the disproportionate dependence of their populations on the state: cuts in state expenditure have been most severely felt in our cities. Dunleavy's (1980) urban political analysis is able, therefore, to harness many 'non-local' processes as necessary components in the study of urban policy. But somewhere along the way the 'urban' concept as it relates to places we know as towns and cities seems to be lost. We need to return to consideration of towns and cities as places, as 'the context in which people live out their daily routine' (Mellor 1975: 277).

In the work of John Urry (1981), the argument has come full circle. He bemoans the neglect of the 'spatial' element in social science. In particular, he criticizes Castells' (1977) argument that space has no meaning outside the social relations that define it. Although space per se can have no independent effect, Urry argues that the spatial arrangement of social objects can affect their social relations. Therefore he dismisses the idea that there can be a general spatial social science as attempted by Chicago human ecologists in the 1920s and quantitative geographers in the 1960s, but there are specific spatial effects that are local in scope. Hence we return to the notion of a locality that in most core countries, and many peripheral countries, will be the daily urban system as the local labour market. Although social classes are mobilized politically at the national scale, their distribution between localities will be uneven. Different mixes of social groups or classes will lead to different patterns of social relations. These differences in experiences will not automatically be evened out by nationalizing processes. In fact, according to Urry (1981), with increasing economic control at the national and international scales, we can expect more politicization of localities as particular labour markets suffer the recession. And this is not just important for reproduction of labour; it can also have local effects on production processes. Urry (1986) links this debate directly to the British localities project, which we discuss below.

We accept Urry's argument in this chapter. This is not a new position, of course, but a restatement of the contextual effect of localities (Filkin and Weir 1972). From our perspective, localities are important since they provide different experiences for their populations, and these will have political implications. But we are not trying to contribute to or develop any new 'urban' theory. The localities we deal with below are largely daily urban systems, but the emphasis is upon the variety of contexts that they provide in terms of class balances and resources rather than as some general notion of 'the city'. Hence we bring back together politics and its spatial context.

Politics: the localities project

In the wake of the deindustrialization of parts of Britain in the late 1970s and early 1980s, there was particular concern for industrial localities that seemed to be losing their economic *raison d'être*. These were not just one or two isolated communities; rather, this change seemed to be engulfing large swathes of the country, especially in the north. In the light of these problems, a large research project was set up in 1985 to compare and contrast how different localities were responding to the economic pressures of the time (Cooke 1989). The 'difficult question' that the research tried to answer was:

> While people's lives continue to be mainly circumscribed by the localities in which they live and work, can they exert influence on the fate of those places given that so much of their destiny is increasingly controlled by global political and economic forces?
>
> (Cooke 1989: 1)

In other words, this prescient project posed the globalization question before it became of widespread concern in the 1990s.

This concern for localities brought politics back into the local equation by considering whether localilities can act as 'a viable base for social mobilization' (ibid.: 3). In this way, local politics was linked to wider geographical scales of activity. Certainly, global trends, which had seen many manufacturing jobs relocated to cheaper overseas labour markets, were an important dimension of locality problems. These trends intersected with national processes, notably the flow of investment out of large cities to smaller towns and the countryside. For some localities, notably cities based upon heavy engineering, these two scale effects operated like a pincer movement, threatening to destroy the local economic fabric of life. But this need not mean the end of that locality as a viable social

Case study

Placing strike action

Where does political action occur? This question is raised to change our thinking about the meaning of 'where'. Instead of identifying 'where' as a particular place, the challenge is to see it occurring within localities, and hence within a structure of inter-related political geographical scales. To illustrate the point we will use Herod's (1997) study of unionized dock workers in the United States between 1953 and 1989.

At the beginning of this period, dock workers and employers negotiated contracts on a port-by-port basis. In other words, there was little inter-locality contact or organization between the different union 'locals' – the American term for the place-based branches of a union. During the 1950s, the dock workers' union (the International Longshoremen's Association, or ILA) became concerned about technological changes in transportation, especially containerization. The strong union base in the New York docks saw such developments as a threat to traditional work practices. Containerization would make it easier for transport companies to use docks with a cheaper labour force, most notably New Orleans and Mobile on the Gulf of Mexico. In addition, the possibilities of reduced manpower through the ease of handling containers urged unions elsewhere to negotiate for the better work practices that had been won by the New York unions. The result was a sustained union campaign to bargain at the state scale rather than regionally or port by port.

After initial opposition to the idea, in 1957 the New York Shipping Association, the most powerful east coast employers' association, was forced to negotiate work conditions for ports from Portland, Maine, to Hampton Roads, Virginia (Herod 1997: 149). After this regional success, the ILA pressed for national work preservation and income maintenance measures. In response to the union's efforts, in 1970 the employers formed a multi-port association (the Council of North Atlantic Steamship Associations) in order to bargain with the ILA's North Atlantic region (ibid.: 149). The union's attempt to forge a national bargaining scheme was eventually successful when, in 1977, a job security programme was adopted in thirty-four ports from Maine to Texas. However, unionized labour soon came under attack as employers began to bring in non-unionized workers. In response to this threat, in 1986 unions in the Gulf coast ports pre-empted national union negotiations and voted for reduced wages, a freeze on benefits and the nullification of some job preservation measures (ibid.: 149). The union's long attempt to increase its leverage through a national bargaining strategy was soon undercut by local-scale experiences in the individual ports.

In sum, the ILA's strategy involved the construction of new scales of negotiation as a response to global technological change and a state-scale political impetus. The union's master contract prevented cheaper work practices initiated by container-ization undercutting unionized labour. In addition, the new contract allowed dockers in other ports to strike legally in support of each other's goals. Such secondary boycotts had been prohibited by the federal Taft-Hartley Act (ibid.: 153). Hence, as we argued in Chapter 1, scales are constructed by political activity in response to particular challenges.

The case study of the ILA illustrates our conceptualization of locality politics rather than urban politics. The decisions of the ILA and its members were made in particular localities with careful consideration of what was occurring in other places and also at other scales. The decisions, for example, of New York dockers were made with an understanding of the situation in Mobile. The politics of unionization was a combination of decisions made based on local experiences, but within a context of what was happening in other localities. The decision of labour activists was not made within a particular urban setting, but within a locality that must interact with other localities. Combined, these localities constitute the capitalist world-economy. Hence, the 'local' decision to participate in national bargaining or not is made within a multifaceted geographical context: the locality itself, the functional connections with other localities, national political decisions, and the dynamics of the capitalist world-economy. Containerization, the arguments within a local union, and the work practices of longshoreman in a dock thousands of miles away are all ingredients in a 'local' decision.

community. The studies found that people in even the worst-hit localities were actively involved in transforming the situation. These were not situations where people just sat back and let things happen to them: much collective social energy went into defining new strategies and futures for each locality studied. These were often based on formal local government policy initiatives, but other private initiatives were also common – the mix depended on the political history of the locality. In this way, locality interacted with other scales, national and international, to restructure parts of the geography of Britain.

The localities project adopted a geographical approach from the opposite direction to that of our world-systems political geography. Whereas we started with the global scale in a 'top-down' manner, the localities research was firmly 'bottom-up', investigating political processes in particular places. We should view these as complementary: you can neither understand the locality without knowing its position in the world-economy and its nation-state, nor understand the world-economy and nation-state without knowing the localities of which they are made up. Starting from different ends of the geographical scale spectrum, we should come to broadly similar conclusions. Cooke's concern for national and global impacts on localities suggests this to be the case. Thus his key finding – that localities are not politically inert – is vital to our world-systems political geography.

Culture and the city

The final element in the construction of contemporary political geography was the 'cultural turn' that developed in the 1990s. Taking the lead from wider trends in the social sciences and humanities, geographers began to question the predominance of political economy approaches, including the world-systems approach, which had dominated the theoretical stance of the discipline in the 1980s. The content and merits of the 'cultural turn' have been vigorously debated (see Barnes 1998). The essential developments were the identification of discourse as a component of social life, the way in which places and regions were represented or 'imagined', and the

recognition that everyday life is embedded within place-specific social settings.

These developments are essential in completing our political geographic understanding of locality. Locations are venues of everyday political actions, and the dominant cultural understandings within a particular locality frame what are understood to be viable political options. Furthermore, the character of a place is not merely a matter of academic conceptualization. Rather it is the product of commonly held 'understandings' that are cultural constructs. For example, in general, residents of Sunderland have a very clear image of Newcastle, and vice versa. The same can be said of rival conceptions of New York and Boston. In this chapter, the material and cultural representation of localities, and the manner in which they frame political activity in the scale of experience are discussed. In the following chapter, we further explore the politics of identity.

Summary

In this section we have surveyed the theoretical foundations for the political geography of localities and:

➤ illustrated the relevance of the concept 'locality' rather than the term 'urban';

➤ introduced the term 'social gatekeepers';

➤ described the 'localities project' as an attempt to understand the political geography of localities and the diversity of the politics of localities;

➤ introduced the politics of the cultural identity of localities and their occupants.

■ Localities matter

The key message of the localities project was that globalization does not mean uniform development or homogenization. Global processes change local areas through processes of worldwide restructuring to be sure, but each case is distinctive. Therefore, rather than eliminating diversity, globalization reorders diversity. Localities are forever changing, but they are certainly not disappearing.

Case study

Gay bars and rural 'values'

The Casa Nova Lounge was a gay bar located in Jenner township in rural Pennsylvania. It was the only gay bar in the county and its existence provoked a political struggle between the owners, resident groups (both supportive and angry), hate groups including the Ku Klux Klan, and the local and federal state. The bar was open from January 1997 to August 2001. Heibel's (2004) narrative of the Casa Nova Lounge illustrates the insecure position of visible homosexual communities in rural America.

This case study exemplifies the way that meaning is attached to particular localities through discourses that are constructed and mobilized for particular purposes. One resident group reacted against homophobic reactions of some neighbours. The Network of Neighbors United Against Hate, while distancing themselves from the Laurel Highlands Gay and Lesbian Alliance, criticized local protest groups in a manner that portrayed rural areas as locations of reactionary, uneducated and 'backward' citizens: 'We are against hate crimes, bigotry, racism and the KKK. We find it unacceptable. This is a rural area, but we're not a lot of rednecks. Many people here are educated professionals' (quoted in Heibel 2004: 118). The connection of a political project of tolerance with an essentialized discourse of rural America was countered by the Concerned Neighbors of Forwardstown, who claimed they were the 'real neighbours': 'We are the neighbors who live 500, 750 and 1,000 feet from the bar. . . . We are not those of you who live one or two miles from the epicenter. . . . You see, we too are made up of a wide array of citizens. . . . And in spite of what certain "neighbors" think, we are educated too' (quoted in Heibel 2004: 118).

These two groups were using representations of 'rural America' as a component of a political battle regarding the material landscape of a particular locality. The one side saw the presence of Casa Nova lounge as a violation of 'rural values' whereas the other invoked a sense of modernity or progress to criticize the actions of some residents.

Socialization in its place

In political geography, the basic text that pleads for localities to be taken seriously politically is John Agnew's (1987) *Place and Politics*. In political science, he notes that it is not so much the global scale that is evoked when neglecting localities but the state scale. It is argued that politics is fundamentally 'nationalized', so any political difference between localities is dismissed as being of marginal interest only. But Agnew argues that national politics is constituted of the politics of localities, a fact ignored especially when dealing with state general elections. After a general election, we are used to hearing such phrases as 'the nation has decided' when in reality it has done no such thing. Voters deliver different verdicts in different localities, and these are aggregated to produce an overall result. For instance, there was no 'nation's verdict' in the United States to elect President George W. Bush in 2000 and 2004 because the country 'changed direction' and opted for a new politics after eight years of a Democrat, President Bill Clinton, in the White House. Rather, different voting districts in different states gave majority support for either the Democrat or Republican candidate, and the aggregate response was calculated to produce a majority winner in each of the fifty states. Finally, the Electoral College system (discussed in Chapter 6) produced a 'national' winner, though a map of results at the county level shows that, at the aggregate level, different localities expressed very different opinions. Even in Italy, with its marked regional differences, Agnew (1988, 1997) finds a strong propensity to ignore diversity and accept what he terms the nationalization thesis. He gives a humorous example, quoting one commentator's explanation for the failure of the Communist Party to enter government on the Italian voters' preference for 'thieves' (in other words,

Christian Democrats), because they were more palatable than the alternative ('reds'). But many localities, in central Italy in particular, regularly voted 'red' and were no less Italian for doing so.

The point is made: we have to take notice of localities to fully understand politics in the state and in the world. In this section, we look at some of the literature that has done just that to provide ideas for the development of a theory of localities in the next section.

For Agnew (1996) there are six processes that interact and combine to frame the local nature of political activity. The six processes are:

1 The social division of labour.

2 Differential access to communications technology.

3 Tensions between localities and central-state authority.

4 The locally specific meanings given to social ethnic and gender divisions.

5 The manner in which localities and regions are represented in political manifestos.

6 The microgeography of everyday life defining the patterns of social interaction.

At the outset, we must stress that these processes are dynamic and not predetermined. The global division of labour is built upon different functions being located in different localities as investment moves and jobs are lost and created. For example, in 2006 there was evidence of back office functions, such as processing airline tickets, moving from towns in India to towns of the Chinese interior (*The Economist* 2006). The geography of communication access results in some localities having access to some pieces of information and not others. Moreover, the same piece of information, a political party's manifesto for example, will be interpreted differently depending upon dominant political attitudes in a particular locality. In the United States, for example, local papers predominate and so editors portray information with their local readership in mind.

Understanding the politics of localities requires an understanding of interaction with states. Some locations may be at the very centre of national politics, the capital city and its environs for example, whereas others may feel oppressed by a central government whose authority is challenged within some localities. Related to central–local tensions are the meanings given to particular social categories. For example, 'working class' could be a matter of local pride, history and identity in one locality, but have negative connotations of, say, inner-city poverty in the culture of another locality. In Britain, for example, travelling supporters of Liverpool Football Club are often greeted with taunts from supporters of London clubs regarding their perceived lifestyles of 'squalor'. Such identities, though based on stereotyping and blind to the facts, can also be built into the messages of political parties. Regional parties, such as the Scottish National Party, portray a positive and distinctive image of Scotland, and lambast the central London-based government for its unfair treatment of Scotland.

Economic well-being, the interpretation of information, and the construction of identity are experienced within localized settings. In combination, it is the dynamic processes of economic political and cultural change of localities within dynamics at the scale of the state and capitalist world-economy that are the frames or venues for political action. This is not to say that politics is isolated within localities. Far from it. Localities are connected to others through trade, migration, and flows of money, goods and information, and localities are nested within a hierarchy of other scales. Although political action is contextualized within localities, the localities are situated within a wider political geography of scale and global differences.

Locality and protest

The dominant ideology in a location is not always unchallenged. Protest politics is one manifestation of locality politics. Other forms of political activity can be related to locality, as many researchers have shown. Tilly (1978) has traced the changing 'repertoire' of protest from the early world-economy to the emergence of liberal democracy. Traditional parades, burning of effigies, sabotage, petitions, riots, strikes, mass demonstrations, insurrection and revolution, all have distinctive social and ideological structures based upon experience of economic relations rooted

Case study

Feminist geographies of everyday life in localities

Socialization in localities is an on-going politics of the everyday. People must negotiate structures of power and dominant representations of groups that can either empower or marginalize them. The politics of 'belonging' or not, and the need to modify one's behaviour to show acceptance of the norm or to dress, speak, and act in certain ways to challenge the norms is a political decision. Furthermore, it is a dynamic politics being negotiated on a daily basis.

Istanbul, the largest city in Turkey, serves as an example of how norms of behaviour are constructed and resisted in localities. The population of Istanbul is approximately 12 million people, about 60 per cent of whom were born outside the city (Secor 2004: 353). Many of these migrants come from the south-east of Turkey where, since the late 1980s, there has been an ongoing conflict between the Turkish state and the PKK (Kurdish Workers Party) fighting for a separate Kurdish state. Kurdish migrants in Istanbul must negotiate a set of norms and representations that are formed within a context of fear of Kurdish terrorism as well as a process of renegotiating understandings of Turkish citizenship. Following the First World War and the establishment of the modern and secular state of Turkey, a civic notion of Turkish citizenship was created that denied ethnic differences. Kurds were labelled 'mountain Turks'. However, since the 1980s a Turkish identity recognizing ethnic and regional differences has begun to emerge (Secor 2004).

Secor's (2004) study of Kurdish women in Istanbul provides insight into the manner in which Kurds living in Istanbul negotiate a dominant understanding of Turkish citizenship and their ethnic identity. The every-day practices of Kurds in Istanbul that Secor's feminist analysis is able to identify illustrate that some Kurds act on a daily basis to resist 'assimilation' and to alter dominant attitudes and norms. In other words, the manner in which a locality socializes, or partially determines, behaviour is fluid and the product of political action.

One such site for renegotiating identity and belonging is schools. As Bahriye, a 21-year-old woman born in Istanbul, identified:

> My younger brother was going to first grade in primary school and one time I looked at his notebook. He had written there, 'The biggest military is our military,' 'My fatherland is Turkey,' etc. I saw these things and I laughed because these are things you are indoctrinated with. Our little sister started to tease my brother, saying, 'Oh are you a Turk? Look here what you have written!' So he said to her, 'At school I am a Turk, it is when I come home that I start saying I am a Kurd.'
>
> (Quoted in Secor 2004: 361)

Bahriye's brother articulates the fact that in Istanbul, as other localities, some spaces suggest certain norms of behaviour. This is further exemplified by Esel, a 36-year-old Kurdish woman who had lived in Istanbul for nineteen years:

> For us to talk about some things comes with a risk. Because of this, people live in two worlds. One is a world that not everyone can enter, a place where you truly belong with the origins of your identity. For example, it is a place where I can unite with other Kurds. But let's say there is someone whose reaction I can't predict. I won't say anything to them on this topic.
>
> (Quoted in Secor 2004: 360)

The politics of negotiating dominant power relations in a locality are not only individual, but a matter of creating groups and organizations. The women in Secor's study were involved in the Kurdish political party HADEP, as well as being active in neighbourhood cultural centres and unions. The existence of such groups can facilitate a display of Kurdish identity in the public spaces of Istanbul. For example, Secor notes that the women's branch of HADEP participated in a Women's Day march, giving them political visibility. However, all such actions require individuals to decide how, where and when to declare their Kurdish identity or hide it. Such everyday decisions connect individual identity to the construction of localities and the dominant meanings they diffuse. The concluding conversation between three women illustrates the everyday politics of creating and negotiating localities:

> *Deren*: There is an Istanbul that belongs to me. In another place, outside of Istanbul, I can say I am an Istanbulite. I have a relationship with Istanbul, a place where I have achieved political and other relations, and I am a creator of this Istanbul myself.
>
> *Nimet*: If you say you are an Istanbulite, your accent says you aren't.
>
> *Deren*: That isn't important. I'm talking about my own Istanbul.
>
> (Quoted in Secor 2004: 365)

in locality. But there is no simple relation between political activity and locality. Tilly argues that material interests alone will not produce protest but that, in addition, organization, mobilization and opportunity are required. Hence even the very plausible 'isolated mass hypothesis' of Kerr and Siegel (1954) that segregated homogeneous workforces such as miners and dockers/longshoreman are prone to high strike levels is shown to be untenable as a general process (Tilly 1978: 67).

Nevertheless, numerous studies have related localities to political activity. Briggs's (1963) critique of Mumford's (1938) characterization of the nineteenth-century city as the appalling 'Coketown' depends upon drawing a distinction between different locales. Although Mumford's model fits Manchester, with its large factories and resulting large social distance between classes, it does not fit Birmingham, with its small workshops and closer contact between classes. This idea that the industrial nature of a locality will be reflected in social relations and hence politics is similarly drawn by Read (1964: 35) when he describes Birmingham and Sheffield as 'cities of political union'

and Manchester and Leeds as 'cities of social cleavage'. The most detailed study of the way in which local social structures are reflected in political activity is the comparison of Oldham, Northampton and South Shields by Foster (1974). These studies of social relations in place are interesting but too specific for our purposes, and we shall not describe them in detail here. Instead, we concentrate on two topics: the relation of size of place to political activity and attempts to plan places to control political activity.

Protest and size of place

To many observers in the nineteenth century, large cities were the centres of agitation and protest. This is certainly the opinion of Engels (1952) and of the pioneer town planners, as we relate below. But recent statistical analysis of protest and organization has shown a remarkable regularity which disputes the militancy of large cities. In nineteenth-century Britain, strikes were more likely in medium-sized towns (Lees 1982). Similarly, in late nineteenth-century America, labour organizations and socialism were found to be stronger in medium-sized towns

Case study

Locating peace protesters

Tim Cresswell's (1996) study of place-specific protest and identity emphasizes that particular places are imbued with certain behavioural norms, or what is deemed appropriate. Protest movements utilize this commonly held understanding by demonstrating in symbolic places (such as the National Mall in Washington, DC). There is also the shock-strategy of acting in ways that are deemed inappropriate in particular places and, therefore, highlighting certain political norms or structures. The Greenham Common peace activists of the 1980s are a good example of this latter strategy. For a

number of years a group of women maintained a camp outside a nuclear missile base in rural England. In this public and militarized space the women conducted everyday activities such as washing, nursing and cooking (Herb 2005: 360). In Cresswell's (1996) language, these acts were 'out of place', being normally associated with the home. The political effect was to make visible feminist concerns over the norms of militarization and nuclear armament and to contrast them with values of nurturing and reproduction. The type of protest in its particular location forced reflection upon the norms that

led to the militarization of the English countryside (Woodward 2004).

Location can, however, also derail peace protest. In the 1980s some local authorities in the United States and Britain declared themselves nuclear-free zones in protest against the deployment of nuclear weapons. One such locality was Cambridge, Massachusetts. The largely middle-class and well-educated residents were unable to sustain the movement though. The desire to disseminate values of peace contrasted with the location's reliance upon the defence industry for jobs (Miller 2000; Herb 2005).

than in large cities (Bennett and Earle 1983). Let us consider the process involved in each case.

Lees' (1982) evidence relates to strikes in Yorkshire, Lancashire, Nottinghamshire and Leicestershire in two 'strike waves', 1842 and 1889–91. He divided these counties into settlements of different populations and computed strike rates for each of five size categories. Strike rates in small (less than 2,000) and large (over 300,000) settlements were consistently the lowest. The highest rates were usually found in his middle category of towns, from 20,000 to 100,000 population. Lees' explanation is that towns of different sizes had qualitatively different forms of relations with political authority, and that this changed over the two periods. In the 1840s, large towns were incorporated and had their own local political authority, including justices and police. In contrast, other towns were controlled by outside (county) authority, with policing carried out by the army. In the incorporated towns, there was more likelihood of mediation and less likelihood of provocation by the authorities. By the 1890s, these contrasts were even more marked. The various organs of mediation in work disputes – general trade councils for labour and chambers of commerce for capital, often combined in conciliation boards – were particularly developed in the larger towns and cities. In Tilly's (1978) terms, the repertoire of protest could be extended towards more 'legitimate' politics and the single weapon of the strike blunted. The strike in nineteenth-century Britain, therefore, seems to have been a phenomenon of the social relations and authorities of medium-sized towns more than other places.

Bennett and Earle (1983) considered the two 'surges' of socialism in the United States in the late nineteenth and early twentieth centuries. In the 1880s, the Knights of Labour reached a membership approaching 2 million, and in 1912 the Socialist Party presidential candidate obtained nearly 1 million votes. In investigating the geography of these two movements, Bennett and Earle found that size of place was important. The Knights of Labour 'displayed surprising strength in unexpected places, notably the small towns and cities of the Middle West' (ibid.: 47). In a regression analysis of the 1912 Socialist Party vote by counties, population of county was found to be a significant variable. For counties with populations under 85,000, socialist vote and population size was positively correlated, whereas for counties over 85,000 the relationship was an inverse one. That is to say, the Socialist Party obtained fewer votes in small and large localities. American socialism was a feature of the medium-sized settlement, localities where paternalistic social relations persisted and interfered with the domination of capital (Gordon 1976).

A political location theory

Gordon (1976) has incorporated the processes described above into a more general location theory of labour control. He argues that there are two forms of efficiency: quantitative efficiency related to improved technology; and qualitative efficiency related to control of the workforce. The latter can be indexed by strike activity. In the process of capital accumulation, firms that combine both efficiencies will be successful competitors in the market. Gordon argues that qualitative efficiency has dominated location decisions for investment. Before 1870 in the United States, for instance, industrial growth was spread among towns of all sizes. After 1870, however, there was a marked concentration of growth in large cities at the expense of medium-sized towns. This reflects the qualitative inefficiency of the latter, as described above. Gordon takes the argument further forward in time. Concentration of workers in large cities eventually led to local commerce losing political control of their cities. This is accompanied by the suburbanization of industry, symbolized by the development of steelworks in Gary, beyond Chicago. Gordon describes the 'sudden' emergence of the decentralization strategy and relates it directly to the rise of corporate capitalism. The late 1890s was a period of phenomenal merger activity, which produced large corporations able to take broad strategic decisions such as decentralization of plant. The qualitative efficiency explanation is far superior to quantitative efficiency explanations, which emphasize changes in transport and land needs. Gordon is able to document that the decision makers at the time were clear in their motives (Gordon 1978: 75). Cities had become 'hotbeds of trade unionism', and corporations were relocating investment to non-union plants in the suburbs. The final stage in this process is the regional shift from the unionized north-eastern United States to the less unionized south and west of the present day. The 'rise of the sun belt' can thus be seen as the third location strategy in US capital's battle to keep control of US labour.

This process has not been documented in detail elsewhere, but the general pattern seems to fit. In Britain, for instance, we have described the militancy of medium-sized towns, and they certainly grew more slowly than large cities in late nineteenth-century Britain. Similarly, decentralization and regional shifts in investment have also been inversely related to unionism in the twentieth century. The qualitative efficiency argument is clearly attractive here too. It is consistent with the 'runaway shop' process that has typified the world-economy from its inception. Wallerstein (1980b: 194) shows how the stagnation of the logistic B-phase was combated in part by a relocation strategy in many industries. Throughout central and western Europe, the power of labour guilds was broken by moving industry to the countryside. This physical dispersion led to weaker labour organization and lower wages. And moving today's runaway shop to peripheral countries – Vietnam, Morocco and so on – is the latest example of the qualitative efficiency strategy. Here we have a political location theory rather than the economic location theory that has dominated geography.

Planning for harmony

The lesson of the above discussion is that ultimately capital, by investment and disinvestment, creates and destroys places. In core countries in the twentieth century, this power of capital has been mediated by the state through the activity of 'town planning'. 'Planning' is sometimes contrasted with 'market' and is then assumed to be anti-capitalist in some way. In fact, in the theory of the state that we developed in Chapter 4, planning occurs as an alternative way to safeguard or promote the interests of dominant classes. Sarkissian (1976) has described how planning has attempted to apply the neighbourhood effect to promote social harmony.

Planners have a long history of promoting the social mixture of communities. Why should social mix be preferred to segregation of classes, which the housing market left to itself would produce? Early planners argued that by mixing classes the behaviour of the lower classes would be raised by their emulating their more affluent neighbours. In this way, social harmony could be created and social tensions reduced. In some ways, this was an anti-urban 'back-to-the-village' movement, but it also encompassed ideas such as equality of opportunity for all classes and provision of leadership for the urban poor. But the emphasis upon emulation meant that social problems were reduced to individual behaviour. It was just a matter of teaching the poor to behave.

The practical application of these ideas can be traced back to George Cadbury's building of Bourneville on the outskirts of Birmingham in the 1880s. This was a paternalistic project – it was a temperance settlement, for example – but social mix was an integral part of the plan to produce an ideal 'balanced community'. The idea of a balanced locality is most developed in Ebenezer Howard's garden city movement, which is probably Britain's major contribution to town planning. Although his book is now well known as *Garden Cities of Tomorrow*, it was originally published in 1898 as *Tomorrow: A Peaceful Path to Real Reform*, which clearly places the movement in a political perspective. This planning in no way challenged the basic forces operating in British society; rather, it steered them into politically safe directions. Howard's motives were explicitly anti-revolutionary, as his original title suggests. In fact, garden city advocates were careful to maintain segregation at the neighbourhood level, because complete integration implied 'equality and hence mediocrity' (Sarkissian 1976: 236).

These ideas became popular after the Second World War. The 'classlessness' of national sacrifice became directed towards reconstruction, and in Britain in particular this involved town planning dominated by garden city ideas. The neighbourhood unit as a small balanced community fitted into this planning ideology and was imported from the United States. The whole process was promoted in both countries by the onset of the Cold War. Balanced neighbourhoods where different classes had equal opportunities became an important element in the 'free', 'democratic' world's claim to continue to be the 'true' 'progressive' force in the world (ibid.: 239). In the United States, this inevitably led to the issue of racial segregation and ultimately to the Supreme Court desegregation victories in education and housing. (See the earlier case study in this chapter regarding the violent street politics that drove racial segregation.) There is a continuity of ideas from Bourneville, England, to Little Rock, Arkansas, centring on opportunity but ultimately based on containing social conflict. First benevolent capitalists with foresight and then the state have used planning in an attempt to produce places of harmony to replace places of strife. It is ironic, therefore, that planning as a reformist solution to 'urban' problems should often appear today to be part of the problem.

Summary

In this section we have focused upon the way localities are arenas of political socialization and are, in turn, the products of political activity. Specifically we:

➤ noted that politics is framed within unique localities rather than being a matter of national norms;

➤ identified the major elements and processes that constitute the socializing role of localities;

➤ illustrated the process of socialization in localities through a focus on protest;

➤ introduced the concept of qualitative efficiency to explain the importance of localities in labour–capital disputes;

➤ shown how localities have been constructed through planning as particular political projects.

■ A new theory of politics in localities

One of the attractions of the concept of political culture has been its use in accounting for long-term continuities in political attitudes and actions. This is why Johnston (1986; Griffiths and Johnston 1991), for instance, has researched the politics of the

Dukeries coalfield during the twentieth century. In 1926, after the collapse of the national strike, the Nottinghamshire miners set up a new local union in opposition to the more militant national union; and in 1985 after a second national strike they repeated this action. This 'moderation' has also been reflected in voting at national, local and union elections. More commonly, continuities in local politics have been traced for 'radical regions'. Cooke (1985), for instance, identifies such localities in South Wales, Provence in southern France and Emilia in central Italy. But the concept of political culture is much less satisfactory when we come to consider political change.

Practical politics

No region or locality is insulated from the world around it. Hence, as in the case of national politics, the ups and downs of the world-economy will generate 'new politics' in the sense that the opportunities and constraints for local people will alter. A strong local culture may smooth out these patterns of new politics. It is more likely, however, that the latter is interpreted within the political culture to give an appearance of continuity when in reality very different political practices occur. In particular, continuity in voting patterns need not indicate a political continuity, since the same political party can represent very different politics at different times. We encountered this with 'new politics' at the national scale in Chapter 6, and there is no reason to think that similar processes do not operate at the local scale. In this context, Savage (1987a) talks about the different 'trajectories' of localities, so that some radical regions may be declining while others are emerging. For instance, Glasgow and Liverpool can be contrasted in this respect, the former having a radical past in the first half of the twentieth century but more moderate in the second half, and the latter having the inverse of this trajectory. We can suggest that even the 'level' trajectory of the Dukeries area may change with the national political decision to eliminate most of the British coal industry.

Jones (1986) has provided us with a particularly clear-cut example of a changing local politics related to wider changes in the world-economy. The Michigan economy is heavily dependent on the automobile industry, and the recent politics of the state has reflected changes in the fortunes of that industry. From 1965 to 1979, a liberal–labour alliance, represented by the Democratic Party, produced a strong welfare state programme on the back of the automobile industry's prosperity. Stagnation during the 1970s was followed in 1979 by a collapse in the industry that totally changed the political scene. Corporations now had the upper hand. In the new competitive politics, Michigan was deemed a high-cost state, and the local welfare state alliance was no longer sustainable. Michigan's 'vigorously progressive political culture' (ibid.: 374) was not able to prevent the state moving along the new pro-business trajectory.

The key concept for linking localities to the processes operating in the wider world is uneven development, which is not an alternative term for the traditional geographical concept of areal differentiation. The latter merely implies diversity over space, whereas uneven development posits a hierarchy of spaces. They can exist globally, as in our core–periphery concepts; within countries, unequal development denotes rich and poor localities or regions. At this scale, places are explicitly made and broken by sequences of investments that are related to Kondratieff waves. The most famous model of this process is Doreen Massey's (1984) geology analogue. Every A-phase of the world-economy is expressed through an upturn in investments. These are not evenly distributed over space, since for any given package of new investments some places will be more suitable – provide a larger profit on the investment – than others. But different places will be favoured in the various rounds of investment over several Kondratieff cycles. Hence every locality can be viewed as having a different pattern of 'layers' of investment, corresponding to the rhythms of the world-economy. For instance, in the traditional industrial zones of northern Britain, many localities have investment 'layers' that signify their nineteenth-century prosperity, but subsequent rounds of investment are underrepresented. Their economic trajectory has taken them from 'boom regions' to 'problem regions'. The politics of all localities concerns how local politicians

and people have coped with the changing world-economy and how change has impinged on their area. Local political cultures may interact with the changing economy, but to fully understand the politics of a locality we need to investigate the actual political practices.

Practical political strategies

The most well-known model of protest activity is Hirschman's (1970) exit–voice–loyalty trilogy of responses to a problem. 'Exit' refers to migration to overcome the problem – for instance, the movement of unemployed people to find jobs. This common strategy is the basis of the making and remaking of all places – boom regions are easily identified by high levels of in-migration, for instance, and out-migration is usually very high in problem regions. 'Loyalty' refers to acceptance of a set of unfavourable circumstances. This need not be because of some deferential ideology or culture; it may be a rational decision on the grounds that any action is likely to fail and be counterproductive. This may represent a very realistic appraisal of the politics of a situation.

Hirschman's concept of 'voice', on the other hand, signifies protest in an attempt to change an unsatisfactory situation. Johnston (1982a: 236–40) describes three types of local protest. First, there is protest through formal channels, such as giving evidence to public inquiries in Britain or fighting an action through the courts in the United States. Second, there is 'generalized protest', which involves taking the case 'to the public', for instance by organizing petitions, public meetings, marches and demonstrations. Third, direct action is to take steps to prevent an unsatisfactory situation beginning or continuing. Sit-in protests, occupying empty houses or closed-down factories are typical examples.

The types of practical politics will change over time and between places. Savage (1987b) provides us with a carefully constructed picture of the practical politics in England in the period before the welfare state. Concentrating on the problems of the working classes, he defines their basic interests to be the reduction of insecurity. This is because in the industrial cities of England they had become separated from the means of subsistence. Every household therefore had to devise a survival strategy centred on the wages of its members in employment. There may have been other supplements to the wages – for example, many employers provided vegetable gardens for their employees – but the survival of the household relied primarily on exchanging wages for subsistence goods. Practical politics was therefore about reducing the insecurity inherent in wage labour.

Savage (1987b) indentifies three types of equally rational response to the problem of insecurity. Mutualistic practical politics involves forming collective organizations to insure against calamities and to keep control, as far as possible, of different aspects of day-to-day subsistence. Many working-class institutions were founded to lessen insecurity in this way, the most well-known being co-operative (retail) societies and friendly (insurance) societies. Economistic practical politics centred upon enhancing security in the workplace through trade union activities. This is not simple wage bargaining but involves negotiating for control over the work process. Demarcation agreements and control of entry into occupations represent examples of economistic solutions to insecurity. Finally, statist practical politics involves lobbying the state to intervene and lessen insecurity by providing social 'safety nets'. The end result of this practical politics was the welfare state.

Savage (ibid.) uses these alternative practical political strategies to show how simple reference to voting patterns can be quite misleading. In the 1920s, for instance, both South Wales and Sheffield became local Labour strongholds, but this support was based on totally different practical politics. This was reflected in the different reactions to the 1919 Housing Act, which promoted the construction of council (public) housing. In Sheffield, a statist politics involved building houses to let at low, subsidized rents. In South Wales, a mutualist politics had resulted in much of the housing stock being built by 'housing clubs'. This 'self-building' tradition plus out-migration in the 1920s meant that council housing was not seen in a redistributive role, and low-rent schemes were uncommon. Hence popular Labour councils in different regions pursued quite different sorts of politics.

But why do different practical politics occur in different localities? Savage's (ibid.) answer is a complex one involving the local social structure facilitating different politics. The pattern of skills is one important factor. Craft occupations in small-scale workshops, for instance, facilitate mutualist political struggle. In contrast, general factory skills in large organizations provide the potential for economistic struggle. Finally, casualized labour will often seek security through statist strategies. These patterns of potentials interact with local gender relations. Where the household is dependent on male wages, economistic struggle is facilitated through trade unions. In contrast, where the household as a whole is involved in wage labour – for instance, 'weaving families' in the textile industry – the potential for mutualistic or statist politics is produced, with the demand for welfare services for the working family.

The interaction between the local social structure and the resulting politics is obviously very complex and produces a mosaic of different politics across a range of localities. Savage (ibid.) illustrates this through a case study of the Lancashire textile town of Preston. Two parts of his story are of relevance here. In the late nineteenth century, Preston, despite its earlier radical reputation, had become a Conservative Party stronghold. Most working people voted Conservative, and Labour candidates made little headway in elections. Savage is at pains to point out that this was not a case of deferential voting but represented a rational response to working-class insecurity in a period when Preston's economic base was declining. The Conservative Party offered a 'politics of regeneration' supported by both unions and employers. Hence working-class support was purely an economistic political strategy: Conservative policy matched the practical politics. The Labour Party was only able to achieve electoral success in Preston in the 1920s when it could tap a new statist practical politics that demanded security through welfare provision. Labour's municipal socialism matched this practical politics and it reaped its electoral reward.

After 1945, the Labour Party in Britain was able to come to power on the basis of a statist political programme that produced the welfare state. This created

the congruent politics of the liberal–social democratic state described in Chapter 6. In recent years,
there has been an increase in the spatial polarization
of voting just as levels of class voting are declining
(Johnston 1985). Savage (1987a) uses a theory of
locality politics to explain this phenomenon. Basically, the uneven development of the downturn of the
Kondratieff wave has produced additional contrasts
in prosperity between localities in Britain. All residents in poorer localities have a common interest in
upgrading the local labour and housing market, and
hence their practical politics can continue to coincide
with Labour's statist approach. In prosperous areas,
on the other hand, both working-class and middle-
class households were gaining from the buoyant
local labour and housing markets, making them
susceptible to the market growth policies of the
Conservatives. The result is that voting contrasts
between localities are cross-cutting the earlier class
cleavage to generate a new congruent politics. Since
poorer areas have been Labour districts and wealthier areas have been Conservative districts, the new
locality-based voting can produce a neighbourhood
effect pattern (accentuating Labour and Conservative
stronghold support) without the need to hypothesize
the vague neighbourhood effect process.

Although Savage's work applies to Britain, the
concepts he has developed would seem to be applicable to a wide range of contexts beyond one country.

Alternative politics of location

So far, our argument has concentrated on the practical politics of households. This is reasonable: since
households are locally dependent, their everyday life
is situated within localities. Let us now add capital to
our local politics. The controllers of capital must also
be concerned with localities, since all capital investments must also be situated. The practical political
strategies of capital can be reduced to three options
(Harvey 1985). First, they can meet competition from
other capitalists by increasing investment in a locality
to produce more efficiently. Second, they can relocate
and invest elsewhere to a more favourable location –
Gordon's political theory of location in the previous
section describes this strategy. Third, capital can build
a coalition for growth in its existing locality, where it

will benefit from the resulting increase in the value
of the locality. Cox and Mair (1988: 310) refer to this
as 'enhancing the flow of value through a specific
locality.' This last strategy is particularly common
in the United States, where locally dependent capital
is encouraged by the anti-monopoly laws whereby
banks and public utilities are forced to remain local.

Growth coalitions represent one of two types of
'politics of turf' that Cox has identified. The class
politics of location is when local politics puts the needs
of labour's consumption fund above capital's profit
fund. This class politics depends on the local changing balance of power between capital and labour. In
Jones's (1986) study referred to previously, the politics of Michigan between 1965 and 1979 could be
referred to as a class politics of location. In a territorial politics of location, on the other hand, people
are mobilized as localized consumers. This enables
territorial coalitions to form between locally dependent capital and households concerned with home
ownership and local services such as education. These
growth coalitions produce a politics of 'boosterism'.
Jones describes just such a new politics in Michigan
after 1979, with the governor in a new role as lobbyist to corporations. Davies (1988, 1991) provides a
description of a very competitive territorial politics
on Tyneside, England, where 'municipal socialism'
has been transformed into 'municipal capitalism'.

Cox (1988) argues that there is currently a dominance of territorial policies in localities, but whether
the outcome is one or other of the location politics
depends upon particular contingencies, both local
and non-local. It may well be that territorial politics
is more common in the United States than in western
Europe (but see Davies 1988, 1991). In Britain, for
instance, local governments have attempted to devise
local economic initiatives that are clearly part of a
class politics (Boddy 1984). These policies bring us
to more formal political practices and the local state.

The local state

The institutions of the modern nation-state exist at
more than one geographical scale. All territorial states
have institutions that operate at the level of the locality. A wide range of functions are typically organized

at this particular scale – education, housing provision, public transport and land-use planning are all typical examples. This state activity is commonly referred to as the local state to distinguish it from the activities at the scale of the state territory, commonly referred to as the central state.

The concept of the local state is an ambiguous one. The use of the term 'state' in this context does not indicate sovereignty, since that is vested with the territorial state. For this reason, some writers have doubted the validity of the concept. It has been pointed out that the phrase 'local government' can be substituted for local state in many cases with no change in meaning (Duncan and Goodwin 1982). Hence the call to drop the concept. This would be a pity. The term 'government' merely implies imposing authority in a locality whereas 'state' suggests a wider set of relations in which to view the formal politics of a locality. It is for this reason that local state has come to be an important concept in the new theory of politics in localities. Let us explore this further.

The nature of the local state

Given the formal power of the nation-state, why is this not instituted as a single-tier government structure? The fact that one-tier states are conspicuous by their absence in the world-economy suggests that extra tiers, and in particular the local state, are a necessary requirement. This need can be derived from two aspects of the modern state. First, the state's institutions consist of a large-scale bureaucracy. All bureaucratic organizations operate through specific spans of control in hierarchical structures. Quite simply, it is far more efficient to decentralize various functions rather than make all decisions from a remote centre: it does not make practical sense to schedule the garbage collection for a provincial city from government offices in the capital city. Second, the government function of the state requires legitimacy, and this can rest to varying degrees on traditions of local autonomy. Such communalism is particularly strong in the United States, but the notion that people living in localities should have some say in the running of their locality is accepted in many countries. The combination of these administrative-efficiency and government-legitimacy

functions produces the local state. The balance between administration and government will vary between places and over time, so that empirically the local state will differ depending on the context. Theoretically, however, the nature of the local state will remain the same.

The concept of the local state was introduced into the literature by Cockburn (1977) in a study of the political practices in a London borough. She used the term to emphasize the fact that the local government she was studying was an integral part of the capitalist state. She employed a Marxist theory of the state, in which the local state is distinguished by the functions it performs, in particular 'social reproduction'. Hence the local state is an instrument of class domination managing the social needs of households for the ultimate benefit of capital. Duncan and Goodwin (1988: 34) criticize this theoretical 'top-down' perspective as being too one-sided. Why is there a history of centre–local tensions within states if the local state is primarily just an agent of the central state?

The major theoretical development since Cockburn (1977) has been Saunders' (1984) 'dual-state thesis', whereby the functions of the two levels of state are conceptualized into two distinct sets of political processes. At the centre, a class politics based upon issues of production is to be found, whereas in the local state a politics of consumption cuts across class lines. Since the central state is concerned with the needs of capital accumulation and the local state focuses more on legitimation needs, then tensions between the two scales are hardly surprising. But this thesis has come under severe criticism. Empirically, it is not very easy to show that this division of functions actually exists. Both production and consumption issues are to be found in the politics at both scales. Theoretically, the allocation of functions to state levels implies a rather static pattern of social relations and the state. The dual-state thesis is ahistorical; we need a longer-term view of the local state than has been the norm in most studies (Duncan and Goodwin 1988).

The most important characteristic of the local state is the ambiguity of its role. Kirby (1987: 20) refers to the 'curious political position occupied by the local state'. This is because it is part of the state apparatus, but it can also be used to oppose the state. Duncan and Goodwin (1988: 46) trace this idea of local state as both agent and obstacle to Miliband (1969). They develop an instrumental theory of the local state as 'a double-edged sword' (Duncan and Goodwin 1988: 41). Once again, uneven development is the key concept that necessitates the state's organizing local agents to manage its territorial diversity. But the very nature of this diversity will mean that different interests from those of the national dominant groups can dominate particular localities. Local groups can, therefore, use the local state to promote their own policies in opposition to those of the central state. Hence for Duncan and Goodwin (ibid.), it is this contradictory role of the local state, and not its particular functions, that enables us to specify the concept. This is the true nature of the local state.

Local state as instrument

The particular balance between the local state as an instrument of the state and as an instrument of the locality will vary widely with the context of the politics. We describe two contrasting examples here before considering more generally the changing potential for conflict.

The tension between central state and local state was a prominent feature of British politics in the last decades of the twentieth century. The Conservative central governments of the 1970s and 1980s had an electoral mandate to reduce public expenditure at the same time as Labour city governments claimed a mandate to resist cuts in services. This has led to a classic centre–local conflict within the state apparatus. Major clashes occurred in all main areas of policy, including education, housing, transport and planning.

The central government's response to this challenge has been varied. The simplest method was to transfer functions to the centre. In education, for instance, a 'national curriculum' was instituted, and schools have been encouraged to 'opt out' of local authority control and become funded directly by central government. The policy of withdrawing national government control over schools continues to be a fault-line in British politics. A second simple solution is to transfer functions from the elected public sector to new appointed public corporations. Non-elected urban development corporations took over land-use

control functions in several cities, for instance. Alternatively, functions can be transferred out of the public sector to the private sector. This privatization strategy has been carried out for public transport and public housing. The local state can also be bypassed by direct appeal to residents, which is part of the policy in education and housing. Far more direct controls are involved with central intervention in local government budgets. The policy of 'capping' expenditure involved placing a limit on the raising of local taxes for selected local governments. Finally, the ultimate central state sanction is abolition. All seven metropolitan governments in England were abolished. These had formed some of the major centres of opposition to the Conservative central government.

The 1980s restructuring of local government in Britain included the construction of a new scale of government between local authorities and the national state. The emphasis was upon both economic efficiency and representation. However, the changes were made with an eye to maintaining the territorial integrity of the state. In other words, although there was to be a semblance of regional government it was viewed as a means of preventing rather than facilitating territorial separation. This was most important for Wales and Scotland.

Beginning in the late 1990s the Labour government under prime minister Tony Blair created a three-pronged strategy of regional governance (Tomaney 2002). First, regional development agencies (RDAs) were established to facilitate integrated regional development that would, in combination, boost the national economy and create full employment. The agencies were criticized for being too dependent upon edicts from the central government. However, they got the ball of regional government rolling. Second, regional chambers, or assemblies, were created consisting of elected local government councillors. Ostensibly, the task of the chambers was to oversee the RDAs. However, they quickly evolved into regional advocates, sometimes clashing with central government (Tomaney 2002). For example, in April 2001 the East of England Regional Assembly (EERA) accused its RDA of avoiding the difficulties of regional development (ibid.: 726). In other words, the EERA was acting as advocate for what it identified as the interests of the region. The third set of institutions is the presence of government offices (GOs) in the regions of Britain, whose task is to coordinate government policy. They have met with mixed success so far (ibid.: 727).

The construction of these three institutions is the manifestation of a regionalization of governance in Britain. There are two important additional matters to consider. The first is geographical scale. Regional government, as practice and idea, was catalysed by the growing regional presence of the European Union. The EU was providing its own funds for regional development projects. In addition, the EU was suggesting an alternative scalar government structure which, arguably, empowered regional government within an overarching context of EU representation and support.

The situation in the United States is very different. Often the local state has been an instrument for middle-class households to prevent national policies impinging on their localities. This is most clearly seen in housing and education policies within metropolitan regions. These areas are highly fragmented into many separate government units. The larger metropolitan regions such as New York and Chicago have more than one thousand governments each. In the United States as a whole, Johnston (1982b) reports 80,000 municipalities and special districts, which, at that time, was one for every 3,000 people!

This political fragmentation is particularly associated with suburbanization. The typical metropolitan region consists of a central city surrounded by numerous suburban municipalities and special districts. The major advantage of separate incorporation is being able to control local land use through zoning. In the Connecticut section of the New York metropolitan region, for example, 75 per cent of the land is zoned for housing plots of one acre or more. Such a policy is designed to produce a social segregation that is the United States' spatial representation of the class conflict. Rich suburbs avoid any cross-subsidization with the poorer central city while at the same time freeloading on city services (Cox 1973). In court cases challenging zoning, these local states have always been able to maintain their discrimination policies (Johnston 1984).

The second major form of discrimination in US cities has been in education. Here the central government has used the courts to force local states to conform to national standards of non-discrimination on racial grounds. Originally, this was a battle in the South as local 'separate but equal' education policies were overturned. In the North, residential segregation meant that neighbourhood schools produced *de facto* segregated education. This was attacked by the central government in the courts by replacing neighbourhood schools with 'mixed' schools using bussing. This precipitated a 'white flight' to the suburbs to avoid the policy. In 1974, Detroit tried to overcome the problem by devising a bussing scheme to incorporate the suburbs. This was defeated in the courts (Johnston 1984).

In these exercises in constitutional politics, the local state has been the prime instrument through which suburban households have defended their localities. This has not really been an option for suburbanites in other countries, since it reflects use of the particularly strong American ideology of communalism. This has provided the ammunition for the local state victories, and it illustrates the importance of the political context in any discussion of local states.

The changing potential for conflict

Intense periods of local–central government tensions, such as those in Britain in the 1980s, raise questions of the temporal–spatial context of locality politics. In fact, we can see that every B-phase in the Kondratieff cycles has been associated with just such a restructuring of the local state. In the 1830s, the new English Poor Law reorganized the dispensing of relief from the parish level to 'unions' of parishes in order to curb 'local excesses'. In the 1880s, the whole local government map was redrawn to take into account the growing problems of cities. In the 1920s, in the earlier 'politics of crisis' identified in Chapter 6, new laws were brought in to curb the activities of various local authorities. And so we come to the new reorganization of local government in the 1970s and the continuing round of legislation to bring local states into line.

The comparison of particular interest is that between the last B-phase and the recent situation.

This is because the first 'politics of crisis' coincided with the rise of the Labour Party as a major party. Although not very successful nationally at this time, the uneven development meant that, inevitably, Labour as a major party began winning control of local governments. The result was centre–local conflict as Labour pursued expenditure programmes that were incompatible with the central state's austerity policies. 'Little Moscows' appeared in several industrial regions (Macintyre 1980). The most serious challenge, however, came in the London borough of Poplar, and local government rebellion is still sometimes referred to as 'poplarism' (Branson 1979). Here the elected members of the local state were far too generous in their dispensing of relief to the poor for the central government's liking. New Acts of Parliament were passed to control expenditure more tightly. New districts were drawn, and some elected authorities suspended and replaced by central government commissioners. Finally, in 1934, in the period we have called the first 'politics of national interest', the abolition solution was finally used, as it has been in the current politics of national interest. A new scheme of social security to be run by civil servants from the central state replaced the local systems of payments. The result was a new system that was 'centrally directed, uniformly fixed and removed from the vagaries and dissensions of party politics' (Runciman 1966: 78–9). The politics of the locality was sacrificed for the national interest then as now.

Contemporary regional development in Britain exemplifies a broader trend in the form of the local state. The local state is increasingly just one of many actors involved in local politics. The move is from local state government to city or local governance. Tomaney (2002: 728) notes that, in the north-east region, 'over 20 "regional" organizations were involved in the preparation of at least 12 regional strategies. . . . Although the RDA, Regional Assembly and, especially, the Government Office were heavily involved in the preparation of regional strategies, many other government departments, non-departmental government bodies, and other agencies were also involved.'

The notion of urban governance has been constructed within the discourses of neo-liberalism we discussed in Chapter 4. Localities are promoted as sites of economic production that need to manage and discipline themselves within a competitive globalized economy. Non-governmental organizations, government offices, local government and the private sector create strategies for cities around the notion of 'growth' that react to and reinforce the dominant discourse of neo-liberalism: competitive wages and financial incentives to attract investment (Imrie and Raco 1999; Ward 2000). Localities are conceptualized as being in competition with other localities across the globe. The desired outcome is a disciplined local population willing to 'compete' within the context of globalization (Wilson, forthcoming).

Just how powerful is the local state, therefore? Formally, the answer seems to be that local governments can be easily disposed of as necessary by sovereign nation-states. Clark's (1984) analysis of local autonomy supports this view. He identifies two sources of local power: the power to initiate policy and the power of immunity from oversight by higher authorities of the state. Whether or not a local unit possesses these powers produces four types of autonomy. Where there is no initiative and no immunity, the situation is one of local administration only. These are literally just agents of the central state pursuing central policies irrespective of the locality. The opposite case, where there is both initiative and immunity, produces an autonomous city-state. This ideal type does not really exist, except as a sovereign state such as Singapore. Real levels of autonomy within territorial states are provided where either initiative or immunity exists. With no initiative but discretion to implement central policies for locality needs with immunity, the system can be described as 'top-down' autonomy. With initiative to pursue policies for a locality but with no immunity from central state interference, we can describe the system as 'bottom-up' autonomy. Clark's (ibid.) application of these criteria to local government in the United States leads him to conclude that, formally, they are closest to the 'local administration' type of autonomy – they have little or no initiative or immunity. We can extend this conclusion increasingly to the situation in Britain as the context of neo-liberalism partially defines locality politics.

This 'pessimistic' view of the local state's power can be easily countered, however. For all the lack of formal autonomy, local states do have a real manoeuvrability that makes them potentially powerful instruments, as our previous discussions have shown. The simplest evidence is the variety of expenditure on different services across local government units. The uniformity implied by local administration is simply not found. In both the United States and Britain there is a huge literature on the variability of local government 'outputs' (Newton 1981). This clearly implies some local 'choices' that are overriding the formal limitations on autonomy. However, the contemporary emphasis upon governance within a context of neo-liberalism (in our framework a particular label given to B-phase restructuring) suggests that the local state is constrained by the involvement of the private sector and the way localities are situated within the capitalist world-economy. Blame is being put upon the amorphous and abstract force of neo-liberalism by some participants in local politics to legitimate emphasis upon pro-capital policies, such as welfare cuts and tax breaks (Wilson, forthcoming).

We can conclude, therefore, that uneven development will inevitably force the central state to organize control of its territory through some local autonomy. Hence abolition of the local state is never a solution to the tension of centre–local relations. In our examples of abolition above, only part of the local state system has been removed. Complete abolition of the local state is unthinkable. And even if this policy were pursued, it would not get rid of locality politics. It would remove an instrument from the locality, but in doing so it would also remove an instrument of the central state. That is the fundamental ambiguity of the local state. Local states are important to central government in many ways. In times of crisis, for instance, political problems can be diverted down the state hierarchy until blame rests with the local state (Dear 1981). By abolishing the local state, the central state would have no link into localities. The state would be less integrated: it would lose part of its legitimacy. Locality politics would continue but take different forms of practical politics, some of which have been described earlier. This might well be more dangerous to the status quo than having localities

firmly integrated into the nation-state through their local states.

We have concluded this section on a contradiction, and that is very appropriate. Local states are not agents of the state or agents of opposition to the state. They are both simultaneously. That is their hallmark. Further, the new theory of politics in localities incorporates other contradictions. This motif fits the politics of the world-economy generally, as the earlier chapters have shown. The world is not a simple machine that we can understand by learning a few formulae. Our world is complex, incorporating severe contradictions that translate into many difficult dilemmas for politicians. The political geography of this world must therefore also be complex. A political geography perspective on the world-economy organized around geographical scales can make sense of our modern world without losing too much of the complexity of that reality.

Since scale is at the heart of our perspective, it is appropriate to turn to a discussion of a class of phenomena that are overtly both local and global in nature. However, we also pose a question regarding what appears to be the attempt to construct two contrasting political geographies. First, we examine the construction of bounded and essentialized territories within the War on Terrorism. We then conclude the chapter by examining world cities, localities within a political geography that challenges bounded territorial states.

Summary

In this section we have related theories of political action in localities to the world-systems approach by:

➤ identifying different forms of practical politics;

➤ identifying different practical political strategies;

➤ conceptualizing the local state and noting the tensions between the local and national state;

➤ situating the timing of practical politics within Kondratieff cycles;

➤ situating the timing of local–national state tensions within Kondratieff cycles.

■ Localities and the War on Terrorism

Since the terrorist attacks of 11 September 2001, localities have been destroyed and constructed in acts of terrorism and the military response of the United States. The attacks of 9/11 dramatically transformed the physical landscape of New York City. The World Trade Center towers were targeted for their symbolic value, representing the economic power and global reach of the United States. However, their destruction had a significant effect upon those familiar with the city's skyline. Their disappearance was a daily reminder that things were no longer the same. The response, the War on Terrorism, has resulted in destruction of locations in Afghanistan and Iraq. The most advanced military technology, the use of satellites to aim depleted uranium ordinance, has been utilized to ensure the most effective annihilation of buildings, neighbourhoods and, of course, their occupants. In stark contrast to memorialization of the victims of 9/11 it is hard even to count the number of deaths from the War on Terrorism (Gregory 2004). Moreover, cities in the United States and other countries have seen physical changes in the name of security, including concrete barriers surrounding buildings, road blocks and pervasive security cameras.

The physical targeting of localities in the War on Terrorism has required the mobilization of political rhetoric to represent both the places and the geopolitical context as one that requires acts of war and the militarization of everyday spaces. Stephen Graham (forthcoming) has summarized the processes and rhetoric transforming localities on both sides of the War on Terrorism. We shall first outline the representation of towns and cities identified as 'terrorist' places before turning to the way 'homeland' cities have been reimagined (Graham, forthcoming).

Urbicide: material and rhetorical destruction of localities

Political geographers' concern with the War on Terrorism has been coupled with traditional interest in the shape of urban areas and the more recent focus

upon the social dynamics of cities. Graham (2004a: 25) coined the term 'urbicide': 'the deliberate denial, or killing, of the city'. The targeting of cities has become a central component of the War on Terrorism. The insurgency after the invasion of Iraq, considered a front in the War on Terrorism in light of US rhetoric rather than actual and proven connections between Sadaam Hussein's regime and terrorist organizations, has emphasized a notion that the United States is fighting a series of urban conflicts. Urbicide places localities at the centre of the United States' contemporary projection of military power across the globe. Such an act of geopolitics requires two simultaneous and related movements: the physical destruction of buildings and neighbourhoods and a series of representations to justify and legitimate the outcome.

Graham (forthcoming) notes that the key rhetorical strategy is to deny a sense of humanity or social life within cities that are being targeted by military actions. Instead of being seen as dynamic and fluid locations of a variety of social groups going about their everyday business, certain localities are portrayed as 'terrorist places'. The goal is to paint a picture of such localities as nothing but a 'nest' of terrorists, both now and throughout history. The richness and diversity of a locality is denied, a cultural urbicide, to legitimate its material destruction.

Targeting the 'terrorist nest'

Four particular, but interrelated, rhetorical tools have been used to portray localities within the broader rhetoric of the War on Terrorism (Graham, forthcoming). The first is the use of satellite-based imagery, reprinted in newspapers and utilized in television broadcasts, to portray cities such as Baghdad as empty cartographic surfaces. In other words they are static depopulated representations to deny the existence of a vibrant and diverse social life. As Derek Gregory notes, localities become:

> letters on a map or co-ordinates on a visual display. Then, missiles rain down on K-A-B-U-L, on 34.51861N, 69.15222E, but not on the eviscerated city of Kabul, its buildings already devastated and its population already terrorized by years of grinding war.
>
> (Gregory 2004, quoted in Graham, forthcoming)

Second, if the localities being targeted are seen to be populated then the rhetoric portrays a singular view of the occupants. Arab cities are portrayed as 'terrorist nests' (Graham, forthcoming). Numerous examples exist. Par for the course is the observation of General Richard Myers, Chair of the US Joint Chiefs of Staff, in April 2004 during fighting in the Iraqi city of Fallujah that left six hundred civilians dead. Myers defined the city as a 'rats' nest' or 'hornets' nest', and so dehumanized the occupants, that needed to be 'dealt with' (News24.com, quoted in Graham, forthcoming). Such portrayals are emphasized by newspaper commentaries. The disturbing voice of Ralph Peters is not atypical. In an article in the *New York Post* entitled 'Kill faster!' Peters (2004) claimed that urban warfare required the rapid and overwhelming use of military technology to wreak havoc before the media arrived. Peters fearing, generously, that the media would offer portrayals of the locality as a location of everyday social life rather than an essentialized 'terrorist nest' (Graham, forthcoming).

The third process involves the virtual simulation of localities. A host of computer games have come on to the market allowing players to manoeuvre US forces around the streets of an Arab urban landscape and blast away 'bad guys.' In popular games such as 'Full Spectrum Warrior' or 'America's Army' the streets are depopulated of citizens going about their everyday life. Again, the locality is dehumanized. The only people to appear are the targets, 'shadowy, sub-human, racialized figures' (Graham, forthcoming). Both of these computer games began as military training videos, blurring the distinction between military and civilian participation in the War on Terrorism. The virtual simulation of Arab localities in computer games allows for a sanitized and safe interactive participation in the War on Terrorism, but in a manner that reinforces a sense of a battle in depopulated locations in the name of 'good'.

The fourth process is the US military's physical construction of model Arab cities for training purposes. Though the morphology of Arab street plans are painstakingly reconstructed, those streets are again empty except for the 'terrorists' to be targeted. The sense of Arab localities as places devoid of social life other than uncivilized terrorists who must be shot is perpetuated.

The four processes of representation are an essential feature of urbicide within the War on Terrorism. They combine to deny targeted localities as cities constituted of a diverse population going about its everyday business. Such an image would not sit well with the knowledge that such cities are being bombed from on high and shelled with depleted uranium ordinance.

Representing homeland localities

The justification for military acts targeting localities in Afghanistan and Iraq requires a complementary, though asymmetric, representation of localities in the United States. Graham (forthcoming) defines four interrelated processes in the reimagining of the 'homeland'. First, is the rhetoric of protecting those who 'belong', or 'non-threatening, full US citizens' (Graham, forthcoming) from demonized Others armed in a variety of ways but never clearly identified or located. The outcome has been the identification of a 'domestic front' and the associated political discourse of creating the United States as a bounded national territory. The manifestation has been increased border control and stricter immigration policy. Second, everyday urban spaces have been redesigned and reconstructed as militarized and 'secure' spaces (Graham, forthcoming). The scale of experience is part of the war:

> To live in America now, at least to live in a port city like Seattle – is to be surrounded by the machinery and rhetoric of covert war, in which everyone must be treated as a potential enemy until they can prove themselves a friend. Surveillance and security devices are everywhere: the spreading epidemic of razor wire, the warnings in public libraries that the FBI can demand to know what books you're borrowing, the Humvee laden with troops in combat fatigues, the Coast Guard gun boats patrolling the bay, the pat-down searches and X-ray machines, the nondescript gray boxes equipped with radar antennae, that are meant to sniff pathogens in the air.
>
> (Raban 2004: 6, quoted in Graham, forthcoming)

Related to the militarization of the scale of experience is a third process, 'the production of permanent

Case study

The physical side of urbicide

Urbicide is not just a matter of representation. The physical destruction of localities is a harsh reality. For example, the Israeli Defence Forces have destroyed homes and olive groves in the name of counter-terrorism (Falah and Flint 2004). In May 2004 Amnesty International reported that

> more than 3,000 homes, hundreds of public buildings and private commercial properties, and vast areas of agricultural land have been destroyed by the Israeli army and security forces in Israel and the Occupied Territories in the past three and a half years. Tens of thousands of men, women and children have been forcibly evicted from their homes and made homeless or lost their source of livelihood.
>
> (Amnesty International 2004: 1, quoted in Falah and Flint 2004: 122–3)

The physical destruction of Palestinian urban spaces has been a feature of Israel's military action against the Palestinians. Done in the name of counter-terrorism the outcome is the denial of urban spaces and the facilitation of Israeli territorial control. The number of physical structures destroyed is staggering. The *Palestine Monitor* (2004) website claims that between September 2000 and February 2002 the Israeli Army destroyed 720 homes and damaged a further 11,552 affecting 73,600 people. Thirty mosques, 12 churches, and 134 water wells were destroyed and damaged and 34,606 olive trees uprooted (Falah and Flint 2004).

The process has been labelled urbicide by bulldozer (Graham 2004b). Falah and Flint (2004) report statements collected by Palestinian human rights groups and newspapers to give voice to the people who lived in these localities. The following quote is typical:

> Around 12:40 A.M., I woke up to the sound of gunfire and the noise of bulldozers. . . . Suddenly, one of the children screamed, 'get out, the Jews are demolishing the houses.' . . . I saw elderly people and women and men carrying their children, leaving their homes. . . . Then I understood that they [the Israeli Army] were demolishing the houses in our area. . . . I rushed to wake up my three brothers and their wives and children, and we went outside without taking anything with us.
>
> (B'TSELEM 2002: 9–10, quoted in Falah and Flint 2004: 124)

anxiety around everyday urban spaces, systems, and events that were previously banalized, taken for granted or ignored in US urban everyday life' (Graham, forthcoming, referring to Luke 2004). Everyday experience in the United States is a negotiation of government colour-coded alerts and endless media portrayal of 'threats', from terrorism, to bird flu and 'freakish' weather events. In sum, the scale of experience is the scale of mass anxiety.

Our political geographic conceptualization of localities situates, or contextualizes, them within the dynamics of the capitalist world-economy. In the discussion of the War on Terrorism, declining US hegemony is a defining process, and helps us explore the tensions within the fourth process that Graham identifies: the rhetorical construction of the United States as a spatially delimited, territorially fixed, and demographically homogeneous nation. Cosmopolitanism has been identified as risky or even threatening (Gilroy 2003). The actual reality of US localities

is denied. Rather than identifying their fluidity and demographic make-up of intertwined migrant groups, they are seen as consisting of distinctive, identifiable and opposed groups – citizens deserving of protection and those who carry a diverse sense of 'threat'. Ironically, it is the identity of the victims of the 9/11 attacks that illuminate the fiction of this rhetoric. The casualty list consisted of people from 41 different countries, though the rhetoric of remembrance denies the diasporas that suffered in the identification of '3,200 American dead' (Graham, forthcoming).

Graham's essay is a powerful identification of the War on Terrorism's material transformation of the scale of experience in the United States and the representations that are employed to justify such changes. Similar to the everyday actions of Kurdish women in Istanbul that we discussed earlier in the chapter, individuals in US cities negotiate the physical spaces of localities and the dominant meanings associated with them. The degree of compliance or resistance is a matter of individual choice, and can vary in different settings. Such individual political actions are made within the context of a militarized scale of experience.

However, our world-systems approach requires the consideration of other scales and processes. Returning to the wider context of the capitalist world-economy, the militarization of US localities is one manifestation of the tensions and contradictions facing the United States as hegemonic power. The United States is negotiating a 'hegemonic dilemma' (Flint 2004). On the one hand, the role of hegemonic power requires the United States to maintain flows of goods, capital and information across borders. On the other hand, the US government is, to some degree, held accountable by its own citizens; it must create a sense that movements across borders are being restricted in a manner that provides 'security'. The 'hegemonic dilemma' does not deny that the US government is itself active in defining threats that must be acted upon. The purpose of the concept is to show that localities in the United States, and other countries, must balance processes of globalization with political concerns of security. By constructing and achieving legitimacy through the concept of 'homeland security' the United States is, to some extent, contradicting its previous role as chief promoter of cross-border flows.

Two questions arise. Is 'homeland security' a sign that states are asserting their authority over the flows of globalization? Or, are alternative political geographies being created by non-state institutions within a network of localities? It is to the latter question we now turn.

Summary

In this section we have discussed the locality politics within the War on Terrorism by:

➤ introducing the concept 'urbicide';

➤ illustrating the geopolitical representation of localities targeted by US forces in the War on Terrorism;

➤ illustrating the geopolitical representation of homeland localities.

■ Political geography of world cities

Our theory of politics within localities has focused upon competition between localities. Within the world-economy as a whole, there is a set of localities that have been great winners of this spatial competition: world cities. They have been termed the '"basing points" for global capital' (Friedmann 1983: 69). These very special localities generally share the following characteristics (King 1990). They have high concentrations of world corporate headquarters and are centres of the world's financial system as indexed by a concentration of foreign banks. They house an international elite of professionals in the transnational producer services sector (law, advertising, insurance, accounting, and so on). They are, in short, the great office centres of the world, and this is reflected in their skyline and local property markets. But this occurs alongside the growth of low-pay employment, producing a highly polarized urban structure of very rich and very poor. World cities are the peak of the 'first world', but they have taken on features of the 'third world' with their homelessness and informal street economy. They are microcosms of the extreme inequalities in the capitalist world-economy as a whole, and the street violence and street crime in many of these cities mirror the increasing instability of our political world. They are special localities for many reasons.

The idea of world cities came to prominence with the introduction of 'the world city hypothesis' by John Friedmann (1983). This was a framework for research in which he linked urbanization to the international division of labour or, in our terms, locality to world-economy. He presented seven theses for exploration:

1 The *integration thesis* states that the opportunities and constraints facing every city (or locality) depend on the nature of its integration into the changing world division of labour. This we have used as the basis of our new theory above.

2 The *hierarchy thesis* states that world cities as 'basing points' for capital in its spatial organization of markets and production can be organized into a world hierarchy of control centres.

3 The *production thesis* states that the global functions of world cities are represented directly in the socio-spatial structure of the locality.

4 The *accumulation thesis* states that world cities are major sites for the concentration and accumulation of international capital.

5 The *migration thesis* states that world cities have become magnets of attraction for international migrants producing an ethnic diversity.

6 The *polarization thesis* states that this social mix interacts with the economic functions to produce spatial segregation in a socially polarized locality.

7 The *social cost thesis* states that the new polarization produces potential for social costs that are beyond the fiscal capacity of the local state, generating crises from which the global capital control functions are insulated.

It is obvious that this 1980s 'hypothesis' fits well with 1990s concern for globalization (Knox 1995). In fact, we can think of world cities as the most visible geographical manifestation of globalization. More recently termed 'global cities', with special reference to London, New York and Tokyo, they have been interpreted as information-rich localities, providing the necessary information milieu for the provision of global corporate services and innovative production complexes (Sassen 2001, 2006). More broadly, they are the nodes of a new 'network society' being built through the combining of computers and telecommunications (Castells 1996).

From this brief discussion, we can see that the concept of world cities is essentially a political economy concept in that politics and economics are indelibly intertwined in its definition. Nevertheless, we can draw out the political implications of these special localities and show how they may be crucial to the future of the capitalist world-economy.

Friedmann (1983) identified world cities as located at the interface between the inter-state system and international capital. As such, they are a contemporary expression of the contradiction between the continuous space in which capital operates and the territorial space of politics. As we discussed in Chapter 4, the latter was created in the 'long sixteenth century', in part through the elimination of the city scale of political organization. Braudel (1984) describes 'world cities' as the centre of international finance in the early modern world – first Antwerp, then Genoa, followed by Amsterdam. In the latter case, however, the world city was part of a new territorial state, and it was the latter political form that prospered. Hence, through to the twentieth century there have been new world cities, but they have been firmly linked into state structures such as London as 'imperial capital' in the late nineteenth century. It is the possibility that world cities may be beginning to derive some new forms of independence from territorial states that makes them so relevant to our political geography (Taylor 1995). This is not to say that territorial states are about to disappear; rather, world cities are becoming new loci of power, which will interact with states in new ways. For instance, Singapore acts as the key site where investment decisions about all the countries of south-east Asia are made and implemented. Miami acts in a similar role for the Caribbean states and even for Latin America. The influences of London, New York and Tokyo are even more transnational in their scope. Hence, in some ways, the map of world cities (see Figure 7.1) is superseding the traditional world political map as the crucial spatial structure in world politics. This is the geographical globalization that we introduced briefly in Chapter 1.

This conclusion is a profound one and is a suitable place to bring this first chapter on localities to a conclusion. It points to a new metageography in the sense of a new way of conceiving the organization of humanity across the Earth's surface. In the modern world, we have been taught to think in terms of countries – a key part of the doctrine of nationalism that we discussed in Chapter 5. This means that we see the world as a map of boundaries, the mosaic of differently coloured countries to be found at the beginning of every atlas. Figure 7.1 provides a very different image of the world of humanity: the squares represent nodes in a worldwide space of flows. Of course, the space of flows has been integral to the modern world-system since its inception, but its existence has tended to be obscured by the dominance of our state-centric thinking (Arrighi 1994). Today, with its manifestation

Political geography of world cities

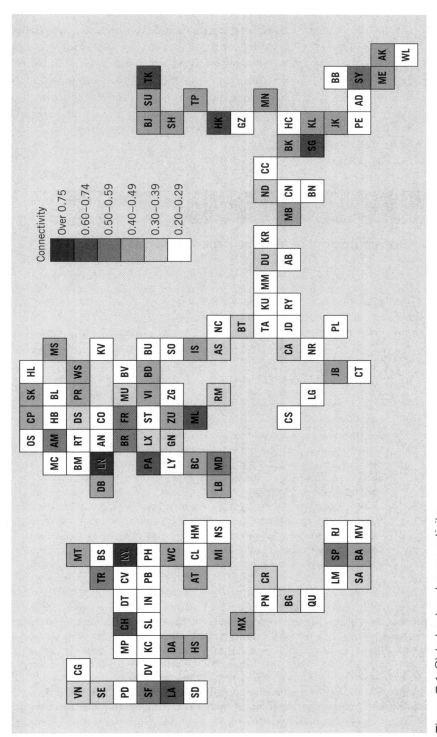

Figure 7.1 Global network connectivity

The cartogram places cities in their approximate relative geographical positions. The codes for cities are: AB Abu Dubai; AD Adelaide; AK Auckland; AM Amsterdam; AS Athens; AT Atlanta; AN Antwerp; BA Buenos Aires; BB Brisbane; BC Barcelona; BD Budapest; BG Bogota; BJ Beijing; BK Bangkok; BL Berlin; BM Birmingham; BN Bangalore; BR Brussels; BS Boston; BT Beirut; BU Bucharest; BV Bratislava; CA Cairo; CC Calcutta; CG Calgary; CH Chicago; CL Charlotte; CN Chennai; CO Cologne; CP Copenhagen; CR Caracas; CS Casablanca; CT Cape Town; CV Cleveland; DA Dallas; DB Dublin; DS Dusseldorf; DT Detroit; DU Dubai; DV Denver; FR Frankfurt; GN Geneva; GZ Guangzhou; HB Hamburg; HC Ho Chi Minh City; HK Hong Kong; HL Helsinki; HM Hamilton(Bermuda); HS Houston; IN Indianapolis; IS Istanbul; JB Johannesburg; JD Jeddah; JK Jakarta; KC Kansas City; KL Kuala Lumpur; KR Karachi; KU Kuwait; KV Kiev; LA Los Angeles; LB Lisbon; LG Lagos; LM Lima; LN London; LX Luxembourg; LY Lyons; MB Mumbai; MC Manchester; MD Madrid; ME Melbourne; MI Miami; ML Milan; MM Manama; MN Manila; MP Minneapolis; MS Moscow; MT Montreal; MU Munich; MV Montevideo; MX Mexico City; NC Nicosia; ND New Delhi; NR Nairobi; NS Nassau; NY New York; OS Oslo; PA Paris; PB Pittsburgh; PD Portland; PE Perth; PH Philadelphia; PL Port Louis; PN Panama City; PR Prague; QU Quito; RJ Rio de Janeiro; RM Rome; RT Rotterdam; RY Riyadh; SA Santiago; SD San Diego; SE Seattle; SF San Francisco; SG Singapore; SH Shanghai; SK Stockholm; SL St Louis; SO Sofia; SP Sao Paulo; ST Stuttgart; SU Seoul; SY Sydney; TA Tel Aviv; TK Tokyo; TP Taipei; TR Toronto; VI Vienna; VN Vancouver; WC Washington DC; WL Wellington; WS Warsaw; ZG Zagreb; ZU Zurich.

Source: Taylor, P. J. (2004).

271

through massive global transactions, it can no longer be relegated to second fiddle. Hence, the image of world cities may well come to rival the old familiar state mosaic as the way we view our world in the very near future.

What are the political implications of this new metageography? Our whole analysis has been premised on the existence of a modern world-system wherein the inter-state system is integral. All previous chapters of the book have provided material on the power of these states and their importance to understanding our world. But the capitalist world-economy is only a historical system; it will not, and at the rate it is using up the world's resources and polluting the world's life support system it cannot, endure for ever. World cities taking functions away from territorial

Case study

Global cities and the construction of a new political geography?

Global cities perform economic and political functions that, in combination, comprise the capitalist world-economy. The territorial organization of these functions and the manner in which states try to control the flows between them constitute one element of the political geography of globalization. Emphasis upon the flows between localities across the globe is the very essence of the globalization thesis, that economic processes transcend states and, to a large degree, escape the attempts of states to regulate them. If we extrapolate the process, the possibility of a dominant political geography organized in global networks of association, rather than 'bounded' within territorial states, is worthy of consideration.

If new networks of political association that challenge the existence of territorial states are emerging then it is logical to see world cities as the crucial sites for activity creating this new political geography. Taylor (2005) conceptualized and identified three such world city networks. Each of the networks is constructed by a particular set of institutions, with different political agendas, that serve as the network makers. First, is an inter-state city network constructed by state departments. Second, is a supra-state city network with UN agencies acting as network makers. Third, is a trans-state city network in which non-governmental organizations (NGOs) act as the network makers.

The inter-state city network serves as a benchmark to compare and understand the geography of the other two networks. This network represents both the diplomatic actions of states between, primarily, capital cities, and the political agency of states to manage relationships between states. It is not, therefore, a challenge to the political geography of states, but the manifestation of diplomatic flows that maintains the inter-state system. The supra-state network is the political geography of global governance. A key actor in the politics of global governance is the UN, especially in its provision of services in 'the fields of health, food, science, labour, development, finance, communications, human rights or refugees' (Taylor 2005: 714). The trans-state network is the manifestation of global civil society, 'a vast, interconnected, and multilayered social space that comprises many hundreds of self-directing or non-governmental institutions and ways of life' (Keane 2002: 23, quoted in Taylor 2005: 716).

In combination, different political actors in the form of different institutions are creating different networks of political association. Comparing the geography of the three networks, Taylor (2005) finds evidence for new political geographies. The NGOs are in the process of creating a network that connects localities in the global north to those in the global south to a greater extent than the inter-state network. The UN agencies are creating a new hierarchical network of power relations that is an alternative expression of power to that of hegemonic states. Taylor's (2005) study is an analysis of political agency in process. It is not a statement of definite outcomes. The significance lies in the identification of institutions as key political actors and their construction of networks of political association. These networks connect localities, with the global cities being important nodes, in hierarchical power relations that are different from the geography of power politics between competing states.

states may be a key indicator that we are indeed at the beginning of a transition to a different world-system with a fundamentally different politics. However, the construction of localities within the context of the War on Terrorism emphasizes the territorial bounding of localities. The tension between a geography of global networks and one of bounded and essentialized places brings us to the role of multiple political actors in the construction of multiple political geographies. To explore this idea we need to consider the politics of identities.

Summary

In this section we have discussed a key subset of localities called world cities by:

➤ noting the different functions that world cities perform;

➤ highlighting the metageography of networks and flows organized by world cities;

➤ discussing the relationship between the political metageography of world cities and the political metageography of nation-states.

Chapter summary

This chapter has focused upon the scale of experience in our political geography framework: the locality. We have:

➤ introduced the concept 'locality' in preference to the term 'urban';

➤ identified the key role that localities play as arenas of political socialization;

➤ exemplified the processes of socialization in place through discussion of protest and planning;

➤ explored theories of political action in localities by discussing practical politics;

➤ discussed the local state, its roles and ambiguous relationship with the national state;

➤ situated practical politics within the dynamics of Kondratieff cycles;

➤ situated local state–national state tensions within the dynamics of Kondratieff cycles;

➤ explored the representation of localities in the War on Terrorism;

➤ identified the concept 'urbicide' and the material and rhetorical destruction of localities;

➤ introduced the concept 'world cities';

➤ discussed the role of world cities in creating a metageography of networks;

➤ discussed the tensions and interactions between the metageography of networks and the metageography of nation-states.

We have identified the scale of experience as the scale of everyday political activity. However, the possibilities and constraints of such activity can be understood only by situating localities in a wider whole of the capitalist world-economy. This we have done by situating practical politics and the politics of the local state within the temporal context of Kondratieff cycles, and also discussing the rhetorical and material destruction of localities in the War on Terrorism. Finally, world cities are key localities constructing an emerging political geography of networks.

Key glossary terms from Chapter 7

administration	decentralization	Kondratieff cycles	power
Annales school of	democracy	(waves)	practical politics
history	dependency	liberal democracy	qualitative efficiency
autonomy	elite	local government	racism
capital city	European Union (EU)	local state	Second World War
capitalism	geopolitics	logistic wave	socialism
capitalist world-	globalization	managerial thesis	sovereignty
economy	government	militarization	space
Christian democracy	hegemony	nation	state
civil society	home	nationalism	structuralist theories of
classes	homeland	nation-state	the state
Cold War	households	neighbourhood effect	third world
collective consumption	human rights	neo-liberal	urbicide
congruent politics	ideology	opposition	world cities
conservative	instrumental theory of	periphery	world-economy
constitution	the state	place	world-system
core	inter-state system	political parties	world-systems analysis

Suggested reading

Agnew, J. (1987) *Place and Politics.* **London: Allen & Unwin.** The seminal text exploring the manner in which places and politics are mutually constructed.

Cresswell, T. (1996) *In Place/Out of Place: Geography, Ideology, and Transgression.* **Minneapolis: University of Minnesota Press.** Intriguing and accessible text exploring how places structure norms of behaviour and how the politics of protest challenges dominant understandings of place.

Taylor, P. J., Derudder, B., Saey, P. and Witlox, F. (eds) (forthcoming) *Cities in Globalization: Practices, Policies and Theories.* **New York: Routledge.** A collection of essays that reviews studies of Anglo-American world city networks and introduces recent European contributions to understanding contemporary inter-city relations.

Activities

1 Type the phrase 'world social forum' into an internet search engine. Note that there is not one central website. How many different localities can you associate with the political network that is the World Social Forum? In what ways is the hierarchy of the capitalist world-economy still evident in this network? In what ways can you see established power hierarchies being challenged?

2 Consider your home town. What institutions and settings played a role in the political socialization of that locality? Was your locality a site of political competition or did one party dominate? How was this situation manifest in the presence of parties in the locality?

3 Using the website(s) of your local government, explore the ways in which the 'local state' relates to higher state levels, especially the national state. Which sorts of programmes are due to local initiatives and which are dependent on outside state support? Is there any way in which the policies of your local government contradict or challenge national policies?

FATHERS 4 JUSTICE PROTESTER AT BUCKINGHAM PALACE, LONDON.
Source: Alisdair Macdonald/Rex Features

Chapter 8

Place and identity politics

What this chapter covers

In the army now

Who wants to be identified as a killer? Thankfully, only the psychotic minority. But yet armies require people to kill and are faced with the problem of turning recruits into soldiers willing and able to slaughter complete strangers. Army training, therefore, requires the construction of a new identity. The citizen must become a soldier prepared to kill. This has not been easy. A famous and influential study of the Second World War (Marshall 1947) claimed that only 25 per cent of US soldiers returned fire. In the wake of the report, the US Army revamped its training. The results are dramatic. According to an article in *Rolling Stone*, in the Korean War (1950–53) the firing rate had risen to 60 per cent, in the Vietnam War it was 90 per cent, the First Gulf War 98 per cent and in the ongoing conflict in Iraq it is pretty much 100 per cent (Tietz 2006).

The US Army initiated a training programme that reconstructed the identity of recruits to enable them to be soldiers and potential killers. The training programme is called Total Control. Bluntly, it is a 'carefully crafted hell that hard-wires kids for combat' (Tietz 2006: 54). In essence, the training programme denies the recruit any identity other than that of soldier under command. Personal autonomy is destroyed. For example, in the canteen when asked what you want to eat by the server the only acceptable reply is 'Ma'am, I am not allowed to say what I want, ma'am' (ibid.: 55). Individual personality is annihilated. Tietz (ibid.: 56) quotes a drill sergeant Randy Shorter: 'The minute you find

a private doing his own thing . . . be it tying a knot his own way, having his boots a certain way, shaving the way he shaved back home – its basically an indication that he's an individual.'

The training programme also constructs the recruit as a member of a new unit to which they owe allegiance. Moreover, if they do not act properly they will endanger the lives of their platoon 'buddies'. With regard to shooting, the recruits are not urged to 'Kill the enemy!', but rather urged to 'Protect your buddy!' and 'Protect the integrity of your unit!' (Tietz 2006: 55). The individual is changed into a soldier, but their identity is also constructed as being a member of the platoon, the US Army, and a nation acting as hegemonic power. At the end of each day recruits recite the Soldier's Creed, the Soldier's Code, the Code of Conduct, the company motto, the seven Army Values, the 'Infantry Song' and the 'Army Song' (ibid.: 56). The Soldier's Code includes the statement, 'I am a protector of the greatest nation on Earth' (ibid.: 56).

The experience of US Army recruits illustrates a number of themes discussed in this chapter. Identity is constructed, by individual actions, by context and by reference to collective identities. This produces a politics of identity that is essentially geographical. It is formed within particular places and spaces and within different geographical scales. Finally, the politics of identity formation has significant implications for future actions – one's own and those of other political actors.

■ Three key concepts of political geography

We have begun each chapter by reiterating some of the seven components of a conceptual framework for political geography. To refresh our memories they are:

➤ Examine the relationship between the material and the representation of the material.

➤ Conceptualize politics geographically.

➤ Challenge long-standing 'silences' to bring in the 'everyday'.

➤ Contextualize politics in spaces and places.

➤ Contextualize politics in the rise and fall of great powers.

➤ Evaluate politics within the larger whole of global politics.

➤ Use geographical scale as a conceptual tool.

In this chapter we engage all seven of the components. In addition, given that this chapter also serves as a conclusion, we shall emphasize three contributions of a world-systems approach to political geography:

➤ Consideration of political actions as simultaneously constructing and being constrained by geographical scale.

➤ Conceptualizing the core–periphery inequalities of the capitalist world-economy within the related metageographies of networks and territorial states.

➤ Contextualizing political actions within the temporal dynamics of the capitalist world-economy.

Simply put, the actions of people, individually and in groups, construct different political geographies. For pedagogical reasons we have presented these geographies in a scalar framework from the broad core–periphery structure of the capitalist world-economy, through the territorial division of the world into nation-states, and local political units. The purpose of this chapter is to emphasize that although such activity occurs in particular localities it is wrong to see it as simply 'local'. For example, the geographical scope of political actions construct geographies of state jurisdiction, the global geopolitical reach of the United States, and the flows of goods, information and money through the networks of the capitalist world-economy. Hence, it is also false to talk of, for example, the 'global' nature of the capitalist world-economy without recognizing the myriad actions of people in particular places, as consumers and producers.

Scale as political product and political arena

We have identified and utilized three geographical scales in our discussion of political geography:

➤ The scale of reality – the structure of the capitalist world-economy.

➤ The scale of ideology – the material and ideological construct of the nation-state.

➤ The scale of experience – the locality as the arena of lived-experience.

In the previous chapter we demonstrated how localities can be understood only by identifying their connections to other localities and scales. Similarly, in this final chapter we emphasize the point that the scales we have identified can be understood only by

conceptualizing them as dynamic structures that are not separate and pre-given scales. Rather, the form of political geographical scale – and the scale at which political activity occurs – is constantly negotiated and requires the construction of linkages to other scales.

In the first chapter we used Schattschneider's analysis to show that political actors are interested in defining the scope of political action. As a rule of thumb, weaker parties seek to broaden the scope by 'jumping scale' to broader levels of political jurisdiction. Stronger parties usually seek to limit the scope of a political struggle. This makes sense in that power is likely to be diluted if other political actors become involved. This is evident in the legal system, where attempts are made to channel actions against states up a hierarchy of courts to, for example, the European Court of Justice which transcends state authority.

Kevin Cox (1997) offers two concepts to help us understand two related processes. First, that the local scale is constructed to facilitate political activity, called the space of dependence. Second, the success of such politics requires a strategic construction of connections to other scales, called spaces of engagement. Cox builds upon his long-standing research interest in the construction of localities as political spaces in which different actors (such as home owners, public utility companies, politicians) have mutual and interlocking interests in the economic growth and social trajectory of a particular locality. They all require realization of long-term investments and/or the generation of a tax base. Cox identifies the geographical expression of such mutual needs as a space of dependence – all actors are dependent on each other to construct and maintain the locality for shared or interlocking needs. The principle of a space of dependence does not apply only to the politics of growth coalitions. For example, protest movements and non-governmental organizations may also cluster in a particular location, such as Geneva, in order to maximize their leverage and utilize shared resources.

But what about strategy to achieve political goals? Cox argues that different political groups faced with different political situations will interact with scales beyond the space of dependence in different ways. These, to some degree, contingent strategies create spaces of engagement. For example, in the previous chapter we noted how the ILA (the dockworkers' union in the United States) constructed a national scale of political engagement from different local situations. Social movements concerned about the environment have constructed networks of connections between localities in a manner that transcends states. Arguably, states would prefer to restrict environmental politics to within their borders. Pollution can then be 'exported' so that as long as 'our' locality is not being polluted it does not matter if another is. A locality could try to lobby its government to restrict pollution and may be satisfied if the source is moved elsewhere. The space of engagement is then restricted to the local and state scales. On the other hand, environmental movements create a global space of engagement so that one locality's gain is not another's loss. Multinational companies would prefer the space of engagement to be restricted to the state scale. If they face protest in one state then they can operate in another. Environmental politics operating in a transnational space of engagement will use the resources of networked local activists to challenge the global operation of capital and its consequences for the planetary ecosystem.

Spaces of engagement and spaces of dependence illustrate that scales are political constructs rather than academic inventions. Three points must be emphasized:

➤ We can identify geographical scales in a manner that is useful for explaining political geography, but it is political actors that actually create the scales.

➤ It is incorrect to see the scales as separate: spaces of engagement are contested constructions that connect, or disconnect, scales as part of political strategy.

➤ The nation-state remains a crucial scale. It is not only a space of dependence and engagement, bringing actors together in the name of territorial sovereignty, but also seen as a key scale of political jurisdiction that may be utilized or must be transcended.

The role of geographical scale is one of the contributions we emphasize in this final chapter. However,

the example of environmental movements illustrates that the politics of scale must be understood in relation to other political geographies. Hence, we turn to networks.

Networks and the capitalist world-economy

In the opening chapter we introduced the structure of the capitalist world-economy, including its core–periphery relations. Interest in globalization has resulted in emphasis upon networks that transcend states and facilitate flows of goods, people, information and money across the globe. We certainly recognize the existence of these networks and the significance of the speed and volume of contemporary flows. In addition, our recognition of the capitalist world-economy as a historical social system requires us to see networks as long-standing political constructs. In our political economy approach, networks facilitating investment and trade create power relations.

One of the important reasons for identifying networks is to deny an identification of core and periphery with swathes of territory. This mistake is often seen in equating the global south with the periphery, or identifying different countries as either core or peripheral states. We need to remind ourselves that the notion of core and periphery relates to different processes (see page 14). In any country, there will be a mixture of core and peripheral processes – though one type may predominate. World cities are locations in which both the core processes of, say, international banking, operate, but often next to low-wage operations, such as sweatshops (Sassen 2006). Networks illustrate that places in the capitalist world economy are connected to maintain the structural relations of core and periphery, but the spatial manifestation of core and periphery processes is differentiated at a finer scale than territorial blocs or states.

Historians of the British Empire have recognized the role of networks in facilitating colonial power, one expression of the core–periphery structure (Lester 2006). For example, Lester identifies three competing colonial projects of the early nineteenth century in the eastern Cape region of South Africa. Each of the projects was attempting to assert power but required a network of connections to facilitate its actions. First, were missionaries who maintained contact with the London Missionary Society and influential sympathizers, such as parliamentary member Thomas Foxwell Buxton. Second, the colonial government transferred information regarding 'effective' colonial government by shipping official documents as well as transferring governors from one location to another. Lester (ibid.: 132) notes Laidlaw's finding that the governance of British colonies relied upon 'informal contacts, patronage, nepotism, and politicking in London'. Third, settlers created a communication network with other colonies to lobby the British government for greater support. Networks of political and economic connections were made to enable territorial control and a relationship of colonial domination.

Networks are political geographic constructs, and operate in tandem with constructs such as scale and territory. They are useful pedagogical tools because

they show that the construction of colonial relationships was not simply a 'core' project: political actions in the colonies and in the metropolitan centres of colonial control were both important. Thinking of networks also requires considering the nodes of a network, either as localities or places. We shall address politics within places in the bulk of this chapter, but before we get there we have one more contribution of our political geography approach to emphasize.

The temporal–spatial context of political action

Our political geography approach emphasizes that we are studying, and living within, a historical social system, namely the capitalist world-economy. We place political actions within a time–space matrix defined by the structure and dynamics of the capitalist world-economy (see pages 15–22). The matrix helps us to understand that all political actions are situated and that geographical location and historical moment combine to offer possibilities for political reaction

---

The politics of pollution

In May 2006 the classic oceangoing luxury liner *SS France* was on its way to be scrapped after forty years of service. As ship historians enthused about the beautiful lines and interior of the ship, and commentators recalled the lavish extravagance of its passengers (Salvador Dali sailed with his pet ocelot, apparently), environmental activists had other things on their minds: the alleged 1,250 tons of asbestos and other toxic materials within the ship. The political geography of scrapping ships is one of differential environmental regulation. The toxins within ships have led governments to export the operation to countries such as India and Bangladesh.

This is a territorial solution for states and business. Political pressure to limit certain activities, such as releasing pollutants into the environment, is sidestepped by transferring to another locality. However, this strategy only works if the protest groups themselves are territorially limited in their

activity and allegiance. Greenpeace is an excellent example of a political group acting within the scope of the global ecosystem and, therefore, transcending state jurisdiction. It is a network of political activity that connects, for example, places in France and Bangladesh. Earlier in 2006, the French government was forced to recall an aircraft carrier bound for Alang, west India, after environmental groups said it contained 1,000 tons of asbestos. The scrapping of *SS France* is being delayed because of similar protests.

Greenpeace is an example of a network of political action that transcends the political geography of nation-states and connects the politics-within-places so that more powerful countries cannot shift their 'dirty work' on to places where peripheral processes dominate.

Source: 'Historic liner heads to scrap yard', CNN.com. http://www.cnn.com/2006/asiapcf/05/09/ghost.ship.ap/index. 9 May 2006. Accessed 16 May 2006.

and construction. Throughout the book we have emphasized some of the consistent features of the capitalist world-economy: notably the inequality of the core–periphery structure and the dynamism of economic restructuring. We have also shown that political geographic features of the system change as the result of political activity. The respective chapters have shown changes in the form and function of the state, national identity and electoral politics, for example.

We balance, on the one hand, the development of new political geographies (the transnational networks formed by institutions in world cities, for example) and the rhythms of Kondratieff waves or the process of hegemonic decline on the other hand. As with our pedagogical use of geographical scale, the trick is to see the dynamics as products of political action rather than as predetermined patterns. So although we have identified a process of British hegemonic decline, the current challenges facing the United States are not indications of a predetermined loss of hegemonic

status. However, we can also learn from the past, and combine analysis of British hegemony (such as Lester's work mentioned above), with the new political geographies being generated within the context of US military action, such as Hardt and Negri's (2000) 'empire'.

The practical usage of our temporal and structural approach is to be able to contextualize current events within broader patterns and processes. In the book we have situated discussion of media stories within the text. The goal has been to illustrate that the daily news reports events in a disconnected or acontextual manner, while the world-systems approach sees them as moments in historical trends. In turn, these trends, or processes, construct political geographies. The events reported in the news are moments of the dynamic political geographies we have presented throughout the book. All events occur in particular places by people acting within certain institutions and political identities. It is to a political geography of places we now turn.

Case study

BIC: the geopolitics of globalization

Brazil, India and China have emerged as important global powers in the wake of economic restructuring and growth. They have been seen as interacting together in international arenas (such as the World Trade Organization) to challenge the power of the United States (Harris 2005). Their rise has led to the identification of the potential for a bloc of states that will challenge the economic assumptions of globalization and neo-liberalism, or the Washington Consensus (Harris 2005). These potential blocs have been named BIC (Brazil, India and China), or perhaps BRIC (to include Russia) and even BRICS (with South Africa).

However, the world-systems approach offers another explanation. The

changes in the national economies of Brazil, India and China are a reflection of economic restructuring at the scale of the capitalist world-economy. BIC is a manifestation of states attempting to negotiate changes in the capitalist world-economy to their advantage. Harris (2005) interprets BIC from the standard social science lens of equating country with society, the multiple society assumption we challenged in Chapter 1. The world-systems approach identifies BIC through its single-society political economy perspective as states trying to improve their relative position within the core–periphery hierarchy. Such a strategy requires a mixture of competition and cooperation. Especially, within the context of globalization

the attempt to capture core processes requires deals with transnational companies at the expense of competing states.

In this case study, we concentrate upon the changes in China, although Harris (2005) describes the dynamics of India and Brazil too. Before we begin, refer back to Figure 1.3 to review the economic rhythm of the capitalist world-economy. Each phase of economic growth was founded upon an economic innovation, and B-phase restructuring set the geographical landscape for which states would capture new industries and core processes and which ones peripheral processes. With that framework in mind, let us outline the recent changes in China.

China accounted for 60 per cent of the world's export growth in 2004 and is integral to global production and finance: China has overtaken the United States as the primary destination for foreign direct investment and 50 per cent of its foreign sales are generated by transnational corporations (Harris 2005: 11). China's economic growth has occurred across all economic sectors. It is a powerhouse in the production and consumption of coal, metals and chemicals, is the largest producer and consumer of steel, and accounts for 40 per cent of the world's demand for oil (ibid.: 12). However, about 20 per cent of China's exports are high-tech, its production of semiconductors and computer chips has risen dramatically. In addition, China has restructured its banking system.

A number of key political geographic developments and implications are raised by the size and manner of China's economic growth. First, is China capturing previous core products and processes rather than new innovations? In other words, will China grow in economic activities that are being peripheralized in this round of Kondratieff B-phase economic restructuring? Steel and coal production may well be necessary to fuel the growth of cities and infrastructure necessary for economic growth in the high-tech and financial sectors. Perhaps the question is whether China's contribution will be to manufacture the hardware for innovations derived in other countries. In other words, will dependence remain? Perhaps the manner in which China has restructured its banking industry is informative. The banking ministry convened a foreign advisory council including the former governor of the Bank of England, the former president of the New York Federal Reserve, and the former general manager of the Bank of International Settlements (Harris 2005: 15).

Second, China's economic restructuring illustrates that the scale of the state is still important, and is a key actor in the processes of globalization. For example, the four biggest state banks hold 70 per cent of the country's banking assets (Harris 2005: 15), and it is the state that is able to offer tax incentives and other inducements to attract Volkswagen, Microsoft and other such companies. The case of China is so interesting because it is a statist communist party that has produced the successes of integrating the country into the networks of globalization. As Harris (ibid.: 9) points out, Chinese leaders have

> Used their control of the government and their statist experience to remould local economic institutions and jettison their communist past without losing their power. In fact, the state-owned sector still produces 68% of GDP and employs hundreds of millions of people.

Third, although the state is still an integral political actor, the goal of state leaders is to situate China's businesses within global economic networks. In the city of Dalian, around 2,500 Japanese companies operate. General Motors, Ford, Honda, Toyota, Hyundai and Daimler all produce inside China. The same story of transnational investment can be found in all sectors. The interaction between global networks and the power of the Chinese state was identified by Yun Yong, chief executive of Samsung:

> Chinese officials are perhaps the most accommodating in the world to foreign investors, because their job performance is evaluated on the amount of foreign capital they attract. . . . You cannot survive in China without becoming a Chinese company. That includes local technology development, product design, procurement, manufacturing and sales.
>
> (see Lee 2004, quoted in Harris 2005: 11)

Fourth, is China another manifestation of the new political geography of world city networks? The country is facing enormous rural-to-urban migration and consequent political strains. Shanghai's GDP of US$80 billion is about the same as Hungary's, Chile's and Pakistan's (Harris 2005: 10).

Fifth, and finally, we must consider the political geography of hegemonic competition. The dramatic growth in China's economic production and trade echoes the similar economic progressions of the Netherlands, Britain, and the United States. But will China capture core innovations and achieve global dominance in finance? Also, is China (in alliance with India and Brazil) challenging the global agenda of the United States? Harris (2005: 15) quotes Ramo (2004) who identifies a Beijing Consensus that challenges the Washington Consensus based on a neo-liberal agenda. For Ramo, the Beijing Consensus takes a more cautious approach to privatization, free trade and capital markets, and aims to nurture an emerging middle class. Ramo (2004, quoted in Harris 2005: 15) argues that the Beijing Consensus has 'the power of a model for global development that is attracting adherents at almost the same speed as the US is repelling them.' We shall see.

Progressive places

We have defined daily urban systems as the localities in which we live our everyday lives. As we saw in Chapter 7, in terms of political geography, this is where we vote for local representatives, protest about local changes, devise our practical politics and confront the local state. In these political practices, everyday environments constitute much more than 'platforms' on which politics unfolds. That is to say, the geography in this local political geography is by no means inert. We provided many examples of this, most overtly in the discussion of socialization in place, and in this chapter we take this argument a stage further.

The everyday urban system is a place. By calling them systems we emphasize their similarities – social scientists have devised generic terms such as central business district, inner city and suburbia for such general comparison. But daily urban systems are each different one from another, singular in their history and geography. Manchester is very different from Liverpool, Charlotte is very different from Atlanta; from such obvious statements many implications follow. Unlike many social scientists, ordinary people understand and value these differences. Most people have an attachment to their birthplace or where they were brought up or where they live today. Such 'sense' of place can be expressed as loyalty to a place. In fact, as we showed in Chapter 5, the rise of the nation-state involved political attempts to eliminate such local attachments as potential threats to the nation. Although reduced in importance in the operation of formal politics, local loyalties can never be simply abolished. As a product of everyday life, they remain regardless of the practices of the state. Alongside attachment to a person's homeland, there may be equally strong ties to their home town. This is a different politics in our scale of experience, what Cope (1996) calls identity-in-place.

How do we define place and, in particular, how does it differ from space? Part of the problem in answering such questions is that in both common language and social science the two words are used interchangeably. But they can be usefully distinguished, and in what follows we draw on Yi Fu Tuan's (1977)

seminal discussion of relations between place and space. His starting point is that '"space" is more abstract than "place"' (ibid.: 6). This is consistent with space being treated as general and place as particular. We might say that space is everywhere, place is somewhere. Further, place has content; the idea of an empty place is eerie, an empty space is merely geometrical. Tuan, however, is concerned for their relations: place as 'humanised space' (ibid.: 54). In this chapter, we are concerned with this specific distinction between space and place.

Tuan (ibid.: 73) has argued that 'when space feels thoroughly familiar to us, it has become place'. This statement leaves space as an impersonal realm, while place is constituted by our everyday behaviour. The most basic of all places is, therefore, the home, which Tuan calls 'an intimate place' (ibid.: 144), and it is for this reason that other places of attachment use this term, as in homeland and home town, mentioned previously. It is these 'home places' to which people feel they belong, and which, therefore, are centrally implicated in the politics of identity.

Our designation of national homeland as place should be noted in terms of scale. Although there is a widespread tendency to equate place with the local scale, places, like spaces, can be designated at several scales. Tuan (ibid.: 149) considers places ranging from a favourite armchair to the whole Earth. There are good reasons why places are often viewed as local relating to the familiarity point above, which is most easily accomplished through micro face-to-face contacts. But there is no need to limit place creation to this one process; as we saw in Chapter 5, the nation is constituted as an 'imagined community', and this is what makes the homeland a place. And green campaigners have been trying to convince us for several decades to view the Earth as humanity's 'home planet'. Although in this chapter we will focus largely upon places at the scale of experience, different scales of place should be kept in mind, since we end the chapter, and this book, by 'jumping scales' back to the global.

Finally, we follow Tuan in arguing that the same locality can be both place and space. It all depends on the perspective of the person and their practices with respect to a given locality. For instance, for the inhabitants of a city their 'home town' is most definitely a place. For a professional urban planner taking a post with the local government as part of a career development, the same city is a space, a locality in which design skills may be used to make the traffic flow more quickly. When the latter practices threaten neighbourhoods with demolition, we have the familiar 'place–space tensions' (Taylor 1998), which have politicized planning in the latter decades of the twentieth century. Clearly, treating localities as a place instead of a space will create a very different politics from the one we described in the previous chapter.

It is not only planners who have tried to dismiss politics based on local attachment to place with disdain or even hostility. Such politics is usually attacked as standing in the way of progress, which leaves it open to a much wider critique. The politics of places is often seen as inherently reactionary in nature. By uniting in 'place coalitions', a selfish politics of beggar thy neighbour can easily be produced. This is most well known as the NIMBY (not in my back yard) syndrome, which avoids general considerations of equity and justice. But place-based politics need not be of this nature. We can take a much more optimistic view of how local politics can evolve, which is indicated by the title of this chapter. As briefly introduced in Chapter 1, the idea of a progressive place is derived from Massey's (1994) work on places and power. Her main point is that viewing local politics as regressive stems from emphasizing the introverted and inward-looking aspects of place. But no place exists in isolation. Places exist in a complex power geometry of flow of people, commodities and information. This is Massey's (1993: 66) 'global' sense of place:

> The uniqueness of a place, or a locality . . . is constructed out of particular interactions and mutual articulations of social relations, experiences and understandings, in a situation of copresence, but where a large proportion of those relations, experiences and understandings are actually constructed on a far larger scale than what we happen to define at that moment as the place itself. . . . Instead, then, of thinking of places as areas with boundaries around, they can be imagined as articulated moments in networks of social relations and understandings. And this in turn allows a sense

of place which is extraverted, which includes a consciousness of its links with the wider world, which integrates in a positive way the global and the local.

It is in such progressive places that new politics of ethnicity, race, gender and class are being forged and which form the subject matter of this chapter. Debates over the meaning given to particular identities have developed as part of contemporary politics. For example, during the civil rights struggle, 'negro' as a label was rejected by political activists, who preferred the perceived dignity given by the term 'black'. Over time, 'black' fell out of favour to be replaced by 'Afro-American' and then again by 'African-American' (Jackson and Penrose 1993: 16). The importance of the debates over these labels lies in the meaning given to them by both the labelled and the labellers. Defining the term by which one is referred to is part of the political struggle to obtain empowerment. In addition, recognition of multiple identities has ignited political debates by problematizing the hegemony of traditional political identities, namely class and nation. The result has been a rise to prominence of *identity politics* as competition over the meaning and import of particular identities has challenged traditional political practices and structures.

The progressive sense of place now being adopted by political geographers reflects the belief that people can forge their own identities and, to a certain extent, futures through the construction of place. A theoretical framework that engages the interaction of structures and agents was, therefore, a necessary component in the new conceptualizations of place. Anthony Giddens' structuration theory provided a means of thinking about how people's behaviour was influenced by structures while, in turn, they created and recreated those structures. In other words, Giddens provided the means by which geographers could delineate how political behaviour within places was both mediated and constrained by the place and, in turn, maintained the particular character of a place. The theoretical background to this chapter is, therefore, a story of how geography has used social theory to place itself in the mainstream of social science.

However, both our materialist theoretical framework and our emphasis upon geographical scales requires us to consider processes beyond the immediacy of specific places. What we aim to show in this chapter is the dynamism of a progressive politics of place and its linkages to dynamics at the global scale. First, we link identity politics in places to dynamics at the global scale by recourse to hegemonic cycles (Chapter 2) and the construction of new modernities by each hegemon. The historical cyclical patterns are related to current notions of identity politics via the concept of reflexive modernization – the idea that everyday decisions in contemporary society are much more individualized than in the recent past. This helps us to answer the question: why identity politics now? The next step is to explain the types of identity politics that are currently prominent. To do this, we resort to the key institutions of the capitalist world-economy and the politics defined by the interactions between them. Once we have laid out our theoretical framework, we shall exemplify it through discussion of contemporary political conflicts.

Because of the way we have used geographical scale to order our argument, it falls upon this chapter to conclude the book. Hence the final section doubles as both chapter and book conclusion. We mentioned earlier that we return to the global, but not in a simple break with what has gone before. Treating economic globalization as essentially spatial in nature and ecological globalization as ultimately about place, we consider this extreme yet decisive place–space tension. The War on Terrorism has led to the creation of new political spaces, such as Camp X-Ray in Guantanamo Bay, Cuba. We conclude the book by contrasting the dehumanized spaces being constructed in the name of the War on Terrorism with the rich humanized places that are the subject of feminist geography and the Annales school of history. The context of American empire provides opportunities posed by the dynamism of identity-in-place (Cope 1996), and illustrates the role of both geographical scale and place in structuring political outcomes.

Summary

This opening section has served as both an overarching synthesis of the whole book as well as a segue from the previous chapter to this one. We have emphasized three interconnected themes and concepts:

➤ the political geography of scale;
➤ the persistence of core–periphery differences;
➤ the usefulness of identifying political actions within their temporal-spatial context.

In addition we have introduced the concept of 'place' to complement the previous chapter's identification of 'locality'. Through a discussion of place we shall focus this final chapter on the politics of identity.

■ Theorizing political action in places

The terms 'turf battles' and social 'movements' imply two things. First, political struggles occur in arenas or, in more formal words, contexts and structures. Second, political struggles are initiated by groups of people or, again in formal terms, by human agency. Influenced by the structuration theory of Anthony Giddens, political geographers came to view places as structures that mediated political activity while, in turn, that political activity continually created and recreated the place. Political geography has seized the opportunity provided by social theorists, such as Giddens, and has positioned itself within the mainstream of social science by emphasizing how political actions are shaped by the places within which they occur and, in turn, continuously constitute the uniqueness of different places. This approach has been termed a 'social constructivist view' of place (Staeheli et al. 1997: xxix) as places are a product of decisions made by managers of capital, the state and civil society. We saw examples of such processes in Chapter 7, as different workplace contexts evoked different political responses.

Structuration theory

Anthony Giddens' structuration theory acted as a catalyst for political geographers wishing to understand political actions within places. Giddens developed his structuration theory over time as a result of his own thoughts as well as responses to critics. The source of the theory was Giddens' frustration with two schools of sociological thought: humanistic sociology, which placed too much emphasis on the freedom of human actions, and Marxist structuralism, which Giddens felt was too deterministic (Cloke et al. 1991: 97). Instead, structuration theory treated individuals as knowledgeable agents while also recognizing the constraining and enabling possibilities of power structures and other social institutions. Giddens developed the notion of social rules to link structures and agents. Social rules fashion the way that people interact in a consistent manner so that regularities and norms of social interaction are established (ibid.: 102). These norms, which are no more than regular human activities, become established as structures. However, as these structures are the product of human actions, they can also be changed in the same way. A simple example is the regularity of family life, with its norms of meal times and 'table manners'. Although these rules may be interpreted as a constraining structure, enterprising adolescents may find ways of changing them.

In later development of his theory, Giddens placed further emphasis on the role of time and space. Following Hägerstrand's (1982) time geography, the idea that people had regular paths in their daily lives was incorporated into structuration theory. Furthermore, these paths were structured by resources and institutions. Hägerstrand's time geography ignored the role of power relations in shaping people's everyday lives, but the addition of structuration theory rectifies this glaring omission. For example, poor women in rural areas of the United States are often frustrated in their efforts to find employment because of their needs to find child care and cheap and convenient public transport. Giddens also noted that not only is the routine of everyday life influenced by people and institutions that the individual has contact with in immediate time and space but also that that interconnectivity includes relationships across broader expanses of time and space. With this, we can relate Giddens' abstract theory to the daily pressures of globalization and Massey's (1993) notion of pro-

gressive places. People's lives are structured by norms and institutions that are obviously features of the places within which they experience their everyday lives. But social practices are also structured by more distant and intangible relationships. For example, the norms and rules of nationalism are embedded in our everyday lives by structures that are quite abstract and dependent upon a long-term view of history and community. In addition, the nature of our everyday life depends upon whether we are lucky enough to be living in the core rather than the periphery of the capitalist world-economy. The time spent on leisure activities rather than searching for household fuel is structured by the three-tier hierarchy of the world-economy.

Structural definitions of place

Our discussion of structuration theory emphasizes its compatibility with our treatment of geographical scales. But much more importantly, the introduction of a structure/agency dynamic has fundamentally changed our understanding of human action within places. Structuration theory showed that human actions are structured in everyday and place-specific contexts *and* how the routinized nature of these actions reproduce these contexts. As a result, context is conceptualized as being 'structural' rather than 'compositional' (Thrift 1983). That is to say, the influence of places was not a simple effect of their content or composition; rather, the nature of places is implicated in understanding social change. For example, in the case of electoral geography, context could no longer be seen as merely the relationship of particular variables (such as, say, the percentage of people in public housing) to a voting pattern. Instead, place was interpreted as a structure, a context that both mediated the voting decision and was a product of that decision. Socialization within place, for instance, predisposed someone to be a supporter of the Conservative or Republican Party and, in turn, the political hue of a place was part of its character, a structure continuously being reproduced, or contested, by predisposed agents. We illustrate this further in a case study of Nazi voting patterns in inter-war Germany below.

John Agnew, in his *Place and Politics* (1987), identified the components of place that lead to an expectation of place-specific political behaviour. In other words, if places are unique then we should expect different forms of political behaviour in different places. The propensity to strike, or to vote Republican or Conservative, or to participate in local government, is likely to vary because of the institutionalized political history of different places. Agnew saw place as comprising three features: location, locale and sense of place. Location is the role that a place plays in the world-economy, its industrial base or geopolitical role, for example. Locale refers to the institutionalized setting of a place (traditions of unionism or Catholicism, perhaps), which permeates most aspects of life, including politics. Sense of place is the essence of a place-specific identity, which gives coherence and meaning to the actions of its inhabitants.

The combination of identity, local institutions and global linkages is also a feature of Massey's (1994) notion of place. Place is the setting for a dynamic set of social interrelationships: relationships that extend beyond the particular place but interact within a particular setting. The specific variety of relationships and the unique way in which they interact produce the particularity of place. In addition, the dynamism of the social relationships suggests that places change while retaining their uniqueness. Massey (1994) reflects upon her place of home, Kilburn in north London. Looking down the high street, she sees the shops and clubs servicing different waves of immigrants, Irish and south Asian. The migrants illustrate the role of linkages across the space of the world-economy that come together to form a particular place.

Case study

Nazi social construction of places, 1924–32

Following Agnew and Massey, place is both dynamic and 'extra-local' in scope. This is a fertile definition of place in that it can be used as a framework for a variety of political geographic analyses. For example, the task of quantitative electoral geography is to conceptualize these notions of place so that they can be included in the statistical analysis of voting behaviour. Electoral geography becomes a dynamic analysis of place and the activities of social groups. What we should expect to see are patterns of voting behaviour that reflect the components of place put forward by Agnew (1987) and Massey (1994), namely place-specific political behaviour derived from 'extra-local' connections.

Analysis of the electoral support for Adolf Hitler's Nazi Party in inter-war Germany exemplifies these points. Initially, the Nazi Party's electoral support was quite insignificant but then grew rapidly as economic conditions deteriorated within the context of the stagnation associated with Kondratieff IIIB (see Table 1.1 in Chapter 1). B-phases are periods of economic restructuring when new core economic processes are replacing the once cutting-edge industries being peripheralized (see Chapter 1). Places whose links with the world-economy are established through the new profitable industries will enable better life opportunities than those dominated by the declining and stagnating industries. Kondratieff IIIB was also a period of geopolitical competition (see Chapter 2) during which the United States and Germany were competing with each other to replace Britain as the hegemonic power. In a global context of economic restructuring and political competition, it is not surprising that the Nazi Party became so popular. The full name of the Nazi Party was the National Socialist German Workers' Party, from which it can be gleaned that nationalism and populist anti-capitalism were important components of its manifesto. Table 8.1 shows the dramatic rise of the party's electoral support over time. The surge in the party's popularity reflects the growing economic dislocation being experienced by the German people, while the earlier periods of minimal electoral support occurred during periods of economic optimism. The broad context set by the political and economic cycles of the capitalist world-economy only explains the Nazi Party's support to a certain extent. Despite its extreme nationalist rhetoric, the attraction of the Nazi Party was never even across the national space. The next step is to explain the place-specific

Table 8.1 Percentage change in the Nazi Party vote between consecutive Reichstag elections

May 1924/December 1924	–3.5
December 1924/May 1928	–0.4
May 1928/September 1930	15.7
September 1930/July 1932	19.0
July 1932/November 1932	–4.2
November 1932/March 1933	10.8

Source: Data from Hamilton (1982).

nature of the Nazi Party vote within Germany.

Figure 8.1 maps the electoral support for the Nazi Party in the Reichstag (parliament) election of 1930. The triangles pointing upwards represent clustering of counties that voted strongly for the party. Triangles pointing downwards represent pockets of counties where the vote for the party was exceptionally low. The larger the triangles, the greater the magnitude of the clustering. By reference to Figure 8.2, we can identify regions of strong support for and strict opposition to the Nazi Party. This pattern is explained by both the role that a place plays in the world-economy, or Agnew's location, and its institutional

setting, or locale. First, the regions' economic linkages to the capitalist world-economy go a long way to explaining the pattern of support. For example, Schleswig-Holstein and Oldenburg were agricultural regions of the north-west, heavily dependent upon livestock farming. The farmers perceived themselves to be adversely affected by the government's agricultural policy, which advocated the importation of food, especially from Denmark, as well as by high interest rates and the high cost of fertilizers and electricity (Brustein 1993: 179). The Nazis' agrarian programme addressed these issues and, consequently, generated rural support. The social context of the voter was also of

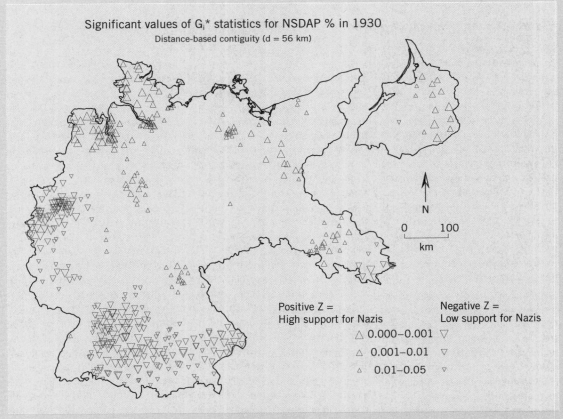

Significant values of G_i^* statistics for NSDAP % in 1930
Distance-based contiguity (d = 56 km)

Positive Z = High support for Nazis
△ 0.000–0.001
△ 0.001–0.01
△ 0.01–0.05

Negative Z = Low support for Nazis
▽
▽
▽

Figure 8.1 Geographical clustering of the Nazi Party vote, 1930.

Source: reproduced by kind permission of Colin Flint from an unpublished PhD thesis at the University of Colorado.

Figure 8.2 Historical–cultural regions of Germany.

importance in determining their vote. Support for the Nazi Party's policy of compulsory inheritance of farmland by the oldest son was strong in the north-west because this custom was already practised in the region, and the National Socialists added the promise of resettlement of disinherited farmers' sons in eastern Germany (ibid.: 69). The intersection of the trade and inheritance issues illustrates how both the material and institutional components of place, and their 'extra-local' connections, fostered a context for Nazi Party support in north-west Germany.

A social constructivist (Staeheli et al. 1997) view of place requires more than an understanding of place-specific concerns and the subsequent relevance of the Nazi Party's manifesto. Party supporters had to generate electoral support, and in doing so they created the geography of the Nazi Party vote (Flint 1998). For example, in Baden the party's support in the Reichstag elections of 1924 was restricted to the northern part of the province. In the initial period of growth, between 1928 and 1930, the party undertook a saturation campaign, with its members disseminating leaflets from lorries and cars and organizing rallies across the whole of Baden. Its intention was to forge a space of political power across all of Baden by attracting supporters from a broad spectrum of social classes. However, the party's support in the Reichstag election of 1930 was geographically bifurcated. In the north of Baden, the Nazis attracted support from blue-collar workers, but in the south its support was white-collar. By 1932, sustained organizational efforts by Nazi activists accomplished the party's goal of cross-class support across the whole of Baden (Flint 1998). The activity of the Nazi Party had, over time, constructed a homogeneous electorate across the whole of Baden by expanding its social appeal through extending its geographical reach.

The example of the Nazi Party in Baden illustrates a number of points. First, by using social theories about the social construction of place in conjunction with a quantitative analysis of voting data, electoral geography becomes a tool in identifying how contexts and places influence the behaviour of individuals. The purpose is not the mapping and counting of votes but to use the geographical distribution of those votes to investigate how places mediate political behaviour. Second, the politics of identity in place is a dynamic process in which the manufacture of political identity requires the construction of spaces of power. Finally, this particular example offers a word of warning about a progressive sense of place. Although progressive political activists may be able to use extra-local linkages to advance their positions, so also may reactionary politicians to replicate past Nazi uses of geography.

Summary

In this section we have discussed:

➤ structuration theory and its use by geographers;

➤ a structural definition of place to explain the role places play in socializing particular identity politics.

The structural definition of place builds upon social theory to identify the elements of places that make them unique and socialize their occupants in unique ways. However, places should not be seen as a separate scale but as situated within a wider whole. It is to the wider whole we now turn.

■ Modernity and the politics of identity

Both Massey and Agnew lead us to consider structures at a variety of scales. Human agency is simultaneously mediated by local institutions and norms, state authorized practices and global imperatives. Our multi-scale framework demands a more explicit linkage to the global structure, the capitalist world-economy. Recourse to the space–time matrix introduced in Chapter 1 suggests that the current growth of interest in identity politics may be a product of global dynamics. In other words, can Kondratieff waves and hegemonic cycles help us to understand why identity politics has become of both great political importance and academic interest? To suggest an answer we need to explore further the nature of the hegemonic cycles we introduced in Chapter 1. Following Wallerstein (1984b), we have thus far emphasized the political-economic aspects of world hegemony, but in other contexts the concept of hegemony has a much more cultural meaning. This 'cultural hegemony' is no less political, of course. The basis of this conception can be found in the work of the Italian Marxist theorist Antonio Gramsci in the 1920s and 1930s. His hegemony was about a dominant group providing political and *intellectual* leadership to enable it to rule largely by consensus rather than coercion. That is to say, the rulers are able not only to set a generally agreed political agenda but also to provide the basic assumptions that define how all others form their politics. Hence the dictum that 'the ruling ideas of society are the ideas of the ruling class'. In world-systems analysis, this model is transferred to the system level, hence 'world' hegemony, so that the ruling ideas of the world-system are the ideas of the hegemonic state, at least during periods of high hegemony. Thus, for instance, it can be said that the twentieth century was the 'American century'. It is this aspect of world hegemony that we develop at this stage of our argument. In the first section, we introduce the idea of prime modernities as the cultural face of world hegemonies. This historical excursion is a necessary backcloth to an understanding of the contemporary nature of modernity – reflexive modernity – which is directly implicated in the rise of the politics of identity.

Hegemony and modernity

Being and becoming modern has been a major force for change by both societies and individuals. Whether construed as a 'mass need' of a people trying to modernize their society (the hallmark of nationalism as we saw in Chapter 5) or a personal need to be seen as 'up to date' by wearing the latest fashion, the idea of the modern is almost universally seen in a positive light. Millions of individuals over many decades, and even centuries, have striven to 'join the modern world', either by migrating to 'new' places – cities or foreign countries – or by changing their home place to make it 'more modern'. It is the major unexamined concept of our world; taken for granted because that is simply what our world is: modern. Popular faith in the modern is so strong that it is usual to treat particularly unpleasant contemporary events as not modern. For instance, in recent years the mass murders in Bosnia and Rwanda have been labelled 'ethnic cleansing' and 'ethnic genocide', respectively, and interpreted as throwbacks to premodern times, presumably on the grounds that modern cannot be 'barbaric'. But if we live in the modern world, these must be modern events (Taylor 1999).

In recent years, social scientists have come to unveil the popular mask of what is the modern.

In path-breaking work, the sociologist Zygmunt Bauman (1989) has taught us that the Holocaust was a modern policy of a modern state based upon modern planning using modern transport, modern chemistry and modern machinery. Evidently, 'modern' incorporates mass murder as well as the latest fashions. It is incredibly ambiguous: 'a double-edged phenomenon' in the words of Giddens (1990). This makes definition very difficult, but we can proceed by focusing on the associated concept of modernity. Modernity may be interpreted as the concrete face of being modern; the condition of modernity describes the nature of modern society (Taylor 1999). The 'double edge' of this condition is simultaneous tendencies towards rapid change and controlled order. To experience the modern is to live with perpetual structural change, but as modern people we are forever trying to control this change to make life liveable (Berman 1983). In short, we try to tame incessant modern change through projects, both personal and political, to build stability. Such projects come in many forms and sizes, and many are territorial in nature. That is to say, the strategy is to protect people within a defined locale. Two well-known examples are national planning and home making. National planning attempts to create stability through protection from the uncertainties of the world market, and creating a home is to make a haven from the stresses and pressures of the labour market.

How does this relate to hegemonic powers? Quite simply, hegemons are the sources of major economic restructurings of the world-economy, which unleash massive social change and uncertainty into the world-system. As new economic developments associated with the hegemon begin to spread across the world, the hegemon also creates a means of dealing with the consequent social changes. Thus, as well as destroying old lifestyles and jobs, new opportunities arise in the form of a different matrix of lifestyles and jobs. In other words, hegemons are inventors of new modernities in terms of both new social changes and new ways to deal with those changes. Hence there is a cultural contribution; hegemonic states are much more than simply the location of the most efficient economies of their time.

Prime modernities

The position we have taken above involves identifying multiple modernities. This is counter to most popular and social science interpretations, which treat modernity as a singular concept: there has been only one modernity, and that is industrial society. In fact, modern society and industrial society are usually used as if they are synonyms. We are trying to overturn this legacy here. Modern societies come in many forms, of which industrial society is just one, albeit a very important one. There have been overt state planning exercises to create new modern worlds, such as in Revolutionary and Napoleonic France and latterly in the Soviet Union. Less planned but much more influential, the modernities that have emanated from the activities of, and within, hegemonic states have ultimately become system-wide. These are therefore the prime modernities of the modern world-system: Dutch-led mercantile modernity, British-led industrial modernity and US-led consumer modernity.

Notice that the three prime modernities have each emphasized one aspect of the overall process of capital accumulation: exchange in mercantile modernity; production in industrial modernity; and consumption in consumer modernity. This is not to suggest that each modernity involved just the one aspect of the process. That would be impossible – to realize any capital investment there must be production *and* exchange *and* consumption. Rather, we identify the cutting edge of the economic restructuring, the process where the key innovations were occurring to make the new modernity a distinctive new form of society. New competences in navigation and trade enabled the Dutch to create mercantile modernity; new competences in machines and engineering enabled the British to create industrial modernity; and new competences in communication and advertising enabled the Americans to create consumer modernity. Of course, these were very complex developments, which involved much more than these processes, but the essence of the changes can be gleaned from this tripartite model of modernities.

The key point about these modernities is that their origins are geographically precise. They are places, locations that come to epitomize how we think of the modern. Thus our images of these different modernities derive concretely from real historical panoramas. For mercantile modernity, the harbour of Amsterdam, full of ships' masts sticking up into the sky, as represented on many early prints, was the wonder of its age and represents what it was to be modern in the seventeenth century. In Britain, the 'dark satanic mills' of the textile industry in northern England, in which new production and new urbanization created the first modern industrial regions, represent how we still view industrial modernity. Finally, the image of consumer modernity is much more wholesome: suburbia was designed as the antithesis of the industrial town, the smoke and grime replaced by houses with gardens and car access to the ubiquitous shopping mall. These three images represent three distinct modern experiences that have dominated lives and lifestyles during successive hegemonic cycles. As such they are the ultimate progressive places of their respective eras.

Places of the future

Defining the hegemonic states as 'laboratories of modernity' (Taylor 1996) implies a cultural position with profound political effects. If being modern is widely seen as a positive attribute, then other states will want to copy the new practices being developed within the hegemonic country. Such emulation is how these modernities convert from being local to system-wide, to becoming prime modernities. But this is not a series of simple preferences to become modern; there is an imperative operating at the centre of the process. The economic success of the hegemon illustrates new opportunities, which other states will ignore at their peril. As arch-progressive places, hegemons represent the future, and this means that other states have to emulate or else miss out on the future. Not to emulate is to remain stagnant in a mire of traditionalism, to be old-fashioned and, above all, to be non-competitive. Hence the invention of a prime modernity leads to a procession of other states attempting to 'catch up'.

Defining the future of others is an immense cultural power. The whole system comes to resemble aspects of the hegemon writ large. As such, it is much more than a political process defined at the state level. Through prime modernities, hegemonic power pene-

trates the everyday lives of people throughout the world-system. There is a common terminology for the way in which this process operated with each hegemon. Emulating the Dutch has come to be called mercantilism; emulating the British industrialization, and emulating the United States is called, more pedantically, Americanization. Let us look briefly at these three processes.

In response to the Dutch golden age of economic success, both the English and the French devised state policies that attempted to stimulate economic growth in their territories at the expense of the Dutch. For instance, Oliver Cromwell's government in 1651 passed the Navigation Acts, which limited shipping in English ports to just carriers from the two countries engaged in the trade. This banned intermediate carriers, notably the great Dutch merchant fleet. However, the best example of emulation at this time came from Russia. Tsar Peter the Great actually worked incognito in a Dutch shipyard in the late seventeenth century in order to learn the secrets of the new world that the Dutch had created. When he returned to Russia, he built St Petersburg to act, in part, as a great new mercantile city on the Baltic, a new Amsterdam as Russia's 'window on the West'. Berman (1983) identifies this as the first example of a 'non-western' country attempting to 'catch up' through a state development policy.

With the Industrial Revolution came industrial espionage. Many of the new industrial practices pioneered in Britain had critical military implications. Without the latest military hardware, states had much more to lose than simply failing to catch up. But as the nineteenth century unfolded, emulation became much more open, culminating in the Great Exhibition of 1851, where Britain laid out its industrial wares for the rest of the world to marvel at. This was a period when industrialists from Europe and the United States visited England, specifically northern England, to learn how to manufacture in the new way. They wanted to build 'new Manchesters' in their own countries. Manchester became the 'shock city' of the age, 'the centre of modern life' (Briggs 1963: 94). According to Arblaster (1984: 260), it came to represent 'the essence of modernity . . . the place to which people came to see the future'.

In the twentieth century, the United States has taken over as world 'role model', but in this case you do not necessarily have to visit the country to see the future. Images of American suburbia have been portrayed in, first, Hollywood films, and then American TV programmes to become known the world over. These have set the standards to which ordinary people aspire, and suburbs have appeared around cities in countries that had long resisted such expansion. When people did visit the United States, they knew exactly what they were seeing. As one Italian immigrant described it:

> What interested me in the United States . . . was the 'modern' concept, in which all things seemed to be done in a revolutionary 'modern' way. Everything was the product of fresh thinking, from the foundations up. Everything had been 'improved', and was continually being improved from day to day, almost from hour to hour. The restlessness, mobility, the increased quest for something better impressed me.
>
> (Barzini 1959: 73)

But alongside this incessant change, there was a new stability. A Swiss observer identified the haven that Americans build for this new modernity:

> This America of everyday life is simple and accessible. . . . It is a modern land where technical ingenuity is apparent at every point, in the equipment of the kitchen as well as of the car, but at the same time a land of gardens, of flowers, of home activities, where a man, away from his office or his work-place, enjoys tinkering at his bench, making a piece of furniture, repainting his house, repairing a fence, or mowing his lawn. A land of luxury but also of simple pleasures.
>
> (Freymond 1959: 84)

This was the America of the 'affluent society, of the good life, the seed and kernel of the future' (Broughton 1959: 263). It is this 'American Dream' as 'world dream' that is consumer modernity.

Ordinary modernity

J. K. Galbraith (1958) described the 'affluent society' as something unique in the history of humanity. The world pioneered by the United States was creating a society where ordinary men and women not only aspired to the good life but actually lived it. In 1950s United States, high wages meant not only that American families could buy the new consumer durables – washing machines, refrigerators, TVs, etc. – but they could also buy the house in which they were to be located. Consumer modernity relied on this congruence between mass production and mass consumption. Although we agree with Galbraith that this was indeed a novel historical situation, the roots of this affluent society are to be found within the two other prime modernities.

We use the term 'ordinary modernity' to describe the common everyday features of the prime modernities. We derive this generic term covering all three hegemons by focusing on the cultural similarities between the three modernities. Put simply, the three prime modernities have incorporated in their make-up a specific cultural celebration of ordinariness (Taylor 1996). We can see this when Dutch genre painting is compared with other art in early modern Europe. Whereas the latter portrays the great and celestial – kings, nobility, Bible scenes, angels – Dutch art depicted scenes of ordinary Dutch life – taverns, burghers, homes, streets. Similarly, the English novel represents another cultural art form that celebrates the trials and tribulations of ordinary lives. From Jane Austen to the great Victorian novelists such as Charles Dickens, their stories continue to resonate with us today because they are so 'modern' in their concerns. And, of course, America has pioneered popular cinema. Not for Hollywood the 'arty' high-brow productions of the European cinema; American films always included a high proportion of images portraying, in a non-patronizing manner, ordinary people living ordinary lives. Their subliminal message was that the good life, suburban living, was possible for all.

Hence we see that the thread of ordinary modernity leads to consumerism. There is a sort of logic to this, but mass consumption is not its final destination. Ordinary modernity was expressed in practice through individual tastes within the home. Invented as a private family space by the Dutch, it was marked from the beginning by an individuality (Rybczynski 1986). That is to say, every home was different in

Case study

Religious fundamentalism as modernity

For most of the twentieth century there has been a common assumption that religion would continue to decline as society became more and more secular. Thus one of the great surprises of the late twentieth century was the resurgence of religious belief in many parts of the world. Within this resurgence there were some who have come to be called fundamentalists, a militant religiosity which, according to Armstrong (2004: ix), rejects many of the positive values of modernity: 'democracy, pluralism, religious toleration, peacekeeping, free speech or the separation of church and state'. Instead they have 'gunned down worshippers in a mosque, have killed doctors and nurses who work in abortion clinics, have shot their presidents, and have even toppled a powerful government', not to mention 9/11. Given their adherence to 'tradition' (see the following case study on fundamentalism and globalization), fundamentalists

are sometimes seen as a throwback to 'medieval' times. Nothing could be further from the truth: they are thoroughly modern.

The spiritual lives of premodern people were distinctively different from those of modern people. Armstrong (2004: xiii) argues that they evolved two ways of thinking and constructing knowledge: *Mythos* and *Logos*. Both were necessary and complemented each other. *Mythos* was timeless knowledge of values, the foundations of culture, to make sense of day-to-day lives. *Logos* was practical knowledge, how to navigate day-to-day lives. *Mythos* did not demand 'proof': it was just how things were. *Logos* was functional and therefore had to be rational. For Armstrong, modernity has lost the sense of *Mythos* and the basis of modern society is *Logos* (ibid.: xiv).

Fundamentalists are modern because they have accepted the scientific

rationality of modernity. If the latter provides 'truth' then the *Mythos* of faith has to be turned into a contradictory *Logos* of faith. The emphasis on the literalism of holy texts – treating them as *Logos* rather than *Mythos* – is wholly a modern reaction to the assault of modernization on religion. Counting back events to find the exact date of the Creation treats the Bible as a text of modern history. Such activity would not be countenanced in a premodern culture, where history was valued for its meanings, portraying timeless realities not specific real events. Thus creationism or intelligent design is as modern as the theory of evolution – debate is conducted on *Logos* grounds. It is in this sense that religious fundamentalisms are modern and integral to contemporary globalization.

The interested reader is directed to Armstrong (2004) for more detailed discussion of the issue.

detail, reflecting a family's preferences in furniture and decor. This individuality came to a head with the Victorian English love of 'nick-nacks' in their homes (Briggs 1988) – we still think of Victorian homes as excessively cluttered. Mass production involves standardized products, and therefore mass consumption did not immediately reflect this individuality. There was some choice between products of different companies, but one suburban home tended to look very much like another. However, consumer modernity has evolved to begin to fully express individuality. Generally, the range of products available to purchase has grown out of all proportion to the recent past, with much niche marketing on top of this. Consumption today is mass in numbers, not in its nature.

On the way, it has even generated a 'consumer movement' to protect consumers as individuals against the large corporations as producers. This is an indicator of a new reflexive modernity, which is, in part at least, the culmination of ordinary modernity.

Reflexive modernization

Reflexive modernity is a concept coined by Ulrich Beck (1994) to represent the uncertainties and fears facing people today, which, in turn, force them to challenge traditional politics and identities. Beck argues that we are currently in a new form of modernity, which he calls the risk society. What sociologists have traditionally called 'industrial society' Beck sees as developing in an unplanned manner, thus ultimately threatening the very survival of our planet. Global warming, toxic waste and nuclear catastrophes are all products of the capitalist world-economy that cause us to reflect upon our roles as producers and consumers. Risks, in this context, are challenges and uncertainties created by the imperative of capital accumulation and the growth of industry. The myriad of risks currently facing us (from the decline of the nuclear family, to crime in the city, to fear of terrorism right up through global ecological catastrophe) forces us to challenge the institutions of industrial society. The magnitude of the perceived threats leads to a questioning of the efficacy of the underlying assumptions of industrial society: business, law and science, for example. As a result of the questioning of established institutions, collective and group meanings based upon the institutions are suffering from exhaustion. For example, questioning the underlying assumptions of business challenges the once universal ideology of progress.

This provocative theory is very relevant to our concerns here. However, it derives from a simpler view of modernity than the one we have advanced. Apart from the contemporary situation, Beck conflates industrial society with modern society. Hence his multiple modernities extend to only two cases: 'old' industrial modernity, which includes our consumer modernity, and the 'new' reflexive modernity of his risk society. This need not stop us using his ideas, however – as we noted in the previous section, we can derive the reflexivity from developments in consumer modernity as a culmination of ordinary modernity. However, we shall focus on the idea of post-traditional derived from Beck's co-author Anthony Giddens (Beck et al. 1994).

The post-traditional condition

The challenges to authority that Beck theorizes occur across a wide range of institutions. The professions are particularly vulnerable to this process because their position relies on their claim to specialized knowledge. But this 'expertise' is no defence, because the 'consumer movement' has extended way beyond retailing: teachers are challenged by parents and pupils, doctors are challenged by patients, lawyers are challenged by their clients, and architects are challenged by everyone. Giddens calls this rise of individual assertion and concomitant decline of tradition authority the 'post-traditional condition'. The limiting case must be contemporary soldier dissent for the war in Iraq. There have always been deserters in wars but this latest war has thrown up a uniquely new phenomenon. In the UK, soldiers refusing to go to Iraq are claiming to be 'conscientious objectors'. The original conscientious objectors were pacifists (for example Quakers, who oppose all wars) in the First and Second World Wars when conscription was used to draft all eligible citizens. Today's serving soldiers, however, are volunteers: they have chosen an army career. Having been in other wars (for example in Bosnia) they are conscientiously objecting to this one particular war in Iraq. Losing in court and going to jail because the judge ruled that armies cannot operate with soldiers choosing which wars to fight, this challenge to the 'expertise' of the state in fundamental defence matters by assertive soldiers is the ultimate sign of decline for traditional sovereign authority. The post-traditional condition appears to be leading to a pick and choose war mentality by the state's most loyal of servants: soldiers.

Individuals are more assertive because they are more reflexive; instead of taking social knowledge for granted, they interrogate it and decide how it relates to their individual needs and concerns before accepting or rejecting an idea. Giddens (1994) calls the resulting social reflexity 'a world of *clever people*' (italics in the original), by which he means not that contemporary people are more intelligent than their forebears but rather that they are more informed and therefore more able to challenge traditional argument, even when dressed up with the authority of 'science', for instance in rejecting new genetically modified food. This situation is called post-traditional because it is not conducive to the reproduction of tradition of any kind. All tradition relies on acceptance of a taken-for-granted truth as specified by some authority such as custom or experts. Social reflexivity requires each tradition, used simply to asserting its position, to justify it instead.

The best example of the effect of this condition is the rise of so-called fundamentalism. According to

Giddens (1994), this term is of recent origin and refers to assertion of traditional truths in a self-reflexive world. Before reflexive modernity, such assertions were either accepted by the faithful or ignored by the rest. Under conditions of globalization, bringing groups together that were geographically separated, the new reflexivity can be particularly subversive. Centuries of asserting truth may now be under threat, not just from the outside but also from the inside as new communications penetrate all. In these circumstances, some traditions have reacted by internal purification and external aggression. This reaction seems to be worldwide and is most familiar as 'religious fundamentalism': Christian, Hindu, Jewish or Muslim. But it occurs in other contexts, such as the American 'patriot' movement and the 'men's movement', which tries to restore the patriarchy of the family. Unable to accommodate to a new world of dialogue, targeting of ordinary people has become acceptable in a politics of hate such as the bombing of US federal buildings and the use of nerve gas in Japanese subways. This is the extreme reactive side of the new politics of identity that reflexive modernization is ushering in.

Case study

Fundamentalism, globalization and post-traditional society

We cannot understand contemporary political geography without having some grasp of what fundamentalisms are. Giddens (1994) provides a useful introduction that fits in well with our ideas on globalization in the modern world-system.

For Giddens (ibid.: 100), fundamentalism is simply 'embattled tradition'. Traditions are the enduring mores and beliefs through which we give meaning to our lives. As 'collective memories' they provide integrity and continuity to help cope with 'the buffeting of change'. Religions are the sacred core of traditional thinking. But in the modern world-system change is especially severe so that modernity is a great eroder of traditions. This has come to a head in the twenty-first century when the erosion has threatened to become fully global. Thus, according to Giddens, we live not in a 'postmodern society' but in a 'post-traditional society'.

Fundamentalism is what tradition becomes in post-traditional society. In the past, practices relating to traditions constructed spaces of places, 'holy places' – Rome, Jerusalem, Mecca – plus the land of the believers – Christendom, the 'land of Israel', Islam. There is mixture of peoples with different traditions, but by and large religions are able to keep to themselves and reproduce their traditions without too much threat. Globalization has changed all this. This is a world where spaces of flows are dominant. Multiple diasporas, new media and communications, cultures of consumption and other 'post-traditional' practices have enhanced the erosion of tradition to intolerable levels for believers. With the modern world 'in your face' as it were, and with nowhere to turn – you cannot get outside globalization – the only recourse is a violent reaction against the perpetrators. They show no respect for tradition, so the fundamentalist shows no respect for modernity: each violates the other.

Note that this is not a 'clash of civilizations' argument: Huntington's (1993) model is a divided world, a space of places. Giddens' globalization is all about the violation of those traditional places by new spaces of flows. Examples of violent fundamentalisms, although post 9/11 associated with Islam, can be found in all traditions and religions in their reactions to the 'wrongs' of modernity. Thus bombing of abortion clinics by Christian fundamentalists is an example, as is the non-religious attack on the Federal Building in Oklahoma City by 'patriots' wanting to return to a simpler traditional life. In gender relations the violent reactions of men against women is another example of violence in a post-traditional society. The demise of the 'traditional family' leaves men losing their 'due respect' expected from a partner. In this case the traditional hierarchical place that was the home is violated by feminist ideas of gender equality.

For more on this argument the interested reader is directed to Giddens (1994).

Social reflexivity and multiple identities

The key process in the rise of identity politics for Beck (1994) is individualization, in which the old certainties of 'industrial society' are replaced by a desire to find and invent new certainties. Specifically, the once standard mainstays of identity (class, family, gender) are replaced by a personal biography, which we choose ourselves. Hence identity no longer becomes something given to us by our assigned statuses (class, family, gender) but becomes a more reflexive matter of choice. This is reflexive modernization. The challenges and uncertainties imposed by the risk society lead to a constant evaluation of political and social choices. Choosing a series of multiple identities is one form of politics aimed at charting a course through the risk society.

In this way, we can see how Beck provides a plausible explanation for the current growth of interest in identity politics. His ideas of risk society and reflexive modernization reflect the concerns for globalization and hegemonic change within world-systems theory. But what of the politics of identity-in-place (Cope 1996)? Beck (1994) goes on to talk about how the 'political vacuity' of the institutions of industrial society is causing a growth of politics in non-institutional settings. A 'sub-politics' is emerging, denoted by the importance of citizens groups and grassroots organizations whose political activities are extra-parliamentary. Furthermore, these new political groups are not tied to classes or parties. However, the move from mainstream political institutions to 'extra-parliamentary' activities is a gradual one. Individuals still participate in the old institutions and forms of politics, but they also engage in new activities based upon new identities.

In other words, people leave their traditional political affiliations in a piecemeal fashion. Hence each individual will possess contradictory and multiple political beliefs and goals. Again, we can see through Beck's ideas why identity politics has become an imperative component of political geography. It is also a form of politics that has global concerns but is grounded in unique places, for sub-politics means shaping society *from below* (Beck 1994: 22) as it is driven by agents outside the political system through both individual and collective acts. It seems that now, more than ever, it is politics within places that construct the broader mosaics of state and global politics.

Summary

In this section we have introduced the terms 'modernity' and 'prime modernity' to situate the politics of identity within the structure and dynamics of the capitalist world-economy. We:

➤ made a connection between hegemony and modernity;

➤ introduced the concept of prime modernity;

➤ noted the relationship between the construction and meaning of particular places and prime modernity;

➤ introduced the concept of reflexive modernization to illustrate that political identity is fluid and often formed in resistance to power structures;

➤ introduced the concept of the post-traditional condition;

➤ noted that social reflexivity makes for dynamic and multiple political identities.

The next task is to make sense of the multiplicity of political identities by identifying the multiple politics of the institutions of the capitalist world-economy.

Identity politics and the institutions of the capitalist world-economy

Now that we have established a connection between global trends and the growth of locally based identity politics, we can use our materialist framework to distinguish some of the parameters of the new politics within places. Such a taxonomy is bound to be incomplete because, as Beck argues, politics is being continually defined by the agent. However, we are currently in a transition period in which traditional political attachments still hold some sway. More importantly, the new politics of identity both challenges and cross-cuts older institutions. By identifying the key institutions in the capitalist world-economy, we can uncover some of the initial battlegrounds of new locally based sub-politics.

As we described in Chapter 1, Wallerstein (1984b) has identified four key institutions in the capitalist world-economy: classes, states, 'peoples' and households. All of the institutions are ambiguous in that they can be both liberating and repressive. For example, the state can crack down on civil liberties but also provide resources for education and social mobility. Such ambiguity has always characterized these institutions, but we can expect it to be particularly enhanced under conditions of reflexive modernization. People can view the institutions as either enabling or constraining, depending upon the particular 'risk' that they are challenging. To make sense of the cacophony of political battle cries and the dynamic mosaic of identities, in the first subsection below we produce a taxonomy of many different political geographies derived from the four institutions (Taylor 1991a). In the second subsection, we use some of these new political geographies to illustrate new politics of identity.

Fourteen political geographies

In many contexts, both popular and academic, politics is conceived of as being limited to practices relating to the state: elections and wars, welfare reform and tax policies, these are the stuff of politics. Such a position was never really tenable, which is why we have adopted the wider view that politics is about power within whatever arena. Today, more than ever, we have to look beyond the state to fully understand the political geography of our times. States still provide much of the 'stuff of politics', but there is much more to politics than this single institution. Using Wallerstein's four key institutions, we can identify fourteen different politics.

First, there are four *intra-institutional politics*, or politics within the separate institutions:

1 Intra-state politics, or the competition for government. This is what political science focuses upon.

2 Intra-people conflict includes a whole array of disputes but often centres upon goals and objectives

(for example independence versus regional autonomy in national politics).

3 Intra-class politics also pertains to strategies and objectives (for example reform versus revolution in early socialist politics).

4 Intra-household politics is dominated by patriarchal and generational disputes brought to the fore through feminist politics.

Second, there are four *inter-institutional politics*, or politics between the institutions:

5 Inter-state politics is the preserve of statesmen and -women and diplomats as studied in international relations.

6 Inter-people politics is manifest in racialism, aggressive nationalism and ethnic cleansing.

7 Inter-class politics is the traditional dispute between capital and labour.

8 Inter-household politics involves local struggles for resources and access; neighbour disputes are an example.

So far, these eight politics remain firmly within the traditional parameters of the politics of 'industrial society'. The complexities of the politics of reflexive modernization are more fully captured by politics between the institutions. As people migrate from traditional politics to new strategies there will be tensions as they simultaneously reject some traditions and institutions while clinging on to others. There are six *politics between institutions*:

9 State–people politics has been dominated by the desire to construct nation-states (Chapter 5). This traditional politics has been accentuated rather than lessened in the wake of contemporary globalization. Contemporary conflicts often revolve around language rights, with minorities claiming equal use of their language in state institutions such as courts and schools.

10 State–class politics has been dominated by the incorporation of classes into state politics by political parties (Chapter 6). Globalization and reflexive modernization have raised new issues regarding the protection of workers as capital has become more mobile.

11 State–household politics has centred upon rights and welfare issues. Increasingly, this politics has become more moralistic and rhetorical as the traditional family is both championed and questioned.

12 People–class politics is concerned with the differential material impact of globalization upon ethnic groups and subsequent problems of political mobilization due to the fragmentation of objective classes into subjective group identities.

13 People–household politics has been concerned with the cultural issues defined by the ideology of the family. Globalization and reflexive modernization heighten these tensions by providing both opportunities and risks that traditional family structures cannot accommodate.

14 Class–household politics has been concerned with the feminist critique of the left's neglect of patriarchy. Globalization intensifies such conflicts through its demand for a feminized labour force and the destruction of traditional work practices, brilliantly satirized in the film *The Full Monty*.

Notice that we have by no means abandoned the state in this fresh review of politics: states appear in five of the fourteen politics. What we have done is open up politics, and with it political geography, to new perspectives. Each of these fourteen politics has a consequent political geography that investigates the spaces and places in which the politics occur at different geographical scales. But more than this, the spaces, places and scales are part of the politics we have previously identified throughout the book (Taylor 2000).

Politics between institutions

We have chosen to focus on politics between institutions because it is here that we expect to find new politics. In this section we describe six case studies which cover each of these politics. What becomes clear from the examples described below is the inadequacy of politics and identities centred upon just one kind of institution. Reflexiveness has resulted in a questioning of the traditional politics within each of the

institutions by reference to the politics within other institutions. The simultaneous use and challenge to the institutions puts an end to simple institutional politics and identity. Instead, we are in a complex and ever-dynamic period of politics defined by the 'internally fractured and externally multiple' (Bondi 1993: 97) nature of identity. In other words, the politics of reflexive modernization is a function of the interacting internal and external components of place (Massey 1994). As our examples show, the contemporary situation is one in which new politics progresses by negotiating and, at times, utilizing the institutions and practices of old politics. For each of our examples, we provide a summary of this juxtapositioning of old and new politics.

State–class politics

The space–place tension we defined at the beginning of this chapter is evident in the relationship between people and the power of governments to limit their mobility. Borders are a crucial geographical manifestation of state control over movement of goods. In addition, the operation and existence of borders are important factors in the way people think of states. Borders may be parts of national identity that define the extent and meaning of 'homeland' for some people (Kaplan 2003; Murphy 2005) and, simultaneously, the violent expressions of state power that restrict movement. Borders are, therefore, products and definers of political identity. Traditionally, such an identity has related to membership of a particular nation and a sense of citizenship tied to rights and duties within a particular state. Increasingly, borders define an identity based upon membership in transnational classes: either the mobility of the cosmopolitan business class or the carceral cosmopolitanism of refugees and people identified as 'security risks' (Calhoun 2003; Sparke 2006). Examination of the differential access to mobility facing regular business travellers on the one hand, and refugees and undocumented migrants on the other, is an illustration of how membership in classes results in differential relationships with the state. Moreover, this case study exemplifies the interaction between the processes of globalization and the War on Terrorism.

Sparke (2006) offers an analysis of North American border regimes from the development of the North American Free Trade Agreement (NAFTA) in the mid-1990s to the current War on Terrorism. The analysis contrasts the construction of fast-track border crossing facilities for 'frequent travellers' who are understood to be businesspeople with the procedures in place for those deemed security risks. Sparke's argument reflects the political geographic contradiction of the 'hegemonic dilemma' we discussed earlier (see page 269) in which states, especially the United States as hegemonic power, are under pressure to facilitate the cross-border flows that define globalization while maintaining a sense of 'national security' that is understood to include secure borders. Sparke (2006) contributes a clear understanding of the differential treatment given to different classes.

First, Sparke (2006) describes a series of expedited border-crossing programmes, such as NEXUS and CANPASS that are designed to facilitate passage across the US–Canadian border without the delays imposed by the production and examination of travel documents. Participants in these programmes are registered and personal details, including biometric information (fingerprints and iris images) are recorded in a database. Pressure was put on the US government to enact such programmes in the context of increased volumes of trade, and also the need to minimize delays. For example, the world's busiest commercial crossing, the Ambassador Bridge between Detroit and Windsor, connects factories operating in the just-in-time car manufacturing industry. Time is money!

More importantly, Sparke (2006) outlines how participants in frequent-crossing programmes are constructed, by the state and their own actions, as members in a particular transnational entrepreneurial class. In the light of the pressures on the US government both to maintain ease of movement across its borders for goods and certain people (the 'frequent travellers') and to enact border security in the wake of the terrorist attacks of 11 September 2001, the notion of a 'smart border' was disseminated. To become a member of the class of people who could pass swiftly across the border was to

become what Sparke (2006: 167) calls 'NEXUS subjects: citizens of nation-states, yes, but with an all important new kind of transnational para-citizenship in a fast lane designed for business class frequent travelers'. Becoming a member of this class did not just entail registering personal and biometric information on a database, it also meant acting as a particular person who would self-regulate their own behaviour in order to 'fit' the behavioural patterns deemed normal, or defining, for the class of frequent business travellers. It was, in the terminology of Michel Foucault, an act of governmentality: the self-monitoring of behaviour that reflects the needs of dominant power relations. In this case, these power relations relate to the needs of transnational capital and the US and Canadian states.

But what of the flip side? Sparke (2006) is careful to remind us that such fast-lanes are not just a matter of the annoyance that the bulk of us feel standing in queues at airports while others walk briskly to their aeroplane, latte in hand. The asymmetry is felt by the world's lowest economic strata and those labelled 'security risks': those deemed to be members in this class face 'volatile mixes of movement and immobility' (ibid.: 169). Members of these classes face the unpleasantness of carceral cosmopolitanism. Sparke identifies two forms in the United States. One is the process of 'expedited removal' by which immigration inspectors were given powers to evict asylum seekers without appropriate documents. The immigration inspectors were invested with sovereign power as the asylum seeker had no recourse to a judge. These powers were extended to all undocumented immigrants and the geographical scope of such power was extended beyond the immediate vicinity of airports and docks. In 2004 the Department of Homeland Security expanded the practice of expedited removal to allow the Border Patrol along parts of the US–Mexican border to evict undocumented aliens caught within 14 days of entry and within 100 miles of the border.

So far, this case study has been situated within the processes of globalization. But carceral cosmopolitanism has its ugliest incidences within the enactment of the War on Terrorism. The term 'extraordinary rendition' has entered our vocabulary; the process by which those labelled as 'terrorists' are forcibly transferred to countries in order to be tortured. The irony is that the victims of this form of cosmopolitanism often travel in the same vehicles as the elite transnational cosmopolitan business class. As Sparke (2006) describes, one victim of extraordinary rendition was able to trace and report his journey by identifying the luxury Gulfstream jet that transported him, while being chained to the luxury leather seats, from Washington, DC, to Maine, via Rome to Jordan. He was then driven across the border into Syria and incarcerated and tortured for ten months. Investigative reporters discovered that the very same jet was often used to transport the manager of the Boston Red Sox baseball team. The same vehicle, but used for a different purpose, for members of different classes, with different group and individual identities, and different relationships with the state.

State–household politics

The changes inherent in globalization have restructured the politics between households and the state. In the United States, the economic impacts of globalization include a twenty-year stagnation in median household incomes and the redistribution of earnings and wealth from low- and middle-income households to the most affluent (O'Loughlin 1997). Although such statistics focus upon class polarization, restructuring has social dimensions too, defined by race, gender and generation (Kodras 1997a). The conjunction of economic restructuring with gender, age and racial discrimination at a variety of geographical scales has produced a social and geographical mosaic of life-chances. The state has steadily reduced its commitment to welfare programmes aimed at ameliorating these social and geographical differences. Increasingly, place-specific institutional contexts – organizational capacity and experience, financial and social capital, and political power – define the capacity of government and non-governmental organizations to deliver social programmes (Kodras 1997a: 88).

The geography of poverty in the United States illustrates the social and geographical complexity of the problem. The national poverty rate rose for the second year running in 2002. The increase in poverty was concentrated among African-Americans, sub-

urban residents and Midwesterners (Weisman 2003). Poverty rates remained steady for most of the population, but rose to 24 per cent for African-Americans. In addition, some other cultural groups were particularly disadvantaged. For example, Puerto Ricans suffered from the economic restructuring associated with Kondratieff IVB because of their concentration in north-eastern central cities experiencing dramatic economic changes, their over-representation in job sectors hit hardest by job losses, and discrimination in the job market (Tienda 1989). Marc H. Morial, president of the National Urban League, identified the decline of US manufacturing industry as a key process:

'The sector has lost 2.7 million jobs since its employment peaked at the end of 2000,' he said, 'and many African Americans in the Midwest, in the industrial heartland, were employed in manufacturing, in autos, steel, glass, and rubber.'
(Quoted in Weisman 2003)

Another line of fragmentation is gender and family composition. The conjunction of global economic restructuring, the subordinate position of women in the workforce and political strategies has pushed female-headed households into poverty (Kodras 1997a, b). As welfare assistance to children in families in which the father has left the home was being reduced, women were forced into low-paying jobs. Such disadvantages were magnified for minority households. For example, in 2004, 18 per cent of children in the United States lived in poverty. The racial differential is staggering: ranging from a rate of 11 per cent for non-Hispanic whites, 29 per cent for Hispanics and 33 per cent for African-Americans (National Center for Children in Poverty 2006).

After the national trauma of the Great Depression in the 1930s, the US federal government began welfare programmes aimed at the most marginalized. The programmatic focus upon the household is illustrated by the name of the most prominent welfare package, Aid to Families with Dependent Children (AFDC), which began as a widow's pension for fatherless families (Cope 1997: 193). However, intensified globalization has changed the post-war contract between the state and the household. Capital has been freed from its national constraints and has become globally mobile. States have restructured in light of this global economic context (see Chapter 4) and so have the relationships between the state and households. The new relationship is between households and localities within the global economy. As the responsibilities of the federal state wane, the well-being of low-income workers and the unemployed becomes a function of the ability of localities to offer social programmes.

The Contract with America produced by the 104th Congress (1994) in the United States optimizes such changes. Its goal was to facilitate the mobility of capital by reducing the strength and responsibilities of the state in a three-pronged strategy of devolution, dismantlement and privatization (Staeheli et al. 1997). A key arena in this strategy has been welfare provisions. The attack on 'big government' involvement in welfare by the 104th Congress was the loudest note in a crescendo of discourse that had been building up in the United States since 1988 (Cope 1997: 194). The growing trend has been towards a 'workfare' system rather than welfare, initiated by the 1988 Family Support Act (FSA). Two principles were encapsulated in the FSA. First, it placed emphasis upon moving welfare recipients into training and jobs through the establishment of the Job Opportunities and Basic Skills (JOBS) training programme. Second, greater freedom was given to individual states to experiment with programmes designed to move welfare recipients into employment. More recently, similar policies were enacted by President Clinton and the Republican Congress (ibid.: 195).

'Workfare' policies suspend welfare payments after a specified period of unemployment. Welfare in the United States has, therefore, ceased to be an entitlement. In a period of increasing global competition, reduced welfare payments have been one means by which the burden of globalization has been placed upon households, notably the poor ones. By reducing federal responsibilities, greater burden is put upon individual states and localities to provide for those marginalized within the capitalist world-economy. However, there is a simultaneous burden upon individual states and localities to become attractive sites for investment. Thus, the new imperative is to reduce

taxes, especially on businesses, to ensure a flow of investment. The result is 'competitive downgrading' (Peck 1996: 252) or a 'race to the bottom' as individual US states cut social programmes to reduce taxes. To make the situation worse, the rhetoric of the fiscal conservatives emphasizes the ability and desirability of charities to provide resources for the poor after government benefits are withdrawn. However, it is far from certain that private charities have the capability for such a large-scale and constant commitment (Wolpert 1997).

The increasing relevance of the locality in welfare provision is also reflected in the political activity of individuals within their household contexts. In a study of the small town of Pueblo, Colorado, Staeheli (1994) found that changing economic conditions intertwined with household context to define the nature of political activism. For the activists interviewed, political and economic restructuring had created an atmosphere of threat and uncertainty about their own futures, as well as those of their family and community. Political restructuring had reduced the perceived efficacy of political activism directed towards the state. Instead, many of the activists in Pueblo were targeting their employers for the provision of initiatives such as scholarships and after-school programmes. The decision to undertake political activism was made within the household context. Households provide a number of resources to facilitate activism, such as financial support from the income of other household members, emotional support, access to social networks and release time from chores (ibid.: 852). Also, decisions to engage in political activism are often made on the basis of threats to the household, especially regarding educational and work opportunities (ibid.: 865).

The intersection of globalization, the retreat of the state from welfare programmes, and the specificity of place have produced a restructuring of the relationships between households and the state. A uniform and stable commitment to entitlements aimed at supporting marginalized households cannot be assumed. Instead, the burden of provision falls upon the locality. The rhetoric of political conservatism calls upon charities to step in where government once stood. However, the limitations to this possibility are clear

(Wolpert 1997). The political strategies of households must become more flexible as they focus upon the private sector and networks of activists to achieve their goals. The geographical counterpart to Beck's reflexive modernization is the flexible use of scale and scope (Kodras 1997a: 92) through which connections are made to other political activists in other places and at other geographical scales, depending upon the goal in question. Hence, similar to Herod's (1997) analysis of trade unions, the construction of scales of activity is an essential part of contemporary political actions.

In summary, although the patriarchal nature of the household remains in both practice and rhetoric, this old politics is being challenged by new feminist politics. Political parties in both the United States and Britain, for example, are reinforcing their rhetoric about the traditional role of the family. Meanwhile, more and more children are being born out of wedlock, and the labour market challenges the traditional role of the 'stay-at-home' wife and mother. Feminists, too, are challenging the traditionally restrictive roles of women, and political activists are creating networks of households and contacts with the private sector to further their goals. The flexible use of scale and scope is the geographical strategy of the new politics, but the terrain that this strategy must negotiate is partially composed of the institutions and assumptions of old politics.

People–class politics

Politics between people and class is a function of competing identities brought to the forefront by socio-economic change and competition. Nagar's (1997) study of the Hindu community in Dar es Salaam, Tanzania, illustrates how caste-based differences fractured ethnic identity and politics. The caste system is much more rigid and formalized than social classes. Despite recent legal proscriptions aimed at relaxing the constraints of the caste system, movement between the classes is practically impossible, and one's caste determines life opportunities, including marriage partner, residence and occupation. Although caste and social class are not exactly the same, Nagar's study is an effective illustration of how socio-economic differences can fragment ethnic identity. The majority of immigrants from the

Indian sub-continent arrived in Tanganyika (in 1964, Tanganyika and Zanzibar merged to become Tanzania) during the time of British colonial rule (1918–61). Although south Asians never comprised more than 1 per cent of the total population, they had greater prominence because of their role in the middle rung of the colonial race–class hierarchy (ibid.: 711). The south Asians were employed as shopkeepers, teachers and civil servants. In 1961, 31,000 south Asians lived in Tanganyika, while in 1993 the south Asian population of Dar es Salaam was estimated to be about 35,000, of whom 10,000 were Hindus.

The Hindu community in Dar es Salaam underwent profound change, beginning in 1960. As the United States achieved hegemonic maturity (see Chapter 2), Britain was forced to grant independence to its colonies. As Tanzania became independent, the country initiated policies against the south Asian community. Many upper- and middle-class Hindus emigrated to the Britain because of Africanization and nationalization policies, which made Hindu businesses less viable. The Hindu emigrants included upper-caste Lohana and Jain business families and brahman civil servants as well as lower-caste labourers and workers, including those from the vanzaa (tailor), mochi (leather worker), dhobi (washer) and prajapati (potter) castes (Nagar 1996). While the upper- and middle-caste populations declined, emigration by the lower castes was balanced by an influx of Hindu families from Zanzibar after the revolution of 1964, in which south Asians were attacked and victimized (Nagar 1997: 711). Particularly significant was an influx of a variety of manual labourers, who came to constitute the largest castes in Dar es Salaam.

A politics focusing purely upon ethnicity would suggest a unified Hindu identity because of the threats faced by the community. Instead, by considering caste identity, the fragmented nature of politics and identity becomes clear. Social labels became geographical: Hindus from India referred to the Jangbaria (Zanzibaris) and the Bharwala (people from the interior). The term Jangbari Gola became a common derogatory term implying that Zanzibari south Asians are culturally closer to Africans than to mainland south Asians (Nagar 1997: 713).

Differences between the newly arriving castes and the established upper and middle castes were heightened as the Hindu manual labourers settled in Kariakoo, denoted as a 'dangerous African area' by upper-class Asians (ibid.: 713).

Before 1961, there was a variety of social institutions and buildings representing the different castes, but after independence the Tanzanian government put pressure upon the Hindu community to create an umbrella organization to represent all its members. The Hindu Mandel, established in 1919, was the oldest Hindu organization in Dar es Salaam. It became the Hindus' umbrella organization, although it was dominated by prosperous and upper-caste Hindu men (ibid.: 717). By overseeing government-mandated tasks, such as the provision of education and the performance of marriage rites, the Hindu Mandel assumed a homogenizing role in its aim to fuse all Hindu castes 'into one entity' (ibid.: 717).

Despite the attempts to create a unified Hindu identity, the Hindu Mandel was perceived by the lower castes and Hindu women as a way of preserving male and upper-caste power and influence. Heightened caste awareness because of the emigration of upper- and middle-caste members and the immigration of lower-caste members inflamed the perception of Hindu Mandel paternalism. Instead of nurturing a unified Hindu community and identity, the various life experiences of the castes fostered an increase in religious and social events for the separate castes within their own halls. Intra-caste socializing, which had been restricted to marriages, births and deaths, became regular bi-weekly events (ibid.: 721). Most lower-caste members saw the Hindu Mandel as serving elite interests to the detriment of the broader Hindu community.

In summary, the caste politics within the Hindu community of Dar es Salaam illustrates how identities are internally fragmented but externally constituted (Bondi 1993). The identity of the lower-caste women Hindus is a function of their class, household and ethnic status. In addition, the juxtaposition of these identities is put into motion by migrations initiated by global changes and linkages. Finally, the form that the intra-caste politics takes is a function of the particularity of place, in this case Dar es Salaam. The

neighbourhoods of the city help to derive differences within the Hindu community, and the interventions of the Tanzanian state led to a specific institutional context. Again we see that the politics of reflexive modernization (Beck 1994) are particular social relations defined by place-specific institutions (here, the Hindu Mandel) and extra-local connections (migratory flows in and out of Dar es Salaam). The old politics of patriarchy, class and nationalism persist in the dominant role of wealthier Hindu men in forging a unified identity. But the fragmentation of Hindu identity and resistance to powerful male figureheads are seen in the informal networks created by lower-caste women and their own use of public space within the Hindu Mandel (Nagar 1997).

People–household politics

Geographers have been very interested in the politics between household and nation because it reflects a tension between private and public space. Feminist geographers have shown how a discourse that attaches women to the private and domestic sphere prevents women's participation in politics and, therefore, maintains patriarchy. Images and rhetoric used to construct the national 'imagined community' are usually the product of masculinized memories and visions (Enloe 1983). Nationalist rhetoric and images promote women as the mothers and nurturers of the nation, a role best accomplished within the household. Alternatively, men are portrayed as warriors and heroes or, in other words, active participants in the public sphere of politics (Dowler 1998). Although nationalism emphasizes the unity of the people's interests, gendered ideology reinforces gender differences. The result is the continued understanding of civil society as two distinct spheres, with women remaining in the less politically relevant private domain and men maintaining their control of the public sphere.

At times of war, the principles of national identity, including the different gender identities, are heightened (Cock 1993). Looking at the conflict in Northern Ireland, the attitudes of male members of the IRA (Irish Republican Army) illustrate the view that women's primary role is to produce children for a future generation of pro-nationalist supporters

(Dowler 1998). Although women are rhetorically integrated into the unified republican-imagined community through their portrayal as heroes of the struggle with the British, such activity is pictured within the household. For example, one Catholic man, Sean, states:

> Me ma Annie was the bravest person I knew. Braver than anyone in the Ra [IRA]. When the soldiers would come to round up the men to be interned, Annie would get into the back of the Saracens [armoured cars] and say 'you're not taking these boys away today' and she'd push us out. The soldiers didn't know what to do. But they never stopped her, they would just leave and round up some other poor bastards.
>
> (Quoted in Dowler 1998: 166)

In a similar vein, another of Dowler's interviewees, Thomas, reflected:

> Aye, I do, the women were the brave ones. When the soldiers would come up the road all the men would be standing in the background with our guns at our sides ready to use, but the women would go right up to the soldiers and yell at them and spit in their faces. They had the balls. The men had the guns, but they had the guts. No one talks about how the women carried messages and guns in the baby buggies. They would walk right by the soldiers while we were hiding in the shadows. Aye, they had the balls.
>
> (Quoted in Dowler 1998: 167)

These two quotes illustrate that women were included in the ideology of a nationalist struggle against an occupying force. However, the language in the second quote equates bravery with masculinity. Also, the women's roles were remembered primarily in relation to their domestic chores. Sean's quote focuses upon the nurturing role of the mother and her defence of the home. Thomas's quote begs the question: why weren't the men pushing the baby buggies?

The women of the republican movement interviewed in Dowler's study illustrate the attempts to renegotiate these traditional views of the relationship between the household and the nation. One of the key media of nationalist memory is Irish resistance songs. Two of the most popular are 'Men in the IRA'

and 'The men behind the wire'. As one republican woman, Peggy, angrily said:

> What of the 'Men in the IRA', they love to sing that one around here. I was in the IRA but that song is not written about me. I also hate the song 'The Men Behind the Wire'. I was in prison for four years, there were women behind that wire too!
>
> (Quoted in Dowler 1998: 170)

As these songs are being sung in a republican prisoners' club, Peggy and her female friends sing loudly to change the words of the song from 'men' to 'women'. Such renegotiation of dominant discourses reflects the efforts to change the relegation of women to the private sphere. Other renegotiations are under way too. The republican women understand the similar trials and tribulations experienced by Protestant women. The possibility of a cross-national identity based upon the politics of the households is witnessed in Maureen's goal of a written history of the conflict:

> I want the women to write their own stories. . . . I would want my mother and her friends to tell their stories about trying to keep everything going while their husbands and sons were in prison. . . . I think we should have some of the women from the Shankill [a Protestant neighbourhood] tell their stories too. Look I'm a republican, I went to prison for being a republican but some things just run deeper than that. If some Protestant women wanted to tell their stories I would say do it. The men wouldn't understand that, they would say we weren't being good republicans. Oh for shite's sake they would say we weren't being good republicans just by writing the book.
>
> (Quoted in Dowler 1998: 170–1)

Maureen's quote returns us to Beck's (1994) reflexive modernization and the ensuing politics between the institutions of the capitalist world-economy. The nation has been a dominant institution in the world-economy yet 'some things just run deeper than that'. The idea that other politics and identities can challenge the ideologies and practices of nationalism suggests that significant changes are afoot. The people of Northern Ireland see challenges other than nationalist struggle.

In summary, there is, therefore, recognition that new politics must be created within the context of old politics. The women of Northern Ireland also wish for patriarchy to be put on the political agenda. A Women's Coalition has been formed in Northern Ireland, and some of its members were elected to the new Northern Ireland Assembly in 1998, a formal participation of women in the public sphere. All members of the Assembly had to declare themselves as 'Nationalist', 'Unionist' or 'Other'. The members of the Women's Coalition chose the label 'Other', eschewing national identities for other politics by prioritizing their gender. In this case, the combination of two old political struggles (feminism and nationalism) has created a new politics of collaboration across traditional divisions.

Class–household politics

In the four previous examples, we have emphasized that new opportunities coexisted with traditional practices and constraints (Beck 1994). Such interinstitutional tensions are clear in the case of female labour participation. Entry into paid labour may reflect desired changes in the status of women, but it is also a feature of a globalized economy in which women are sought as cheaper workers. What is clear throughout the world-system is the gender inequalities in the labour market, as women are consistently paid less and face barriers to promotion. Efforts, often union-led as in the case of Britain (Lawrence 1994), to equalize pay and opportunity illustrate how once class-based institutions have adapted to include feminist strategies. However, the twin ideologies of patriarchy and capitalism still predominate, and gender inequalities persist. Women's subordinate status in the household mirrors the difficulties that women face in the labour market. Similar to the discussion of state–household politics, the private domain of the household is a traditional structure that can serve to restrict women's participation in the public domain.

Social scientists have taken two approaches to explaining women's inequality within the labour market. First is the argument that inequalities stem from relationships within the household, namely child care responsibilities. These factors, and the heavier domestic workloads placed upon women, all reflect women's subordination in the household. However, theorists have yet to agree on how to apportion blame between patriarchy and capitalism for gender discrimination (Hanson and Pratt 1995: 5). The second take on gender discrimination has been to focus upon the segmented nature of labour markets. Labour markets are split between primary and secondary segments. In the primary segment, wages and benefits are determined through institutionalized negotiations, whereas in the secondary segment market forces play a greater role. Not surprisingly, workers in the primary segment usually obtain higher wages, better benefits and greater job security. Patriarchal practices based upon stereotyping and gender discrimination have been blamed for the relative exclusion of women from the primary segment (ibid.: 6).

Clearly, both of these explanations are partial. Geographers have sought to weld them together by emphasizing the role of place in mediating social relationships. Localized social networks ameliorate the constraints of women's status in the household. Although the nature of everyday contacts and experiences is restricted by commitments to the home, they also provide a means of finding jobs. A longitudinal study of the town of Worcester, Massachusetts (ibid.), shows the linkages between social contacts established through the household and women's employment. Most people found their jobs through personal contacts or by seeing 'Help Wanted' signs during their everyday activities: 77 per cent of men and women in professional and managerial jobs and 70 per cent of skilled manual workers (ibid.: 194). Such social contacts determine work choices in a number of ways. First, socialization within households and related social networks defines the image of what is a possible or desirable work path. What is defined as a 'good job' is voiced through conversations around the dinner table. In different households, a professional career for a daughter may either be assumed or be considered beyond the realms of possibility. Particular cultures, such as traditional Hindus and Muslims, may not even consider non-domestic paid work for women. Second, information about jobs is shared through social networks. An advertisement on the bus in State College, Pennsylvania, touts a job-seeking network administered by the Pennsylvania State University Alumni Association by ridiculing the traditional word of mouth. Of course, it is the professional and managerial classes that will have greater access to commercial networking. In many cases, women's ties to the household restrict their mobility and, therefore, make them more dependent upon localized contacts structured by their everyday activities (ibid.).

Hanson and Pratt's study found that women and men find employment through different networks. The contacts through which women find work tended to be more family- and community-related than men's. In addition, the channelling of job information is gendered, with men finding their work through other men, and women through women. Clearly, these findings have implications for the type and location of jobs found by women. The use of gendered information flows to find jobs perpetuates gender divisions within the job market. In addition, the use of family and community contacts by women increases the likelihood that they will find work close to home, thus reducing their mobility. If a woman finds news of a job from another woman who is herself working close to home then the chances are that the new job will be local too.

Hanson and Pratt's study shows how place-based institutions and networks of contacts structure women's participation in the workforce. Attempts to gain advancement in the public domain are structured by subordination in the household. Patriarchal assumptions regarding women's roles at home may restrict their job search capabilities to gendered and localized personal contacts. Such a limited, but in its own way enabling and effective, strategy perpetuates discrimination and stereotyping in the workplace as 'men's jobs' remain a male preserve. However, particular households can also enable women's participation in the public domain. Households may contain professional women who can act as role models. The ability for households to negotiate domestic responsibilities and income pooling can provide women with the time and financial security to achieve their goals. In other words, households can be enabling instead of constraining.

In summary, the example of class–household politics illustrates many of the geographical components of reflexive modernization. The changes in female employment opportunities stem from globalization and the changing work practices that it demands. Women have greater opportunity to participate in the workforce, but the pressures of globalization demand a cheap and flexible workforce. Therefore, female workforce participation is a double-edged sword. More significantly for our argument, new opportunities for women in the public domain may be restricted by traditional practices in the household. The tools that women must use to counter these constraints are the structured contacts of their everyday experiences. How women take advantage of the dynamics of the capitalist world-economy is a function of their immediate geography.

State–people politics

We have left this example to last because it covers political relations we have previously devoted a whole chapter to (Chapter 5). States, nations and their association with territory lie at the heart of traditional political geography. Their importance most certainly has not diminished under reflexive modernization; rather, they are contested in new and more complex ways. It is, therefore, in this inter-institutional politics more than any other that we can see how political geography is changing right to its core as both practice and discourse.

Nationalism has been the dominant form of group identity since the end of the eighteenth century. As discussed in Chapter 5, national identity has been constructed over time by recourse to ethnic ties framed by a dominant state discourse. The old politics of nationalism has become a means of legitimating the power of the state by reference to a territorially defined (home)land. Given the importance of the nation for the existence of the state, it is not surprising that state institutions, such as schools and the armed forces, have played an active role in defining the ingredients and referents of national identity. Moreover, states have emphasized particular territorial claims and historical events in order to legitimize inter-state disputes and rivalries (Dijkink 1996). In other words, nationalism as old politics sees the nation as a strategy to obtain, use and legitimate state power (Breuilly 1993). Above all, it simplified politics to a basic dual, us (the nation-state) and them (all foreigners). However, in a context of reflexive modernization (Beck 1994), we expect to see challenges to the authority of singular and imposed national identities issued by the state. Indeed, the geographical counterpart to the myriad of issues being processed by individuals is a complex imagined geography that sees many boundaries and spatial interconnections that do not reflect state borders. The result is a new politics of nationalism that originates from people's localized social settings and challenges the spatial framework of inter-state disputes.

Globalization is the context for the more complex politics of nationalism. Contemporary levels of flows of capital, goods and people across state borders under the auspices of globalization can create new transactional identities that challenge the old dominance of the state. The political projects of the European Union and a post-nationalist Ireland were discussed in Chapter 5 as new identities initiated by 'cross-border' interaction. The following example of contemporary Ecuador specifically focuses upon the tensions between state-defined nationalist discourse and alternative geographies of identity formulated by citizens who have a different geographical imagination from that promoted by the state.

In January 1995 there was a series of incidents along the disputed border between Ecuador and Peru in which patrols exchanged fire and, allegedly, helicopters attacked border posts (Radcliffe 1998). The precise path of the border had been in dispute for some time. After a conflict in 1941, the United States, Brazil, Chile and Argentina had summoned Ecuador and Peru to Rio de Janeiro to sign an agreement over the location of the border. However, the border was based upon contemporary knowledge of watersheds in a remote area, and when surveys provided new information in 1960, Ecuador declared the 1942 Rio Protocol invalid. The Ecuadorian government had been active in constructing a national identity in which the Peruvian border featured prominently. Via state school texts, geographical institutions, military maps, and literary and visual images, the Ecuadorian government constructed a national identity that contained anti-Peruvian sentiments while emphasizing patriotism and 'sacrifice for the territorial integrity, defense of honour, decorum and national glories' (García Gónzalez 1992: 212–13, quoted in Radcliffe

1998: 281). The Ecuador/Peru border lies in the area of the Amazon basin, known in Ecuador as the Oriente. The Ecuadorian government imbued the Oriente with a prominent role in national identity by promoting images of the region's resource and development potential within a discourse of national progress (Radcliffe 1998: 276). The importance of the border and the Oriente in which it lies has been combined in the government's consistent promotion of a national Amazonian territory '"truncated" by an "invalid" border' (ibid.: 278).

Using the state institutions at its disposal, the government of Ecuador constructed a national identity which promoted the maintenance of state borders and the image of Peru as an evil rival, or in some texts 'the Cain of the South' (ibid.: 281). However, alternative images of national identity exist in Peru. The Shuar and Achuar indigenous groups resident in the Oriente region experience a complex relationship with Ecuadorian nationalism. During the 1995 conflict, the recruitment of Shuar–Achuar men into the army increased the ties with both state and nation. The state conspicuously identified indigenous soldiers as national war 'heroes'. However, simultaneously, the Federation of Shuar–Achuar Centres (FCSA) had been calling on the government for decentralized decision making, collective land rights and recognition of cultural differences (ibid.: 286). These tensions between indigenous identity and Ecuadorian national identity are intimately tied to the designation of the state boundary. While accepting that the boundary defines Ecuadorian citizenship, for the indigenous people it also divides them between Ecuadorians and Peruvians. Catholic missionaries attempted to integrate the Shuar–Achuar into the Ecuadorian nation. However, the FCSA holds a jaundiced view of these efforts: '[Missionaries] taught us to respect authorities which were not Shuar, that we were part of a different nation, so we could not visit our brothers who remained on Peruvian territory' (CONAIE 1989: 89, quoted in Radcliffe 1998: 287). The war brought these differences to a head:

Such was the history of the wars between Ecuador and Peru, we were forced to kill ourselves and they declared victories and each time there are conflicts, the first victims are us, the Shuar–Achuar, and our sons [in their capacity as soldiers] are obliged to build trenches against their other brothers Shuar, Achuar, Awajun, Wampis on the Peruvian side. . . . They call our Federation leaders subversives, enemies of the nation [patria], who threaten the integrity of the state by proposing a parallel state.

(FCSA 1992, quoted in Radcliffe 1998: 287)

There was also a second challenge to the state during the 1995 conflict. Sixteen women's rights groups from both Ecuador and Peru issued a joint statement calling for a ceasefire (Radcliffe 1998: 228). Rejecting the 'trap of war', they promoted a social justice that transcended the space of 'republican brotherhood'. They referred to women 'being guided by the love for this tormented Latin America'. This cultural–regional–continental identity directly challenged the state's basic function as 'war machine'.

In summary, these two very different challenges are examples of new politics that do not see state imperatives as the most important politics. Here we have a most traditional form of old politics, a border conflict, generating new responses based on identities that challenge nation as defined by state. Places are invoked at different scales, local indigenous cross-border 'homeland' and Latin America as a place of identity, but with the same effect of subverting the government's unitary view of national identity, which was challenged by the indigenous peoples of the Oriente, who see the border as unjust rather than invalid. Different social settings produce different patterns of interactions and transactions, which, in turn, lead to different views of group and national identity. The old politics of state-defined national identity has emphasized identity with the nation-state in the context of inter-state geopolitical rivalries. The new politics of nationalism is more reflexive as transactions across state borders come to define group identity. The coexistence of the old and new politics of nationalism means that new boundaries and identities will be shaped within the context set by old ones. Regional identities within, across and above state borders will be negotiated by challenging and, at times, utilizing the state and the nation-state.

Consistent across these six examples of politics between the institutions is the way that the institutions and practices of old politics provide both opportunities and constraints for new politics. While new politics are forging a new agenda, the arena is still partially defined by old politics. Although new politics may use a flexible strategy of scope and scale (Kodras 1997a: 92), dominant institutions and identities still carry over from the old politics. The difference is that instead of politics being a matter of who controls the institutions (therefore perpetuating their existence), the new politics is renegotiating the power and role of these institutions by fusing them in new political struggles. This new political strategy has the potential for more fundamental change as the institutions themselves are being challenged. As such changes progress, new political geographies will be created as more efficacious scales, institutions and identities are created.

Summary

In this section we have identified the multiplicity of political identities and agendas in the capitalist world-economy by:

➤ relating political identity to the institutions of the capitalist world-economy;

➤ illustrating the tensions and interconnections between the politics of these institutions;

➤ conceptualizing identity politics in the capitalist world-economy through the identification of fourteen political geographies.

Beyond globalization and 'empire'

As we come to the end of the book, it is time to reflect upon the political implications of our materialist perspective. Our goal has been to provide a perspective that helps readers to understand the dramatic and often bewildering changes at different geographical scales up to and including the global while also understanding people's political responses by reference to their place-specific setting up to and including the nation-state. Our subject matter has ranged from the geopolitical models that organized politico-military decisions in the Cold War, through the renegotiation of state power in the face of globalization, to women's local strategies to change their daily political and economic settings. Both geographical scale and the dynamism of place and space are implicated in all of the politics we have addressed.

This final section of our text takes on a dual role: we provide a conclusion both for this chapter and for the book as a whole. We write both conclusions under the heading 'beyond globalization' because we feel that globalization is an ideology with a tendency to be debilitating for critical thought and politics. Whatever else we may have achieved in this book, we hope readers will have come to appreciate that for all the globalizing going on, the basic structures of the capitalist world-economy remain in place. Hence there is an uneven globalization, with some places highly favoured and others virtually abandoned. The geography of these patterns is often more complex than in the past, but economic disparities are probably greater today than at any other time in the history of the modern world-system. To combat 'blanket' interpretations of globalization, our conclusions will, first, focus upon place–space tension to show how political conflicts over place and space and the construction of geographical scale are integral components of future struggles regarding the trajectory of the capitalist world-economy; and, second, examine the 'ultimate' place–space tension at the global scale, where the sustainability of the Earth's ecosystems puts all other politics in the shade.

Place–space tensions

The definition of geographical scales and both places and spaces is always a political act. This is the basic premise of our political geography. For example, the state originated as a political space with the related connotations of abstract power. States administered spaces through medieval legacies of power and tribute, with additional military power. For the majority of its inhabitants, the state deserved no loyalty. Instead, the state was a remote institution demanding

taxes. Over the past two hundred years, the state has been transformed into a place rather than a space. The construction of the nation-state has merged the space of sovereign territory with the sacred place of the homeland. This was a political struggle in that as state space became nation-state place, subjects became citizens. Citizens could put demands towards the state, resulting in the welfare state. To illustrate the 'homely' familiarity of this new place, the welfare state was to care for its citizens 'from the cradle to the grave'. But in the process, it produced large anonymous state bureaucracies that alienated the very people they were supposed to help. For many, the state became part of the problem rather than the solution.

But our political geography is not only about the state. A similar political story can be told about the household. In the creation of particular modernities within the modern world-system by the hegemons, there were redefinitions of the household. The Dutch invented the modern home by separating the upper floors of their town houses for the family (Rybczynski 1986). Women were isolated from the public and economic spaces of the lower floors while being portrayed by (male) Dutch artists as being content in their new domestic places. During the period of British hegemony, lower-middle- and working-class women created the Victorian home, which acted simultaneously as a haven for male workers but an oppressive workplace for women (Mackenzie and Rose 1983). More recently, the suburban household was an essential component of the United States' redefinition of modernity. Suburbanization increased the isolation of women by distancing them physically as well as socially from public space.

The image of the American suburban household as locus of consumption and bliss was broadcast across the globe through programmes such as *Bewitched*, *Leave it to Beaver* and *The Brady Bunch*. As an interesting aside, contemporary views of the same period (for example *The Wonder Years*) reflect angst rather than security. The current television imagery of the American family suggests cynicism and failure rather than optimism and inevitable success. The world outlook of *The Simpsons* and *Desperate Housewives* is very different from that of the *Brady Bunch*. Although the home-place component of modernity may be challenged during periods of hegemonic decline and uncertainty, a consistent feature of the creation of the home has been the different gender-specific perceptions of the household. For men, the household space has been a haven from their engagements with public space; for women, it has been an oppressive space designed to reduce access to the male-dominated public space. Renegotiations of traditional practices and identities have been restructuring the household and role norms within it.

One feature that these two examples encompass is a basic place–space tension. As state and household, they can both be defined as delimited spaces with controlled access. These are their constraining sides, sometimes expressed through prison imagery: the bureaucratic state as 'iron cage' and housewives as 'captive'. Simultaneously, but not necessarily for the same people, as nation and home they can be defined as secure places, havens from a turbulent world. These are their enabling sides. Hence there is a 'haven–cage' tension in these institutions viewed as places and spaces. It is such tensions that define the political geographies within the struggle over the institutions of the capitalist world-economy and the social relations that they mediate. The contested nature of institutions as places and spaces translates into their capabilities to enable one politics by constraining another.

Geographical scales are equally contested, as we emphasized in Chapter 1. Herod's (1997) analysis of the longshoremen's union, from Chapter 7, was premised upon the construction of new scales of collective bargaining, which enabled the dockers to secure their working conditions despite pressures from globalization. Labour politics has always been interested in geographical scale in recognition of the limitations of restricting strikes and negotiations to single factories.

Since its inception, the workers' movement has realized that the global scale is the most efficacious arena for its struggle: *The Communist Manifesto* of 1848 ends with the ringing declaration:

The proletarians have nothing to lose but their chains. They have a world to win. *Working men of all countries unite*!

(Marx and Engels n.d.: 103; italics added)

Of course, this prescription has always been easier to declare than to practise, not least when consumer modernity ensured that workers, male and female, did indeed have much more to lose than their 'chains'. The key point is that since the inception of the capitalist world-economy, capital has always been able to move across the world – globalization is an enhancement of a long-standing process. Quite simply, the organization of labour at the local scale can be sidestepped by investment in other places within the space of the capitalist world-economy. On Marx's prompting, labour's response was to construct a scale of cooperation and activity through the Workers' Internationals. In 1862, French and English trade unionists organized the International Working Men's Association. One of the International's goals was the provision of cross-national support as a strategy against foreign 'black-leg', or 'scab', labour (Taylor 1987: 293). In other words, the International was a scale of political activity designed for efficacious labour politics within the space of the capitalist world-economy. However, the history of the International was brief, and its demise can also be explained by a partial repudiation of the initial scale politics. By 1875, the First International was effectively finished, to be finally replaced in 1889 by the Second International. This, however, was a very different type of organization, an umbrella grouping for nationally organized workers' parties. The initial cross-border (or trans-state) strategy was diverted into an inter-state strategy of national parties turning oppressive state spaces into better places for the working class. Although both trans-state and inter-state dimensions of the organization were destroyed by the First World War, when workers fought against one another on behalf of their nations, domestic success finally arrived with the establishment of welfare states in the core of the world-economy by the mid-twentieth century. It is just these welfare states that have been prime targets of proponents of contemporary globalization. Thus 'international' organization is on the labour agenda again. The construction of new capitalist spaces such as the North American Free Trade Agreement has helped to initiate cross-border worker cooperation in a process dubbed '(re)politicizing the global economy' (Rupert 1995). As in 1862, today's international labour movements are an attempt to construct a global place for democratic and equitable politics out of a global capitalist space.

As in the rest of the book, the discussion above treats issues of geographical scale separately from questions of place and space. This is wholly a pedagogical decision, which we can no longer sustain. Political constructions of different scales and political constructions of places and spaces are in no way autonomous. The political stories we have just briefly told are in no sense separate from one another. In the world-systems approach we have adopted, they are both part of the larger story of hegemonies and modernities (Taylor 1996, 1998). However, they do not have to be linked only in this theoretical manner; there are clear connections of a practical nature. For example, the rise of suburbia in the United States was not unrelated to state policies and initiatives. The best example is the Hoover Report of 1931, where government, bankers, manufacturers and builders agreed that suburbanization through single-family dwelling units (home-households) would be a long-term solution to the Depression because it maximized consumption (Hayden 1981: 23). And so it worked out in the 'post-war boom' – the 1950s were the decade of greatest suburbanization in the history of the United States. But as a solution to 'under-consumption', it created other more fundamental problems. If hegemony is about emulating the hegemon, the Earth is not big enough for the world to live 'the American way'.

The 'ultimate' place–space tension

One particular manifestation of contemporary globalization is 'ecological globalization', the concern that current social trends will outstrip the Earth's capacity to survive as a living planet. Consideration of 'ecological globalization' differs from concentration upon the economic or political manifestations of globalization in two ways. First, ecological concerns pre-date the rise of globalization theses by two or three decades. Second, and more importantly, from a geographical perspective, ecology is the prime way in which the global is represented as a place. It is our

'home planet', the 'home of humanity', which we destroy, literally, at our peril. In contrast, economic and political globalizations treat the world as an action space, an abstract platform on which to perform, for instance the 24-hour-a-day financial space of world cities. We ask of a place: how sustaining is it? We ask of a space: how efficient is it? Hence there are totally contrasting concerns for, for instance, saving tropical forest biodiversity and maintaining the competitiveness of, say, London as a world city player.

The facts are quite stark. The United States has secured its hegemonic role through an ideology of consumerism. The good life promised by the United States to the rest of the world is a suburban lifestyle. Can this promise be kept? Some years ago, Watt (1982: 144) estimated that the carrying capacity of the world, assuming an American standard of living, is 600 million people: a figure exceeded in 1675, a century before the United States even existed. Today, the world's population is about 6 billion and will rise to somewhere between 10 billion and 14 billion in the next century. It is easy to see why people argue that current trends are not ecologically sustainable. It is this unsustainability of secular economic growth that underlies the tensions defined by Beck (1994) as the 'risk society'. The global ecological crisis is, therefore, the ultimate place–space tension between making planet Earth a habitable home and using it as an exploited resource space.

One example will suffice to show the urgency of the situation: communist China is falling for the irresistible seductiveness of consumer modernity. Until recently, its cities operated largely through a mixed system of private bicycles and cheap mass public transport. For very large cities, and there are many in China, this simple mixed transport system is relatively efficient as a low pollution solution to the traffic problem. In contrast, western cities with their high car densities are becoming pollution sinks and inefficient at moving traffic. Nevertheless, it is exactly the latter traffic model that the Chinese government is imposing on its cities. In fact, one of the ironies of this example is that most city planners in the West are beginning to promote bicycle travel and predict a necessary return to mass transit in the wake of unacceptable pollution and gridlock. However, contemporary consumer modernity has the motor car at its core, and therefore the 'modernizing' Chinese government sees conversion of its many cities into 'car cities' as necessary for 'catching up' with the West. By 2010, China plans to be the third largest car maker in the world, with 40 million cars on its streets (R. Smith 1997). This is just a start; to finally catch up with the United States will require another tenfold increase in cars, but China is now on a trajectory to that end. This is the ultimate example of Americanization: adding another billion ordinary shoppers to the great mall that is the world market (Zhao 1997). The Chinese leaders think that they are recreating China as a new progressive place, but nothing could be further from the truth; in their political failure to see other than an American future, they illustrate the absurdity of contemporary politics for the future of the Earth as a sustainable home for humanity.

'Empire' and the opportunity to create places or spaces

Earlier in the chapter we introduced the difference between spaces and places, and noted that the same geographical entity could be either a space or place depending upon the point of view. If we return to the geopolitical constructions of localities in the War on Terrorism discussed in the previous chapter, it is evident that the places of Iraq, Afghanistan or Palestine are simultaneously identified as spaces to be targeted. As we conclude this book, it is poignant to contrast the different views of the same locations on the map. Specially, we contrast the construction of dehumanized spaces and the related category of *homo sacer* with the feminist commitment to a humanist perspective. The latter extols a methodology that enables us to understand the complexities of people and the multitude of power relations they negotiate in their everyday lives. It aims to investigate and demonstrate the richness and diversity of humankind, and the geography of places, by focusing on the scale of the individual. We have exemplified this approach throughout the book.

But let us conclude with the contemporary geopolitics of the War on Terrorism and the geography of spaces that is being constructed by contemporary

representations and practices. Gregory (2004) builds upon the historical category of *homo sacer*, originally associated with the Roman Empire. *Homo sacer* translates, roughly, as sacred man. It was a title given to enemies who were deemed neither sacred, and so were beyond divine law, nor within the scope of the law, and hence deserving of legal protection and process. They were, in other words, categorized as individuals who could be killed with impunity. In contrast to the project of feminist geography, the geopolitical construction of *homo sacer* denies the humanity of individuals. To do so requires their construction as essentialized beings, with no complexity of layers: no roles of mother, father, husband, wife, brother, sister, and so on.

The term *homo sacer* has come to prominence through the contemporary work of Italian philosopher Giorgio Agamben (1995), and has been utilized by Gregory and other scholars to explore the political geography of the War on Terrorism. One manifestation is the dehumanizing of the places that come under attack in the War on Terrorism, as discussed in Chapters 2 and 7. Iraq, Afghanistan and Palestine, for example, are deemed empty – the occurrence of everyday life is denied. As part of this construction, the inhabitants in such places are reduced to being 'terrorists'. The category is a double construction, everyone in a particular place is seen to have but one role. The term 'terrorist' is a contemporary *homo sacer* and so can be targeted and killed with impunity, often from a distance with depleted uranium shells and missiles.

But *homo sacer* is not just about the representation of individuals, places and countries. In the War on Terrorism it has also entailed the construction of material spaces beyond the reach of the law, or the compassion of humanity. Camp X-ray in Guantanamo Bay is the most obvious example. People from across the globe have been incarcerated here with no recourse to the law. They are beyond any national legal jurisdiction or international court. They have been labelled 'unlawful combatants', a category that did not exist until the administration of President George W. Bush invented it. Camp X-ray is a space that defines its inmates as *homo sacer* – and beyond the jurisdiction of established political geography.

It remains despite censure from the United Nations and governments. Even the Attorney General of Britain, the United States' chief ally, declared it should be closed.

Camp X-ray is a space of the War on Terrorism. It is one of the many material spaces that have been constructed and justified within representations of 'terrorist places' (Graham, forthcoming). It is the world with which we are faced. Two competing political geographies exist to help us understand it and build the future. One compartmentalizes the world into simple spaces with abstract people deemed threats who need to be controlled or killed. The other is a political geography of places within the political and economic structures of the capitalist world-economy. It aims to understand the complexity of people and places across the globe and facilitate empathy. The latter is the one we recommend.

Summary

In this section we have challenged the reader to think about the possibilities of political action within the contemporary contexts of globalization and 'empire' by introducing the notion of place–space tensions. We noted that different political outcomes and agendas come about by thinking of localities and scales as either spaces or places.

Chapter summary

This chapter has served as a conclusion to the whole book. It has done so in two ways. First, we have summarized the concepts within the book and discussed their political implications. At the beginning and end of the chapter we emphasized three key concepts that have been discussed throughout the book:

➤ the political geography of scale;

➤ the persistence of core–periphery differences;

➤ the temporal–spatial context of political action.

At the end of the chapter we conceptualized politics as space–place tensions, a concept that may be applied to the political geography of the past, present, or future.

Second, we have conceptualized the form and practice of identity politics by:

➤ introducing the concept of prime modernity;

➤ illustrating the dynamism of identity politics though the concept of reflexive modernization;

➤ emphasizing the nature of contemporary identity politics by introducing the concept of the post-traditional condition;

➤ situating identity politics within fourteen political geographies related to the institutions of the capitalist world-economy.

We concluded by emphasizing that although identity politics is framed within the structure, dynamics and institutions of the capitalist world-economy it is also a matter of individual and group agency.

Key glossary terms from Chapter 8

administration	European Union (EU)	local state	risk society
Annales school of history	First World War	minorities	scope of conflict
	free trade	nation	semi-periphery
autonomy	fundamentalism	nationalism	sovereignty
boundary	geopolitics	nation-state	space
capitalism	globalization	neo-liberal	spaces of flows
capitalist world-economy	government	patriarchy	spaces of places
	hegemony	peoples	state
civil liberties	home	place	structural power
civil society	homeland	place–space tension	United Nations
classes	households	pluralism	Washington Consensus
Cold War	ideology	political parties	world cities
conservative	inter-state system	politics of identity	world-economy
core	Islam	post-traditional society	world market
democracy	Kondratieff cycles (waves)	power	world-system
elite		practical politics	world-systems analysis
'empire'	local government	reflexive modernity	

Suggested reading

Flint, C. (ed.) (2004) *Spaces of Hate: Geographies of Discrimination and Intolerance in the USA.* New York: Routledge. A collection of essays describing how hate activity, and resistance towards it, creates places.

Gregory, D. and Pred, A. (eds) (2006) *Violent Geographies: Fear, Terror, and Political Violence.* New York: Routledge. A collection of essays demonstrating how places shape political violence, and the scalar connections between the 'global' and the 'everyday'.

Taylor, P. J. (1999) *Modernities: A Geohistorical Interpretation.* Minneapolis: University of Minnesota Press. A concise and accessible essay on the way in which hegemonic powers define the 'modern' and the impact upon places and identity.

Activities

1 Think about a story that has been in the news for a while. The length and ongoing nature of the story should allow you to access a number of media reports on the topic. Conceptualize the story using the fourteen political geographies we identified in the chapter. (Hint: you will have to translate the language used in the media to identify the concepts we have used.) How many of the fourteen politics could you see as a component of the current affair you chose? In what ways are they related?

2 Considering a contemporary news story, list three ways in which the world-systems approach helps you explain the event and three ways in which a feminist approach is illuminating. In what ways do the two approaches complement each other to further your understanding of the 'real-world' event?

3 Do you live and behave in a post-traditional world? Think of how you relate to 'authority figures' (professionals) and discuss this with a member of an older generation. Note the similarities and differences in your use of 'experts' – are you a member of 'the world of clever people' that Giddens refers to?

Glossary

absolutism A form of rule in which the rulers claim complete power. It is usually applied to the politics of the 'absolutist states' that developed in Europe in the seventeenth and eighteenth centuries.

administration The implementation of government policy by ministers and civil servants.

anarchy of production The sum of the investment and disinvestment decisions of many entrepreneurs in a free market or capitalist system. The essence of the process is that there is no overall planning.

Annales school of history Named after a journal, the *Annales*, this French school of historians emphasize the day-to-day social and economic processes in opposition to the traditional political history of major events. Its leading recent exponent was Fernand Braudel, one of the originators of world-systems analysis.

ANZUS A defence and security pact for the Pacific area comprising Australia, New Zealand and the United States, formed in 1952. In 1985, the United States suspended its defence obligations to New Zealand.

apartheid An organization of society that keeps the races apart. It was introduced in South Africa by the National Party after 1948 as a means of ensuring continued white political dominance and was dismantled after 1989.

Arab League A political organization formed by eight Arab states in 1945. It now has twenty-one members.

aristocracy The traditional upper class, whose power is based upon land ownership.

ASEAN The Association of South East Asian Nations, formed in 1967, is a regional association of non-communist states.

autarky A policy of economic self-sufficiency based upon protectionism and the creation of large economic blocs.

authoritarian A form of rule where the rulers impose their policies without any effective constraints. Military dictatorships in the 'third world' are usually described thus.

autonomy The situation when a territory has self-government but not full sovereignty.

balance of power A theory of political stability based upon an even distribution of power between the leading states.

bloc A group of countries closely bound by economic and/or political ties.

boundary The limits of a territory; the boundary of a state defines the scope of its sovereignty.

bourgeoisie The urban middle class, the original political foes of the aristocracy. In Marxist analysis, members of the capitalist class, which owns the means of production.

capital city The chief site of the most important elements of the state apparatus.

capitalism A system of economic organization based upon the primacy of the market, where all the key decisions are to maximize profit. In Marxist analysis, it is defined as a mode of production by the existence of wage labour, the proletariat, exploited by owners of the means of production, the bourgeoisie.

capitalist world-economy The modern world-system based upon ceaseless capital accumulation.

centralization Concentration of power in the hands of the central government at the expense of other levels of government.

centre/centrist A political position between right-wing and left-wing orientations. Centrist politics claims to be non-ideological.

centrifugal forces Political processes that contribute to the disintegration of the state.

centripetal forces Political processes that contribute to the integration of the state.

Christian democracy A common political label and ideology of right-wing parties in Europe and Latin America. Originally derived from Catholic political movements, today it is associated with a more collectivist approach among conservative parties.

civil liberties The fundamental rights of citizens, typically violated by authoritarian regimes but respected in liberal democracies.

civil society The sum of all the voluntary associations through which a social system operates, it was devised as a concept to represent society outside the activities of the state. Sometimes, it appears as a dual (state/civil society), and sometimes as part of a trilogy including economy (state/civil society/ economy).

classes In world-systems analysis, one of the four key social institutions. These are the economic strata of the world defined, as in Marxism, in relation to the mode of production.

Cold War A geopolitical world order that lasted from 1946 to 1989. It pitted the communist world led by the Soviet Union against the United States and its allies.

collective consumption The consumption of public services and goods, especially as associated with urban areas.

colonialism The occupation of foreign territory by a state for the purposes of settlement and economic exploitation. It is another term for formal imperialism.

colony A territory under the sovereignty of a foreign power.

Comecon The Council for Mutual Economic Assistance (CMEA) was formed by communist countries in 1949 to promote socialist economic integration and the planned development of the economies of member states. Dissolved in 1990, its members were Bulgaria, Cuba, Czechoslovakia, Hungary, Mongolia, North Korea, Poland, Romania, Vietnam and the Soviet Union.

commonwealth A loose political grouping of states, most of which are former members of the British Empire.

communism A social system based upon the communal ownership of all property; it is usually used to mean the state-controlled social systems that were set up in former Soviet bloc countries.

congruent politics When the politics of support broadly matches the politics of power. It is the basis of liberal democracies.

conservative Originally a political ideology against social change, now a general term for right-wing politics.

constitution The fundamental statement of laws that define the way in which a country is governed.

containment The name given to the family of geopolitical codes devised by US governments against the Soviet Union during the Cold War.

contradictory politics When the politics of support is the opposite of the politics of power; an unstable political situation.

core One of three major zones of the world-economy – the others are periphery and semiperiphery – in world-systems analysis. It is characterized by core processes involving relatively high-wage, high-tech production.

core area of states An area where a state originates and around which it has gradually built up its territory.

coup d'état A change of government by unconstitutional means, usually involving a rebellion by the armed forces.

decentralization Dispersion of power away from central government to other levels of government.

decolonization The transfer of sovereignty from a colonial power to the people of the colony at the time of political independence.

democracy A form of government where policy is made by (direct democracy) or on behalf of (indirect democracy) the people. As indirect democracy, it usually takes the form of competition between political parties at elections.

dependency An economic or political relationship between countries or groups of countries in which one side is not able to control its destiny because of oppressive links with the other side.

derivationists Theorists of the state who attempt to derive the nature of the state from Marx's writings on capitalism.

détente A phase of the Cold War when accommodation policies were pursued, notably 1969–79.

development of underdevelopment The economic processes that occur in the periphery of the world-economy that are the opposite of the development which occurs in the core. The phrase was coined by Gunder Frank of the Dependency School of Development to show why poor countries were failing to catch up economically. ·

diplomacy The art of negotiating between countries. Diplomats conduct the foreign policies of states short of war.

disconnected politics When there is no relation between the politics of support and the politics of power.

domino theory A Cold War model of how countries become communist; once one country 'falls', it precipitates further communist takeovers among its neighbours.

economism The Marxist theory that all non-economic processes can be traced back to the economic base of society.

elite A small group in a society that has a disproportionate influence on events.

empire A political organization comprising several parts, one of which is the centre of power to which the rest are subordinate.

'empire' Hardt and Negri's (2000) concept of a new form of sovereign power based not on territorial sovereign states but a number of power relations and institutions, such as racism and multinational companies.

error of developmentalism The idea that all countries follow the same path of development.

European Union (EU) A group of states comprising most of western Europe that is carrying out a variety of policies ultimately leading to economic and political integration. Formerly known as the European Community (EC).

executive The branch of government that carries out the policies and implements the rules agreed by the legislature. The 'chief executive' is usually the prime minister or president.

faction A distinctive political group with its own policy, which preceded modern political parties in legislatures.

fascism An ideology developed by the Italian dictator Benito Mussolini. It is associated with the glorification of the state and its leader, militant anti-communism and military expansion.

federation A state where power is shared between two levels of government: a central or 'federal' government and a tier of provincial or 'state' governments.

feudalism A form of society based upon landlords collecting dues from the agricultural producers or serfs in return for military protection. This hierarchical society of mutual obligations preceded capitalism in Europe.

First World War A major war (1914–18) pitting Germany, Austria-Hungary and the Ottoman Empire (Turkey) against Britain, France and Tsarist Russia, and later the United States.

formal imperialism The political control of territory beyond a state's boundary.

franchise The voting rights in a country, for instance the universal adult franchise used in elections in most modern states.

free trade The policy of allowing commodities into a country from all other states without prohibitive tariffs, in order to maximize trade.

frontier The zone at the edge of a historical system where it meets other systems.

functionalism   An argument that you can understand an institution through analysis of what it does, as in functional theories of the state.

fundamentalism   Tradition defending itself within the difficult circumstances of reflexive modernity.

geopolitical code   The operating code of a government's foreign policy that evaluates places beyond its boundaries.

geopolitical transition   The short period of rapid change between one geopolitical world order and the next.

geopolitical world order   A stable pattern of world politics dominated by an agenda set by the major powers.

geopolitics   The study of the geographical distribution of power among states across the world, especially the rivalry between the major powers.

gerrymander   A general term indicating political bias in the operation of an electoral system; more specifically, it means the manipulation of constituency boundaries to favour a particular candidate or party.

globalization   A contentious term that is used to describe contemporary society. It has two main dimensions: (1) it denotes an up-scaling of human activity to the global scale; and (2) it refers to the expansion of transnational relations in a global space of flows. As such, globalization challenges the primacy of the state in political, economic and social processes.

government   The primary political institution in a state, responsible for making and implementing laws and policies.

heartland–rimland thesis   A development of the heartland theory that allows the sea power to balance the land power's strategic position by controlling the area between them.

heartland theory   A geostrategic theory devised by Halford Mackinder that gives the land power in control of central Asia ultimate strategic advantage over sea power in competition for control of the world.

hegemony   A position held by a state or a class when it so dominates its sphere of operation that other states or classes are forced to comply with its wishes voluntarily. States are defined as hegemonic at the scale of the world-system, classes at the scale of the state.

home   The locale of the household, it is seen as the epitome of socially constructed place. It is interpreted as both a place of security and of constraint.

homeland   A place of belonging and identity, it is associated with nationalism through the conversion of state territory as space into national homeland as place.

households   One of the four key social institutions in world-systems analysis. These are the 'atoms' of the system, where small groups of people share a budget.

human rights   Fundamental rights that all persons should be able to expect, such as humane treatment in prison and a fair trial.

iconography   Symbolic representation, as in nationalist use of images and flags. It is a strong centripetal force in integration theories of the state.

idealism   In international relations, an approach to world politics that emphasizes cooperation and believes that the inter-state system can be organized peacefully.

ideology   A world view about how societies both do and should work. It is often used as a means of obscuring reality.

imperialism   The process whereby one country dominates another country, either politically or economically.

informal imperialism   Dominance of a territory outside a state's borders but without political control; a process inherent in the economic structures of the world-economy.

instrumental theory of the state   A theory in which the state apparatus is available for control by competing groups. In the Marxist and world-systems version this competition is hollow, since the dominant class retains control of the state as their political instrument.

inter-state system   The political organization of the capitalist world-economy as a multiple set of polities.

Islam   The community of believers in the Muslim religion.

isolation   A policy of keeping a country apart from foreign entanglements such as alliances and wars.

judiciary   The branch of government responsible for interpreting the laws.

Kondratieff cycles (waves)   The cyclical pattern of the world-economy of about fifty years' duration and consisting of an A-phase of growth and a B-phase of stagnation.

League of Nations   Precursor of the United Nations, established by the Treaty of Versailles as a form of world government to prevent future war, it quickly failed as the US senate failed to ratify the Covenant of the League.

Lebensraum   German word for living space; it became the name given to the policy of expansion of Nazi Germany.

left-wing   A general term to denote anti-establishment political views. It is specifically used as a label for socialist or communist parties.

legislature   The branch of government responsible for enacting laws.

liberal   A political ideology that promotes individual freedom.

liberal democracy   A form of state in which relatively fair elections are held regularly to decide the multi-party competition for government.

liberal democratic interlude   A period of elected government in a state where government formation is also carried by other means.

liberation movement   A political organization aiming at overthrowing a non-representative government, usually foreign.

local government   The lowest tier of government, dealing with the particular affairs of individual cities, towns and communities.

local state   The state apparatus that deals with localities.

logistic wave   Very long cycles of material change (150–300 years) with an A-phase of growth and a B-phase of decline or stagnation.

longue durée   Fernand Braudel's concept of the gradual change through the day-to-day activities by which social systems are continually being reproduced.

malapportionment   An electoral abuse where electoral districts are unequal in population size, thus favouring some parties or groups over others.

managerial thesis   An argument about the distribution of goods in urban areas that emphasizes the power of urban managers or 'gatekeepers' to influence the allocation.

Marxism   The system of thought derived from Karl Marx in which politics is interpreted as a struggle between economic classes. Since it promotes communal ownership of all property when it is practised, it is frequently termed communism.

mercantilism   A state policy and economic theory devised and implemented in the seventeenth and eighteenth centuries. It used protectionist measures to control trade, with the ultimate purpose of concentrating bullion within the state.

militarism   An ideology and practice that believes the armed forces must necessarily play the dominant role in foreign policy and, in turn, sees the military as serving its own needs rather than the state and the needs of the civilian population.

militarization   Institutionalized process of socialization that embeds militarism within all aspects of society, such as education and national identity.

mini-systems   Small-scale historical societies that no longer exist and that were based upon a reciprocal-lineage mode of production.

minorities   A category of the world-systems institution of peoples, it consists of a minority group of persons in a society who view themselves as separate from the rest of society on ethnic, language or religious grounds.

mode of production   In world-systems analysis, the overall organization of the material processes in a historical system, including production, distribution and consumption.

Monroe Doctrine   Statement made by US president James Monroe in 1823 that established Latin America within a US sphere of influence and warned European powers against interfering in newly independent Latin American countries.

multinational corporations   The most common term to describe large companies whose activities straddle state boundaries.

multi-party system   A situation where several parties compete for votes in an election.

Napoleonic Wars   Constituted a major war (1795–1815) in which the French drive to European and world leadership was decisively beaten, leaving Britain supreme on the world stage.

nation   A group of persons who believe that they consist of a single 'people' based upon historical and cultural criteria and therefore should have their own sovereign state.

national determinism   The allocation of persons to a national group on some ascribed basis, usually language and territory.

national self-determination   The right of all peoples to determine their own government.

nationalism   An ideology and political practice that assumes all nations should have their own state, a nation-state, in their own territory, the national homeland.

nation-state   As an ideal state, the situation where all the inhabitants belong to one nation. Although most states claim to be nation-states, in practice almost all of them include sizeable minority groups outside the dominant nation.

NATO   The North Atlantic Treaty Organization is a military alliance formed in 1949 under US leadership and consisting of North American and west European states. Its prime purpose was to counter the perceived Soviet threat to the latter group of states.

neighbourhood effect   The tendency for persons to conform to the political norms of the area in which they live, usually illustrated by voting patterns.

neo-conservative   Political ideology promoting limited state intervention and social spending, and, in the case of the United States, militaristic foreign policy.

neo-liberal   Political ideology and practice promoting free trade and the free movement of capital, and a limited role of the state.

neutralism   The policy of refusing to join any military alliances.

non-aligned movement   A grouping of third world countries that refused to align themselves with either the United States or Soviet Union during the Cold War.

non-decision making   The power to control the political agenda so that there is no challenge to the status quo, i.e. unwanted questions are not allowed to arise, so no decisions have to be made about them.

OAS   The Organization of American States, formed in 1948 with its headquarters in Washington, DC, is a political association of American states that is dominated by the United States.

OAU   The Organization of African Unity, formed in 1963, is the major political association of African states and has campaigned for decolonization and against apartheid.

opposition   Generally a position of disagreement, it is formalized in many liberal democracies as the role and title of the major party not in government, i.e. the government in waiting.

pan-region   A large division of the world, relatively self-sufficient and under one dominant state.

partition   The division of a state into two or more territories, which constitute new states.

patriarchy   A form of gender politics whereby men dominate women.

peoples   One of the four key social institutions in world-systems analysis. These are culturally defined status groups, the most politically potent being the nation.

periphery   One of three major zones in the world-economy – the others are the core and semi-periphery – in world-systems analysis. It is characterized by peripheral processes consisting of relatively low-wage, low-tech production.

place   Often used interchangeably with space, in geography place indicates a more humanized locality through which we live our lives.

place–space tension   Although place and space are commonly defined as different categories, since they are social constructs it follows that the same tract of the Earth's surface can be constructed as a place by one social group (e.g. as homeland) while being treated as a space by a different social group (e.g. as target map for bombing). This is a contingent relation and therefore a tension rather than a dialectic.

plebiscite   A vote by an electorate to change its constitutional status, such as for the secession of a territory from a state.

pluralism   A viewpoint that political power should be distributed among a wide range of groups and interests in a society.

pluralist theories of the state   An argument that contemporary states act as neutral umpires in deciding between the claims of numerous interest groups.

political parties   Political organizations whose goal is to win control of the state apparatus through combining a centre organization of political elites with branches of members through all or part of the country.

politics of contentment   A phrase coined by J. K. Galbraith to describe the politics of core states when right-wing parties continue to be elected to implement policies of reducing taxes for the rich and cutting services for the poor.

politics of failure   A description of unstable politics in the material conditions beyond the core where no government is able to construct a viable support base, so that every government is doomed to failure.

politics of identity   A more individualized politics relating to choices of lifestyle within reflexive modernity.

politics of power   The party politics associated with interests supporting and funding the party and the policies implemented to favour those interests.

politics of support   The party politics of attracting voters, both long-term through socialization and short-term through campaigning.

populism   A very loose ideology whose common theme is that the ordinary person is the source of virtue against the greed of special interests. It is commonly associated with rural protest movements against corrupt 'big city' politics, but in Latin America it has been an ideology of urban-based protest movements.

post-traditional society   An interpretation of present social conditions in which individual reflexivity combined with globalization makes the maintenance of old traditions difficult and the creation of new ones potentially impossible.

power   The ability to be successful in a conflict; this may be overt through force or the threat of force, or covert through non-decision making or structural advantage.

practical politics   The day-to-day politics of survival, which may not always involve the formal politics of the state.

proportional representation   A way of organizing elections so that each party's proportion of legislative seats is approximately the same as its proportion of votes won in the poll.

protectionism   A trading policy that discriminates against rival states by placing prohibitive tariffs on their goods entering the country.

protectorate   A small state where a foreign power controls its politics, so that it is effectively a colony.

qualitative efficiency   A phrase coined by David Gordon to describe capital's control over the work process, which may be more important than mere quantitative efficiency of costs.

racism/racialism   Hostility to persons or groups merely on account of their physical features or cultural traditions.

realism In international relations, a position that believes the inter-state system to be inherently competitive and always threatening to every state.

reflexive modernity An interpretation of current modernity as a new situation in which individuals demand the right to make choices about their own identity and its meaning in the wider world.

relative autonomy of the state A position that rejects economism and argues that every state has political processes that are partly autonomous with respect to its economy.

right-wing A general term to indicate pro-establishment political views; as a label, it is usually given to the main party in a country, either Christian democrat or conservative, that opposes the socialist party.

risk society Ulrich Beck's description of the result of reflexive modernization, in which risk replaces production as the measure of society.

scope of conflict The total collection of interests brought into a conflict at the point of its resolution.

secession The act of separating a territory from a state.

Second World War A major war (1939–45) pitting Germany, Italy and Japan against Britain, France, the Soviet Union and the United States.

sectionalism A regional bias in a political party's support.

semi-periphery The middle category of the three-zone division of the world in world-systems analysis. It is characterized by a mix of both peripheral and core production processes.

Social Darwinism The application of evolution theory, especially the notion of the survival of the fittest, to social situations to justify material inequalities.

social democracy A form of state with historically high state expenditure on services and welfare paid for by progressive taxation.

social imperialism A policy of using state surpluses from imperial ventures to pay for social welfare.

socialism A political ideology that in theory means the social ownership of economic production but in practice has meant the redistribution of income through state welfare policies.

sovereignty The ultimate power of the state, the legal source of its unique right to physical coercion within its territory. Sovereignty is not just proclaimed; it has to be recognized by other members of the inter-state system.

space In a common interpretation, it is our three-dimensional world abstracted down to its basics (geometry), but in political geography it is a social construction that impinges crucially on our behaviour.

spaces of flows A social construction of space based upon productions of multiple flows in networks and chains. Coined by Castells (1996), he argues that this is the nature of contemporary social space.

spaces of places A social construction of space based upon productions of multiple distinctive places. Coined by Castells (1996), he argues this was the usual form of space before the contemporary era.

state Defined by their possession of sovereignty over a territory and its people, states are the primary political units of the modern world and together constitute the inter-state system.

structural power Power based upon structural location within the world-economy.

structuralist theories of the state The argument that the modern state is inherently capitalist because it is part of the capitalist system.

suffrage The right to vote.

superpower A term used to describe the United States and Soviet Union indicating their immense political power relative to that of all other states. After the Cold War, the United States is often referred to as the 'lone superpower'.

territoriality Behaviour that uses bounded spaces to control activities.

third world A Cold War term used to describe the poorer countries of the world in opposition to the first (western) world and second (communist) world.

Thirty Years War First major war (1618–48) involving nearly all European powers. It confirmed the decline of Habsburg power, France as the strongest continental power and Dutch sovereignty.

Treaty of Tordesillas The division of the non-European world between Spain and Portugal in 1494.

Treaty of Versailles Discussed and signed in 1919 after the conclusion of the First World War. Led to the break-up of the Austro-Hungarian Empire and the establishment of new nation-states. Indicated United States as emerging hegemonic power.

Treaty of Westphalia Brought to an end the Thirty Years War in 1648. It is usually interpreted as the first major agreement in international law.

Truman Doctrine Statement by President Harry Truman in 1947 to oppose communist expansion. It is often interpreted as marking the beginning of the Cold War.

unequal exchange A mechanism of the world-economy based upon labour cost differentials between core and periphery reflected in commodity prices. The effect is that core goods are overpriced relative to periphery goods, to the huge advantage of the former.

unitary state A country where there is one prime level of government, the central government.

United Nations A world organization of states, formed in 1945, to which nearly every country in the world belongs.

urbicide The destruction or denial of the material and cultural aspects of urban spaces.

Warsaw Pact A military alliance in eastern Europe under Soviet leadership; formed in 1955 to counter NATO, dissolved in 1990.

Washington Consensus Term coined to describe the neo-liberal policies imposed upon states in the name of globalization and economic growth.

world cities Large urban centres with global connections. The term was coined by John Friedmann to describe the major control centres of capital in the contemporary world-economy.

world-economy A form of world-system based upon the capitalist mode of production (ceaseless capital accumulation).

world-empire A form of world-system based upon a redistributive–tributary mode of production (a small military–bureaucratic ruling class exploiting a large class of agricultural producers).

world market The system-wide price-setting mechanism for commodities in the world-economy.

world-system A historical social system where the division of labour is larger than any one local production area.

world-systems analysis The study of historical systems, especially the modern world-system or capitalist world-economy.

Bibliography

Abbott, C. (1999) *'Political Terrain': Washington DC from Tidewater Town to Global Metropolis*. Chapel Hill, NC: University of North Carolina Press.

Abraham, D. (1986) *The Collapse of the Weimar Republic*, 2nd edition. New York: Holmes & Meier.

Abrams, P. (1978) 'Towns and economic growth: some theories and problems' in P. Abrams and E. A. Wrigley (eds) *Towns in Society*. Cambridge: Cambridge University Press.

Agamben, G. (1995) *Homo Sacer: Sovereign Power and Bare Life*, trans. D. Heller-Roaxen. Stanford, CA: Stanford University Press.

Agnew, J. A. (1987) *Place and Politics*. London: Allen & Unwin.

Agnew, J. A. (1988) '"Better thieves than Reds"? The nationalization thesis and the possibility of a geography of Italian politics', *Political Geography Quarterly* 7: 307–22.

Agnew, J. A. (1996) 'Mapping politics: how context counts in electoral geography', *Political Geography* 15: 129–46.

Agnew, J. A. (1997) 'The dramaturgy of the horizons: geographical scale in the "Reconstruction of Italy" by the new political parties, 1992–95', *Political Geography* 16: 99–121.

Agnew, J. A. (2003) *Geopolitics: Re-visioning World Politics*, 2nd edition. New York: Routledge.

Agnew, J. A. (2005) *Hegemony: The New Shape of Global Power*. Philadelphia: Temple University Press.

Agnew, J. A. and Corbridge, S. (1995) *Mastering Space: Hegemony, Territory and International Political Economy*. London: Routledge.

Alavi, H. (1979) 'The state in post-colonial societies' in H. Goulbourne (ed.) *Politics and State in the Third World*. London: Macmillan.

Allen, J. (2003) *Lost Geographies of Power*. Oxford: Blackwell.

Amin, S. (1987) 'Democracy and national strategy in the periphery', *Third World Quarterly* 9: 1129–56.

Amin, S., Arrighi, G., Gunder Frank, A. and Wallerstein, I. (1990) *Transforming the Revolution: Social Movements and the World-System*. New York: Monthly Review Press.

Amnesty International (2004) 'Israel and the Occupied Territories, under the rubble: house demolition and destruction of land and property'. http://web.amnesty.org/library/index/engmde150332004.

Anderson, B. (1983) *Imagined Communities*. London: Verso.

Anderson, J. (1986) 'Nationalism and geography' in J. Anderson (ed.) *The Rise of the Modern State*. Brighton: Wheatsheaf.

Annan, D. (1967) 'The Ku Klux Klan' in N. Mackenzie (ed.) *Secret Societies*. New York: Holt, Rinehart & Winston.

Appadurai, A. (1991) 'Global ethnoscapes: notes and queries for a transnational anthropology' in R. G. Fox (ed.) *Recapturing Anthropology*. Sante Fe, NM: School of American Research Press.

Arblaster, A. (1984) *The Rise and Decline of Western Liberalism*. Oxford: Blackwell.

Archer, J. C. (2002) 'The geography of an interminable election: Bush v. Gore 2000', *Political Geography* 21: 71–7.

Archer, J. C. and Taylor, P. J. (1981) *Section and Party: A Political Geography of American Presidential Elections from Andrew Jackson to Ronald Reagan*. Chichester: Wiley.

Armstrong, K. (2004) *The Battle for God: A History of Fundamentalism*. New York: Ballantine.

Arquilla, J. and Ronfeldt, D. (2001) 'Osama bin Laden and the advent of Netwar', *New Perspectives Quarterly* 18: 23–33.

Arquilla, J., Ronfeldt, D. and Zanini, M. (1999) 'Networks, netwar and Information-Age terrorism' in I. O. Lesser, B. Hoffman, J. Arquilla, D. Ronfeldt and M. Zanini (eds) *Countering the New Terrorism*. Santa Monica: Rand.

Arrighi, G. (1994) *The Long Twentieth Century*. London: Verso.

Arrighi, G., Silver, B. J. and Brewer, B. D. (2003) 'Industrial convergence, globalization, and the persistence of the North–South divide', *Studies in Comparative International Development* 38: 3–31.

Artibise, A. (1996) 'Cascadian adventures: shared visions, strategic alliances, and ingrained barriers'.

Paper presented at the Pacific Northwest History Meetings, Seattle.

Bacevich, A. (2002) *American Empire: The Realities and Consequences of US diplomacy*. Cambridge, MA: Harvard University Press.

Bachrach, P. and Baratz, M. (1962) 'Two faces of power', *American Political Science Review* 56: 947–52.

Banks, A. S. and Textor, R. B. (1963) *A Cross-Polity Survey*. Cambridge, MA: MIT Press.

Barnes, T. J. (1998) 'Cultural twists and turns', *Environment and Planning D: Society and Space* 16: 631–4.

Barnett, R. J. and Muller, R. E. (1974) *Global Reach*. New York: Simon & Schuster.

Barraclough, G. (ed.) (1998) *The Times Atlas of World History*, 5th edition. London: Times Books.

Barratt Brown, M. (1974) *The Economics of Imperialism*. London: Penguin.

Bartlett, C. J. (1984) *The Global Conflict, 1800–1970*. London: Longman.

Barzini, L. (1959) 'From Italy' in F. M. Joseph (ed.) *As Others See Us*. Princeton, NJ: Princeton University Press.

Bassin, M. (1987) 'Race contra space: the conflict between German geopolitik and National Socialism', *Political Geography Quarterly* 6: 115–34.

Bauman, Z. (1989) *Modernity and the Holocaust*. Ithaca, NY: Cornell University Press.

Beard, C. A. (1914) *An Economic Interpretation of the Constitution of the United States*. New York: Macmillan.

Beaverstock, J., Smith, R. G. and Taylor, P. J. (1999) 'The long arm of the law', *Environment and Planning A* 31: 1857–76.

Beck, U. (1994) 'The reinvention of politics: towards a theory of reflexive modernization' in U. Beck, A. Giddens and S. Lash (eds) *Reflexive Modernization: Politics, Tradition and Aesthetics in the Modern Social Order*. Stanford, CA: Stanford University Press.

Beck, U., Giddens, A. and Lash, S. (1994) *Reflexive Modernization: Politics, Tradition and Aesthetics in the Modern Social Order*. Stanford, CA: Stanford University Press.

Bennett, S. and Earle, C. (1983) 'Socialism in America: a geographical interpretation of failure', *Political Geography Quarterly* 2: 31–56.

Bergesen, A. J. and Lizardo, O. A. (2005) 'Terrorism and hegemonic decline' in J. Friedman and C. Chase-Dunn (eds) *Hegemonic Declines: Present and Past*. Boulder, CO: Paradigm.

Bergesen, A. and Schoenberg, R. (1980) 'Long waves of colonial expansion and contraction, 1415–1969' in A. Bergesen (ed.) *Studies of the Modern World System*. New York: Academic Press.

Bergman, E. F. (1975) *Modern Political Geography*. Dubuque, IA: William Brown.

Berman, M. (1983) *All that is Solid Melts into Air: The Experience of Modernity*. New York: Simon & Schuster.

Best, A. C. G. (1970) 'Gaborone: problems and prospects of a new capital', *Geographical Review* 60: 1–14.

Bhabha, H. (1990) *The Nation and Narration*. London: Routledge.

Billig, M. (1995) *Banal Nationalism*. Thousand Oaks, CA: Sage.

Billington, J. H. (1980) *Fire in the Minds of Men*. London: Temple Smith.

Blake, C. (ed.) (1987) *Maritime Boundaries and Ocean Resources*. London: Croom Helm.

Blaut, J. M. (1975) 'Imperialism: the Marxist theory and its evolution', *Antipode* 7(1): 1–19.

Blaut, J. M. (1980) 'Nairn on nationalism', *Antipode* 12(3): 1–17.

Blaut, J. M. (1987) *The National Question*. London: Zed Books.

Blondel, J. (1978) *Political Parties*. London: Wildwood House.

Blouet, B. W. (1987) 'The political career of Sir Halford Mackinder', *Political Geography Quarterly* 6: 355–68.

Boddy, M. (1984) 'Local economic and employment strategies' in M. Boddy and C. Fudge (eds) *Local Socialism?* London: Macmillan.

Bogdanor, V. (1992) 'Founding elections and regime change', *Electoral Studies* 9: 288–94.

Boggs, C. (2005) *Imperial Delusions: American Militarism and Endless War*. Lanham, MD: Rowman & Littlefield.

Bondi, L. (1993) 'Locating identity politics' in M. Keith and S. Pile (eds) *Place and the Politics of Identity*. Minneapolis: University of Minnesota Press.

Boogman, J. C. (1978) 'The raison d'état politician Johan de Witt', *The Low Countries History Yearbook 1978*: 55–78.

Boot, M. (2002) *The Savage Wars of Peace: Small Wars and the Rise of American Power*. Basic Books: New York.

Booth, K. (1991) 'Security and anarchy: utopian realism in theory and practice', *International Affairs* 67: 527–42.

Boulding, K. (1990) *Three Faces of Power*. Newbury Park, CA: Sage.

Bowle, J. (1974) *The Imperial Achievement*. London: Secker & Warburg.

Bowman, I. (1948) 'The geographical situation of the United States in relation to world policies', *Geographical Journal* 112: 129–42.

Branson, N. (1979) *Popularism 1919–1925*. London: Lawrence & Wishart.

Brass, P. R. (1975) 'Ethnic cleavages in the Punjabi party system 1952–1972' in M. R. Barnett et al. (eds) *Electoral Politics in the Indian States*. Delhi: Manohar.

Braudel, F. (1973) *Capitalism and Material Life 1400–1800*. London: Weidenfeld & Nicolson.

Braudel, F. (1984) *The Perspective on the World*. London: Collins.

Brenner, N. and Theodore, N. (2002) 'Cities and the geographies of "actually existing neoliberalism"', *Antipode* 34: 349–79.

Brenner, N., Jessop, B., Jones, M. and Macleod, G. (2003) *State/Space: A Reader*. Oxford: Blackwell.

Breuilly, J. (1993) *Nationalism and the State*, 2nd edition. Manchester: Manchester University Press.

Brewer, A. (1980) *Marxist Theories of Imperialism: A Critical Survey*. London: Routledge & Kegan Paul.

Briggs, A. (1963) *Victorian Cities*. London: Penguin.

Briggs, A. (1988) *Victorian Things*. London: Batsford.

Broughton, M. (1959) 'From South Africa' in F. M. Joseph (ed.) *As Others See Us*. Princeton, NJ: Princeton University Press.

Brown, L. R. (1973) *World without Borders*. New York: Vintage.

Brucan, S. (1981) 'The strategy of development in eastern Europe', *Review* 5: 95–112.

Brustein, W. (1993) *The Logic of Evil: The Social Origins of the Nazi Party, 1925–1933*. New Haven, CT: Yale University Press.

B'TSELEM (2002) 'Israel's policy of house demolitions and destruction of agricultural land in the Gaza strip'. http://www.btselem.org/English/Publications/Summaries/Policy_of_Destruction.asp.

Buchanan, K. (1972) *The Geography of Empire*. Nottingham: Spokesman.

Bukharin, N. (1972) *Imperialism and the World Economy*. London: Merlin.

Bull, H. N. (1977) *The Anarchical Society: A Study of World Order*. London: Macmillan.

Burch, K. (1994) 'The properties of the state system and global capitalism' in S. Rosow et al. (eds) *The Global Economy as Political Space*. Boulder, CO: Lynne Rienner.

Burger, J. (1987) *Report from the Frontier*. London: Zed Books.

Burghardt, A. F. (1973) 'The bases of territorial claims', *Geographical Review* 63: 225–45.

Burnett, A. and Taylor, P. J. (eds) (1981) *Political Studies from Spatial Perspectives*. Chichester: Wiley.

Burnham, W. D. (1967) 'Party systems and the political process' in W. N. Chambers and W. D. Burnham (eds) *The American Party System*. New York: Oxford University Press.

Busteed, M. A. (1975) *Geography and Voting Behaviour*. London: Oxford University Press.

Butler, J. (1990) *Gender trouble, feminism and the subversion of identity*. London: Routledge.

Calhoun, C. (2003) 'The class consciousness of frequent travelers: towards a critique of actually existing cosmopolitanism' in D. Archibugi (ed.) *Debating Cosmopolitics*. New York: Verso.

Canovan, M. (1981) *Populism*. London: Junction.

Castells, M. (1977) *The Urban Question: A Marxist Approach*. Cambridge, MA: MIT Press.

Castells, M. (1978) *City, Class and Power*. London: Macmillan.

Castells, M. (1996) *The Rise of the Network Society*. Oxford: Blackwell.

Ceaser, J. W. (1979) *Presidential Selection: Theory and Development*. Princeton University Press: Princeton, NJ.

CEH (Commission for Historical Clarification) (1999) *Guatemala: Memory of Silence*, Report of the Commission for Historical Clarification, Guatemala, Conclusions and Recommendations. http://shr.aaas.org/guatemala/ceh/report/english/toc.html (English language executive summary).

Chappell, R. G. and Chappell, G. E. (2000) *Letters from Korea Corpsmen*. Kent, OH: Kent State University Press.

Chase-Dunn, C. K. (1981) 'Interstate system and capitalist world-economy: one logic or two?' in W. L. Hollist and J. N. Rosenau (eds) *World System Structure*. Beverly Hills, CA: Sage.

Chase-Dunn, C. K. (1982) 'Socialist states in the capitalist world-economy' in C. K. Chase-Dunn (ed.) *Socialist States in the World-System*. Beverly Hills, CA: Sage.

Chase-Dunn, C. K. (1989) *Global Formation*. Oxford: Blackwell.

Chase-Dunn, C. K. and Anderson, E. N. (eds) (2005) *The Historical Evolution of World-Systems*. New York: Palgrave Macmillan.

Chase-Dunn, C. and Hall, T. D. (1997) *Rise and Demise: Comparing World-Systems*. Boulder, CO: Westview Press.

Chouinard, V. (2004) 'Making feminist sense of the state and citizenship' in L. A. Staeheli, E. Kofman and L. J. Peake (eds) *Mapping Women, Making Politics*. New York: Routledge.

Clark, G. L. (1984) 'A theory of local autonomy', *Annals, Association of American Geographers* 74: 195–208.

Clark, G. L. and Dear, M. (1984) *State Apparatus*. Boston: Allen & Unwin.

Cloke, P., Philo, C. and Sadler, D. (1991) *Approaching Human Geography: An Introduction to Contemporary Theoretical Debates*. New York: Guilford Press.

Coates, D. (1975) *The Labour Party and the Struggle for Socialism*. Cambridge: Cambridge University Press.

Cobban, A. (1969) *The Nation State and National Self-determination*. London: Collins.

Cock, J. (1993) *Women and War in South Africa*. Cleveland, OH: Pilgrim Press.

Cockburn, C. (1977) *The Local State*. London: Pluto.

Cohen, S. (1973) *Geography and Politics in a World Divided*, 2nd edition. New York: Oxford University Press.

Cohen, S. B. (2003) *Geopolitics of the World System*. Lanham, MD: Rowman & Littlefield.

Colley, L. (1992) *Britons*. London: Pimlico.

Collyer, M. (2006) 'Determinants of political transnationalism'. Paper presented at Political Pre-Conference, Political Geography Specialty Group, Champaign, IL, 6 March.

CONAIE (1989) *Las Nacionalidades indígenas en el Ecuador: nuestro proceso organizativo*, 2nd edition. Quito: Abya-Yala.

Cooke, P. (1985) 'Radical regions' in G. Rees (ed.) *Political Action and Social Identity*. London: Macmillan.

Cooke, P. (1989) *Localities*. London: Unwin Hyman.

Cope, M. (1996) 'Weaving the everyday: identity, space, and power in Lawrence, Massachusetts, 1920–1939', *Urban Geography* 17: 179–204.

Cope, M. (1997) 'Responsibility, regulation, and retrenchment: the end of welfare?' in L. A. Staeheli, J. E. Kodras and C. Flint (eds) *State Devolution in America: Implications for a Diverse Society*. Thousand Oaks, CA: Sage.

Cope, M. (2004) 'Placing gendered political acts' in L. A. Staeheli, E. Kofman and L. J. Peake (eds) *Mapping Women, Making Politics*. New York: Routledge.

Corbridge, S. (1997) 'Editor's introduction: India, 1947–1997'. *Environment and Planning A* 29: 2091–8.

Coulter, P. (1975) *Social Mobilization and Liberal Democracy*. Lexington, MA: Lexington Books.

Cox, K. R. (1969) 'The voting decision in a spatial context', *Progress in Geography* l: 81–117.

Cox, K. R. (1973) *Conflict, Power and Politics in the City: A Geographic View*. New York: McGraw-Hill.

Cox, K. R. (1979) *Location and Public Problems: A Political Geography of the Contemporary World*. Chicago: Maaroufa.

Cox, K. R. (1988) 'The politics of turf and the question of class' in M. Dear and J. Wolch (eds) *Territory and Reproduction*. Beverly Hills, CA: Sage.

Cox, K. R. (1997) 'Spaces of dependence, spaces of engagement and the politics of scale, or: looking for local politics', *Political Geography* 17: 1–23.

Cox, K. R. and Mair, A. (1988) 'Locality and community in the politics of local economic development', *Annals, Association of American Geographers* 78: 307–25.

Cox, M. (1986) 'The Cold War as a system', *Critique* 17: 17–82.

Cox, R. (1981) 'Social forces, states and world orders: beyond international relations theory', *Millennium* 10: 126–55.

Cresswell, T. (1996) *In place/out of place: geography, ideology and transgression*. Minneapolis: University of Minnesota Press.

Crump, J. (2004) 'Producing and enforcing the geography of hate: race, housing segregation, and housing-related hate crimes in the United States' in C. Flint (ed.) *Spaces of Hate*. Routledge: New York.

Dalby, S. (1990a) *Creating the Second Cold War*. London: Pinter.

Dalby, S. (1990b) 'American security discourse: the persistence of geopolitics', *Political Geography Quarterly* 9: 171–88.

Dalby, S. and Ó Tuathail, G. (1996) 'Editorial introduction: the critical geopolitics constellation', *Political Geography* 6/7: 451–6.

Danielson, M. N. (1976) *The politics of exclusion*. New York: Columbia University Press.

Davies, J. G. (1988) 'From municipal socialism to . . . municipal capitalism?' *Local Government Studies*, March/April: 19–22.

Davies, J. G. (1991) 'Letter from Tyneside, in the semiperiphery of the semicore. A UK experience', *Review* 14: 437–50.

Davis, H. B. (1978) *Towards a Marxist theory of Nationalism*. New York: Monthly Review Press.

Dear, M. (1981) 'A theory of the local state' in A. D. Burnett and P. J. Taylor (eds) *Political Studies from Spatial Perspectives*. Chichester: Wiley.

Dear, M. and Clark, G. (1978) 'The state and geographic process: a critical review', *Environment and Planning A* 10: 173–83.

De Blij, H. J. (1967) *Systematic political geography*. New York: Wiley.

Deighton, A. (1987) 'The "frozen front": the Labour government, the division of Germany and the origins of the Cold War, 1945–7', *International Affairs* 63: 449–65.

Delaney, D. and Leitner, H. (1997) 'The political construction of scale', *Political Geography* 16: 93–7.

Delpar, H. (1981) *Red and Blue, The Liberal Party in Colombian Politics, 1863–1899*. Alabama: University of Alabama Press.

Dent, M. (1995) 'Ethnicity and territorial politics in Nigeria' in G. Smith (ed.) *Federalism: the Multiethnic Challenge*. Longman: London.

Deutsch, K. W. (1961) 'Social mobilization and political development', *American Political Science Review* 55: 494–505.

Deutsch, K. W. (1981) 'The crisis of the State', *Government and Opposition* 16: 331–43.

Dijkink, G. (1996) *National Identity and Geopolitical Visions: Maps of Pride and Pain*. London: Routledge.

Dikshit, R. (1975) *The Political Geography of Federalism*. Delhi: Macmillan.

Dikshit, R. D. and Sharma, J. C. (1982) 'Electoral performance of the Congress Party in Punjab (1952–1977): an ecological analysis', *Transactions, Institute of Indian Geographers* 4: 1–15.

Dittmer, J. (2005) 'Captain America's Empire: reflections on identity, popular culture, and post-9/11 geopolitics', *Annals, Association of American Geographers* 95: 626–43.

Dix, R. H. (1984) 'Incumbency and electoral turnover in Latin America', *Journal of Interamerican Studies* 26: 435–48.

Dodds, K. (1997) *Geopolitics in Antarctica*. New York: Wiley.

Dodds, K. and Atkinson, D. (eds) (2000) *Geopolitical Traditions: A Century of Geopolitical Thought*. London: Routledge.

Dowler, L. (1998) '"And they think I'm just a nice old lady": women and war in Belfast, Northern Ireland', *Gender, Place and Culture* 5: 159–76.

Dowler, L. and Sharp, J. (2002) 'A feminist geopolitics?' *Space and Polity* 5: 165–76.

Dua, B. D. (1979) *President's Rule in India*. New Delhi: Chand.

Duncan, S. S. and Goodwin, M. (1982) 'The local state: functionalism, autonomy and class relations in Cockburn and Saunders', *Political Geography Quarterly* 1: 77–96.

Duncan, S. and Goodwin, M. (1988) *The Local State and Uneven Development*. Cambridge: Polity Press.

Dunleavy, P. (1980) *Urban Political Analysis*. London: Macmillan.

Duverger, M. (1954) *Political Parties*. London: Methuen.

Easton, D. (1965) *System Analysis of Political Life*. New York: Wiley.

Economist, The (2006) 'Watch out, India', 6 May, pp. 69–70.

Edwardes, M. (1975) *Playing the Great Game: A Victorian Cold War*. London: Hamilton.

Ek, R. (2000) 'A revolution in military geopolitics?' *Political Geography* 19: 841–74.

Emmanuel, A. (1972) *Unequal Exchange: A Study of the Imperialism of Trade*. New York: Monthly Review Press.

Engels, F. (1952) *The Condition of the Working Class in England in 1844*. London: Allen & Unwin.

Enloe, C. (1983) *Does Khaki Become You? The Militarisation of Women's Lives*. Boston, MA: South End Press.

Enloe, C. (1989) *Bananas, Beaches, and Bases: Making Feminist Sense of International Politics*. Berkeley, CA: University of California Press.

Esposito, J. L. (1998) *Islam: The Straight Path*, 3rd edition. Oxford: Oxford University Press.

Falah, G-W. and Flint, C. (2004) 'Geopolitical spaces: the dialectic of public and private space in the Palestine–Israel conflict', *The Arab World Geographer* 7: 117–34.

Falah, G-W., Flint, C. and Mamadouh, V. (2006) 'Just war and extraterritoriality: the popular geopolitics of the United States' war on Iraq as reflected in newspapers of the Arab world', *Annals, Association of American Geographers* 96: 142–64.

Featherstone, M., Lash. S. and Robertson, R. (eds) (1995) *Global Modernities*. Sage: London.

FCSA (Federacíon de Centros Shuar-Anchuar) (1992) 'Shuara Achuara Nunke, Tierra Shuar–Achuar'. Mimeo, Sucua, July.

Feldman, N. (2003) *After Jihad: America and the Struggle for Islamic Democracy*. New York: Farrar, Straus & Giroux.

Fifer, J. V. (1976) 'Unity by inclusion: core area and federal state at American independence', *Geographical Journal* 142: 402–10.

Fifer, J. V. (1981) 'Washington, DC: the political geography of a federal capital', *Journal of American Studies* 15: 5–26.

Filkin, C. and Weir, D. (1972) 'Locality' in E. Gittus (ed.) *Key Variables in Social Research*, Vol. 1. London: Heinemann.

Fincher, R. (2004) 'From dualisms to multiplicities: Gendered political practices' in L. A. Staeheli, E. Kofman and L. J. Peake (eds) *Mapping Women, Making Politics*. New York: Routledge.

Fishman, R. (1987) *Bourgeois Utopias: The Rise and Fall of Suburbia*. New York: Basic Books.

Flint, C. (1998) 'Forming electorates, forging spaces: the Nazi Party vote and the social construction of space', *American Behavioral Scientist* 41: 1282–303.

Flint, C. (1999) 'Changing times, changing scales: world politics and political geography since 1890', in G. Demko and W. B. Wood (eds) *Reordering the World*, 2nd edition. Boulder, CO: Westview Press.

Flint, C. (2001) 'Right-wing resistance to the process of American hegemony: the changing political geography of nativism in Pennsylvania, 1920–1998', *Political Geography* 20: 763–86.

Flint, C. (2002) 'Political geography: globalization, metapolitical geographies and everyday life', *Progress in Human Geography* 26: 391–400.

Flint, C. (2003a) 'Political geography II: terrorism, modernity, governance and governmentality', *Progress in Human Geography* 27: 97–106.

Flint, C. (2003b) 'Dying for a "P"? Some questions facing contemporary political geography', *Political Geography* 22: 617–20.

Flint, C. (2003c) 'Terrorism and counterterrorism: geographic research questions and agendas', *The Professional Geographer* 55: 161–9.

Flint, C. (2004) 'The "war on terrorism" and the "hegemonic dilemma": extraterritoriality, reterritorialization, and the implications for globalization', in J. O'Loughlin, L. Staeheli and E. Greenberg (eds) *Globalization and its Outcomes*. New York: Guilford Press.

Flint, C. (forthcoming) 'Mobilizing civil society for the hegemonic state: the Korean War and the construction of soldiercitizens in the United States' in D. Cowen and E. Gilbert (eds) *War, Citizens, and Territory*.

Flint, C. and Falah, G-W. (2004) 'How the United States justified its war on terrorism: prime morality and the construction of a "just war"', *Third World Quarterly* 25: 1379–99.

Flint, C. and Shelley, F. M. (1996) 'Structure, agency and context: the contributions of geography to world-systems analysis', *Sociological Inquiry* 66: 496–508.

Fluri, J. and Dowler, L. (2004) 'House bound: women's agency in white separatist movements' in C. Flint (ed.) *Spaces of Hate*. New York: Routledge.

Foeken, D. (1982) 'Explanation for the partition of sub-Saharan Africa, 1880–1900', *Tijdschrift voor Economische en Sociale Geografie* 73: 138–48.

Foster, J. (1974) *Class Struggle and the Industrial Revolution*. London: Weidenfeld & Nicolson.

Frank, A. G. (1977) 'Long live transideological enterprise! The socialist economies in the capitalist division of labour', *Review* 1: 91–140.

Frank, A. G. (1978) *Dependent Accumulation and Underdevelopment*. New York: Monthly Review Press.

Frank, A. G. (1984) *Critique and Anti-critique*. New York: Praeger.

Freymond, J. (1959) 'From Switzerland' in F. M. Joseph (ed.) *As Others See Us*. Princeton, NJ: Princeton University Press.

Friedmann, J. (1983) 'The world city hypothesis', *Development and Change* 17: 69–83.

Fukuyama, F. (1992) *The End of History and the Last Man*. New York: Free Press.

Gaddis, J. L. (1982) *Strategies of Containment*. New York: Oxford University Press.

Galbraith, J. K. (1958) *The Affluent Society*. Boston, MA: Houghton Mifflin.

Galbraith, J. K. (1992) *The Culture of Contentment*. Boston, MA: Houghton Mifflin.

Gallagher, J. and Robinson, R. (1953) 'The imperialism of free trade', *Economic History Review* (2nd series) 6: 1–15.

Galtung, J. (1971) 'A structural theory of imperialism', *Journal of Peace Research* 8: 81–117.

Galtung, J. (1979) *The True Worlds*. New York: Free Press.

García Gónzalez, L. (1992) *Resumen de Geografía, historia y cívica*. Quito: Editorial Andina.

Gellner, E. (1964) *Thought and Change*. Chicago: University of Chicago Press.

Gellner, E. (1983) *Nations and Nationalism*. Oxford: Blackwell.

Giddens, A. (1981) *A Contemporary Critique of Historical Materialism. Vol. 1. Power, Property and the State*. London: Macmillan.

Giddens, A. (1990) *The Consequences of Modernity*. Stanford, CA: Stanford University Press.

Giddens, A. (1994) 'Living in a post-traditional society' in U. Beck, A. Giddens and S. Lash (eds) *Reflexive Modernization: Politics, Tradition and Aesthetics in the Modern Social Order*. Stanford, CA: Stanford University Press, pp. 56–109.

Gilmartin, M. and Kofman, E. (2004) 'Critically feminist geopolitics' in L. Staeheli et al. (eds) *Mapping Women, Making Politics*. London: Routledge.

Gilroy, P. (2003) '"Where ignorant armies clash by night": homogenous community and the planetary aspect', *International Journal of Cultural Studies* 6: 271–6.

Glassner, M. I. (ed.) (1986) 'The new political geography of the sea', *Political Geography Quarterly* 5: 5–71.

Godlewska, A. and Smith, N. (eds) (1994) *Geography and Empire*. Oxford: Blackwell.

Gold, D. A., Lo, C. Y. H. and Wright, E. O. (1975) 'Recent developments in Marxist theories of the capitalist state', *Monthly Review* 27: 29–42 and 28: 36–51.

Goldfrank, W. L. (1979) 'Introduction: bringing history back in' in W. L. Goldfrank (ed.) *The World-System of Capitalism: Past and Present*. Beverly Hills, CA: Sage.

Goldstein, J. S. (1988) *Long Cycles: Prosperity and War in the Modern Age*. New Haven, CT: Yale University Press.

Gordon, D. M. (1976) 'Capitalist efficiency and socialist efficiency', *Monthly Review* 28: 19–39.

Gordon, D. M. (1978) 'Class struggle and the stages of American urban development' in D. C. Perry and A. J. Watkins (eds) *The Rise of the Sunbelt Cities*. Beverly Hills, CA: Sage.

Gordon, D. M. (1980) 'Stages of accumulation and long economic cycles' in T. K. Hopkins and I. Wallerstein (eds) *Processes of the World System*. Beverly Hills, CA: Sage.

Gottmann, J. (1973) *The Significance of Territory*. Charlottesville, VA: University Press of Virginia.

Gould, P. R. and Leinbach, T. R. (1966) 'An approach to the geographic assignment of hospital services', *Tijdschrift voor Economische en Sociale Geographie* 57: 203–6.

Graham, S. (2004a) 'Introduction: cities, warfare, and states of emergency' in S. Graham (ed.) *Cities, War, and Terrorism*. Oxford: Blackwell.

Graham, S. (2004b) 'Constructing urbicide by bulldozer in the occupied territories' in S. Graham (ed.) *Cities, War, and Terrorism*. Oxford: Blackwell.

Graham, S. (forthcoming) 'Inter-city relations and the "War on Terror"' in P. Taylor, B. Derudder, P. Saey and F. Witlox (eds) *Cities in Globalization: Practices, Policies, and Theories*.

Grano, O. (1981) 'External influences and internal change in the development of Geography' in D. R. Stoddart (ed.) *Geography, Ideology and Social Concern*. Oxford: Blackwell.

Grant, R. and Agnew, J. (1996) 'Representing Africa: the geography of Africa in world trade, 1960–1992', *Annals, Association of American Geographers* 86: 729–44.

Grant, R. and Nijman, J. (1997) 'Historical changes in US and Japanese foreign aid to the Asia-Pacific region', *Annals, Association of American Geographers* 87: 32–51.

Gregory, D. (2004) *The Colonial Present*. Oxford: Blackwell.

Gresh, A. and Vidal, D. (2004) *The New A–Z of the Middle East*. London: Tauris.

Griffiths, M. J. and Johnston, R. J. (1991) 'What's in a place?' *Antipode* 23: 185–215.

Habermas, J. (1975) *Legitimation Crisis*. Boston, MA: Beacon Press.

Hägerstrand, T. (1982) 'Diorama, path and project', *Tijdschrift voor Economische en Sociale Geografie* 73: 323–39.

Halliday, F. (1983) *The Making of the Second Cold War*. London: Verso.

Hamilton, R. (1982) *Who Voted for Hitler?* Princeton, NJ: Princeton University Press.

Hanson, S. and Pratt, G. (1995) *Gender, Work, and Space*. London: Routledge.

Harbutt, F. J. (1986) *The Iron Curtain: Churchill, America and the Origins of the Cold War*. New York: Oxford University Press.

Hardgrave, R. L. (1974) *India: Government and Politics in a Developing Nation*. New York: Harcourt Brace & World.

Hardt, M. and Negri, A. (2000) *Empire*. Cambridge, MA: Harvard University Press.

Harris, J. (2005) 'Emerging Third World powers: China, India, and Brazil', *Race and Class* 46: 7–27.

Hartmann, H. (1980) *Political Parties in India*. New Delhi: Meenakshi Prakasha.

Hartshorne, R. (1950) 'The functional approach in political geography', *Annals, Association of American Geographers* 40: 95–130.

Harvey, D. (1973) *Social Justice and the City*. London: Arnold.

Harvey, D. (1982) *The Limits to Capital*. Oxford: Blackwell.

Harvey, D. (1985) *The Urbanization of Capital*. Baltimore, MD: Johns Hopkins University Press.

Harvey, D. (2005) *The New Imperialism*. Oxford: Oxford University Press.

Haseler, S. (1976) *The Death of British Democracy*. London: Elek.

Hayden, D. (1981) *The Grand Domestic Revolution*. Cambridge, MA: MIT Press.

Hechter, M. (1975) *Internal Colonialism. The Celtic Fringe in British National Development, 1536–1966*. Berkeley, CA: University of California Press.

Heibel, T. (2004) 'Blame it on the Casa Nova? "Good scenery and sodomy" in rural southwestern Pennsylvania' in C. Flint (ed.) *Spaces of Hate*. New York: Routledge.

Helin, R. A. (1967) 'The volatile administrative map of Rumania', *Annals, Association of American Geographers* 57: 481–502.

Henderson, G. and Lebow, R. N. (1974) 'Conclusions' in G. Henderson et al. (eds) *Divided Nations in a Divided World*. New York: McKay.

Henige, D. P. (1970) *Colonial Governors from the Fifteenth Century to the Present*. Madison, WI: University of Wisconsin Press.

Henkel, W. B. (1993) 'Cascadia: a state of (various) mind(s)', *Chicago Review* 39: 110–18.

Henrikson, A. K. (1980) 'The geographical mental maps of American foreign policy makers', *International Political Science Review* 1: 495–530.

Henrikson, A. K. (1983) '"A small, cozy town, global in scope": Washington, DC', *Ekistics* 50: 123–45.

Hepple, L. (1986) 'The revival of geopolitics', *Political Geography Quarterly* 5 (Supplement): 21–36.

Herb, G. H. (2005) 'The geography of peace movements' in C. Flint (ed.) *The Geography of War and Peace*. Oxford: Oxford University Press.

Herod, A. (1997) 'Labor's spatial praxis and the geography of contract bargaining in the US east coast longshore industry, 1953–89', *Political Geography* 16: 145–69.

Herz, J. H. (1957) 'Rise and demise of the territorial state', *World Politics* 9: 473–93.

Heske, H. (1986) 'German geographic research in the Nazi period', *Political Geography Quarterly* 5: 267–82.

Heske, H. (1987) 'Karl Haushofer: his role in German geopolitics and Nazi politics', *Political Geography Quarterly* 6(1): 35–44.

Hewison, R. (1987) *The Heritage Industry. Britain in a Climate of Decline*. London: Methuen.

Hinsley, F. A. (1966) *Sovereignty*. London: Watts.

Hinsley, F. A. (1982) 'The rise and fall of the modern international system', *Review of International Studies* 8: 1–8.

Hirsch, A. R. (1983) *Making the Second Ghetto: Race and Housing in Chicago, 1940–1960*. Cambridge: Cambridge University Press.

Hirschman, A. O. (1970) *Exit, Voice and Loyalty*. Cambridge, MA: Harvard University Press.

Hobsbawm, E. J. (1977) 'Some reflections on "The Breakup of Britain"', *New Left Review* 105: 3–24.

Hobsbawm, E. J. (1987) *The Age of Empire, 1875–1914*. London: Guild.

Hobsbawm, E. J. (1990) *Nations and Nationalism since 1780*. Cambridge: Cambridge University Press.

Hobson, J. A. (1902) *Imperialism: A Study*. London: Allen & Unwin.

Hobson, J. A. (1968) 'Sociological interpretation of a general election' in P. Abrams (ed.) *The Origins of British Sociology 1834–1914*. Chicago: University of Chicago Press.

Hoffman, B. (2001) 'Change and continuity in terrorism', *Studies in Conflict and Terrorism* 24: 417–28.

Holdar, S. (1994) 'From American hegemony to pan-regional dominance? The changing geography of the western aid regime, 1966–1990', *Tijdschrift voor Economische en Sociale Geografie* 85: 236–48.

Holloway, J. and Picciotto, S. (eds) (1978) *State and Capital: A Marxist Debate*. London: Arnold.

Holm, H-H. and Sorenson, G. (1995) *Whose World Order? Uneven Globalization and the End of the Cold War*. Boulder, CO: Westview Press.

Horrabin, J. F. (1942) *An Outline of Political Geography*. Tillicoultry, Scotland: NCLC Publishing Society.

Hoyle, B. S. (1979) 'African socialism and urban development: the relocation of the Tanzanian capital', *Tijdschrift Economische voor en Sociale Geografie* 70, 207–16.

Hudson, B. (1977) 'The new geography and the new imperialism: 1870–1918', *Antipode* 9: 12–19.

Hunt, A. (1980) 'Introduction: taking democracy seriously' in A. Hunt (ed.) *Marxism and Democracy*. London: Lawrence & Wishart.

Huntington, S. P. (1991) *The Third Wave: Democratization in the Late Twentieth Century*. Norman, OK: University of Oklahoma Press.

Huntington, S. P. (1993) 'The clash of civilizations', *Foreign Affairs* 72: 22–49.

Huttenback, R. A. (1976) *Racism and Empire*. Ithaca, NY: Cornell University Press.

Hyndman, J. (2000) *Managing Displacement: Refugees and the Politics of Humanitarianism*. Minneapolis, MN: University of Minnesota Press.

Hyndman, J. (2001) 'Towards a feminist geopolitics', *Canadian Geographer* 45: 210–22.

Hyndman, J. (2004) 'Mind the gap: bridging feminist and political geography through geopolitics', *Political Geography* 23: 307–22.

Ignatieff, M. (2003) 'The challenges of American imperial power', *Naval War College Review* Spring 2003.

Ikporukpo, C. O. (1986) 'Politics and regional politics: the issue of state creation in Nigeria', *Political Geography Quarterly* 5: 127–40.

Imrie, R. and Raco, M. (1999) 'How new is the new local governance? Lessons from the United Kingdom', *Transactions of the Institute of British Geographers* NS 24: 45–63.

Isaacs, A. (1948) *International Trade: Tariff and Commercial Policies*. Chicago: Irwin.

Itzigsohn, J. (2000) *Developing Poverty: The State, Labor Market Deregulation, and the Informal Economy in Costa Rica and the Dominican Republic*. Philadelphia: Pennsylvania State University Press.

Jackson, R. H. (1990) *Quasi-states; Sovereignty, International Relations and the Third World*. New York: Cambridge University Press.

Jackson, P. and Penrose, J. (1993) 'Introduction: placing "race" and nation' in P. Jackson and J. Penrose (eds) *Constructions of Race, Place and Nation*. London: University College London Press.

Jahnige, T. P. (1971) 'Critical elections and social change', *Polity* 3: 465–500.

James, A. (1984) 'Sovereignty: ground rule or gibberish?' *Review of International Studies* 10: 1–18.

James, D. (2006) 'US intervention in Venezuela: a clear and present danger'. http://www.globalexchange.org/countries/americas/venezuela/USVZrelations.pdf. Accessed 13 April 2006.

Jefferson, M. (1939) 'The law of the primate city', *Geographical Review* 34: 226–32.

Jessop, B. (1982) *The Capitalist State*. Oxford: Robertson.

Jessop, B. (2002) *The Future of the Capitalist State*. Cambridge: Polity Press.

Johnson, C. A. (2004) *The Sorrows of Empire: Militarism, Secrecy and the End of the Republic*. New York: Metropolitan Books.

Johnston, R. J. (1973) *Spatial Structures*. London: Methuen.

Johnston, R. J. (1979) *Political, Electoral and Spatial Systems*. London: Oxford University Press.

Johnston, R. J. (1980) 'Electoral geography and political geography', *Australian Geographical Studies* 18: 37–50.

Johnston, R. J. (1982a) *Geography and the State*. London: Macmillan.

Johnston, R. J. (1982b) *The American Urban System*. Harlow: Longman.

Johnston, R. J. (1984) *Residential Segregation, the State and Constitutional Conflict in American Urban Areas*. London: Academic Press.

Johnston, R. J. (1985) *The Geography of English Politics: The 1983 General Election*. London: Croom Helm.

Johnston, R. J. (1986) 'The neighbourhood effect revisited: spatial science or political regionalism?' *Environment and Planning D: Society and Space* 4: 41–55.

Johnston, R. J., O'Neill, A. B. and Taylor, P. J. (1987) 'The geography of party support: comparative studies in electoral stability' in M. J. Heller (ed.) *The Logic of Multiparty Systems*. Munich: Springer-Verlag.

Johnston, R. J., Shelley, F. W. and Taylor, P. J. (1990) *Developments in Electoral Geography*. London: Routledge.

Johnston, R. J., Gregory, D., Pratt, G. and Watts, M. (2000) *The Dictionary of Human Geography*, 4th edition. Oxford: Blackwell.

Jones, B. D. (1986) 'Government and business: the automobile industry and the public sector in Michigan', *Political Geography Quarterly* 5: 369–84.

Jones, S. B. (1959) 'Boundary concepts in the setting of place and time', *Annals, Association of American Geographers* 49: 241–55.

Kamath, P. M. (1985) 'Politics of defection in India in the 1980s', *Asian Survey* 25: 1039–54.

Kaplan, A. (2003) 'Homeland insecurities: reflections on language and space', *Radical History Review* 85: 82–93.

Kaplan, R. (1994) 'The coming anarchy', *Atlantic Monthly*, February.

Kashyap, S. C. (1969) *The Politics of Defection*. National: Delhi.

Kautsky, K. (1914) 'Ultra-imperialism', *Die Neue Zeit*, September 1914. Available at http://www.marxists.org/archive/kautsky/1914/09/ultra-imp.htm. Accessed 20 April 2006.

Keane, J. (2002) 'Global civil society? Introducing global civil society', in H. Anheier, M. Glasius and M. Kaldor (eds) *Global Civil Society 2001*. Oxford: Oxford University Press.

Kennedy, P. (1988) *The Rise and Fall of the Great Powers*. New York: Random House.

Kerr, C. and Siegel, A. (1954) 'The inter-industry propensity to strike' in A. Kornhauser (ed.) *Industrial Conflict*. New York: Wiley.

King, A. D. (1990) *Global Cities*. London: Routledge.

Kirby, A. (1987) 'State, local state, context and spatiality: a reappraisal of state theory', Working Paper 87–07, Institute of Behavioral Science, Boulder, CO.

Kirby, A. (1993) *Power/Resistance: Local Politics and the Chaotic State*. Bloomington, IN: Indiana University Press.

Klare, M. T. (1995) *Rogue States and Nuclear Outlaws: America's Search for a New Foreign Policy*. New York: Hill & Wang.

Klare, M. T. (2004) *Blood and Oil: The Dangers and Consequences of America's Growing Petroleum Dependency*. New York: Metropolitan Books.

Kleppner, P. (1979) *The Third Electoral System: 1853–1892*. Chapel Hill, NC: University of North Carolina Press.

Knight, D. B. (1982) 'Identity and territory: geographical perspectives on nationalism and regionalism', *Annals, Association of American Geographers* 72: 514–31.

Knight, D. B. (1988) 'Self-determination, for indigenous peoples: the context for change' in R. J. Johnston, D. B. Knight and E. Kofman (eds) *National Self-determination and Political Geography*. London: Croom Helm.

Knox, P. L. (1995) 'World cities and the organisation of global space' in R. J. Johnston, P. J. Taylor and M. J. Watts (eds) *Geographies of Global Change*. Oxford: Blackwell, pp. 232–48.

Kobayashi, A. (1997) 'The paradox of difference and diversity (or why the thresholds keep moving)' in J. P. Jones, H. Nast and S. Roberts (eds) *Thresholds in Feminist Geography; Difference, Methodology and Representation*. Lanham, MA: Rowman & Littlefield.

Kobayashi, A. and Peake, L. (2000) 'Racism out of place: thoughts on whiteness and an antiracist geography in the new millennium', *Annals, Association of American Geographers* 90: 392–403.

Kodras, J. E. (1997a) 'Restructuring the state: devolution, privatization, and the geographic redistribution of power and capacity in governance' in L. A. Staeheli, J. E. Kodras and C. Flint (eds) *State Devolution in America: Implications for a Diverse Society*. Thousand Oaks, CA: Sage.

Kodras, J. E. (1997b) 'Globalization and social restructuring of the American population: geographies of exclusion and vulnerability' in L. A. Staeheli, J. E. Kodras and C. Flint (eds) *State Devolution in America: Implications for a Diverse Society*. Thousand Oaks, CA: Sage.

Kolko, J. and Kolko, G. (1973) *The Limits of Power*. New York: Harper & Row.

Koopman, S. (2006) 'Being solidarity activists: white middle-class women in the movement to close the School of the Americas'. Unpublished paper, Department of Geography, University of British Columbia.

Koves, A. (1981) 'Socialist economy and the world-economy', *Review* 5: 113–34.

Kristof, L. D. (1959) 'The nature of frontiers and boundaries', *Annals, Association of American Geographers* 49: 269–82.

Langhorne, R. (1981) *The Collapse of the Concert of Europe*. London: Macmillan.

Laski, H. J. (1935) *The State in Theory and Practice*. London: Allen & Unwin.

Laurie, N. and Calla, P. (2004) 'Development, postcolonialism, and feminist political geography' in L. A. Staeheli, E. Kofman and L. J. Peake (eds) *Mapping Women, Making Politics*. New York: Routledge.

Lawrence, E. (1994) *Gender and Trade Unions*. London: Taylor & Francis.

Lee, B. J. (2004) 'Gotta be Chinese', *Newsweek*, 28 June, p. E8.

Lees, L. H. (1982) 'Strikes and the urban hierarchy in English industrial towns, 1842–1901' in J. E. Cronin and J. Schnear (eds) *Social Conflict and the Political Order in Modern Britain*. London: Croom Helm.

Leib, J. (1995) 'Heritage versus hate: a geographical analysis of the Georgia Confederate flag debate', *Southeastern Geographer* 35: 37–57.

Lester, A. (2006) 'Imperial circuits and networks: geographies of the British Empire', *History Compass* 4: 124–41.

Leuchtenburg, W. E. (1958) *The Perils of Prosperity, 1914–32*. Chicago: University of Chicago Press.

Levinson, C. (1980) *Vodka Cola*. Horsham: Biblios.

Levy, J. S. (1983) *War in the Modern Great Power System, 1495–1975*. Lexington, KY: University Press of Kentucky.

Lewis, B. (2002) *What Went Wrong? Western Impact and Middle Eastern Response*. Oxford: Oxford University Press.

Libby, O. G. (1894) 'The geographical distribution of the vote of the thirteen states of the federal constitution, 1787–1788', *Bulletin of the University of Wisconsin* 1: 1–116.

Lichtheim, G. (1971) *Imperialism*. London: Penguin.

Lijphart, A. (1971) 'Class voting and religious voting in European democracies', *Acta Politica* 6: 158–71.

Lijphart, A. (1982) 'The relative salience of the socio-economic and religious issue dimensions: coalition formation in ten western democracies, 1919–1979', *European Journal of Political Research* 10: 201–11.

Lowenthal, D. (1958) 'The West Indies chooses a capital', *Geographical Review* 48: 336–64.

Luard, E. (1986) *War in International Society*. London: Tauris.

Luke, T. (2004) 'Everyday techniques as extraordinary threats: urban technostructures and nonplaces in terrorist actions', in S. Graham (ed.) *Cities, War, and Terrorism*. Oxford: Blackwell.

Macintyre, S. (1980) *Little Moscows*. London: Croom Helm.

Mackenzie, S. and Rose, D. (1983) 'Industrial change, the domestic economy and home life' in J. Anderson et al. (eds) *Redundant Spaces in Cities and Regions?* London: Academic Press.

Mackinder, H. J. (1904) 'The geographical pivot of history', *Geographical Journal* 23: 421–42.

Mackinder, H. J. (1919) *Democratic Ideals and Reality: A Study in the Politics of Reconstruction*. London: Constable; New York: Holt.

MacKinnon, C. (1989) *Toward a Feminist Theory of the State*. Cambridge, MA: Harvard University Press.

McColl, R. W. (1969) 'The insurgent state: territorial bases of revolution', *Annals, Association of American Geographers* 59: 613–31.

McCormick, R. P. (1967) 'Political development and the second party system' in W. N. Chambers and W. D. Burnham (eds) *The American Party Systems*. London: Oxford University Press.

McCormick, R. P. (1974) 'Ethno-cultural interpretations of American voting behaviour', *Political Science Quarterly* 89: 351–77.

McKay, D. V. (1943) 'Colonialism in the French geographical movement, 1871–1881', *Geographical Review* 33: 214–32.

McPhail, I. R. (1971) 'Recent trends in electoral geography', *Proceedings of the 6th New Zealand Geography Conference* 1: 7–12.

Mamadouh, V. (2005) 'Geography and war, geographers and peace' in C. Flint (ed.) *The Geography of War and Peace*. Oxford: Oxford University Press.

Mann, M. (1986) *The Sources of Social Power*, Vol. 1. New York: Cambridge University Press.

Mann, M. (2003) *Incoherent Empire*. New York: Verso.

Marshall, D. D. (1996) 'National development and the globalization discourse', *Third World Quarterly* 17, 875–901.

Marshall, S. L. A. (1947) *Men against Fire: The Problem of Battle Command in Future War*. New York: William Morrow.

Martin, H-P. and Schumann, H. (1997) *The Global Trap*, trans. P. Camiller. New York: St Martin's Press.

Massam, B. (1975) *Location and Space in Social Administration*. London: Arnold.

Massey, D. (1984) *Spatial Divisions of Labour*. London: Macmillan.

Massey, D. (1993) 'Power geometry and a progressive sense of place' in J. Bird et al. (eds) *Mapping the Futures*. London: Routledge.

Massey, D. (1994) *Space, Place, and Gender*. Minneapolis, MN: University of Minnesota Press.

Mayer, T. (2004) 'Embodied nationalisms' in L. A. Staeheli, E. Kofman and L. J. Peake (eds) *Mapping Women, Making Politics*. New York: Routledge.

Mellor, R. (1975) 'Urban sociology in an urbanized society', *British Journal of Sociology* 26: 276–93.

Mercer, D. (1993) '*Terra nullius*, Aboriginal sovereignty and land rights in Australia', *Political Geography* 12: 299–318.

Mercer, D. (1997) 'Aboriginal self-determination and indigenous land title in post-*Mabo* Australia', *Political Geography* 16: 189–212.

Mikesell, M. W. (1983) 'The myth of the nation state', *Journal of Geography* 82: 257–60.

Miliband, R. (1969) *The State in Capitalist Society*. London: Quartet.

Miliband, R. (1977) *Marxism and Politics*. London: Oxford University Press.

Miller, B. (2000) *Geography and Social Movements: Comparing Antinuclear Activism in the Boston Area*. Minneapolis, MN: University of Minnesota Press.

Minghi, J. V. (1963) 'Boundary studies in political geography', *Annals, Association of American Geographers* 53: 407–28.

Modelski, C. (1978) 'The long cycle of global politics and the nation state', *Comparative Studies of Society and History* 20: 214–35.

Modelski, G. (1981) 'Long cycles, Kondratieffs and alternating innovations' in C. Kegley and P. McGowan (eds) *The Political Economy of Foreign Policy Behaviour*. Beverly Hills, CA: Sage.

Modelski, G. (1987) *Long Cycles of World Politics*. London: Macmillan.

Modelski, G. and Thompson, W. R. (1995) *Leading Sectors and World Powers*. Columbia: University of South Carolina Press.

Mohan, G. (2002) 'The disappointment of civil society: the politics of NGO intervention in northern Ghana', *Political Geography* 21: 125–54.

Moisio, S. (2002) 'EU eligibility, central Europe, and the invention of the applicant state narrative', *Geopolitics* 7: 89–116.

Momsen, J. H. and Townsend, J. (1987) *Geography of Gender in the Third World*. London: Hutchinson.

Morris, J. (1968) *Pax Britannica*. London: Faber & Faber.

Mosse, G. L. (1975) *The Nationalization of the Masses: Political Symbolism and the Mass Movements in Germany from the Napoleonic Wars through the Third Reich*. New York: Howard Fertig.

Mosse, G. L. (1993) *Confronting the Nation: Jewish and Western Nationalism*. Hanover: Brandeis University Press.

Mouzelis, N. P. (1986) *Politics in the Semi-periphery*. London: Macmillan.

Muir, R. (1981) *Modern Political Geography*, 2nd edition. London: Macmillan.

Mumford, L. (1938) *The Culture of Cities*. London: Secker & Warburg.

Murphy, A. B. (1990) 'Historical justifications for territorial claims', *Annals, Association of American Geographers* 80: 531–48.

Murphy, A. B. (2005) 'Territorial ideology and inter-state conflict: comparative considerations', in C. Flint (ed.) *The Geography of War and Peace*. Oxford: Oxford University Press.

Nagar, R. (1996) 'The South Asian diaspora in Tanzania: a history retold', *Comparative Studies of South Asia, Africa and the Middle East* 16: 62–80.

Nagar, R. (1997) 'The making of Hindu communal organizations, places, and identities in postcolonial Dar es Salaam', *Environment and Planning D: Society and Space* 15: 707–30.

Nairn, T. (1977) *The Break-up of Britain*. London: New Left Books.

National Center for Children in Poverty (2006) 'Low income children in the United States, 2004', May 2004. http://www.nccp.org/pub_cpf04.html. Accessed 22 May 2006.

Nelund, C. (1978) 'The national world picture', *Journal of Peace Research* 315: 273–8.

Newton, K. (ed.) (1981) *Urban Political Economy*. London: Pinter.

Nielsson, G. P. (1985) 'States and "nation-groups": a global taxonomy' in E. A. Tiryakian and R. Rogowski (eds) *New Nationalisms of the Developed West*. London: Allen & Unwin.

Nijman, J. (1992) 'The dynamics of superpower spheres of influence: US and Soviet military activities, 1948–1978' in M. D. Ward (ed.) *The New Geopolitics*. Philadelphia: Gordon & Breach.

Nolin Hanlon, C. and Shankar, F. (2000) 'Gendered spaces of terror and assault: the testimonio of REMHI and the Commission for Historical Clarification of Guatemala', *Gender, Place and Culture* 7: 265–86.

Nugent, N. (1991) *The Government and Politics of the European Community*. Durham, NC: Duke University Press.

Nye, J. (2004) 'Soft power and American foreign policy', *Political Science Quarterly* 119: 255–70.

O'Brien, R. (1992) *Global Financial Integration: The End of Geography*. London: Pinter.

O'Connor, J. (1973) *The Fiscal Crisis of the State*. New York: St Martin's Press.

O'Loughlin, J. (1997) 'Economic globalization and income inequality in the United States' in L. A. Staeheli, J. E. Kodras and C. Flint (eds) *State Devolution in America: Implications for a Diverse Society*. Thousand Oaks, CA: Sage.

O'Loughlin, J. (2001) 'The regional factor in contemporary Ukrainian politics: scale, place, space, or bogus effects', *Post-Soviet Geography and Economics* 42: 1–33.

O'Loughlin, J. (2005) 'The political geography of conflict: civil wars in the hegemonic shadow', in C. Flint (ed.) *The Geography of War and Peace*. Oxford: Oxford University Press.

O'Loughlin, J. and Bell, J. E. (1999) 'The political geography of civic engagement in Ukraine', *Post-Soviet Geography and Economics* 40: 233–66.

O'Loughlin, J. and Van der Wusten, H. (1990) 'The political geography of panregions', *Geographical Review* 80: 1–20.

O'Loughlin, J., Ward, M. D., Lofdahl, C., Cohen, J. S., Brown, D. S., Reilly, D., Gleditsch, K. S. and Shin, M. (1998) 'The diffusion of democracy, 1946–1994', *Annals, Association of American Geographers* 88: 545–74.

O'Sullivan, P. (1986) *Geopolitics*. New York: St Martin's Press.

Ó Tuathail, G. (1992) 'Putting Mackinder in his place: material transformations and myths', *Political Geography* 11: 100–18.

Ó Tuathail, G. (1996) *Critical Geopolitics*. Minneapolis, MN: University of Minnesota Press.

Ó Tuathail, G. (2000) 'The postmodern geopolitical condition', *Annals, Association of American Geographers* 90: 166–78.

Ó Tuathail, G. and Agnew, J. (1992) 'Geopolitics and discourse: practical geopolitical reasoning in American foreign policy', *Political Geography* 11: 190–204.

Ó Tuathail, G. and Dalby, S. (eds) (1998) *Rethinking Geopolitics*. London: Routledge.

Ohmae, K. (1995) *The End of the Nation-State: The Rise of Regional Economies*. New York: Free Press.

Orridge, A. (1981a) 'Varieties of nationalism' in L. Tivey (ed.) *The Nation-State*. Oxford: Robertson.

Orridge, A. (1981b) 'Uneven development and nationalism I and II', *Political Studies* 24: 1–5 and 181–90.

Osei-Kwame, P. and Taylor, P. J. (1984) 'A politics of failure: the political geography of Ghanaian elections, 1954–1979', *Annals, Association of American Geographers* 74: 574–89.

Paasi, A. (1997) *Territories, Boundaries and Consciousness: The Changing Geographies of the Finnish–Russian Boundary*. New York: Halsted Press.

Paddison, R. (1983) *The Fragmented State: The Political Geography of Power*. Oxford: Blackwell.

Pahl, R. E. (1970) *Whose City?* London: Longman.

Painter, J. (1995) *Politics, geography, and 'political geography'*. London: Arnold.

Painter, J. (2004) 'Prosaic states'. Paper presented at the Annual Conference of the Association of American Geographers, Denver, CO.

Palestine Monitor (2004) 'Palestine fact sheets'. http://www.palestinemonitor.org/factsheet/ palestinian_intifada_fact_sheet.htm.

Park, R. L. and de Mesquita, B. B. (1979) *India's Political System*. Englewood Cliffs, NJ: Prentice-Hall.

Parker, W. H. (1982) *Mackinder. Geography as Aid to Statecraft*. Oxford: Clarendon Press.

Paterson, J. H. (1987) 'German geopolitics reassessed', *Political Geography Quarterly* 6: 107–14.

Peck, J. (1996) *Work-place: The Social Regulation of Labor Markets*. New York: Guilford Press.

Peters, R. (2004) 'Kill faster', *New York Post*, 20 May. www.nypost.com.

Phillips, P. D. and Wallerstein, L. (1986) 'National and world identities and the inter-state system', *Millennium* 14: 159–71.

Potts, D. (1985) 'Capital relocation in Africa', *Geographical Journal* 151: 182–96.

Poulantzas, N. (1969) 'The problem of the capitalist state', *New Left Review* 58: 119–33.

Poulsen, T. M. (1971) 'Administration and regional structure in east-central and south-east Europe' in G. W. Hoffman (ed.) *Eastern Europe*. London: Methuen.

Pounds, N. J. G. (1951) 'The origin of the idea of natural frontiers in France', *Annals, Association of American Geographers* 41: 146–57.

Pounds, N. J. G. (1954) 'France and "les limites naturelles" from the seventeenth to the twentieth centuries', *Annals, Association of American Geographers* 44: 51–62.

Pounds, N. J. G. (1963) *Political Geography*. New York: McGraw-Hill.

Pounds, N. J. G. and Ball, S. S. (1964) 'Core areas and the development of the European states system', *Annals, Association of American Geographers* 54: 24–40.

Prescott, J. R. V. (1965) *The Geography of Frontiers and Boundaries*. London: Hutchinson.

Prescott, J. R. V. (1969) 'Electoral studies in political geography' in R. Kasperson and J. V. Minghi (eds) *The Structure of Political Geography*. Chicago: Aldine.

Raban, J. (2004) 'Running scared', *The Guardian* 21 July, pp. 3–7.

Radcliffe, S. (1998) 'Frontiers and popular nationhood: geographies of identity in the 1995 Ecuador–Peru border dispute', *Political Geography* 17: 273–93.

Ramo, J. C. (2004) 'China has discovered its own economic consensus', *Financial Times* 7 May.

Rapkin, D. P. (ed.) (1990) *World Leadership and Hegemony*. Boulder, CO: Lynne Rienner.

Ratzel, F. (1897) *Politische Geographie oder die Geographie der Staaten, des Verkehres und des Krieges*. Munich and Leipzig.

Read, D. (1964) *The English Provinces c. 1760–1960. A Study in Influence*. London: Arnold.

Reich, R. (1998) 'When naptime is over', *New York Times Magazine*, 25 January, pp. 32–4.

Research Working Group (1979) 'Cyclical rhythms and secular trends of the capitalist world-economy: some premises, hypotheses and questions', *Review* 2: 483–500.

Rieber, J. N. and Cassaday, J. (2002a) 'Enemy Chapter Two: One Nation' from *Captain America* 2: 1–23. New York: Marvel Comics.

Rieber, J. N. and Cassaday, J. (2002b) 'Enemy Chapter Three: Soft Targets' from *Captain America* 2: 1–22. New York: Marvel Comics.

Roberts, S. (2004) 'Gendered globalization' in L. A. Staeheli, E. Kofman and L. J. Peake (eds) *Mapping Women, Making Politics*. New York: Routledge.

Robertson, G. S. (1900) 'Political geography and empire', *Geographical Journal* 16: 447–57.

Robinson, K. W. (1961) 'Sixty years of federation in Australia', *Geographical Review* 5: 1–20.

Robinson, R. (1973) 'Non-European foundations of European imperialism: sketch for a theory of collaboration' in R. Owen and B. Sutcliffe (eds) *Studies in the Theory of Imperialism*. London: Longman.

Robinson, R., Gallagher, J. and Denny, A. (1961) *Africa and the Victorians*. London: Macmillan.

Rodriguez, N. P. (1995) 'The real "New World Order": the globalization of racial and ethnic relations in the late twentieth century' in M. P. Smith and J. R. Feagin (eds) *The Bubbling Cauldron: Race, Ethnicity, and the Urban Crisis*. Minneapolis, MN: University of Minnesota Press.

Rokkan, S. (1970) *Citizens, Elections, Parties*. New York: McKay.

Rokkan, S. (1980) 'Territories, centres and peripheries: towards a geoethnic–geoeconomic–geopolitical model of differentiation within western Europe' in J. Gottman (ed.) *Centre and Periphery*. Beverly Hills, CA: Sage.

Ronfeldt, D. and Arquilla, J. (2001) 'Networks, netwars, and the fight for the future', *First Monday* 6. http://firstmonday.org/issues/issue6_10/ronfeldt/index.html. Accessed 3 January 2003.

Runciman, W. G. (1966) *Relative Deprivation and Social Justice*. London: Routledge & Kegan Paul.

Rupert, M. E. (1995) '(Re)politicizing the global economy: liberal common sense and ideological struggles in the US NAFTA debate', *Review of International Political Economy* 2: 658–92.

Russett, B. M. et al. (1963) *World Handbook of Political and Social Indicators*. New Haven, CT: Yale University Press.

Russett, B. M. (1967) *International Regions and the International System: A Study in Political Ecology*. Chicago: Rand McNally.

Rustow, D. A. (1967) *A World of Nations: Problems of Political Modernization*. Washington, DC: Brookings Institute.

Ryan, H. B. (1982) *The Vision of Anglo-America*. Cambridge: Cambridge University Press.

Rybczynski, W. (1986) *Home. A Short History of an Idea*. London: Penguin.

Sadasivan, S. N. (1977) *Party Democracy in India*. New Delhi: Tata McGraw-Hill.

Said, E. W. (1979) *Orientalism*. New York: Vintage Books.

Sandner, G. (1989) 'Historical studies of German political geography', *Political Geography Quarterly* 8: 311–400.

Sarkissian, W. (1976) 'The idea of social mix in town planning: an historical review', *Urban Studies* 13: 231–46.

Sassen, S. (1996) *Losing Control? Sovereignty in an Age of Globalization*. New York: Columbia University Press.

Sassen, S. (2001) *The Global City: New York, London, Tokyo*. Princeton, NJ: Princeton University Press.

Sassen, S. (2006) *Cities in a World Economy*, 3rd edition. Thousand Oaks, CA: Pine Forge Press.

Sauer, C. O. (1918) 'Geography and the gerrymander', *American Political Science Review* 12: 403–26.

Saunders, P. (1984) 'Rethinking local politics' in M. Boddy and C. Fudge (eds) *Local Socialism?* London: Macmillan.

Savage, M. (1987a) 'Understanding political alignments in contemporary Britain: do localities matter?' *Political Geography Quarterly* 6: 53–76.

Savage, M. (1987b) *The Dynamics of Working Class Politics*. Cambridge: Cambridge University Press.

Scase, R. (1980) *The State in Western Europe*. London: Croom Helm.

Schattschneider, E. E. (1960) *The Semi-Sovereign People*. Hinsdale, IL: Dryden.

Schlesinger, P. (1992) 'Europeanness: a new cultural battlefield', *Innovations* 5: 12–22.

Secor, A. (2004) '"There is an Istanbul that belongs to me": citizenship, space, and identity in the city', *Annals, Association of American Geographers* 94: 352–68.

Serres, M. and Latour, B. (1995) *Conversations on Science, Culture, and Time*, trans. R. Lapidus. Michigan: University of Michigan Press.

Seton-Watson, H. (1977) *Nations and States*. London: Methuen.

Sharp, J. (2000) *Condensing the Cold War: Reader's Digest and American identity*. Minneapolis, MN: University of Minnesota Press.

Sharp, J. (2004) 'Doing feminist political geographies' in L. A. Staeheli, E. Kofman and L. J. Peake (eds) *Mapping Women, Making Politics*. New York: Routledge.

Shelley, F. M. (2002) 'The electoral college and the election of 2000', *Political Geography* 21: 79–83.

Shelley, F. M. and Archer, J. C. (1994) 'Some geographical aspects of the American presidential election of 1992', *Political Geography* 13: 137–59.

Shideler, J. H. (1973) 'Flappers, philosophers, and farmers: rural–urban tensions of the twenties', *Agricultural History* 47: 283–99.

Short, R. J. (1982) *An Introduction to Political Geography*. London: Routledge & Kegan Paul.

Skinner, Q. (1978) *The Foundation of Modern Political Thought*, Vol. 2. Cambridge: Cambridge University Press.

Sklair, L. (2002) *The Transnational Capitalist Class.* Oxford: Blackwell.

Small, M. and Singer, J. D. (1982) *Resort to Arms: International and Civil Wars 1816–1980.* Beverly Hills, CA: Sage.

Smith, A. D. (1979) *Nationalism in the Twentieth Century.* Oxford: Robertson.

Smith, A. D. (1981) *The Ethnic Revival in the Modern World.* Cambridge: Cambridge University Press.

Smith, A. D. (1982) 'Ethnic identity and world order', *Millennium: Journal of International Studies* 12: 149–61.

Smith, A. D. (1986) *The Ethnic Origins of Nations.* Oxford: Blackwell.

Smith, A. D. (1995) *Nations and Nationalism in a Global Era.* Cambridge: Polity Press.

Smith, D. (1978) 'Domination and containment: an approach to modernization', *Comparative Studies in Society and History* 20: 177–213.

Smith, G. (1995) 'Mapping the federal condition: ideology, political practice and social justice' in G. Smith (ed.) *Federalism: the Multiethnic Condition.* London: Longman.

Smith, G. E. (1985) 'Ethnic separatism in the Soviet Union: territory, cleavage and control', *Environment and Planning C: Government and Policy* 3: 49–73.

Smith, N. (1984) 'Isaiah Bowman: political geography and geopolitics', *Political Geography Quarterly* 3: 69–76.

Smith, N. (1993) 'Homeless/global: scaling places' in J. Bird et al. (eds) *Mapping the Futures.* London: Routledge.

Smith, N. (2003) *American Empire: Roosevelt's Geographer and the Prelude to Globalization.* Berkeley, CA: University of California Press.

Smith, R. (1997) 'Creative destruction: capitalist development and China's destruction', *New Left Review* 222: 3–42.

Smith, T. (1981) *The Pattern of Imperialism.* Cambridge: Cambridge University Press.

Sparke, M. (2005) *In the Space of Theory: Postfoundational Geographies of the Nation-State.* Minneapolis, MN: University of Minnesota Press.

Sparke, M. B. (2006) 'A neoliberal nexus: economy, security and the biopolitics of citizenship on the border', *Political Geography* 25: 151–80.

Spate, O. H. K. (1942) 'Factors in the development of capital cities', *Geographical Review* 32: 622–31.

Staeheli, L. (1994) 'Restructuring citizenship in Pueblo, Colorado', *Environment and Planning A* 26: 849–71.

Staeheli, L., Kodras, J. E. and Flint, C. (1997) 'Introduction' in L. A. Staeheli, J. E. Kodras and C. Flint (eds) *State Devolution in America: Implications for a Diverse Society.* Thousand Oaks, CA: Sage.

Staeheli, L., Kofman, E. and Peake, L. (2004) *Mapping Women, Making Politics: Feminist Perspectives on Political Geography.* New York: Routledge.

Stilgoe, J. R. (1982) 'Suburbanites forever: the American dream endures', *Landscape Architecture* 72: 89–93.

Storper, M. (1997) 'Territories, flows, and hierarchies in the global economy' in K. R. Cox (ed.) *The Spaces of Globalization.* New York: Guilford Press.

Strayer, J. R. (1970) *On the Medieval Origins of the Modern State.* Princeton, NJ: Princeton University Press.

Sumartojo, R. (2004) 'Contesting place: antigay and lesbian hate crime in Columbus, Ohio' in C. Flint (ed.) *Spaces of Hate.* New York: Routledge.

Swyngedouw, E. (1997) 'Neither global or local: "glocalization" and the politics of scale' in K. Cox (ed.) *Spaces of Globalization.* New York: Guilford Press.

Szymanski, A. (1982) 'The socialist world system' in C. K. Chase-Dunn (ed.) *Socialist States in the World-System.* Beverly Hills, CA: Sage.

Tabb, W. K. and Sawers, L. (eds) (1978) *Marxism and the Metropolis.* New York: Oxford University Press.

Taibbi, M. (2006a) 'Mr. Republican', *Rolling Stone*, 6 April.

Taibbi, M. (2006b) 'How to be a lobbyist without even trying', *Rolling Stone*, 6 April.

Takeyh, R. (2003) 'Iran at a crossroads', *Middle East Journal* 57: 42–56.

Taylor, C. and Hudson, M. (1971) *World Handbook of Political and Social Indicators.* New Haven, CT: Yale University Press.

Taylor, M. and Thrift, N. (1982) *The Geography of Multinationals.* London: Croom Helm.

Taylor, P. J. (1978) 'Progress report: political geography', *Progress in Human Geography* 2: 153–62.

Taylor, P. J. (1981) 'Geographical scales within the world-economy approach', *Review* 5: 3–11.

Taylor, P. J. (1982) 'A materialist framework for political geography', *Transactions, Institute of British Geographer* NS7: 15–34.

Taylor, P. J. (1984) 'Accumulation, legitimation and the electoral geographies within liberal democracies' in P. J. Taylor and J. W. House (eds) *Political Geography: Recent Advances and Future Directions.* London: Croom Helm.

Taylor, P. J. (1986) 'An exploration into world-systems analysis of political parties', *Political Geography Quarterly* 5 (Supplement): 5–20.

Taylor, P. J. (1987) 'The paradox of geographical scale in Marx's *Politics*', *Antipode* 19: 287–306.

Taylor, P. J. (1989) 'The world-systems project' in R. J. Johnston and P. J. Taylor (eds) *World in Crisis?* Oxford: Blackwell.

Taylor, P. J. (1990) *Britain and the Cold War: 1945 as Geopolitical Transition*. London: Pinter.

Taylor, P. J. (1991a) 'Political geography within world-systems analysis', *Review* 14: 387–402.

Taylor, P. J. (1991b) 'The crisis of the movements: the enabling state as quisling', *Antipode* 23: 214–28.

Taylor, P. J. (1991c) 'The changing political geography' in R. J. Johnston (ed.) *The Changing Geography of the United Kingdom*. London: Methuen.

Taylor, P. J. (1992a) 'Understanding global inequalities: a world-systems approach', *Geography* 77: 10–21.

Taylor, P. J. (1992b) 'Tribulations of transition', *Professional Geographer* 44: 10–13.

Taylor, P. J. (1992c) 'Nationalism, internationalism and a "socialist geopolitics"', *Antipode* 24: 327–36.

Taylor, P. J. (1993a) 'Geopolitical world orders' in P. J. Taylor (ed.) *Political Geography of the Twentieth Century*. London: Belhaven.

Taylor, P. J. (1993b) 'States in world-systems analysis: massaging a creative tension' in B. Gills and R. Palan (eds) *Domestic Structures, Global Structures*. Boulder, CO: Lynne Rienner.

Taylor, P. J. (1994) 'The state as container: territoriality in the modern world-system', *Progress in Human Geography* 18: 151–62.

Taylor, P. J. (1995) 'Beyond containers: internationality, interstateness, interterritoriality', *Progress in Human Geography* 19: 1–15.

Taylor, P. J. (1996) *The Way the Modern World Works: World Hegemony to World Impasse*. Chichester: Wiley.

Taylor, P. J. (1998) *Modernities: A Geohistorical Interpretation*. Cambridge: Polity Press.

Taylor, P. J. (1999) 'Places, spaces and Macy's: place–space tensions in the political geography of modernities', *Progress in Human Geography* 23: 7–26.

Taylor, P. J. (2000) 'Havens and cages: reinventing states and households in the modern world-system', *Journal of World-Systems Research* 11: 544–62.

Taylor, P. J. (2004) *World City Network: A Global Urban Analysis*. London: Routledge.

Taylor, P. J. (2005) 'New political geographies: global civil society and global governance through world city networks', *Political Geography* 24: 703–30.

Taylor, P. J. and Johnston, R. J. (1979) *Geography of Elections*. London: Penguin.

Terlouw, K. (1992) *The Regional Geography of the World-System: External Arena, Periphery, Semi-Periphery and Core*. Nederlandse Geografische Studies, 144 Utrecht: Faculteit Ruimtelijke Wetenschappen, Rijksuniversiteit Utrecht.

Therborn, G. (1977) 'The rule of capital and the rise of democracy', *New Left Review* 103: 3–42.

Thompson, E. P. (1968) *The Making of the English Working Class*. London: Penguin.

Thompson, E. P. (1985) *The Heavy Dancers*. New York: Pantheon Books.

Thrift, N. (1983) 'On the determination of social action in space and time', *Environment and Planning D: Society and Space* 1: 23–57.

Thrift, N. (2000) 'It's the little things' in K. Dodds and D. Atkinson (eds) *Geopolitical Traditions: A Century of Geopolitical Thought*. London: Routledge.

Tienda, M. (1989) 'Puerto Ricans and the underclass debate', *Annals, American Academy of Political and Social Sciences* 501: 105–19.

Tietz, J. (2006) 'The killing factory', *Rolling Stone*, 20 April.

Tietz, M. (1968) 'Towards a theory of urban public facility location', *Papers of the Regional Science Association* 21: 35–51.

Tilly, C. (1975) 'Reflections on the history of European state-making' in C. Tilly (ed.) *The Formation of Nation States in Western Europe*. Princeton, NJ: Princeton University Press.

Tilly, C. (1978) *From Mobilization to Revolution*. Reading, MA: Addison-Wesley.

Tilly, C. (1990) *Coercion, Capital, and European States, AD 990–1990*. Cambridge, MA: Blackwell.

Tivey, L. (1981) 'States, nations and economies' in L. Tivey (ed.) *The Nation-State*. Oxford: Robertson.

Tomaney, J. (2002) 'The evolution of regionalism in England', *Regional Studies* 36: 721–31.

Trevor-Roper, H. (1983) 'The invention of tradition: the highland tradition of Scotland' in E. Hobsbawm and T. Ranger (eds) *The Invention of Tradition*. Cambridge: Cambridge University Press.

Tuan, Yi Fu (1977) *Space and Place*. London: Edward Arnold.

Urry, J. (1981) 'Localities, regions and social class', *International Journal of Urban and Regional Research* 5: 455–73.

Urry, J. (1986) 'Locality research: the case of Lancaster', *Regional Studies* 20: 233–42.

Walker, R. B. (1993) *Inside/Outside: International Relations as Political Theory*. Cambridge: Cambridge University Press.

Wallace, W. (1991) 'Foreign Policy and national identity in the United Kingdom', *International Affairs* 67: 65–80.

Wallerstein, I. (1974a) *The Modern World System: Capitalist Agriculture and the Origins of the European World-Economy in the Sixteenth Century*. New York: Academic Press.

Wallerstein, I. (1974b) 'The rise and future demise of the capitalist world system: concepts for comparative analysis', *Comparative Studies in Society and History* l6: 387–418.

Wallerstein, I. (1976) 'The three stages of African involvement in the world-economy' in P. C. W. Gutkind and I. Wallerstein (eds) *The Political Economy of Contemporary Africa*. Beverly Hills, CA: Sage.

Wallerstein, I. (1979) *The Capitalist World-Economy*. Cambridge: Cambridge University Press.

Wallerstein, I. (1980a) 'Maps, maps, maps', *Radical History Review* 24: 155–9.

Wallerstein, I. (1980b) *The Modern World-System II. Mercantilism and the Consolidation of the European World-Economy 1600–1750*. New York: Academic Press.

Wallerstein, I. (1982) 'Socialist states: mercantilist strategies and revolutionary objectives' in E. Friedman (ed.) *Ascent and Decline in the World-System*. Beverly Hills, CA: Sage.

Wallerstein, I. (1983) *Historical Capitalism*. London: Verso.

Wallerstein, I. (1984a) *The Politics of the World-Economy*. Cambridge: Cambridge University Press.

Wallerstein, I. (1984b) 'Long waves as capitalist process', *Review* 7: 559–75.

Wallerstein, I. (1991) *Unthinking Social Science*. Cambridge: Polity Press.

Wallerstein, I. (2003) *The Decline of American Power*. New York: W. W. Norton.

Wallerstein, I. (2004) *World-Systems Analysis: An Introduction*. Durham, NC: Duke University Press.

Ward, K. (2000) 'A critique in search of a corpus: revisiting governance and re-interpreting urban politics', *Transactions of the Institute of British Geographers* NS 25: 169–85.

Ward, K. (2005) 'Making "flexible" Manchester: competition and change in the temporary staffing industry', *Geoforum* 36: 223–40.

Waterman, S. (1984) 'Partition – a problem in political geography' in P. J. Taylor and J. W. House (eds) *Political Geography: Recent Advances Future Directions*. London: Croom Helm.

Waterman, S. (1987) 'Partitioned states', *Political Geography Quarterly* 6: 151–70.

Watson, J. W. (1970) 'Image geography: the myth of America in the American scene', *Advancement of Science* 27: 71–9.

Watt, K. E. F. (1982) *Understanding the Environment*. Boston: Allyn & Bacon.

Weber, E. (1976) *Peasants into Frenchmen*. London: Chatto & Windus.

Webster, G. R. (2002) 'The US presidential election and the Bush v. Gore Supreme Court decision', *Political Geography* 21: 99–104.

Weisman, J. (2003) 'US incomes fell, poverty rose in 2003', *Washington Post*, 27 September. http://www.washingtonpost.com/ac2/wp-dyn/A4586-2003Sep26?lang.html. Accessed 23 May 2006.

Werz, N. (1987) 'Parties and party systems in Latin America' in M. J. Holler (ed.) *The Logic of Multiparty Systems*. Munich: Springer-Verlag.

Wesson, R. (1982) *Democracy in Latin America*. New York: Praeger.

White, S. (1992) 'Democratizing eastern Europe', *Electoral Studies* 9: 227–87.

Williams, C. H. (1986) 'The question of national congruence' in R. J. Johnston and P. J. Taylor (eds) *World in Crisis?* Oxford: Blackwell.

Williams, O. (1971) *Metropolitan Policy Analysis*. New York: Free Press.

Wilson, C. (1958) *Mercantilism*. London: Historical Association.

Wilson, D. (forthcoming) *Race and Cities: America's New Black Ghetto*. London: Routledge.

Wolpert, J. (1997) 'How federal cutbacks affect the charitable sector' in L. A. Staeheli, J. E. Kodras and C. Flint (eds) *State Devolution in America: Implications for a Diverse Society*. Thousand Oaks, CA: Sage.

Woodward, R. (2004) *Military Geographies*. Oxford: Blackwell.

Wright, E. O. (1997) *Class Counts*. Cambridge: Cambridge University Press.

Yang, D. J. (1992) 'Magic mountains: attracted by pristine mountain beauty, the Pacific Northwest's high-tech wizards are aiming at conquering world markets', *The New Pacific* Autumn: 19–23.

Young, C. (1982) *Ideology and Development in Africa*. New Haven, CT: Yale University Press.

Zhao, B. (1997) 'Consumerism, Confucianism, communism: making sense of China today', *New Left Review* 222: 43–59.

Zolberg, A. R. (1981) 'Origins of the modern world system: a missing link', *World Politics* 33: 253–81.

Index

Numbers in **bold** refer to the Glossary

foundation elections 237, 238
France
  bankrupt 9, 154
  boundaries 128
  Colbertism 102, 124
  concept of the state and 137
  core 21
  decline 21
  defeat by Britain 90, 168
  defeat by Germany 60
  departments 131
  elections 196, 209, 217
  European Union and 131
  hegemonic challenge 50, 54, 56
  imperialism 83, 90, 91–2, 95, 99
  irregular transfer of power 202–3
  Kondratieff cycles and 20–1
  medieval state 122, 125
  mercantilism 102, 124
  modernity 291
  patriotism 166
  revolution 161–2, 164, 168, 176, 180, 181, 197, 290
  separatism 171
  as unitary state 131, 132
  Valois dynasty 9
franchise **317**
Frank, A. Gunder
  core processes 15
  development and development of underdevelopment 11, 87, 129
  modernization 7–8
  nationalism 180
  periphery processes 15, 103, 105, 150
  semi-periphery processes 103, 150
  Soviet Union 64–5
  trade 102
free trade 47, 51, 52–3, 53, 101–2, 139, 184–5, 202, 214, 220, **317**
Friedman, Milton 147
Friedmann, John 269–70
frontiers 126–31, 185, **317**
functionalism **318**
fundamentalism 293, 294–5, **318**
funerals 118

Galbraith, J. K. 35, 108, 222, 292
Galicia 171
Galtung, Johan 67, 70, 86

Gandhi, M. K. (Mahatma) 100
GATT (General Agreement on Tariffs and Trade) 145
Gaulle, Charles de 202–3
GDP *see* gross domestic products
gender 6, 283, 295
  class–household politics 305–6
  globalization and scale 109
  households 310
  inequalities 24
  people–households politics 305
  relations 110
  roles 110, 189–90
  state–households politics 300
gendering of geopolitics 3
General Agreement on Tariffs and Trade (GATT) 145
geo-economics 44, 75
geographical conceptualizing of politics 2
geographical scales 4, 38–9
  empire and 31
  Europe in 1500: 121
  experience 243–4
  identity politics 289
  ideology separating experience from reality 28–31
  localities 250, 265
  place–space tensions 309–11
  as political product and political arena 277–8
  politics of 26
  scope as 26–8
  state–household politics 300, 302
  state–people politics 308
  structuration theory 285
  three-scale analysis 28–30, 38
geometrical boundaries 128
geopolitical codes 45–6, 69, **318**
geopolitical transitions 56, 61–2, **318**
geopolitical world orders 45–57, **318**
geopolitics 3, 5, 32, 38, 42–3, **318**
  codes and world orders 45–57
  critical 69–73
  feminist, interrogation of security and 73–6
  globalization and 51, 280–1
  meaning, past, present and future 43–5
  situating codes and theories 57–65
  War on Terrorism 65–73

Germany
  Communist Party 146
  core 21
  defeat 21, 60
  elections 209, 217, 238
  expansion 5
  federalism 133
  finance capital 84
  geopolitical codes 45
  geopolitics 43–4, 59–60
  Heartland Theory and 58–9
  hegemonic challenge 50, 54, 56–7, 286
  historical-cultural regions 288
  imperialism 91
  invasion of Soviet Union 64
  Kondratieff cycles and 17, 20–1
  nationalism 167, 169, 175, 178
  Nazi Party and Nazism 5, 43, 59–60, 146–7, 168, 178, 188–9, 286–8
  pan-regions 60
  partition 136, 170
  protectionism 102–2
  reunification 136
  rise of 58–9
  as semi-periphery 21, 65, 102–3, 146
  Social Democratic Party 213
  unemployment 16
  unification 56, 169
  USA competition 51
  Weimar Republic 146–7, 222
  world power 56–7
gerrymandering **318**
Ghana 91, 103–4, 105, 150, 209, 228–9
Gibraltar 125
Giddens, Anthony 283, 284–5, 290, 294–5
Gilmartin, M. 2–3, 26, 44
glasnost 65
global inequalities 106–7
global-level geopolitical codes 45
global order 48
global scale of reality 29–31
globalization 308–9, **318**
  as Americanization 12
  business services 18
  democratization and 208
  ecological 311–12
  economy 182